The Scientist in the Early Roman Empire

Richard Carrier

Pitchstone Publishing
Durham, North Carolina

Pitchstone Publishing
Durham, North Carolina
www.pitchstonepublishing.com

Printed in the USA

10 9 8 7 6 5 4 3 2 1

Library of Congress Cataloging-in-Publication Data

Names: Carrier, Richard, 1969-
Title: The scientist in the early Roman Empire / Richard Carrier.
Description: Durham, North Carolina : Pitchstone Publishing, [2017] |
Includes bibliographical references and index.
Identifiers: LCCN 2016057375| ISBN 9781634311069 (pbk. : alk. paper) | ISBN
9781634311083 (epdf) | ISBN 9781634311090 (mobi)
Subjects: LCSH: Science, Ancient. | Scientists—Europe—History—To 1500. |
Science—Europe—History—To 1500. | Science—Philosophy—History. |
Religion and science—Europe—History—To 1500. | Rome—Civilization.
Classification: LCC Q124.95 .C3674 2017 | DDC 509.37—dc23
LC record available at https://lccn.loc.gov/2016057375

Contents

1. Introduction 7
   1.1 Problem 7
   1.2 Focus 19
   1.3 Method 28
2. The Natural Philosopher as Ancient Scientist 35
   2.1 Defining the Natural Philosopher 35
   2.2 Aristotle's Idea of a Scientist 39
   2.3 On Stone & Papyrus 44
   2.4 'Natural Philosophers' as the Presocratics 47
   2.5 The Roman Conception of the Scientist 49
   2.6 The Methods of Roman Scientists 56
   2.7 Mathematics & Causation 68
   2.8 Summary & Conclusion 93
3. The Roman Idea of Scientific Progress 97
   3.1 The Growth of Ancient Science 99
   3.2 Scientific Medicine up to the Roman Era 107
   3.3 Scientific Astronomy up to the Roman Era 129
   3.4 Scientific Physics up to the Roman Era 155
   3.5 Other Sciences 173
   3.6 Technological Progress 190

3.7 Was Roman Science in Decline? 240
3.8 Ancient Tales of Decline 270
3.9 Ancient Recognition of Scientific Progress 307
3.10 Summary & Conclusion 335

4. In Praise of the Scientist 341
4.1 Philosophers for Science 341
4.2 Literary Praise 349
4.3 Evidence of Elite Interest 367
4.4 Seneca and the Aetna 379
4.5 The Scientist as Hero in the Roman Era 391
4.6 The Scientist as Craftsman in the Roman Era 409
4.7 Lack of Institutional Support? 447
4.8 Evidence of Non-Christian Hostility to Science 455
4.9 The Path to Christian Values 461
4.10 Summary & Conclusion 467

5. Christian Rejection of the Scientist 471
5.1 Clement of Alexandria (c. 200 A.D.) 475
5.2 Tertullian (c. 200 A.D.) 483
5.3 Lactantius (c. 300 A.D.) 489
5.4 Eusebius (c. 300 A.D.) 496
5.5 Christian Anti-Intellectualism? 500
5.6 Evidence in the New Testament 503
5.7 Evidence from Christian Writers 521
5.8 Assessment of Christian Hostility 526
5.9 Exceptions That Prove the Rule 529
5.10 Medieval Christianity 535

6. Conclusion 543
6.1 Results 543
6.2 Applications 545
6.3 Speculations 550

Appendix A on Ancient Exploration 553

Appendix B on Science before Aristotle 555

Appendix C on the Books of Sextus Empiricus 559

Bibliography 561

Index of Ancient Inventions 625

General Index 629

About the Author 647

1. Introduction

The present study demonstrates that Christianity in its first three centuries was almost uniformly hostile or dismissive of the value of studying nature, while over the same period there was a significant contingent of influential pagans who embraced and expressed exactly the opposite attitude. Though there were also a variety of negative attitudes among the pagans at all levels of society, the early Christians shared nothing like the positive attitudes found among their pagan peers. The evidence for this includes not only straightforward surveys of direct and indirect expressions of Christian and pagan attitudes in extant literature, but also a survey of the actual and ideal status of 'science' in Roman education (which I treated in a previous volume, *Science Education in the Early Roman Empire*), as well as the appreciation and expectation of 'scientific progress' among Roman intellectuals. For context we will also survey what a 'natural philosopher' was imagined to be and do, and how they are an ancient analog to the modern notion of a scientist. The present chapter introduces the issue by explaining why that question concerns us, and what my focus and methods shall be.

1.1 Problem

One of the outstanding questions in history is why the Scientific Revolution occurred so late in the history of civilization. The state of science and philosophy in Greco-Roman society was remarkably advanced, more so than most people realize. Such a level was not achieved anywhere else in the world, nor again until the 16th century.[1] So why did the ancients not experience a more sweeping revolution in the methods and social role of science, despite seeming to have all the right ingredients in place for almost a thousand years? Why did that revolution only finally happen over the course of the 17th century?

There are two kinds of answer one can give to this question. Either it is all just blind luck—such a revolution could have happened in either era, but in the 17th century we just got lucky, the right individuals simply chanced upon the right discoveries at the right time—or certain necessary social-historical causes converged in the 17th century but not before. A third possibility would be some combination of both, which may be more probable.[2] There are good candidates for 'happenstance', but also indications of broad social forces. For example, the coincidental discovery of the telescope, the printing press, and the New World (not to mention gunpowder and the compass) are the most obvious catalysts many scholars credit for helping launch the Scientific Revolution, yet none of these developments were the planned outcome of the work of scientists but were the product of nonscientists with different goals working independently of the scientific community and each other. Yet at the same time there were broad trends influencing even these developments, such as a rising passion for experimentation and inventiveness among craftsmen, and a prolonged large-scale military and commercial competition among independent states sharing the same seas.[3] The Scientific Revolution seems to have gathered steam even before

---

1. Not even in China (despite comparable technological development): cf. Lloyd 1996a and 2002.

2. For the best and most complete discussion of these options, including the following examples and more, see Cohen 1994, who surveys a diverse range of scholarship on the issue. For examples of recent individual treatments of the Scientific Revolution, synthesizing a diverse range of theories, see Henry 2008 and Shapin 1996 (especially in light of Collins 1998: 523–69). For defenses of the very idea of a Scientific Revolution see Yerxa 2007.

3. Telescope: Sluiter 1997 and Van Helden 1977 (though obsolete but still useful is C.J. Singer 1921, who also treats the parallel history of the microscope, more thoroughly discussed in Disney et al. 1928 and Ford 1985). Printing press: Kapr 1996 and Eisenstein 1980. New World: Phillips & Phillips 1992 and Barrera-Osorio 2006. Compass: Aczel 2001 and Kreutz 1973. Gunpowder: Partington 1960 and Kelly 2004. All these studies exhibit the general trends noted, but for a broader

its zeitgeist was articulated (Francis Bacon, sometimes credited as a father of the Scientific Revolution, actually wrote fifty to a hundred years after a shift in the role and methods of science had already begun), yet happened so quickly (in less than two centuries the methodology and social role of science had radically changed, despite some resistance), and involved so many intellectuals converging on similar ideas all at once (many of whom had no direct contact with each other or with anyone we could call 'the match who lit the fire'), that the most credible explanation must surely include at least some broad socio-historical causes. Something must have been different about 16th and 17th century European society.

One issue that often comes up in attempts to resolve this question is the social position of the 'scientist', who before the close of the Scientific Revolution was only known as one or another variety of 'philosopher'. How respected and socially supported, or how marginalized or opposed, was this sort of person and their work? The present study provides the bulk of the answer to that question for the ancient period, particularly the last stretch of it, by which time any social, cultural, or ideological factors that would have converged to produce a scientific revolution should have had their effect. Whether the social position of the natural philosopher was actually different before or after the Renaissance is a question that must be left for future study. But some scholars insist there was a difference, and though their claims about the early modern period might also be questionable, only their claims about the ancient period will be thoroughly examined here.

In his broader survey of the historiography of the Scientific Revolution, H. Floris Cohen summarized past attempts at explaining why that revolution did not happen in the ancient world.[4] Cohen shows how the explanations vary considerably, but all amount to arguing that something was wrong with how the Greeks and Romans thought, which would not be corrected for another thousand years—either they lacked some ideological assumption that was required, or embraced some ideological assumption that got in the way. At least two scholars in his survey, G.E.R. Lloyd and Joseph Ben-David, proposed it had something to do with, in Lloyd's words, "the weakness of the social and ideological basis of ancient science," in the sense that "there was

---

picture see P. Smith 2004 and 2006.

4. Cohen 1994, esp. §4.2 "Why Did the Scientific Revolution Not Take Place in Ancient Greece?" (pp. 241–60).

no acknowledged place in ancient thought, or in ancient society, for science, or for the scientist, as such," because "the investigators performed different social roles as doctors or architects or teachers" than as researchers, while "the men who engaged in what we should call science had always been a tiny minority who faced the indifference of the mass of their contemporaries at every period," so as a result "the conditions needed to insure the continuous growth of science did not exist, and were never created, in the ancient world."[5]

This same argument had previously been made by Ludwig Edelstein, who claimed that ancient society "on the whole remained completely indifferent" to the value of science, and that this lack of public support hindered scientific progress by ensuring only very few would pursue it.[6] Edelstein finds evidence for both points in the fact that natural philosophy was hardly represented in mainstream education, which I have found is only qualifiedly true.[7] Likewise, Edelstein argues that this "lack of social recognition" was responsible for the "lack of permanent and stable forms of organization" for scientists or scientific research, and this in turn hindered the sciences. Edelstein and Lloyd are probably correct that these factors slowed scientific progress—its pace in antiquity was indeed slow, as we shall see—but it is not immediately clear how a slower pace would prevent a scientific revolution, rather than only make it take longer to happen. Their evidence does establish that natural philosophy, a category of endeavor that included what we now call 'science' (see section 1.2.III below), was to *some* extent marginalized in ancient society. However, this leaves open the question of *how* marginalized, and ultimately of whether there is any significant difference between science's marginalization in antiquity and its social position immediately before the modern Scientific Revolution.

Though Lloyd and Edelstein make many correct observations about the social status of the natural philosopher in antiquity, these facts become problematic when turned into explanations. Certainly in antiquity there was no distinct social category of the 'scientist' *per se*, but neither had there been in the 16th or early 17th centuries—the creation of a distinct and recognized role for science and scientists was clearly a *consequence* and not

---

5. Quotations assembled from Lloyd 1973: 178, 176, 170, 174.

6. Edelstein 1952: 598–99.

7. See my book *Science Education in the Early Roman Empire* (= Carrier 2016).

a cause of the Scientific Revolution, for it seems only to have *followed* the conceptual separation of speculative from 'experimental' philosophy. Cohen recognizes that this is a serious problem with the theory.[8] The same problem undermines Edelstein's assertion that "the rhetorical character of many scientific books is an indirect indication of the insecurity of the position of science" because it demonstrates a desperate need "to gain the approval of public opinion," but the exact same features are found in scientific books right through the 17th century, exactly when Edelstein and Lloyd assume the social position of science had changed.[9] In contrast, the rhetorical nature of *ancient* scientific treatises was most often aimed at rival scientists, not opponents of science, and was a direct consequence of the particular nature of ancient education and discourse, which was inherently agonistic, and thus the rhetorical character of ancient books does not indicate anything peculiar about ancient science as such, since it was a general characteristic of ancient society as a whole.[10] Lloyd even argues (probably correctly) that this fact was essential to the rise and success of Hellenistic science.[11]

It is certainly true that the number of scientific investigators in antiquity was never large and their social position did vary among different social groups and periods, yet scientific knowledge and methodology continuously improved between 400 B.C. and 200 A.D. (from Aristotle and his predecessors to Ptolemy and Galen), as Lloyd himself admits.[12] Hence he specifies that (now adding my emphasis) "the conditions needed to *insure* the continuous growth of science did not exist." In other words, the conditions for growth existed, but not the conditions that could prevent the

8. Cohen 1994: 257–59, in respect to Ben-David's elaboration of Lloyd's thesis. I discuss this further in chapter 5.2.

9. Edelstein 1952: 600; one need only skim the works of Copernicus, Gilbert, Bacon, and Galileo to find them packed with essentially the same defensiveness—in fact, arguably more so.

10. For instance, this is shown (using Vitruvius as an example) in J.-M. André 1987 and (with a variety of examples) in Barton 1994a. For a detailed analysis of the structure and intent of science writing in antiquity, see Asper 2007.

11. Lloyd 1990 and 2001: 202–07.

12. The number of published research scientists in any given generation in antiquity is estimated at around a hundred (which would mean there were at least five hundred over the three centuries of the early Roman empire): see Carrier, *Science Education* (2016), pp. 29–31.

widespread social embrace of superstitious and antiscientific thinking, as Lloyd argues happened after 200 A.D. In other words, the flower of science in antiquity was growing, but easy to kill.[13]

How this relates to social perceptions of the natural philosopher is articulated in more detail by Joseph Ben-David, but his analysis is fraught with even greater problems. He claims that in antiquity "scientists were regarded as philosophers interested in a particularly esoteric and impractical branch of knowledge."[14] But he does not identify *who* thought this, even though there were many different segments of the population with different attitudes and influences. As we shall see in later chapters, Ben-David's assertion does not hold up against considerable evidence to the contrary. Whatever the case, his overall theory is that ancient science never underwent a scientific revolution because it developed in a "slow and irregular" pattern due to the "absence of the specialized role of the scientist and the nonacceptance of science as a social goal in its own right." But, he argues, "in order to become accepted by others and perpetuated, people have to fulfill a recognized social function," and therefore:

> Before science could become institutionalized, there had to emerge a view that scientific knowledge for its own sake was good for society in the same sense that moral philosophy was. Something like this idea had apparently occurred to some natural philosophers. But in order to convince others that this was so, they had to show some moral, religious, or magical relevance of their insights. As a result, the scientific content of natural philosophy was either lost or concealed by the superstitions and rituals of esoteric cults.[15]

Like Lloyd and Edelstein, Ben-David never demonstrates that a notable rise in the rate and regularity of scientific discovery was a *cause* rather than an *effect* of the Scientific Revolution (Ben-David's analysis notably lacks careful attention to chronology), but whatever the case may be regarding that, his assertions about antiquity are far of the mark.

First, by the Roman period, doctors, astronomers, and engineers certainly had specialized and recognized social roles, as did the natural

13. I will discuss this in chapters 2.5 and 5.3.

14. Ben-David 1991: 301.

15. Ben-David 1984: 31.

philosopher generally. Though still strongly associated with other branches of philosophy, and with particular philosophical schools, we shall see how natural philosophers as a class were nevertheless recognized with a distinct name (*physicus* in Latin, *physikos* in Greek) and there was explicit discourse about their function and value in society. Likewise, while he concedes that the "idea" of science's value "had apparently occurred to some natural philosophers," Ben-David claims they buried science in the very attempt to convey its value to society. But this is not believable. It is hard to find what "scientific content of natural philosophy" was actually "lost" during the ancient period—by this process or any other—rather than being lost or buried *in the middle ages*, when inattention, disinterest, and the limited preservation of scientific knowledge was far more typical, and "superstition and ritual" more widely prevalent in eclipsing interest in scientific research, quite literally represented by Christian monks scraping the ink of the scientific treatises of Archimedes off the page of his book and writing hymns to God in their place.[16]

Ben-David also never demonstrates that science was *ever* pursued only "for its own sake." Even Francis Bacon, widely regarded as the paradigmatic defender of an increased social status for science and a key player in the development of the Scientific Revolution, never argued that science should be pursued "for its own sake," but always for some moral or utilitarian end. Specifically, in fact, for the "fame" and "true glory" of the king of England, and for the benefit of "charity" and "use." Notably, Bacon asserted these justifications for science specifically to counter opponents of scientific research among contemporary priests, aristocrats, and scholastics, thus demonstrating that Bacon was not representing or responding to a shift in the social status of the scientist, but joining in the attempt to *cause* one. In fact, contrary to Ben-David's thesis, Bacon struggles at length to justify science by articulating its 'moral and religious relevance', the very thing Ben-David claims supposedly doomed science in antiquity, though there is no evidence of that.[17] We shall see that many among the Roman elite

16. Netz & Noel 2007.

17. Bacon 2001: 3–42 (quotes from 5 and 10; opposition from priests, aristocrats, and scholars: 5–38; religious relevance: 38–43; political, military and economic utility: 43–56; moral value: 57–62). Though this was originally published in 1605, I cite the modern edition for convenience. An excellent discussion of the context

valued 'science' in its ancient sense, and for the same reasons Bacon did (and argued others should). Though it was probably true that most educated people in antiquity "were not very interested in empirical science," it is still unclear how that differed from Bacon's day, when most educated society instead comprised Bacon's opponents or swelled the ranks of disinterested bystanders. Likewise, given the evidence examined later, we shall find it hard to maintain Ben-David's additional claim that an inability to fit scientific findings into some specific philosophical framework divorced scientists from philosophers.[18]

The only point Ben-David certainly gets right is that natural philosophers never achieved the same status enjoyed by moral philosophers, or even orators, poets, and other literati, and this is essentially the same point made by Edelstein and Lloyd.[19] But again it is hard to see how the situation differed in Bacon's day, when priests, artists, and scholars outside the sciences continued to enjoy greater social prestige. Surely, even well after the 16th century, parents preferred to see their sons in the clergy, military, or law, rather than working as scientists. Even in the early 19th century, when Jane Austen crafted Edward Ferrars' monologue on his family's failure to convince him to take a profession, only the church, army, navy, and the law win any mention as the preferences of himself or his family. Doctor, engineer, or research scientist never even come up, and it is incredible to imagine they ever would have.[20] Indeed, "at what time in the world's history has the attitude of the upper and controlling classes been different?"[21]

---

of a broader Christian condemnation of 'curiosity' against which Bacon is arguing is provided in P. Harrison 2001; and Neil 2004: 99–138, concluding, "on the whole" Christian institutions "condemned curiosity" and resisted efforts to rehabilitate it even in the 17th century (ibid., p. 157). And on Christian (not necessarily religious) hostility to science and scientists during the 17th century, which was arguably more severe than any non-Christian hostility known in Roman times, see Lougee 1972 (esp. pp. 45–60) and Crouch 1975 (esp. pp. 37–90).

18. Ben-David 1984: 39.

19. This is the entire thrust of his argument in Ben-David 1984: 39–44.

20. Jane Austen, *Sense and Sensibility* 1811, vol. 1, chapter 19 (= Austen 1996: 92–93).

21. D. Lee 1973: 70–71; Carrier 2016: 8. Even Peter Green, who otherwise makes much of finding this aristocratic attitude in antiquity, nevertheless admits it can be found in every era, even in the 18th century writings of David Hume: P. Green

So the Scientific Revolution had certainly elevated the status of the scientist, but not so high as Ben-David seems to imagine. Hence observing the same or similar situation in antiquity, of natural philosophers occupying a lower social status than other revered groups but not lacking in social status altogether, does not get us very far. As we shall see, to be a natural philosopher guaranteed a certain degree of respect and prestige among the elite, at least as much as it did in the years before Bacon argued they deserved even more. It is thus ironic to see Ben-David claiming that the social marginalization of scientists was demonstrated by the fact that under the Ptolemies they were "simply parts of the entourage of the court," apparently unaware of the fact that there was hardly any higher status to be had among the elite in ancient society. And yet even under the Romans, scientists were not cast into the streets.

It is also important to observe that what aristocrats *said* did not always correspond to what they *did*. A relevant analogy is the status of fine arts under the early Roman empire. An open hostility to the study and practice of music, far greater than any that can be found against science, is easily seen among the writings of the Roman elite. Yet this did nothing to prevent music from being widely learned, practiced, and enjoyed—*especially* by the elite.[22] Likewise for painting and sculpture—a career as a painter or sculptor was looked upon by the elite with open disdain, and their work was often condemned as an immoral luxury, and yet painters and sculptors continued to earn fame and wealth, and their work was always in demand and often of high quality.[23] Sure, Roman aristocrats would never deign to become a sculptor, and looked down on sculptors as beneath them, and sometimes even railed against the decadence of their work, yet it was their passion for highly skilled art and their bottomless bank accounts that sustained a prosperous industry of superb sculptors across the empire who produced beautiful works exhibiting an exceptional knowledge and skill that would not be seen again until the Renaissance. If the presence of negative attitudes among various segments of the pagan elite did not stop art, it could hardly

---

1990: 470–73 (cf. 470) and 855 (notes 38 and 39). For another example and further discussion see chapter 4.6.1.

22. M. Clarke 1971: 52–53. On music education in antiquity, see Carrier 2016 (index).

23. Blagg 1987; Rawson 1985: 88–89. See also chapter 3.8.1.

have impeded science, which was considerably more respectable.

Of course, the vast majority of the population in antiquity was poor and uneducated and did not share the interests or values of the upper and middle classes.[24] But in all ages before the 18th or 19th centuries the uneducated masses probably held no appreciable value for science or were even suspicious of the elitism and impiety of scientists. But since in antiquity these groups did not control any significant economic or political institutions that could affect the outcome of science, either to advance or oppose its promotion or progress, their attitudes toward it were probably as insignificant in antiquity as they were in the 17th century. It was only when those embracing such anti-elitist attitudes found rapid advancement in the Christian Church, and then were elevated to positions of real political and social power when the Christian Church became an official state religion, that their hostility or indifference to science actually succeeded in all but killing it. As we shall see, before the rise of Constantine the attitudes of the authorities of the Christian Church were almost uniformly hostile or notably indifferent to scientific research, so their elevation to power would have predictable results.

Hence this subsequent rise to dominance is the reason for our attention to Christianity's formative years. For this may go a long way toward explaining why the decisive rise of the Christian Church in the 4th century A.D. secured nearly a thousand year delay in the advance of theoretical science, which only the weakening or outright shattering of church power and control appears to have ended.[25] Though it is becoming increasingly popular to deny this, no one to date has presented evidence of any *significant* advances in the sciences being made at any time between 300 and 1200 A.D. Rodney Stark, for instance, fails to muster a single example in his entire survey of medieval "accomplishments."[26] The trivial or incidental

---

24. For what I mean by "upper" and "middle" class in Roman society see chapter 4.6.

25. On the breakdown of Church power and authority as a necessary step toward the Scientific Revolution see P. Harrison 1998.

26. Attempts to argue that science continued unabated during the middle ages are summarized by Stark 2001 and 2005 (and see chapter 5.10). For his survey of medieval 'accomplishments' see Stark 2001: 130–34 and 2005: 38–54. Some are not medieval but in fact ancient (see chapter 3.6). For more detailed attempts to document advances in medieval technology see: L. White 1962 and Gies & Gies

does not count (such as minor modifications to waterwheel technologies that had already been developed and employed in Roman times), nor does the repetition of prior achievements (such as the rediscovery of alternative theories of motion or vision already developed in antiquity), nor mere inventions unconnected with any formal science (like the development of the stirrup or compass), since in general all three phenomena occur in all ages in all cultures and thus do not distinguish any culture or era from any other, so there is nothing meaningfully 'scientific' about them. In matters of *genuine* scientific progress, during the middle ages there is only silence. In astronomy there is hardly anything significant between Ptolemy and Copernicus—indeed, very little progress was made even by Copernicus, who merely resurrected an alternative theory that some of Ptolemy's astronomical colleagues and predecessors had already been advancing. Real progress in astronomical theory, discovery, and explanation would have to await the work of men in subsequent generations, like Brahe, Kepler and Galileo. In medical science there is nothing noteworthy between Galen and Vesalius—and Vesalius merely picked up essentially where Galen left off, leaving major theoretical advances for men like Harvey, whose own methods were not all that far from Galen's. Even in physics there is nothing truly novel to be found between the time of Hero or Ptolemy and the works of Gilbert or Galileo.[27] As we shall see in chapter three, the picture is the

---

1994 (though both rely on obsolete scholarship and are frequently wrong; e.g. White's conclusions regarding water power are now well refuted: e.g. Walton 2006 and the many sources on ancient water power cited in chapter 3.6). There is a concise but thorough refutation of Stark in Carrier 2010.

27. Nicolaus Copernicus articulated his heliocentric theory during the first half of the 1500's, Tycho Brahe improved astronomical data in the second half of the 1500's, and Johannes Kepler's significant work began in the first decades of the 1600's. Andreas Vesalius started achieving and publishing his improved anatomical results in the middle of the 1500's, William Harvey in the early 1600's. Galileo's work occupied the first third of the 1600's, and he was advancing new and experimental work started by others no earlier than the mid-1500's. William Gilbert completed his own work in the late 1500's and published in 1600. On all these basic details see Henry 2008 and Shapin 1996. Of course some notable contributions were made by Muslims in the interim (e.g. Hill 1993), but these were fleeting, scarce, and of limited significance. Most of the advances they are alleged to have made were actually recoveries of lost ancient knowledge and thus not in fact an example of progress. Likewise, some of John Philopon's commentaries in the 6th century A.D.

same in every scientific field.

But the absence of significant scientific development in the middle ages has never been hard to explain. What requires explanation is why science began to be avidly and successfully pursued again in the 15th and 16th centuries, and why it then roared ahead of ancient accomplishment already by the 17th. After surveying the paucity of significant advances throughout the early middle ages, Crombie then links the rise of modern science with the 'rediscovery' in the 12th through 14th centuries of theoretical and conceptual ideas that had already been extant in antiquity, and locates in those centuries the first stages of repetition or corroboration of experimental and theoretical work already done in antiquity. As Crombie's evidence shows, by the 15th century, Europe was roughly back at the same stage of scientific understanding that had been achieved by the early 3rd century A.D. *Then* in only two centuries Europe went on to surpass ancient science in a revolutionary way.[28]

Did ancient attitudes toward the natural philosopher have anything to do with preventing this same advancement under the Roman Empire? Or was Roman science right on the same track, only two centuries away from seeing its own scientific revolution, but instead shot down by the collapse of Roman economic and political institutions in the 3rd century, followed by the rise to power of the Christian Church shortly thereafter?[29] This question

---

are claimed as an exception, though he doesn't say he conducted the experiments he refers to, so we can't say for sure that he wasn't just repeating what he'd read in earlier scientific treatises—since he only reproduced conclusions already reached by Strato or Hipparchus long before the days of Ptolemy and Hero (see discussion in chapter 3.4; as noted in Cohen & Drabkin 1948: 217, n. 1, "It is difficult to say to what extent Philoponus is original in these views and to what extent he is recording an anti-Aristotelian tradition"); either way, Philopon's work does not contain any novel physics (indeed, his work is hardly even scientific at all). Philopon's claims were also dismissed or ignored just as Strato's were. In Philopon's case, this might have been due in part to his being declared a heretic; in Strato's case, it might have been due in part to his being an atheist. See discussion in chapter 3.5, with: Russo 2003: 351; Drake 1989; Wolff 1987; *ODCC* 896; *EANS* 436–37 and 765–66; *OCD* 1135 and 1406; *DSB* 7.134–39 and 13.91–95; *NDSB* 4.51–52 and 6.540.

28. Crombie 1959 (even more thoroughly confirmed in Crombie 1994).

29. On the collapse of the Roman political and economic institutions in the 3rd century A.D., and the subsequent rise of Christian power in the 4th century, see Drinkwater 2005; Southern 2001; Michael Grant 1999; Rathbone 1997; Cameron

is too big to be answered here. But we cannot even begin to answer it without an accurate understanding of the essential pieces to the puzzle, and one such piece is how the scientist and his work was perceived in the ancient world before its fall, particularly whether any significant and influential segment of society held them in esteem, and whether the triumphant Church would inherit an ideology that was favorable or unfavorable to the scientific enterprise. Hence the purpose of this study.

1.2 FOCUS

Our concern is to analyze attitudes toward the 'natural philosopher' before the rise of Constantine, the first Christian emperor, especially as this will inform any connection between such attitudes and the Scientific Revolution. This requires narrowing our focus (I) by chronological period, (II) by general cultural category, and (III) by the specified subject of 'natural philosopher'. Within these parameters, for reasons already explained above, the bulk of our attention will be paid to the two most importantly contrasting groups: Christians and pro-science pagans.

I. Chronological Focus

Though the concept of the *physicus* or 'natural philosopher' remained largely unchanged throughout antiquity, the early Roman period from 100 B.C. to 313 A.D. provides us with the widest diversity of authors using and discussing the word.[30] Their reception and treatment of the concept reflects

1993; Brown 1992; MacMullen 1988 (with 1984 and 1997); and Williams 1985. Chapter three will explore the idea that Roman science may have been on the right track just immediately preceding these events. Two plagues also struck (the first in 165–180 A.D. and the second, widely regarded as substantially worse, in 251–266 A.D.) which may have further damaged or disrupted the social and economic system in the 3rd century (see Jackson 1988: 172–73, who estimates losses could have been as high as 25% of the population in each case, though it was probably less).

30. Note that I shall employ the traditional convention of B.C. and A.D. in lieu of the culturally neutral B.C.E. and C.E. because the original notation is more familiar and there is no good reason to change it. Both indicate the same division of eras, which was the invention of a Christian and only makes sense as such. Changing

the particular interests of this period, which is an important one in the history of science, lying right on the threshold of the middles ages, marking essentially the end of significant scientific progress for centuries to come. Hence the chronological scope of this study shall encompass the last major phase of ancient science, the period after the end of the Ptolemaic patronage of the sciences in Egypt, when the dominance of the Roman Empire over the Mediterranean was most secure, and, for the first and last time, the Western World (as then known) was essentially united under a strong, universal government. This began in the 1st century B.C., then started to fall apart in the 3rd century A.D., and was well in decline by the 4th.

Significant signposts at each end help define our period of interest, which begins shortly after 100 B.C. with the converging circumstances, first, of Rome's conquest of the Mediterranean, when every major nation came under the direct or indirect control of Roman leadership, setting the stage for what is called the Pax Romana or "Roman Peace," and, second, the boldest and most notable promotion of natural philosophy in the Latin language by Lucretius, through his famous epic poem *On the Nature of Things*.[31] Our period then ends with the dawn of the era of Constantine, when the chaos of the 3rd century was partly and tentatively 'solved' by adopting Christianity as the semi-official religion of the Empire shortly after Constantine's rise to power in 313 A.D., thus marking the beginning of a very different political and intellectual atmosphere than had existed before. The period from 100 B.C. to 313 A.D. also happens to mark the era that molded and produced the last great scientists of the ancient world: Hero, Ptolemy, and Galen. It includes their unique and relatively stable social circumstances during the phase of ancient history called the 'Second Sophistic' (which is typically

---

the acronyms does nothing to conceal that fact and therefore serves no purpose. Analogously, calling the sixth day of the week 'Saturday' (literally "Saturn's Day") does not entail embrace of a Eurocentric worldview or belief in the God Saturn. It's just clear English.

31. The exact date of publication for the Lucretian poem is debated, but it must have been sometime between 100 and 40 B.C. Before Lucretius, Epicurean natural philosophy had been introduced into Latin more clumsily by Gaius Amafinius, but at any rate, both authors lived through the early 1st century B.C., as did a few other Epicureans writing in Latin. See Rawson 1985: 284; Farrington 1946: 88–91; with *OCD* 67, 291, 863–65 (s.v. "Amafinius, Gaius"; "Catius, Titus"; "Lucretius (Titus Lucretius Carus)"), *DSB* 8.536–39 (s.v. "Lucretius"), and *EANS* 512–13.

dated from 50 to 235 A.D.) as well as the century immediately preceding and thus producing it, and the century immediately following and thus marking its decline.[32]

II. Cultural Focus

Culturally, we shall concern ourselves with Greco-Roman society as a broad category, since it is only in that cultural context that 'natural philosophers' lived and interacted in any relevant or meaningful sense. Other cultural or linguistic groups within the Roman empire or on its borders are thus mostly outside the scope of this study. Since Greek and Latin societies were more similar than different in their customs, values, and beliefs, and in our period of interest were increasingly integrated, we shall use the word "Roman" to designate everyone fluent in either Latin *or* Greek living within the borders of the Roman empire, regardless of an author's actual language or citizenship. Otherwise, actual differences in language shall be indicated with the terms "Latin" and "Greek," and differences in cultural outlook will be noted when relevant. Most intellectuals during the period in question were essentially bilingual anyway, or at least were expected to be, while most illiterate inhabitants of the empire probably spoke or understood some Latin or Greek.[33] So the term "Roman" is employed here more as a political and chronological category than a cultural one, but even as such it encompasses the common and interacting elements of the Greek and Latin cultures of the time.

III. Subject Focus

The history of ancient science begins with the convergence of two phenomena: a rising interest in acquiring a theoretical understanding of *why* things are as they are or act as they do, specifically in terms of a causal system rather than a mythology of divine or supernatural agency, followed by a rising consciousness of methodology and the importance of

---

32. On the derivation and meaning of the phrase 'Second Sophistic' and on dating when it began and ended, see Bowersock 1974, Anderson 1993, Whitmarsh 2005, and *OCD* 1337–38 (s.v. "Second Sophistic"). For what it involved, especially in respect to Roman science, see von Staden 1995 (using Galen as an example) and Brunt 1994.

33. This point is discussed in more detail in chapter three of Carrier 2016.

epistemological debate. Modern science is a perfection of both endeavors, and thus ancient science falls short of it only in degree. Hence there are both parallels and differences between modern and ancient science.[34] Before the Scientific Revolution, 'science' was not as dependent on experimentation or the hypothetico-deductive method that has proven so successful today, although it did not do without them. It was also either subservient to philosophy or heavily influenced by philosophical speculation. Nevertheless, Ptolemy's rigorous use of mathematics to describe planetary motion and the propagation of sound and sight, and his testing of theories against observations, was by any measure scientific, as was Galen's insistence upon exploratory anatomy and the need to develop a physiological theory in accord with observations, in both cases emphasizing the unification of theoretical reason with empirical observation—and, incidentally, both emphasizing the essential importance of mathematics in such endeavors. In Ptolemy's case this is too obvious to require demonstration. All of Ptolemy's treatises mathematize nature, in optics, harmonics, geography, astronomy, even a lost work in mechanics. He did not produce any major theory of natural phenomena that he did not attempt to describe mathematically and demonstrate empirically. Galen's position is perhaps more surprising, since medical science had not been properly mathematized in any significant way (and would not be until after the Scientific Revolution). Nevertheless, Galen argued explicitly that all empirically-confirmed mathematical descriptions of natural phenomena were superior to philosophical speculation, and that the same rigor and principles of mathematical reasoning must be employed as much as possible even when empirically demonstrating theories in medicine and physiology.[35] Such examples demonstrate that the *idea* of

---

34. For a broad yet brief summary of both the differences and similarities see Edelstein 1952. On the problem created by these issues for identifying any pre-modern activity as 'science' see Sharples 2005: 1–7, Lloyd 1992b and 2004, and Rihll 2002: 7–9, with Dear 2005 and *OCD* 560–81 and 717–18 (s.v. "experiment" and "hypothesis, scientific"). For a more complex approach to defining science (not adopted here): Russo 2003: 15–30. For something simpler: Healy 1999: 100–01. In contrast, there is no use or merit in a complete rejection of modern categorization of the sort voiced in French 1994: ix-xxii (which is aptly criticized in Healy 1999: 115 n. 1).

35. We will summarize Ptolemy's work in chapters 3.3 and 3.4, but on Galen's respect for mathematical method, see relevant discussion and note in chapter

science (as we now know it) was growing in antiquity, though it had not yet flowered into the methodological revolution that characterized the 17th century. Nevertheless, ancient science presaged modern science in often startling ways, in both knowledge and method, and it certainly had a causal-historical role in the development of modern scientific thought.

Studying these connections requires identifying what ancient word, if any, designated the practitioners of ancient science. In older English translations of ancient texts, the noun *mathêma* and its adjective *mathêmatikos* have often been translated as "science" and "scientific" or "scientist," respectively. But this is not a consistently sound practice. Such words had two connotations, and one was far too broad, and the other far too narrow, to correlate with the modern English words "science," "scientific," or "scientist." In their broader connotation, *mathêma* and *mathêmatikos* meant any or all academic subjects, education, and learning—representing the whole scope of the sciences and humanities combined, or vaguely defining any field in that category.[36] In this sense, *mathêma* is far closer in meaning to the modern word "education" or "higher education" while *mathêmatikos* is far closer in meaning to the modern word "academic."[37] In their narrower connotation, these words meant "mathematics," "mathematical," and "mathematician," whether applied or abstract, and even when thus employed in reference to the mathematical sciences (like astronomy or mechanics), this was mostly by metonymy, due to the heavy employment of mathematics in those

3.6.VI here (and chapter seven of Carrier 2016). On the empiricism and scientific method of both Galen and Ptolemy, see relevant notes and discussions here in chapter three (especially 3.7).

36. *LSG* 1072 (s.v. "*mathê, mathêma*" § 1–2; cf. "*mathêmatikos*" § 1). This was always the meaning of the Latin word *scientia*, despite being the etymological source of our word "science." In fact, *scientia* was often even broader in connotation, designating any and all knowledge of any kind, cf. *OLD* 1703 (s.v. "*scientia*"). Likewise, words like *ars* or *technê* (designating *any* art, skill, or science that can be articulated systematically), or *epistêmê* and its cognates (designating the epistemologically well-grounded knowledge of *any* skill or subject), were also either too broad or too specialized in connotation, even though they also sometimes properly translate as 'science' (see *LSG* 660, s.v. "*epistêmê*," "*epistêmonikos*," "*epistêmos*," etc.; *LSG* 1785, s.v. "*technê*," "*technikos*," "*technitês*," etc.; and, e.g., *OLD* 175, s.v. "*ars*").

37. As in 'scholarly', to be distinguished from the use of "academic" in a modern pejorative sense, or from the completely different use of "Academic" as a proper noun in reference to one of the ancient schools of philosophy.

arts. Such a use did not in itself denote the specifically scientific—that is, *empirical*, or even theoretical—aspects of those same arts, and certainly did not denote what we mean by 'science' in any *general* sense.[38] Though in appropriate contexts 'science' would be a fair translation of *mathêma* and 'scientist' would be a fair translation of *mathêmatikos*, this would be so only in those contexts where the terms do happen to designate what we would mark with those words in English.[39] Since these words translate as "science," "scientific" or "scientist" only in certain contexts, and only in connection with a limited range of sciences, they clearly are not the closest thing the ancients had to our words "science," "scientific" or "scientist."

In contrast, in ancient texts the words *physika* and *physikos*, which in their broad connotation meant "natural" in nearly every sense of the modern English word, and in their commonly narrow connotation translate as "natural philosophy" and "natural philosopher" respectively, always denoted the content or study of nature, and in that latter sense always encompassed all theories of nature and all methods of testing or rejecting them, as well as the facts or conclusions thus obtained.[40] These words are therefore as broad and nearly as narrow as our words "science" and "scientist" today, and thus make a far closer fit than *mathêma* and *mathêmatikos*. The words *physika* and *physikos* are as broad because they did not designate only certain fields of inquiry but *all* branches of the study of nature, just as our words "science" and "scientist" do today. And they are almost as narrow, because they never encompassed or denoted subjects in the humanities, and were only broader in connotation than our words "science" and "scientist" for the simple reason that *all* methods of "studying" nature and *all* "conclusions" thus reached, whether sound or ridiculous by modern standards, were denoted by those words, whereas, being on the receiving end of the Scientific

38. *LSG* 1072 (s.v. "*mathê, mathêma*" § 3–4; cf. "*mathêmatikeuomai*" and "*mathêmatikos*" § II.1–2). This was always the meaning of the Latin word *mathematicus*: see *OLD* 1084 (s.v. "*mathematicus*[1]" and "*mathematicus*[2]").

39. The same applies to ancient words for 'engineer' (on which see Donderer 1996: 16–24), and one could build a similar case around ancient words for doctor (not always designating a scientific healer) or astronomer (which sometimes meant astrologer).

40. I will demonstrate this in chapter two. But for an introductory discussion of the contrast between *mathematicus* and *physicus* see Russo 2003: 187–94. See also chapter 2.2 and 2.7.

Revolution, we now narrow the range of methods appropriately designated "science" to what is strictly and soundly empirical, and narrow the range of conclusions appropriately designated "scientific" to what has actually been demonstrated by those methods. Accordingly, a modern "scientist" is someone who employs those kinds of methods to demonstrate those kinds of conclusions. But apart from this narrowing of focus, the ancient words *physika*, as "natural philosophy," and *physikos*, as "natural philosopher," were essentially identical to our words "science" and "scientist," at least in aims, interests, and subject matter.

A passage in the Latin author Aulus Gellius exemplifies this distinction. Writing in the late second century A.D., Gellius describes (according to legend) what used to be done in the first real "school" of philosophers, that established by Pythagoras (notably in Italy, not Greece). Students first had to pass a stage of keeping silent and listening for two or more years, during which they were called *akoustikoi*, "auditors." Then they advanced to the next stage—and:

> During this stage they were called *mathêmatikoi*, obviously from those arts they were then learning and practicing, because the ancient Greeks called geometry, gnomonics, music, and other higher disciplines *mathêmata* (although commoners call *mathêmatikoi* those who should be called by their ethnic name, Chaldaeans [i.e., Babylonian astrologers]). After that, once equipped with a skill in these studies, they advanced to observing the operation of the universe and the principles of nature, and that was when they were finally called *physikoi*.[41]

The observation is then made that in Gellius' day students did not respect this process and simply skipped the listening part and the mathematical studies and insisted instead on being taught whatever subjects they were interested in, even though they were "entirely without preparation, education, or a knowledge of geometry."[42] This must mean students were

41. Aulus Gellius, *Attic Nights* 1.9.6–7.

42. Aulus Gellius, *Attic Nights* 1.9.8, using (though writing in Latin) the Greek words *atheoretoi*, *amousoi*, *ageometretoi*, meaning "without having observed or contemplated," "without culture or education," and "without a knowledge of geometry," respectively—or in other words, "unprepared, uncouth, and mathematically illiterate." See *LSG* 31–32 (s.v. "*atheôrêtos*"), 85 (s.v. "*amousos*"), and 346 (s.v. "*geômetrêtos*").

not pursuing a full course of preparatory training in the mathematical and contemplative arts, but these are clearly not 'science' in the sense of empirical study of the natural world. Gellius understands the latter to be a separate activity, which ideally *mathêmata* only prepared one for. Thus he does not regard the *mathêmatikos* as a scientist in any sense we would recognize, but he clearly sees the *physikos* as such, or as near to it as anyone would have been in his day. And the context clearly represents this as the common view of his time.

Therefore, the focus of this study is the *physicus* as 'natural philosopher'. The social role of the *physicus* was the closest the ancients came to the social role of 'scientist' today, representing in many ways the sociohistorical precursor to the modern scientist (more evidence of which I will present in chapter two). So we will focus on natural philosophers, and as much as necessary on what they did (their methods, interests, and ideals), but we shall emphasize those natural philosophers that most resemble or anticipate what would eventually become modern scientists, since our greatest interest lies in those particular natural philosophers who adopted empirical values and engaged in at least some empirical research toward resolving questions about nature—even though there were also natural philosophers with little interest in either. For this reason, the words 'science' and 'scientist' will be used throughout this study (and have already been used above) to indicate this distinction between the increasingly empirical (and thus proto-scientific) natural philosopher and all natural philosophers generally. So when used of the ancients, the word 'scientist' will denote those natural philosophers who are identifiable precursors, in both interests and methodology, to what we now mean by 'scientist', while 'science' shall denote their most empirical or empirically-directed activities. In contrast, the term 'natural philosopher' (and the Latin and Greek equivalents) shall denote the entire class of ancient theorizers about nature.

This definition of 'science' and 'scientist' allows us to recognize the differences between the subcategories of 'ancient scientist' and 'modern scientist' without excessive anachronism or obscurity. The distinction thus formed between a 'scientist' in our qualified sense and the broader class of 'natural philosopher' (and between 'science' and the broader category of 'natural philosophy') did not exist in antiquity, but that does not mean the existence of such distinctions in antiquity are a modern fiction. There was indeed a difference between the more empirical *physicus* and their more speculative colleagues, and between their more empirical conduct and their more speculative. This difference was simply not yet recognized or given the proper appreciation. Such recognition and appreciation would eventually become a defining feature of the Scientific Revolution. Indeed, it may have been in the early stages of being recognized by the time of Ptolemy and Galen, but subsequent history thwarted any progress in that direction for over a thousand years.

In taking this position I do not mean to imply that only the more 'scientific' of ancient natural philosophy is worthy of interest. Rather, it is merely the most relevant to this study's present concern, which is to aid in explaining one element of the rise of modern science. The less empirical side of ancient natural philosophy, and everything that would eventually be abandoned as unscientific, is certainly worthy of attention (and in fact it will not be entirely neglected here), but a detailed study of the nature of ancient natural philosophy as a whole, on its own terms, would be a different project.[43] Nor do I consider the rise of modern science as the inevitable end result of any process begun in antiquity. Rather, I see modern science as only a contingent result of events and conditions both in and after antiquity, but one that is peculiar, and of considerable significance to understanding ourselves and our society, and therefore deserving attention in its own right as a historical problem. But I do adopt as a controlling assumption throughout that modern science has produced more, and more accurate, knowledge of the true facts of the world, and therefore any system of methods that approaches those now known to increase scientific knowledge, in this sense, is 'better' than any system of methods that does not perform as well, and likewise the *results* of such 'better' methods are themselves 'better' in the limited sense of being more accurate or correct. And I believe antiquity can be judged by these standards, as long as we are sympathetic to their reasons for falling short of them. I take this view because I embrace the improvement of knowledge as a fundamental value, and identify such improvement by its evident success in practice. For example, if modern science were not 'better' at identifying the true facts of the world it could never have landed a man on the moon or harnessed the power of the atom. And yet the story of how that became possible begins in antiquity.[44]

43. For examples of such an approach see Lehoux 2012 and French 1994.

44. My position thus does not correspond to any of the stereotyped battlelines

1.3 METHOD

Since our objective is to identify social attitudes toward a particular category of person, almost all our relevant evidence will be found in what ancient writers said.

In terms of actions, we shall see that very little was done on a social scale that indicated any particular value or disdain for natural philosophers or natural philosophy in general. In broad social terms, ancient society was neutral or indifferent toward them. Obviously, there were no scientific research institutes funded by the Roman government or even by private benefactors, nor were there any research universities in the modern sense (though something akin to them as educational institutions did exist). There were also no other social acts or institutions that promoted science or natural philosophy specifically (though there were organized social and academic societies for scientists). On the other hand (at least in our period of interest) there were no laws passed that opposed or hindered scientific research or speculation, and no outraged mobs tearing scientists limb from limb.[45] The astronomer Hypatia would not suffer that fate until the early 5th century, and the development of laws and acts designed to control and limit intellectual authority did not begin in any significant sense until the late 4th century, under Christian rule, all well after our period of interest.[46]

in the debate over a so-called "Whig interpretation of history," which is in my view a terrible anachronism (originally having to do with a specific question in the political history of Britain). I follow and agree with the analysis of Brush 1995 and Mayr 1990, that one can validly be presentist and progressivist without being "Whiggish," especially in the history of science.

45. On the status of intellectual freedom under the Roman empire see Breebaart 1976. On ancient equivalents of universities, and ancient academic societies, see chapter eight of Carrier 2016.

46. For Hypatia see Carrier 2016 (index), with *NDSB* 3.435–37 and *EANS* 423–24. Ironically, active disdain for natural philosophy was more evident in earlier Athenian history than in Roman, e.g. in the satire of Aristophanes' *The Clouds* (see French 1994: 6–10 and Olson 1984), in the fears expressed by Plato (e.g. *Laws* 10.886d-887a) and in the trials of Protagoras, Anaxagoras, and Socrates, which never saw their like again (see A.E. Taylor 1917 and Dodds 1951: 179–206, with Plutarch, *Pericles* 32.3 and *Nicias* 23.2–4, and Diogenes Laertius, *Lives and Opinions of Eminent Philosophers* 2.12–14). See also *OCD* 227–28, 465 and 739–40 (s.v.

Even the *idea* of a state suppression and policing of 'heresy' did not evolve in any coherent form before the 3rd century, when pagan opposition to Christianity became more organized and more concerned with controlling ideology, and even then such behavior did not specifically affect or concern natural philosophers, until the same tactics were adopted and magnified by Christians in subsequent centuries.[47]

As a result, there is little physical evidence to examine and few actions to analyze. Though there is some important epigraphic evidence in regard to medicine and engineering, and ample modern discussion of the social status of doctors, including studies of doctors and medicine in ancient art, actual 'science' or natural philosophy (hence medical *research* as distinguished from practice) gets little or no mention in inscriptions or any physical medium, possibly because it was practiced by so few or subordinated to

---

"belief," "Diopeithes, decree of" and "intolerance, intellectual and religious"). Later came the merely threatened prosecution of Aristarchus for the same reason (see P. Green 1990: 186 and Plutarch, *On the Face that Appears in the Orb of the Moon* 6 (= *Moralia* 923a)). Also dated near the same period would be the (probably mythical) murder of the discoverer of irrational numbers (according to Pappus, *Commentary on Book 10 of Euclid's 'Elements'* 64). But as Edelstein 1952: 589–94 shows, such active interference with scientific freedom was never repeated in pagan antiquity, and as Dover 1976 argues (slightly corrected by O'Sullivan 2008), much of it had already been exaggerated to begin with.

47. On Christian suppression and control of thought and ideology see MacMullen 1984 and 1997 and relevant sections of Hopkins 1998 and Watts 2006. On the evolution of pagan persecution of Christians see Janssen 1979, Hopkins 1998, Rives 1999, and De Ste. Croix 2006 (and for the converse: Salzman 1987). Even pagan persecution of Christians focused more on controlling actions than beliefs, but the involuntary burning of doctrinal books, which begins in the 3rd century, indicates that control of ideology was now on the radar. This contrasts somewhat with the previous, very limited, and sporadic suppression of 'malevolent' (as opposed to benevolent and hence approved) magic and divination, where it was always *behavior* that was the target of control more than thought, speculation, or belief. Even the burning of magical and astrological books was not an attempt to control thought, doctrine, or ideology, but an attempt to eliminate 'dangerous' behavior, such as the production of poisons or claims to supernaturally 'predict' the fate of reigning emperors. On this limited suppression (alongside widespread acceptance) of magic and divination see Ankarloo & Clark 1999: 243–66 and Dickie 2001: 142–61 (on later Christian efforts: 251–72). For astrology see Swan 2004: 280–81, Africa 1967: 73–81, and Barton 1994a: 1–94 and 1994b.

a career as a doctor, philosopher, or engineer.[48] Since the status of doctors as healers (or architects as builders) provides very little information about the status of empirical research, which (as we shall see) was the particular province of the *physicus*, scholarship on attitudes towards doctors will be of only marginal use. Similar attempts to identify the socially distinctive characteristics of ancient mathematical scientists (such as astronomers and engineers) are conspicuously inconclusive and therefore no more useful.[49] One thing these studies have established, however, is that though empirical scientists were men of wealth and respect, they did not typically come from the aristocratic elite, but were usually a level below in social status and prestige (a fact we will examine in chapter 4.6). On the other hand, though medical, mathematical, astronomical, and other scientific subjects have been found in papyri, this is typically of a specialized nature that does not reveal much about general attitudes.[50] We shall nevertheless examine the very few occasions where the Greek term *physikos* appears in inscriptions or papyri. And we will examine the few actions taken by emperors and others that we can find in the literary sources, which indicate something discernible about

48. Surveys of inscriptions mentioning doctors: Gummerus 1932; Gourevitch 1970; Nutton 1977; Korpela 1987; Meunier 1997; Rémy 2010; and Samama 2003. Scholarship on the social status of doctors (though not always distinguishing the scientifically educated from other *medici*): Scarborough 1970; Kudlien 1976 and 1986; J. André 1987; Horstmanshoff 1990: 187–96; van der Eijk et al. 1995; Nutton 1985, 1995, and 2013: 167–68; Mattern 2008: 21–27; and Israelowich 2015: 11–44. Studies of doctors and medicine in ancient art: Grmek & Gourevitch 1998; Hillert 1990. Medical subjects on ancient coins: Penn 1994. The doctor in ancient literature: Amundsen 1974 and 1977.

49. For example Netz 2002 and the relevant sections of Cuomo 2001 and 2007. R. Taylor 2003: 9–14 and Donderer 1996: 68–76 survey the social status of architects in the Roman period (conclusion: comparable to doctors). Donderer also produces a large number of Roman-era inscriptions mentioning architects or engineers, probably the largest and most complete collection in print. See also Cohon 2010. On the (perhaps lower) social status of surveyors, B. Campbell 2000: xlv-liii.

50. For example: *EANS* 612–24; with: W.H.S. Jones 1947; Sarton 1952: 293–94; Cauderlier 1978; Nutton 1984b: 318; Evans 1998: 345–47; A. Jones 1999; Bagnall 2009: 338–57; Netz 2011. For surveys of medical papyri: Marganne 1981, 1998, 2001; with further discussion in Nicholls 2010. For papyrus fragments of an ancient geographical treatise: Gallazzi & Settis 2006.

attitudes toward the natural philosopher.

Likewise, I have not found representations in ancient art to be of much use. Though a comprehensive search of all extant artifacts is impossible, from what I could find it appears likely that little or no art can be linked in any relevant way to natural philosophy or natural philosophers. For even what might derive from natural philosophy (such as the representation of the cosmos as a globe) or represent its practitioners (such as busts of famous philosophers) does not inform us about social attitudes directed toward that particular class of activity, since there are many other reasons why such images would be created or enjoyed, and the attitudes and intentions of the artist, audience, or owner can rarely be known as precisely as we would need in the present case.

To illustrate the problem, consider what would have been a notable exception from our period: a mosaic allegedly recovered from the excavated library at Herculaneum in the 19th century, which depicts Archimedes at the dramatic moment before his famous death (on which see chapter 4.6.I). This would be exceptional for two reasons that illustrate why most artistic evidence appears to be unusable for our purposes. First, it is a depiction of Archimedes, who, unlike other philosophers represented in art, wrote *only* on subjects in mathematics and natural philosophy, and thus, unlike other philosophers we know, his depiction could not have been inspired by his theories or accomplishments in moral philosophy or any other intellectual field, such as fame as a poet or healer. Second, it uniquely represents Archimedes actually engaged in scientific or at least mathematical work (drawing diagrams in a portable sandbox), and, just as uniquely, it was clearly intended to evoke the tragedy of a scientist's murder by a careless soldier. Unfortunately, the authenticity of this mosaic is almost unanimously rejected. It is now regarded as a 19th century fake.[51] Though its authenticity

51. For image, summary, and sources: Bol 1983. For discussion of its history and assessment: Goethert 1931 and Paul 1962. Similarly, some claim a gold ring seal depicting an astronomer at work is an ancient representation of Aratus (or Eudoxus or Eratosthenes), but it is more widely regarded as a forgery of the 16th century: Henig 1994: 314–15 (§ 656); vs. Schefold 1997: 296–97, 525–26 (Abb. 173). This judgment was reached for several reasons, including "the unusual subject matter" (vs. the common use of the same motifs in art of the 16th century). There *are* authentically ancient depictions of Aratus, as a poet inspired by Urania (the Muse of Astronomy), but their motifs suggest poetry rather than science

may have been rejected on invalid grounds, at present we can only follow the consensus of experts and exclude it from our evidence. I have not found any other artwork that comes even close to being as relevant or useful as this mosaic would have been, with the exception of some unusual coins celebrating the astronomer Hipparchus, which we will examine in chapter 4.3, and the unique discovery of what *appears* to be a visual depiction of human dissection in an early 4th century catacomb painting, which is so enigmatic, and has so many varying interpretations, as to be useless to the present inquiry.[52]

Overall, material evidence is not very helpful. Hence the sources for our study are almost entirely literary. But a reliance on literary evidence presents at least two methodological problems.[53] First, when examining such evidence we must pay attention to the literary and historical context of every passage, which often leaves a lot of room for interpretation. Second, literary studies are limited to what has survived. Yet numerous works that we know were written by and about scientists are no longer extant, and many more may have been written unknown to us. Important examples of this lost literature include the *Lives of Doctors and Their Schools and Works* by Soranus (written in the mid-2nd century A.D.), a book on the astronomy of eclipses by the Roman consul Gaius Sulpicius Gallus (written in the 2nd century B.C.), Varro's encyclopedia on the sciences (the *Disciplinae*, written

---

is the honored subject: Schefold 1997: 286–89, 418, 524–25 (Abb. 163–66 and 300; Schefold suggests some of these images are not of Aratus but of actual astronomers like Eudoxus, but in every case the visual context suggests a poet, which can only be Aratus).

52. See Hillert 1990: 232–34; Ferrua 1991: 121–23 (on date and context: 157–59); and Grmek & Gourevitch 1998: 193–96, who identify in extant scholarship no fewer than ten completely different interpretations of the scene, with most taking it as religio-symbolic and thus absent of any scientific meaning. Since no medical instruments are shown (not even a scalpel) and there are (by one count) twelve disciples in attendance, a religious or symbolic meaning seems more likely than a medical or scientific one. The image also falls outside our period of interest (at least probably—experts date it no earlier than 320 A.D., though it could be earlier).

53. Apart from the more fundamental problem of establishing the accuracy and authenticity of extant textual traditions, but since it is beyond the scope of this study to test the conclusions of modern textual critics, modern critical editions of ancient texts shall be trusted.

in the late 1st century B.C.), most of a similar but superior encyclopedia from Aulus Cornelius Celsus (the *Arts*, written in the 1st century A.D.), and most if not all the works of Hypatia, one of the few women known to have written on science in antiquity (in the late 4th century A.D.).

Nevertheless, over two hundred scientific or quasi-scientific texts survive from the early Roman period, while the scattered references to the sciences and scientists that we have from ancient literature comprise a fairly large body of evidence. The *physicus* in particular is named at least a thousand times in extant Greek and Latin texts, and it is from the period of the early Roman Empire that the largest body of relevant literature survives.[54] Analysis of these and other such literary references will occupy the bulk of this study. We shall begin by examining what the words *physikos* and *physicus* commonly meant in the early Roman period.

---

54. Count determined from a close analysis of comprehensive search results derived from the *Thesaurus Linguae Graecae* (www.tlg.uci.edu). The results of a corresponding analysis from several Latin text databases were similar for the word *physicus* with the same connotation, including the complete *Brepols Library of Latin Texts* (www.brepolis.net) and the *Patrologia Latina*, *Bibliotheca Iuris Antiqui*, and *Packard Humanities Institute Demonstration Disks*.

2. The Natural Philosopher as Ancient Scientist

Before asking what people thought about the natural philosopher, we must first answer what people thought a natural philosopher was or did. That question is answered here.

2.1 Defining the Natural Philosopher

The Latin *physicus* is a loan word from Greek, Romanizing *physikos*, an adjective meaning "natural" or in its adverbial form (*physicê* in Latin, *physikôs* in Greek) "by nature, naturally." It derives from the Greek word *physis* ("nature") and is correctly translated "natural" and *not* "physical" in the sense of solid or material (which would be *materialis* in Latin, *hylikos* in Greek).[55] Though many philosophers did end up defining the natural as the material, when they emphasized that something was "natural" they were less concerned with its having mass or being composed of matter than with its standing in contrast to the artificial and man-made. As Plato put it, the three most fundamental causes recognized in antiquity were *physis*, *tychê*, and *technê*: nature, chance, and design. Those who excluded any role for the gods in ordering the world regarded the interaction of *physis* and *tychê* to

55. Besides being obvious, the derivation of *physikos* from *physis* is directly asserted by the 2nd century A.D. grammarian Aelius Herodianus (or a Roman author assuming his name), in frg. 222 of *On the Modification of Words* (*Peri Pathôn*, suppl.). Also: *OCD* 1001 (s.v. "nature").

be the cause of all *technê*, since the latter then existed only as the product of human or animal intelligence, while others (like Socrates and his most famous student Plato) held exactly the opposite view, placing divine *technê*, in the sense of supernatural intelligent design, as the cause of all *physis* (and sometimes all *tychê* as well). But even here the division of all causes into *physis*, *tychê*, and *technê* is maintained, and by Plato's time (late 5th and early 4th century B.C.) the effective root meaning of *physis* had become whatever was neither *tychê* nor *technê*, which can only mean what Plato calls the *aitia automatê*, the "innate causal" powers and properties of things, which are neither intelligent nor random.[56]

Following this understanding of "nature" as a subject and "natural" as a category, the idea of "physics" was then derived from the Greek word *physikê* (sc. *philosophia* or *epistêmê*, "natural philosophy" or "natural knowledge") or *physika* (sc. *pragmata*, "natural things, matters, studies") and the Latin equivalent *physica* (variously in the neuter plural or feminine singular). In antiquity this was the branch of philosophy concerned with natural things—all things, not just what we delimit by the word "physics" today. It was then distinguished from logic and mathematics (which concerned reason, held to be necessarily prior to the study of nature) or ethics and politics (which concerned how people ought to act, hence with human choice rather than

56. Plato, *Laws* 10.888e-892c, who also links this distinction to a related dichotomy between *physis* and *nomos*, that which is true "by nature" and that which is true "by convention." In *Sophist* 265c, Plato says everything comes into being "either as the product of God's workmanship," whether directly or indirectly, "or else nature produces things from some innate cause without intelligent purpose," which he says is what "many claim and believe." Plato sides with the former, declaring in *Sophist* 265e that all *physis* is "made" by divine *technê*. This is also the gist of Galen's division of causes in *On the Natural Faculties* 1.1 (= Kühn 2.1–2) into *psychê* and *physis* (soul and nature), the *psychê* in this case meaning intelligent, deliberative causes (including divine, human, and animal intelligence), which corresponds to what Plato means by *technê*, although Galen, just like Plato, also understood *physis* as a *technê*-like cause in the special and limited sense that it was rationally designed by God (*On the Natural Faculties* 1.12 = Kühn 2.26–30; and see 1.6 and 3.13 = Kühn 2.15 and 2.199), and he was not alone (see von Staden 1996: 95–96). On similar distinctions in Aristotle and other authors, see Schiller 1978–1979, von Staden 1997b: 187–92, Heinemann 2005b, and Cuomo 2007: 7–40. Not all ancient scientists believed in divine design, however (such as the Epicureans or Strato the Aristotelian); nor materialism (*OCD* 910, s.v. "materiality").

the immutable behavior of nature).[57]

In the masculine, *physicus* and *physikos* became a substantive noun meaning, according to Lewis & Short, "a natural philosopher, naturalist," according to Liddell & Scott, "an inquirer into nature, natural philosopher," and according to the *Oxford Latin Dictionary*, "a natural scientist."[58] Its substantive meaning appears to derive from the actual conjunction of *physikos philosophos*, "natural philosopher," or *physikos anêr*, "man of physics," which both held essentially the same meaning.[59] There was even a verbal cognate in Greek: *physikeuomai*, "to be or speak like a natural philosopher."[60] Of these many definitions, most appropriate is 'natural philosopher', as one who philosophized about nature. Although "naturalist"

---

57. Heinemann 2000; Leisegang 1941; cf. Aristotle, *Meteorology* 338a20–339a9, *On the Heavens* 268a1–7; etc. (see section 2.2 below). For overall context: *OCD* 1145–46 (s.v. "physics"). For an example of a Roman-period survey of the divisions of philosophy: Seneca, *Moral Epistles* 89. For a detailed modern discussion see Hadot 1979 and (more briefly) Kidd 1988: 349–55. On the origin and evolution of the meaning of the word *physis* see: Naddaf 1992 and 2005; Patzer 1993; R.M. Grant 1952: 3–40; Burnet 1930; and Hardy 1884. See also "The Invention of Nature" in Lloyd 1991: 417–34; also Heinemann 2001, though this is currently incomplete without his promised second and third volumes. Aristotle produces the clearest ancient definition of *physis* in *Physics* 193a-194a and *Metaphysics* 1014b-1015a, fully analyzed in Buchheim 2001 and Heinemann 2005a. On the Latin equivalent (*natura*) see pertinent studies in Lévy 1996.

58. *LSL* 1373 (s.v. "*physica*" and "*physicus*"); *LSG* 1964 (s.v. "*physikos*") and 1964–65 (s.v. "*physis*"); *OLD* 1376 (s.v. "*physicê*," "*physicus*[1]," and "*physicus*[2]"). The equivalent word *physiologia* and cognates (both Latin and Greek) carried the same meaning ("*physio-*," natural, "*-logia*," reasoning, discussion, theorizing) but in the Roman period was less common, and was occasionally employed in the more specific sense of "physiology."

59. So the phrase *physikos anêr* and the solitary word *physikos* (when applied to persons), always means the same thing as the phrase *physikos philosophos*. For *physikos philosophos*, by analogy to *êthikos philosophos*, "moral philosopher," see: Sextus Empiricus, *Against the Professors* 7.7, 10.255, 10.351 (late 2nd century A.D.); Plutarch, *Themistocles* 2.4 (late 1st century A.D.); Eusebius, *Preparation for the Gospel* 14.4.8, 14.13.9, 14.16.11 (early 4th century A.D.); etc. For *physikos anêr* see: Galen, *On the Combinations and Effects of Individual Drugs* Kühn 11.460–61; and Eusebius, *Preparation for the Gospel* 10.14.14.

60. For *physikeuomai* see Galen (quoting Julianus), *Against What Julianus Said about the Aphorisms of Hippocrates* Kühn 18a.255–56.

is simpler, it misleadingly conjures up the modern and more limited sense of studying only natural history (e.g. geology, botany, zoology). On the other hand, 'natural scientist' can be too anachronistic, for as scientific as some may have been, as a class the *physici* were still philosophers first and foremost. But every natural philosopher in antiquity could be called a *physicus* and every *physicus* could be called a natural philosopher. Hence we shall treat the phrase 'natural philosopher' as interchangeable with *physicus* or *physikos*.

Though not completely 'scientific' in the modern sense, the *physicus* was still often expected to actually go and look at the facts rather than merely speculate or pass on a tradition (as we shall see), and this could involve the *physicus* in science-like activities, sometimes in a nearly modern sense. Likewise, anyone in the Roman period whom we would call a predecessor to modern scientists could have been called a *physicus* without objection, while the *physici* (*physikoi*) as a class bore the strongest overlapping correlation to the modern genus of "scientist" (as argued in chapter one, though in the present chapter we shall see more evidence confirming this). Still, given their differing methodological assumptions, not everyone who could be called a *physicus* then would be called a "scientist" today, even in the anachronistic sense still applied to men of the 17th and 18th centuries.[61]

By the end of the Middle Ages the Latin word *physicus* had taken on a much more limited meaning as a scientific doctor, as opposed to the word *medicus*, which by that point had come to signify more often a lay medical worker lacking a real education. This was a stark deviation from the original meaning of these words. In antiquity the *physicus* was a general researcher and philosopher of all natural phenomena, while *medicus* most commonly (though not always) indicated an educated rather than a folk healer—often someone, in fact, who carried out his medical practice informed by the methods and findings of the *physici* (as we shall see). But these connotations came to be forgotten over the course of the middle ages, and were ultimately transformed by the influence of Arabic medicine and the rediscovery of Greek medical authors in the 12th century. By the 18th century the new meanings had largely replaced the old. This is how we came by our word

61. Contrary to common assumption, the word "scientist" was not coined until the 19th century (Ross 1964). Thus, "scientists" in a strictly literal sense did not exist even in the days of Galileo or Newton or Harvey or Lavoisier.

"physician" today.[62] But we must deal with the way things originally were, in ancient Greece and Rome.

2.2 Aristotle's Idea of a Scientist

Though we are only concerned with who the *physicus* was during the early Roman empire, the starting point for any discussion of philosophy must be Aristotle, who wrote much earlier, in the 4th century B.C. The earliest known use of *physikos* as "natural philosopher" appears in Aristotle, and this is where the word is first explicitly defined and explained. Aristotle was also a major influence on the sciences of the early Roman Empire, especially in the case of our paradigm examples, Ptolemy and Galen. Thus, we must begin with an understanding of what Aristotle said about the *physicus*.

Aristotle argued that everything natural involves motion (*kinêsis*) and matter (*hylês*), and therefore *physics* concerns only these objects, whereas *mathematics* concerns objects that have no motion or matter.[63] Numbers and relations in the abstract, for example, are eternal and unchanging and have no substance or location. Though he does argue (against Plato and Pythagoras) that mathematical objects only exist when manifested in a material, the mathematician studies them only in abstraction, in the mind, separate from their appearance in nature.[64] Aristotle also names *physics* as one of the three "theoretical philosophies" (*philosophiai theôrêtikai*), the others being mathematics and theology (*mathêmatikê* and *theologikê*), distinguishing these from the "practical philosophies," like ethics and politics, which have to do with how we should act, what we should do, not

62. On this entire process see Schipperges 1970 and 1976.

63. Aristotle, *Metaphysics* 6.1.1026a.

64. On mathematical objects: Aristotle, *Physics* 193b and *Metaphysics* 1026a, 1059b, 1064a, etc. On what the mathematician does: Aristotle, *Physics* 193b-194a, *On the Soul* 403b14–16, *Metaphysics* 1059b. Analogously, just as mathematical objects only exist when manifested in a material, Aristotle also regarded the soul as the form and function of the body, and hence the soul exists only when the body is alive, cf. Aristotle, *On the Soul* 2.1. For discussion of these and similar passages see Distelzweig 2013, H. Lang 2005, and Modrak 1989.

with what "is" *per se*.[65] He regarded the theoretical as more important, and theology as the most important of all, which he called the "first philosophy" because it concerns Being itself, the immovable, unchanging, and immaterial basis for all else (which he says must surely be divine). Though "physics is indeed a kind of wisdom, it is not the first" in prominence, Aristotle says, unless there is nothing else but matter, which he did not believe, though many natural philosophers did, before and after him.[66] So he calls physics the "second philosophy," still giving it a higher place than any other field of inquiry besides theology.[67]

Consequently, the *physikoi* "alone intend to look into the whole of Nature as well as Being" but "since Nature is just one genus of Being" it is not as a *physikos* but as a philosopher *per se* that one studies Being itself.[68] Nevertheless, the *physikos* must study not only matter but also reason, and that "especially" since the analysis of things requires it.[69] Though logicians (*dialektikoi*) differ in their aims from natural philosophers, the latter are inevitably concerned with the same things.[70] But though a *physikos* was thus expected to be a complete philosopher, not 'just' a natural philosopher, "the whole business of the *physikos* is what has in itself a principle of motion and rest" since "the *physikos* is concerned with every function and property of whatever is a body and whatever is matter." This he held in contrast, again, with math and logic, which study abstractions apart from their material instances, or the practical or productive arts, which study how the student himself should move to carry out the art correctly. Aristotle gives medicine, gymnastics, and carpentry as examples of the latter.[71] In other words, the natural philosopher studies motion in other things for mere knowledge, not the motion we must initiate in ourselves to achieve some particular

---

65. Aristotle, *Metaphysics* 6.1.1026a.

66. On Aristotle's disapproval of their view, see Alexander of Aphrodisias, *Commentary on Aristotle's 'Metaphysics'* 70 and 72 (early 3rd century A.D.).

67. Aristotle, *Metaphysics* 1005a-b.

68. Aristotle, *Metaphysics* 1005a-b.

69. Aristotle, *Metaphysics* 1037a.

70. Aristotle, *On the Soul* 403a30–403b9.

71. Aristotle, *Metaphysics* 1059b and *On the Soul* 403b14.

end.[72] Yet psychology is still a part of physics, since "it is up to the natural philosopher to look at some issues regarding the soul, whatever is not free of matter," and so "to examine the soul, either the whole thing or to a limited extent, is the natural philosopher's business."[73]

Thus, Aristotle contrasts the *physikoi* with Parmenides and Melissus because the latter denied the existence of motion or change, yet the *physikos* studies only what is subject to motion or change.[74] Not everyone would agree. Centuries later Plutarch regarded Melissus as a *physicus*, and Eusebius assumed the same of Parmenides, probably because they propounded doctrines about the natural universe and, perhaps, because the Presocratics came to be arbitrarily lumped together (a point we shall examine below).[75] But in the Roman period, Aristotle's classification of these philosophers as *aphysikoi* was still acknowledged by Sextus Empiricus in the late 2nd century A.D., who says Aristotle called them this "because nature is the origin of motion, which they abolished by asserting that nothing moves."[76]

Aristotle sometimes treats astronomy as a branch of mathematics, but more emphatically as a branch of physics.[77] This brings out an important distinction long held between the *physicus* and the *mathematicus* in the specific sense of a mathematician.[78] Astronomy was certainly a part of

72. Aristotle, *Metaphysics* 1064a.

73. Aristotle, *Metaphysics* 1026a, with *On the Soul* 403a28–29.

74. Aristotle, *Physics* 184b17. Parmenides wrote a poem entitled *On Nature* in the 5th century B.C. arguing like a *physicus*, treating the same material, but attacking physical doctrines as absurd. Hence Aristotle argued, "just as the geometer no longer has anything to say to one who denies his first principles—for that belongs either to some other science or to a science common to all—so, too, for the natural philosopher," and therefore a *physikos* has nothing to say to men like Parmenides (*Physics* 185a).

75. Plutarch, *Themistocles* 2.3 (late 1st to early 2nd century A.D.); Eusebius, *Preparation for the Gospel* 14.3.6, 14.16.13.

76. Sextus Empiricus, *Against the Natural Philosophers* 2.45–46 (see Appendix C on the title and numeration of the extant books of Sextus Empiricus).

77. For mathematics: *Metaphysics* 1026a; for physics: *Physics* 193b. In practice, Aristotle certainly treated astronomy as part of his program to develop a completed "knowledge of nature" (*Physics* 184a-b; *Meteorology* 338a-339a; *On the Heavens* 268a).

78. On the Latin term *mathematicus* and its Greek equivalent see discussion in

physics, for "if it is up to the *physikos* to know what the sun or moon are," as Aristotle says in his *Physics*, "it would be out of character for him to know nothing of their properties, especially given the fact that writers on nature also bring to light the appearance of moon and sun and whether the earth and the cosmos are spherical or not."[79] He recognized that there were studies "more like physics than mathematics" yet that use math, "such as optics, harmonics, and astronomy," which use arithmetic or geometry but not in the way a pure mathematician does. As he puts it, "geometry examines natural lines, but not as natural objects, while optics examines mathematical lines, not as mathematical objects but as natural ones."[80] In other words, optics is the study of the geometry of natural phenomena, while pure geometry is the study of geometry alone, without reference to anything in nature. That's what distinguished the *physicus* from the *mathematicus*. It follows, Aristotle reasons, that the *physicus* must study both matter (*hylê*) and the patterns and shapes it takes, and thus must take up mathematics and logic, though in a more applied or descriptive way than the pure mathematician or logician.[81]

Finally, medicine was also not outside the purview of the natural philosopher. Though the art of healing was a practical matter, not theoretical, the theoretical part of medicine belonged to physics, and of course a doctor could be both a *physikos* and a *iatros* (in Greek; in Latin: *medicus*), wearing different hats for different activities. In fact, Aristotle held the two occupations to be almost inseparable, though distinct. As he puts it:

---

chapter 1.2.III.

79. Aristotle, *Physics* 193b.

80. Aristotle, *Physics* 194a; see also *On the Soul* 403b15–19.

81. Aristotle, *Physics* 194a-b; *On the Soul* 403b5–19. In *Posterior Analytics* 1.13–14 (78b30–79a30), Aristotle distinguishes pure mathematics from pure physics and then describes, in effect, a middle science of mathematical physics to mediate between them, adding that even medicine encounters mathematical problems (giving the example of using geometry to solve the problem of why circular wounds take longer to heal than gashes; a proof later accomplished by Herophilus: von Staden 1996: 90). He implies the same for mechanics, optics, harmonics, and astronomy (ibid. 1.9–10, esp. 76a23–76b12). See also Ps.-Aristotle, *Mechanics* 1.847a, on how 'mechanics' results where mathematics and natural philosophy meet. For more on the role of mathematics in Aristotle's natural philosophy see Hussey 2002 and Taub 2003: 106–15.

> It is also the business of the *physikos* to know the first principles concerning health and disease. For neither health nor disease are possible for what has been deprived of life. For this reason most of those interested in nature and those doctors who follow their art more philosophically, are similar: the one group ends at the study of medicine, while the other group, concerning their medicine, begins from the study of nature.[82]

And:

> Concerning health and disease it is not only the business of the doctor but also the *physikos* to discuss their causes, up to a point. But the way in which they differ and the way in which they investigate different things must not escape our notice, given that their activity is at least the same up to a certain point, as the facts are witness to. For all those doctors who are brilliant or inquisitive have something to say about nature and think it fitting to take their principles from that, and the most accomplished of those who occupy themselves with nature practically end up with medical principles.[83]

So Aristotle appears to have imagined good, smart doctors as doing physics (and basing their medicine on their research as a *physikos*), and accomplished *physikoi* as studying everything having to do with nature, right up to medicine itself. Still, the distinction he maintains is that the doctor practices an art, seeking a particular end (health), whereas the *physikos* pursues only knowledge, seeking out the nature and causes of things (like health). This may be the earliest explicit demarcation between theoretical and applied science.[84]

Aristotle's analysis was definitive, since the concept of the *physicus* remained essentially unchanged throughout antiquity. But the Roman period from 100 B.C. to 313 A.D. still provides us with the widest diversity of authors using and discussing the word. Armed now with our Aristotelian

---

82. Aristotle, *On Sense and Sensibles* 436a17–436b2.

83. Aristotle, *On Respiration* 480b21–30. My translation of this and the previous passage is more literal than provided by Lennox 2005: 66–68, though his discussion of them supplements mine.

84. Owens 1991 provides a full and detailed examination of this division of pure and applied science in Aristotle. Galen reiterates it, as we shall see in section 2.7, but he describes the principle more generally in *To Thrasybulus* 30 (= Kühn 5.861).

background, we can survey the evidence of this later but crucial era.

2.3 On Stone & Papyrus

The word *physicus* hardly appears in extant inscriptions, from any period.[85] For example, it is found twice in a copy of a chronicle inscribed (and probably composed) between 16 and 20 A.D., as an epithet for the Presocratic philosophers Anaximander and Xenophanes; similarly an inscription of around the same date naming the Presocratic philosopher Parmenides.[86] It also occurs on an undecipherable fragment of the famous Epicurean inscription of Diogenes of Oenoanda erected in the 2nd century A.D.[87] And there are two fragmentary inscriptions at Delphi, honoring a certain Diogenes Aristokleides, "a *physikos* by profession" (*physikon epistêmên*),[88] around 120–130 A.D., and another man of uncertain name (perhaps [Pe] rses or [Epithe]rses), "a *physikos phi*[*losophos*]," around 119–124 A.D.[89] Most other occurrences of the word in inscriptions and papyri usually have a different connotation, as an indication of blood relation ("natural"

85. As determined from a search of the *Packard Humanities Institute Demonstration Disks*, the *Database of Roman Inscriptions* (containing the volume of the *Corpus Inscriptionum Latinorum* for the city of Rome), and the indexes of numerous epigraphic collections in print.

86. Burstein 1984; and *I.Velia* 21, cf. *Apollo* [Musei provinciali del Salernitano] 2 (1962):125–36.

87. Diogenes of Oenoanda, *Epicurean Inscription*, frg. 114 (col. 1, line 6) = M.F. Smith 1996: 174 (for context and references see *OCD* 457, "Diogenes (5)").

88. For rendering the word "profession" here I follow *LSG* 660 (s.v. "*epistêmê*" §I.2), as in a trained skill that was regularly practiced, whether for gain (monetary or otherwise) or not.

89. *Fouilles de Delphes* 3.4.2: § 83 & 110, both fragmentary but among lists of names of individuals to whom (and to whose children) the Delphians granted "citizenship" and various other honors (such as the right to consult the oracle first, the right to have cases heard at court first, a release from civic duties, and so on). It is probable Aristokleides and Perses received the same or similar honors—for what reason is unknown, although two doctors (Dio the *iatros* and Metrophanes "of the medical profession," *iatrikên epistêmên*) are also among the honorees (ibid. §87 and 108).

children as opposed to children "by adoption").[90]

One unusual case is a bilingual inscription found in a field between Atinas and Volcei, Italy, which is probably from the 1st century B.C.[91] This identifies a certain "Lucius Manneius" [son or freedman] of Quintus, as a "*medicus*" in Latin, then translated into Greek with the phrase "and, by birth, Menekrates of Tralles, son of Demetrius, a *physikos oinodotês*," or "a winegiving natural philosopher." This Menekrates may have been adopted or freed by a Roman citizen and thus taken a Roman name, inscribed here in Latin.[92] Commentators have reasonably assumed that *medicus* is meant to translate *physikos oinodotês*.[93] Although *medicus vinarius* (or perhaps *vinidator*) would be expected, there was no space left even to write out the whole of the word *medicus*, so "*medic*" must have been intended to abbreviate the entire phrase. On the other hand, we would normally expect in the Greek *iatros oinodotês*, in reference to doctors renowned for employing wine in their treatment, a practice introduced in Italy in the same century by the famous doctor Asclepiades.[94] Since Menekrates

---

90. For example: *Die Inschriften von Klaudiu Polis* (= *Inschriften griechischer Städte aus Kleinasien* Bd. 31) §160 (*physiko patri*, "biological father"); *P.Mil.Vohl.* 2.73:8 (*physika tekna*, "birthchild"); *IAph2007* §12.1109 (*physikôn teknôn*, "children by birth"); *P. Oxy.* 44.3136:20 (*physikê th[u]gatêr*, "daughter by birth"), 44.3183:24 (*huioi physikoi*, "sons by birth"); *P.Lips.* 28:18 (*huion gnêsion kai physikon*, "legitimate son by birth"). A search of the *Duke Databank of Papyri* and the indexes of various papyrological collections produced no clear uses of *physicus* or *physikos* in any other sense.

91. *CIL* 10.388 (= *IRN* 236 = *CIL* 1.1256 and 1^2.1684 = *IG* 14.666 = *ILLRP* 799). The inscription was erected by Menekrates to his deceased wife, "Maxsuma Sadria."

92. This would explain why his distinctively Greek name "by birth" (*phusei*) is different from his Latin name. The expected abbreviation for a freeborn Roman, *Q. f.* ("son of Quintus") does not appear, but only a lone *Q* in the Latin. All the interpreters (see note above) take this as *Quinti*, i.e. "of Quintus" and therefore either "son of Quintus" by adoption or "freedman of Quintus" (the interpreters do not agree on which).

93. There are numerous doctors in the epigraphic record named Menecrates, suggesting a possible medical dynasty (e.g. see Korpela 1987: 167–68 and Rawson 1985: 85).

94. On Asclepiades launching the "winegiving" practice in Italy: Galen, *To Thrasybulus* 24 (= Kühn 5.846); cf. also *IG* 14.666 and *Anonymi Londinensis* 24.30 (cf. W.H.S. Jones 1947). Modern scholarship: Jouanna 1996, Garzya 1999, Touwaide 2000, and observations and sources in B.T. Lee 2005: 179; Rawson 1985: 174–75;

chose *physikos* instead of *iatros*, this suggests that he understood an overlap between the roles of doctor and natural philosopher, and regarded the latter as more worth communicating to Greek readers. There are no instances in extant Roman literature of *physicus* used as a literal equivalent for *medicus*—as in Aristotle, the two words are always distinguished in terms of research vs. practice. However, a well-educated and hence 'scientific' doctor was expected to be a *physikos* as well as a *iatros*. So we should probably read this inscription as confirming that Menekrates was a "natural philosopher" who practiced medicine for a living, applying natural philosophy to his practice as Aristotle recommended.[95] At that early date, Menekrates might have assumed readers who only knew Latin (and thus had no education in Greek) would not understand such a nuanced meaning of the word *physicus*, and so chose *medicus* instead as the Latin equivalent (or this choice was dictated by the need to abbreviate, since *medic* would be more readily understood than *physic*).

With this in mind we can look again at the two Delphic inscriptions mentioned earlier. Since Diogenes Aristokleides is listed as being a *physicus* "by profession," perhaps we should understand this to mean that Diogenes was a physician, who thought of himself as a properly-educated and methodologically-grounded doctor in the Aristotelian tradition (though not necessarily an adherent of Aristotelian philosophy). That would mean this Diogenes was not *merely* a doctor, but in some sense a medical researcher and investigator, or at least a medical intellectual and theorist. We should probably assume the same of Menekrates. However, since Perses (if that is the correct reconstruction of his name) is specifically called a natural "philosopher," this could mean he was not a doctor but someone with even broader interests.

With only three definite appearances of the word, all of them vague, the epigraphic evidence for the *physicus* is scanty. To gain any real understanding we must turn to the textual tradition, where we find an enormous treasury of evidence.

---

and sources and discussion in chapter 3.2.

95. Rawson 1985: 85 argues Menekrates was thus boasting he was a student of the medical sect of Asclepiades, but either way, as Rawson correctly notes, "such careful claims concerning origin and training cannot be paralleled in this period epigraphically, for any other profession."

2.4 'Natural Philosophers' as the Presocratics

It was an occasional idiom in the Roman period to refer to many of what we today call 'the Presocratics' as 'the *physici*'.[96] Unless otherwise qualified, this use should be taken as a proper noun, "the Natural Philosophers," as a particular honored category of natural philosophers generally, which applies to philosophers preceding Socrates in the late 5th century B.C. Most of the philosophers before Socrates were regarded as concerned solely with physics, rather than ethics or logic. Hence the ancient love of all things old secured these early "physicists" an honored place in the history of philosophy (as then understood).

As Sextus says, "some place physics as the first of philosophy's three parts, since chronologically the study of physics is the oldest, and even up to now the first who philosophized are called the *physikoi*."[97] Diogenes Laertius put it more extremely in the early 3rd century A.D.:

> Archelaus, the pupil of Anaxagoras and the teacher of Socrates, was the first to transfer natural philosophy from Ionia to Athens. And he was called the *physikos*, because natural philosophy also ended with him—when Socrates introduced ethics.[98]

Sextus avoids this hyperbole by rightly noting that some believed ethics began earlier, perhaps with Heraclitus.[99] And certainly natural philosophy never 'ended'. Quite the contrary, it grew enormously after Socrates, most famously in the hands of Aristotle half a century later. But the general idea is that philosophy was primarily if not solely about physics until Socrates,

---

96. For instance: Cicero, *Timaeus* 1.2, *Prior Academics* (= *Lucullus*) 2.17.55 (1st century B.C.); Sextus Empiricus, *Against the Professors* 7.89, 7.141; Diogenes Laertius, *Lives and Opinions of Eminent Philosophers* 10.134; Hippolytus, *Refutation of All Heresies* 1.pinax (early 3rd century A.D.); Eusebius, *Preparation for the Gospel* 14.2.1, 14.3.6.

97. Sextus Empiricus, *Against the Professors* 7.20.

98. Diogenes Laertius, *Lives and Opinions of Eminent Philosophers* 2.16. Hippolytus also says natural philosophy extended from Thales to Archelaus in *Refutation of All Heresies* 1.10.

99. Sextus Empiricus, *Against the Professors* 7.7.

and thus the philosophers preceding him were first and foremost *physici*, and so deserved the name as a special epithet.[100] It is for this reason that Thales of Miletus, who lived in the 6th century B.C., was widely regarded in the Roman era as the first *physicus*.[101] Sometimes this special use of the word is indicated by qualifying it as "the old natural philosophers" or "the first natural philosophers," or something similar.[102] Eusebius occasionally distinguishes them as the *physikoi* "who came before Plato."[103] One might also infer such a special connotation from the phrase "those called" the *physici*, although such a phrase could refer to *physici* from later periods.[104]

The methods and interests of the Presocratic *physici*, and the resources, background knowledge, and intellectual tools available to them, all differed considerably from the *physici* of the Roman period. And for this study we are only concerned with the latter, so we will not discuss the Presocratics further (except to occasionally mention what Romans said or believed about them, regardless of what was actually the case), nor will we survey the developments in the intervening Hellenistic period that evolved into the situation under the Romans (beyond a brief history of science provided in chapter three). Over the course of the four centuries of our concern (from

---

100. So: Eusebius, *Preparation for the Gospel* 14.2.1.

101. For instance: Tertullian, *Apology* 46 (early 3rd century A.D.); Clement of Alexandria, *Stromata* 2.4.14.2 (late 2nd century A.D.); Sextus Empiricus, *Against the Professors* 7.89; Eusebius, *Preparation for the Gospel* 10.14.10, 10.14.16; Ps.-Plutarch, *Tenets of the Philosophers* [*Moralia*] 883e (composed sometime between the 2nd and 4th centuries A.D.).

102. As 'first' (with the phrase *prôtoi physikoi*): Eusebius, *Preparation for the Gospel* 14.14.7. Similarly: Cleomedes, *On the Heavens* 18, 74, 201 (on dating Cleomedes see note in chapter 3.3, but probably 1st century B.C. to 2nd century A.D.). As 'old' (with various phrasing): *veteres physici*: Cicero, *Prior Academics* (= *Lucullus*) 2.5.13, cf. 2.27.87; *archaioi physikoi*: Posidonius in the early 1st century B.C. (via Strabo, *Geography* 17.1.5, completed early in the 1st century A.D.); Alexander, *Commentary on Aristotle's 'Metaphysics'* 178; *palaioi physikoi*: Diodorus Siculus, *Historical Library* 18.1.1 (late 1st century B.C.); *presbyteroi physikoi*: Eusebius, *Preparation for the Gospel* 14.13.9.

103. Eusebius, *Preparation for the Gospel* 14.4.8, 14.4.12.

104. Referring to the Presocratics: Plutarch, *Themistocles* 2.4; Eusebius, *Preparation for the Gospel* 14.2.1; Diogenes Laertius, *Lives and Opinions of Eminent Philosophers* 10.90. But including later philosophers: Eusebius, *Preparation for the Gospel* 14.13.9.

100 B.C. to 313 A.D.), there were no major differences among the *physici* in regard to their available resources, background knowledge, and intellectual tools. To one extent or another, individual *physici* differed from each other in those and other respects, but not in any identifiable chronological pattern. Consequently, what follows represents a survey of what was typical or common among *physici* throughout the early Roman empire, or what was commonly thought or said about them in that period.

2.5 The Roman Conception of the Scientist

Galen explains the connection between the Presocratics and all *physici* in the early 3rd century A.D.: "some of the old philosophers," he writes "had their name changed a bit, and were called natural philosophers because they would explain everything under the label 'nature' [*physis*]."[105] This represents the Roman conception of the *physicus* generally. Aulus Gellius identified the *physicus* in the 2nd century A.D. as one who investigates nature and its principles; Varro reported in the 1st century B.C. that "those who discuss the whole of nature are called *physici* for that very reason"; and Diogenes wrote in the early 3rd century A.D. that "some take the name of *physikos* from their investigation of nature," just as "others take that of *êthikos* [ethicist] because they discuss ethics, while those who are occupied with verbal jugglery are styled *dialektikoi* [logicians]."[106] In the late 2nd century A.D. Sextus singled out four fields of 'knowledge' (*epistêmai*) on which one could write: the "natural, mathematical, medical, and musical" (*physika, mathêmatika, iatrika, mousika*),[107] and of these, "whoever devotes himself to natural subjects simply must be a natural philosopher." Hence Sextus devoted two books specifically to refuting the *physikoi*.[108] Drawing on a different classification, Diogenes Laertius reports that in the 3rd

105. Galen, *Commentary on Hippocrates' 'On the Nature of Man'* Kühn 15.2.

106. Aulus Gellius, *Attic Nights* 1.9.6–8 (quoted and discussed in chapter 1.2.III); Varro, *On the Latin Language* 10.55.4; Diogenes Laertius, *Lives and Opinions of Eminent Philosophers* 1.17.

107. Sextus Empiricus, *Against the Professors* 1.300.

108. Sextus Empiricus, *Against the Professors* 9 and 10 = *Against the Natural Philosophers* 1 and 2 (see chapter 4.8 and Appendix C).

century B.C. the Cynic Menippus had written a treatise against the "natural philosophers, mathematicians, and grammarians," thus dividing knowledge into the study of language, the study of numbers, and the study of nature, the special province of the *physicus*.[109]

It was common for the *physicus* not merely to study nature but to embrace nature as a central philosophical principle, often forming a worldview similar to modern metaphysical naturalism, wherein all existence is understood as a single interconnected system of natural objects and causes, "especially those *physici* who say everything that exists is one," usually meaning, made of one substance, in endless configurations.[110] Tacitus, for example, can avoid using the word *physicus* when referring to 'natural philosophers' in the early 2nd century A.D. by instead using the periphrasis, "those who believe fate is not guided by the stars, but instead according to the principles and conjunction of natural causes."[111] Hence, according to Alexander of Aphrodisias in the early 3rd century A.D., most *physici* "held that there is nothing except natural things and those in motion," sometimes in contrast with Pythagoreans who "included more among the things that are 'insubstantial' than did natural philosophers," like taking numbers and axiomatic principles as immaterial, abstract objects, which many *physici* did not accept, believing only natural objects existed and nothing else.[112]

Because they were devoted to finding natural causes and explanations for everything, the *physici* were understood as the ultimate experts on natural facts and phenomena—scientifically and philosophically. Cicero introduces them in the 1st century B.C. as "those whom the Greeks call *physikoi*,"[113] and describes them as "the hunters and explorers of nature," perceiving them more as what we would call scientists today, rather than

---

109. Diogenes Laertius, *Lives and Opinions of Eminent Philosophers* 6.101.

110. Cicero, *On Divination* 2.33 (1st century B.C.). See Carrier 2005a for the contemporary meaning of 'metaphysical naturalism' and a modern example of a naturalist worldview.

111. Tacitus, *Annals* 6.22: *non e vagis stellis, verum apud principia et nexus naturalium causarum*. Such avoidance of common expressions through periphrasis is typical of Tacitean style.

112. Alexander, *Commentary on Aristotle's 'Metaphysics'* 72, 76, 264–265.

113. Cicero, *On the Orator* 1.49.217.

mere philosophers.[114] Galen agrees, declaring that "the man who wants to leave none of nature's works unknown, certainly he alone shall rightly be called a natural philosopher," for "he is not yet completely a natural philosopher" who "is still ignorant of some of the works" of nature.[115] In fact, "if we were really natural philosophers, we would thoroughly understand" why all things are the way they are.[116] Though referring to the functions of human organs, it is clear Galen's sentiment was generalized to all the subjects of ancient physics, just as Galen says, declaring that "the natural philosopher, *just as in all other matters*, also attempts to discover the causes" in human physiology.[117] Consequently, Galen favorably refers to the *physikoi* generally as the sort of authorities people should consult on the nature of things.[118] Likewise, according to Plutarch the *physikos* was an individual who in general "discovers the causes that operate inevitably in nature" and who "studies material and instrumental principles" to that end.[119] And when Cicero defines "the philosopher" as "one who strives to know the significance, nature and causes of everything divine or human," and to know and follow right from wrong, the latter of course refers to the moral philosopher, leaving the former to define the natural philosopher.[120]

The interest of the *physicus* was known to be theoretical rather than practical, his goal being to *know*, not to do, although these were not exclusive pursuits, as Aristotle had already said centuries before (and we will see more evidence below).[121] But as Philo put it in the early 1st century A.D., the *physici* are "those for whom the theoretical side of life is the focus

---

114. Cicero, *On the Nature of the Gods* 1.83.

115. Galen, *On the Uses of the Parts* 11.18 (= Kühn 3.922 = M.T. May 1968: 541); and Galen, *On the Uses of the Parts* 12.14 (= Kühn 4.56 = M.T. May 1968: 577).

116. Galen, *On the Uses of the Parts* 15.1 (= M.T. May 1968: 658).

117. Galen, *On Mixtures* Kühn 1.624 (emphasis added).

118. Galen, *On the Combinations and Effects of Individual Drugs* Kühn 11.460.

119. Plutarch, *Questions at a Party* 8.3.1 (= *Moralia* 720e).

120. Cicero, *On the Orator* 1.49.212. That natural philosophy included elements of what we now call theology will be shown below.

121. See also Aspasius, *Commentary on the 'Nichomachean Ethics'* 35 (early 2nd century A.D.).

of their diligence."[122] Hence Galen distinguishes a purely medical from a more 'scientific' interest in the action of drugs when he says "it is not now set before us [doctors], as it is for the natural philosopher, to seek out the *causes* of every effect, but merely to observe the effects of medicines, so we can use them well," meaning that doctors really only needed to know what drugs do, to help them in medical practice, but the natural philosopher's job was to find out *why* drugs did what they did.[123] Accordingly, among "doctors" Pliny the Elder repeatedly cites Phanias "the natural philosopher" as an authority on the medicinal properties of flowers, trees, and plants.[124]

Following Aristotle again, Galen also discusses the distinction between the *physicus* and the *medicus* in respect to the study of anatomy and physiology. "Inasmuch as it is *natural* our body is the subject of natural science," Galen says, but "inasmuch as it is to be *healed*, it is the subject of medicine" in the same way that "inasmuch as it is *well-conditioned*, or receptive of good condition, it is the subject of gymnastics." But even more important than the material subject of an art, according to Galen, is the *goal* of that art, since "these starting-points do not function in the same way for a doctor as for a natural philosopher, a difference which is itself found out by reference to the goal of each science," since the goal is really the ultimate criterion of any art. Thus, "for mathematics, astronomy, or natural philosophy, the aim consists in contemplation alone," whereas medicine has a practical objective.[125]

---

122. Philo of Alexandria, *On the Special Laws* 3.117.

123. Galen, *On the Combinations and Effects of Individual Drugs* Kühn 11.426 (also 11.401 and 11.427).

124. Pliny the Elder, *Natural History* 1.21c.9, 1.23c.7, 1.24c.8, 1.25c.9, 1.26c.9, 22.35.1 (1st century A.D.); similarly, Apuleius, *Defense* 45.14 (2nd century A.D.), regarding the medical effects of minerals. Phanias (of Eresus, sometimes sp. Phaenias or Phainias: *EANS* 641) was a colleague of Theophrastus, second head of Aristotle's school during the late 4th century B.C., and wrote on numerous subjects, including history and logic—but most pertinently in this case, he conducted and published research in botany (though none of his works survive). Over sixty fragmentary references to or from this Phanias can be found in the *Thesaurus Linguae Graecae* (as Phanias or Phainias), e.g. Athenaeus, *The Dinnersages* cites two works by Phanias, *On Plants* (*Peri Phytôn*) at 2.44, 2.59, 2.83, etc., and *Botanical Studies* (*Ta Phytika*) at 2.52, which are probably the works Pliny consulted.

125. Galen, *To Thrasybulus* 28, 29, 30 (= Kühn 5.857, 859, 861). Galen similarly links

In much the same way, in the 1st century B.C., the *physicus* was contrasted with the astronomer by Posidonius, and by Strabo (possibly following Posidonius, whom he knew, read, and admired), on the grounds that the latter dealt with math and the former with physical causes, and not, strictly speaking, the other way around (a distinction we will examine in section 2.7). Yet like Aristotle's ideal doctor, Posidonius wore both hats. Though clearly writing as a *physicus*, he also made astronomical observations and constructed a famous apparatus reproducing the circular motions of the heavens.[126] Accordingly, Plutarch assumes that mastering astronomy requires becoming both a geometer *and* a natural philosopher.[127]

The *physicus* could also be positioned between the theologian (or 'first philosopher') and the mathematician in the Middle Platonic view, as briefly stated in the *Handbook* of Alcinous:

> Of theoretical philosophy, that part which is concerned with the motionless and primary causes and things divine is called theology; that which is concerned with the motion of the heavenly bodies, their revolutions and periodic returns, and the constitution of the visible world is called physics; and that which makes use of geometry and the other branches of mathematics is called mathematics. So ....
>
> Theoretical knowledge has three parts ... theology, physics, and mathematics. The aim of the theologian is knowledge of the primary, highest, and originative causes. The aim of the natural philosopher is to learn what the nature is of the universe, and what kind of living thing is man, and what place he has in the cosmos, and if god exercises providence over all things, and if other gods are ranked beneath him, and what is the relation of men to gods. The aim of the mathematician is to examine the nature of existence in two and three dimensions, and the phenomena of change and locomotion.[128]

---

astronomers and natural philosophers in *On Critical Days* Kühn 9.937.

126. Cicero, *On the Nature of the Gods* 2.34 (Cicero was a good friend and avid pupil of Posidonius: cf. Kidd 1999: 38–40 and 1988: 23–27). We might have actually recovered one of these machines (see discussion in chapter 3.3).

127. Plutarch, *On the Face that Appears in the Orb of the Moon* 26 (= *Moralia* 942b).

128. Alcinous, *Epitome of Platonic Doctrine* (= *Didaskalikos*) 3.4 and 7.1. See commentary in Dillon 1993: 57–60 and 86–89. This treatise also shows Aristotelian and Stoic influences. A previous attribution of it to Albinus has been rejected by recent scholarship, but a date in the 2nd century A.D. is still accepted. For a complete translation and commentary see Dillon 1993. See also *OCD* 53 (s.v.

Of course, it was expected that a philosopher would master and use all three fields of inquiry, and in fact Alcinous says one should use mathematics as an aid to progress in the other two divisions, theology and physics.[129] But of particular note here is the overlap between a natural philosopher and what we would *today* call a theologian. Atticus, a Roman Platonist of the late 2nd century A.D., wrote that a philosopher must study all branches of philosophy, and of these 'physics' is "the knowledge of things divine and of the actual first principles and causes, and all other things that result from them."[130] In such a fashion theology was sometimes subordinated to physics, and 'the gods' subordinated as a part of nature. For example, theology was treated in the first book of the *Physics* by Chrysippus in the 3rd century B.C. and the 2nd book of the *Physical Discourse* by Posidonius in the early 1st century B.C.,[131] while Eusebius in the early 4th century A.D. claimed Plato subdivided physics into two fields, "the observation of things perceptible to the senses" and "the contemplation of incorporeal things," which would have included metaphysics and theology.[132] In the 2nd century A.D., Sextus attests to a continuing debate whether theology should be subordinated or added to physics, and concluded that the "best" *physici* should be counted on to address issues in theology.[133]

Finally, the natural philosophers as a recognized class were common enough for Lucian to make fun of them in the 2nd century A.D. Lucian portrays the moon's annoyance at being constantly observed, discussed, and picked apart by men below, so she begs Zeus to "destroy the *physikoi*, muzzle the *dialektikoi*, raze the Stoa, burn down the Academy, and stop the lectures in the Peripatos," for only then can she "rest and cease to be surveyed by

---

"Alcinous (2)"). Note that the division of theoretical philosophy into theology, physics, and mathematics is already in Aristotle (see section 2.2 above).

129. Alcinous, *Epitome of Platonic Doctrine* (= *Didaskalikos*) 7.2–4.

130. Quoted in Eusebius, *Preparation for the Gospel* 11.2.1 (cf. 11.2.5); Aristocles, an Aristotelian philosopher of the 1st or 2nd century A.D., said essentially the same thing about the Platonist view (Eusebius, ibid. 11.3).

131. According to Diogenes Laertius, *Lives and Opinions of Eminent Philosophers* 7.134.

132. Eusebius, *Preparation for the Gospel* 11.1.1 (compare with 11.2.1 and 11.7.1).

133. Sextus Empiricus, *Against the Professors* 9.12.

them every day."[134] To be a *physicus* could also be a badge of honor. "Strato," says Diogenes, "is held in the highest regard and nicknamed 'The Natural Philosopher' because he took the greatest care and consumed himself more than anyone with natural theory."[135] Even the philosophical emperor Marcus Aurelius lamented the fact that he could not become one, writing to himself in the 2nd century A.D., "because you have lost your hope of becoming a *dialektikos* and a *physikos*, do not for this reason renounce" other hopes.[136] The *physici* were also to some extent held in higher regard than sophists and logicians, if we take as a common sentiment the remark of Alexander of Aphrodisias that "whenever the mathematician and the natural philosopher make an argument they never lie, as the logician [*dialektikos*] and the sophist do, for the logician deliberately tries to win at all costs and sometimes takes to lying, while the sophist always does."[137]

Indeed, in mocking this high opinion of the *physici* Eusebius only confirms it when he describes "the *physikoi*" as "those who wander the wide earth and make the discovery of the truth the most important thing," noting at the same time that they also carefully study the views of thinkers before them, even those in other cultures.[138] A particularly amusing example of the Roman attitude toward the *physicus* is provided by a mock trial in a textbook for pleading cases in court, in which a "Cynic" is the son of an "eloquent speaker" who takes his son to court in order to disown

---

134. Lucian, *Icaromenippus* 21.

135. Diogenes Laertius, *Lives and Opinions of Eminent Philosophers* 5.58 (something similar is said of Strato in Seneca, *Natural Questions* 6.13.2). Strato of Lampsacus (of the early 3rd century B.C., discussed in chapter three), third head of Aristotle's school, was frequently paired with this epithet in ancient literature: e.g., Cicero, *On the Nature of the Gods* 1.35; Tertullian, *On the Soul* 15; Strabo, *Geography* 1.3.4; Galen, *On Semen* Kühn 4.629 and *On Trembling, Palpitation, Convulsion, and Shivering* Kühn 7.616; Plutarch, *On the Cleverness of Animals* 3 (= *Moralia* 961a) and *On Tranquility of Mind* 13 (= *Moralia* 472e); Sextus Empiricus, *Against the Professors* 7.350, 8.13, 10.155, 10.177, 10.228, 10.229 and *Outlines of Pyrhhonism* 3.32; Porphyry, *On Abstinence* 3.21.8 (late 3rd century A.D.); etc.

136. Marcus Aurelius, *Meditations* 7.67.1. This and many other sentiments of Aurelius on this subject are discussed in chapter 4.2.

137. Alexander of Aphrodisias, *Commentary on Aristotle's 'Metaphysics'* 646.

138. Eusebius, *Preparation for the Gospel* 14.9.4 (also implied by Tertullian, *To the Nations* 2.2).

him, on account of his countercultural way of life. "Indeed," the father pleads:

> Every discussion of philosophy is foreign to the customs of our state. Even so, if it still pleases you, are not other sects [besides the Cynics] more justified? If you were considering the natural philosophers, you would be investigating whether fire was the beginning of things or whether it was brought forth from tiny and mobile elements, or whether this world is eternal or mortal.[139]

Apart from revealing the sort of questions the *physicus* entertained, the gist of the father's speech, here and afterward, is that it was more dignified to be a *physicus* than a Cynic.[140] Not everyone agreed, of course. A dislike of the *physici* is rampant throughout Christian writers like Eusebius and Tertullian, a fact we will examine in the next section (and in detail in chapter five).

2.6 The Methods of Roman Scientists

The *physicus* can be further understood by studying what were, or were thought to be, their methods. The *physici* were conscious of the question of method, which was often linked to the Hellenistic concept of the "criterion" (sc. "of truth"). Sextus reports:

> It is taught that the natural philosophers since Thales were the first to begin an examination into the criterion. For once they had condemned

139. Quintilian, *Minor Declamations* 283.4.3 (see also: 268.4 and 268.9). This work might be the unauthorized publication of class material from Quintilian's lectures by some of his students (cf. Quintilian, *Education in Oratory* 1.pr.7 and 7.2.24), or possibly not even his, though it would still date from his era (cf. *OCD* 421, s.v. "*Declamationes pseudo-Quintilianeae*").

140. This "case" comes from the 1st century A.D. The earliest extant references in Latin to the *physici* as a class appear in the 2nd century B.C. in works that introduced Greek genres or material into Latin: Lucilius, *Satires*, frg. 26.635.64, says "all the *physici* say that from the very beginning man draws his existence from a soul and a body," and a *physicus* was a character in the lost play *Chryses* by Pacuvius, as reported in Cicero, *On Divination* 1.131.

> sensation as in many cases untrustworthy, they set up reason as the judge of the truth in existing things, and starting out from this they arranged their doctrines of principles and elements and the rest, apprehension of which is gained through the faculty of reason.[141]

Nevertheless, all natural philosophers believed the universe could be best understood by being observed. In the late 3rd century A.D. Lactantius said that "against their critics, natural philosophers defend knowledge by deriving from what is observable the argument that everything can be known."[142] As Sextus says, "the natural philosopher is satisfied that what lies around us is logical and intelligible."[143] One only needed the right 'criterion' to distinguish true opinions from false.

There was certainly widespread disagreement among the *physici* about which criteria to employ, and no firm consensus developed, but there were some shared principles of method among them. For example, before late antiquity such men were regarded first and foremost not as transmitters of tradition, but as investigators and freethinkers. As Cicero eloquently put it, "should not the *physicus*—that is, the hunter and explorer of nature—be ashamed to consult minds soaked with habit for evidence of the truth?"[144] Instead, Cicero says:

> In every field, continued observation over a long time brings extraordinary knowledge, which may be acquired even without the intervention or inspiration of the gods. Because repeated observation makes it clear what follows any given cause, and what precedes any given event....so predicting the future from the operation of nature does not require divine inspiration, just human reason.[145]

Empirical ideals like this appear to have been a mainstay of the *physici*, especially in opposition to their critics, at least until the 4th century A.D. Though it is sometimes assumed that ancient natural philosophy was

141. Sextus Empiricus, *Against the Professors* 7.89; see also ibid. 7.126 and *Outlines of Pyrrhonism* 1.178.

142. Lactantius, *Divine Institutes* 3.6.

143. Sextus Empiricus, *Against the Professors* 7.127.

144. Cicero, *On the Nature of the Gods* 1.83.

145. Cicero, *On Divination* 1.109, 1.111.

entirely comprised of *a priori* argument, in fact it was more fundamentally empirical and *a posteriori*, differing from modern science only in the quality of its methods.

As a result, although observation and independent thought was part of what was expected of a *physicus* in the early Roman Empire, an equally large part was the application of more philosophical reasoning, often in defense of a comprehensive worldview. For example, writing in the 1st century B.C. Varro said that *physici* "starting from nature as a whole reason backwards to what the first principles of the world might be"[146] and then, according to Vettius, they would work the other way around, "the natural philosophers defending their position by drawing on a first principle."[147] For this reason Cicero says there is no class of person "more presumptuous" than the *physicus*, because he often goes beyond the evidence of the senses and pronounces opinions about the nature of the world on insufficient evidence, for which he says they ought to be "ashamed." For example, on the question of earthquakes, Cicero remarks of the *physicus*:

> It is not too shameless, I think, that they dare to say *after* an earthquake has happened what force brought it about. But do they really see *in advance* that one is going to happen from the color of a spring's water? Many things of that sort are taught in schools, but you know you need not believe everything.[148]

His criticism is apt. Though the *physici* did typically reason from physical facts and observations, they relied much more on logic and analogy than experiments or mathematical laws (though, as we shall see in the next chapter, they *also* relied on experiments and mathematical laws). A typical example is provided by Plutarch's record of a discussion of vision and hearing, where observations play a central role in the arguments advanced, but not in a very scientific way.[149] Accordingly, by the 2nd century A.D.

---

146. Varro, *On the Latin Language* 10.55.4.

147. Vettius Valens, *Anthologies* 248.

148. Cicero, *On Divination* 2.29–33.

149. Plutarch, *Questions at a Party* 8.3.1 (= *Moralia* 720e); armchair methodology (albeit based on real observations in hunting, fishing, and agriculture) also pervades Plutarch, *Natural Questions* (= *Moralia* 911c-919). As another example of unscientific behavior, in his *Life of Apollonius of Tyana* 2.30, Flavius Philostratus claims the *physici*

Galen warned that many *physici* claim more than they should, and that even those who did not should be read more carefully, to avoid mistaking them as having said more than they did.[150]

Still, despite the philosophical presumptions of the *physici*, which he believes are sometimes unjustified or even wrong, Cicero admits that observation and evidence held an important place in their work, and he cites a wide range of fact-gathering by the Stoic *physici* as examples (as we will see in the next section). Cicero was of the opinion that evidence should supersede all speculation, a rather modern point of view. But this was not typical. Most *physici* were not averse to philosophizing and did not have too great a care for the distinction between that and empirical research, or for their relative merits. Hence what many *physici* thought were sound methods of finding and confirming the truth were often quite flawed by modern standards, though they were no worse than anything else that preceded the Scientific Revolution.[151]

Cicero also says "it is not characteristic of a *physicus* to believe that something has a smallest part" and even Epicurus "surely would never have believed this, if he would have had his friend Polyaenus teach him geometry rather than having Polyaenus unlearn it," which implies Cicero thought it was characteristic of a *physicus* to learn and apply advanced mathematical principles.[152] In fact, only die-hard Epicureans maintained that a *physicus* could do without using math, and several Epicureans even

---

could deduce a person's character from an inspection of the features of their eyes and face (similar to the 19th century fad of reading the bumps on someone's head). However, this is the only reference in our period to *physici* as a class embracing the pseudoscience of physiognomy (on which, see Barton 1994a: 95–132), and it appears only in the context of a depiction of Indian high society that is pure fantasy.

150. Galen, *On the Combinations and Effects of Individual Drugs* Kühn 11.547.

151. Per discussion and sources in chapter one.

152. Cicero, *On the Boundaries of Good and Evil* 1.20. In this case, Cicero has in mind the mathematical understanding of infinite divisibility (possibly deriving in part from the work of Archytas in the early 4th century B.C.) which led to the method of exhaustion (a precursor to calculus, employed by the time of Archimedes in the 3rd century B.C.). This partly resolved the paradoxical analyses of Parmenides and Zeno, which the atomism of Epicurus had resolved in a different way by declaring the infinite division of matter to be impossible. See Johansen 1998: 54–58, 65–74, 432–44, and Hussey 2002: 221–25.

studied the subject extensively.[153] Likewise, though Epicurus also rejected logic (*dialectica*), making 'physics' (*physica*) the sum of all things, even Epicurus held that physics was a form of knowledge (*scientia*) through which "one can explore the meaning of words, the nature of language, and the principles of correspondence or contradiction."[154] Thus physics for him went beyond mere exploration of the facts of nature. To an important extent even logic fell under its purview, just as Aristotle had said. It seems, in fact, that the Epicurean disdain for dialectics was born less from a disinterest in logic as from a dislike of jargonizing—in other words, according to Diogenes Laertius, Epicureans "reject dialectic as redundant," since "natural philosophers use ordinary language," which is sufficient.[155] The reaction here is against something Cicero had also observed: that just like writers on logic, geometry, music, grammar, and rhetoric, in pursuing their art natural philosophers had developed a highly specialized vocabulary that was alien even to native speakers of the same language (just as modern scientists have

---

153. See Mueller 1982: 92–95 and 2004: 62–63, Verde 2013, and *OCD* 1174 (s.v. "Polyaenus (1)"). Epicurus nevertheless used geometry in his natural philosophy (cf. Taub 2003: 133), and like the Skeptics who rejected even physics yet nevertheless studied the hell out of it just to debunk it, several Epicureans became obsessed with mathematical study. Hermarchus, like Polyaenus a pupil of Epicurus himself, wrote an entire treatise *On Mathematics* (Diogenes Laertius, *Lives and Opinions of Eminent Philosophers* 10.24–25). In the late 3rd century and early 2nd centuries B.C. the Epicurean philosophers Philonides and Eudemus of Pergamum were practicing mathematicians who shared advanced mathematical treatises with Apollonius of Perga (on whom see chapter 3.3), and fragments of a biography of Philonides recovered from Herculaneum reports that he corresponded with several other mathematicians (see: Crönert 1900; *DSB* 1.179, in s.v. "Apollonius of Perga"; *OCD* 122–23, in s.v. "Apollonius (2)"; *EANS* 659 and *OCD* 1135, s.v. "Philonides (2)"). In the early 1st century B.C. the Epicurean philosopher Zeno of Sidon composed sophisticated Epicurean criticisms of mathematics, which came near to anticipating non-Euclidean geometry and modern theories of induction (*EANS* 847; *DSB* 14.612–13; *OCD* 1588, s.v. "Zeno (5)"). Mueller also links the Epicureans Basilides (*EANS* 190), Protarchus (*EANS* 702), and Demetrius (EANS 233) with serious work in mathematics (on the latter, including fragments of his mathematical work recovered from Herculaneum, see De Falco 1923). See, however, Netz 2015 (who argues Epicureans only wrote against mathematics, and the few men identified as Epicurean mathematicians might not have been Epicureans).

154. Cicero, *On the Boundaries of Good and Evil* 1.63.

155. Diogenes Laertius, *Lives and Opinions of Eminent Philosophers* 10.31.

done).[156] But this Epicurean distrust of logic was not universal. The Stoics, for example, had no such qualms, and argued that the formal study and use of logic were essential to the methods of the *physicus*.[157] And Galen declared that "logic is useful in both medical practice and natural theory" and "no one more rightly deserves to be considered stupid and insane than someone who believes in investigating the nature of things through anything other than logic and reason."[158]

As a result, most scientific progress in antiquity (as now confirmed by modern science and recognized by some even then) was made by natural philosophers who embraced formal logic or even mathematics as essential tools of their trade. And those who carefully documented observations, aimed at precision in measurement, and conducted experiments. The method exhibited in the best works of Ptolemy, Galen, and Hero, as with Archimedes and other scientists before the Roman era who were among the most revered in the Roman era, most commonly involved starting with a set of premises, often drawn from observations, deriving predictions from those premises by deductive logic, and then testing those predictions against further observations to verify if the model (their set of premises) is correct. We see this process in various forms, for example, in the *On Floating Bodies* of Archimedes, the *Catoptrics* of Hero, the *Syntaxis* of Ptolemy, and the *On the Natural Faculties* of Galen.

Cicero provides further insight when he argues that "nothing is more disgraceful for a *physicus* than to say that something happens without a cause," attacking the 'atomic swerve' proposed by Epicurus in the late 4th or early 3rd century B.C.[159] It was also claimed that the most common

156. Cicero, *On the Boundaries of Good and Evil* 3.1.4. A similar complaint appears to underlie remarks in Ptolemy, *On the Criterion* 4–6; as well as Galen, *On the Difference among Pulses* Kühn 8.588, *On the Natural Faculties* 1.1 (= Kühn 2.1–2), and *On the Therapeutic Method* 1.5.5–7. See related commentary in Hankinson 1991a: 132–33.

157. For example: Diogenes Laertius, *Lives and Opinions of Eminent Philosophers* 7.83. The Platonists certainly agreed (e.g. Alcinous, *Epitome of Platonic Doctrine* 4–7), and we have already seen the Aristotelians did. The Stoics had advanced formal logic well beyond Aristotle, cf. e.g. Galen, *Education in Logic* and Russo 2003: 218–21.

158. Galen, *Commentary on Hippocrates' Sixth Book on Epidemics* Kühn 17b.306 and *On Examinations by which the Best Physicians Are Recognized* 8.4.

159. Cicero, *On the Boundaries of Good and Evil* 1.19.

metaphysical premise thought to be defended by all *physici*, setting them apart from other intellectuals, was a denial of creation *ex nihilo*. If, as Galen said, the *physikoi* are those who "explain things under the label 'nature,'" then the *physikoi* should not believe something happens other than by nature. For example, Cicero says, if organs or features of organs appear and disappear, as the diviner's art supposed, it follows that:

> The appearance and demise of all things is not brought about by nature, and something will exist that either arises from nothing or suddenly vanishes into nothing. What natural philosopher has ever said this? Diviners say this. But do you think you ought to believe them instead of the natural philosophers? What!?[160]

Indeed, he says elsewhere, "all the natural philosophers would laugh at us if we said anything happens without a cause," for if "something happens without a cause, then it follows that something comes out of nothing, which neither Epicurus nor any natural philosopher accepts."[161]

This is no doubt connected with their opinion, as Cicero put it, that "the nature of things is continuous and united as one harmonious whole, which I see is the popular view of the *physici*, especially those who have said 'all that exists is one.'"[162] Alexander agrees it is "more or less a common dogma of all the natural philosophers" that "nothing ever comes into being from nothing, but everything comes from something else," or as he says again, "the common doctrine of all the natural philosophers is that [only] nothing comes from the nonexistent."[163] Plutarch even jokes about this when he says lenders "surely have a laugh at the natural philosophers, who say nothing arises out of nothing. After all, for the lenders, interest arises from what no longer exists nor has any foundation!"[164]

Since the *physici* believed everything has a cause (and usually only a natural, inevitable cause), their primary research interest lay in discovering

160. Cicero, *On Divination* 2.37 (cf. Aristotle, *Physics* 187a28).

161. Cicero, *On Fate* 25.1 and 18.14.

162. Cicero, *On Divination* 2.33.

163. Alexander of Aphrodisias, *Commentary on Aristotle's 'Metaphysics'* 652 and 719.

164. Plutarch, *On Why You Should Not Borrow Money* 5 (= *Moralia* 829c).

the causes of things (as we shall see in the next section), whether or not they employed what we now regard as sound methods of discovering those causes. Aristotelians, of course, expected natural philosophers to investigate *all* of the Aristotelian causes, which by Roman times were five in number: material, efficient, formal, instrumental, and final. As Galen says, "if they are *really* natural philosophers, we will expect them to give an answer to each kind" of these five causes "when dealing with all the parts of a living being," and no doubt, we may presume, in all other studies as well.[165] This concept of 'cause' was not just temporal (each cause preceding an effect in time, like the Aristotelian 'efficient' and 'final' causes) but also ontological (like the Aristotelian 'material' and 'formal' causes).[166]

As a result, the *physicus* was apparently expected to be a reductionist, in the sense of figuring out not only all the causes of things, but what elements everything can be reduced to, in both causes *and* composition. As Sextus observes, "those in the camp of Pythagoras say that the real natural philosophers, when investigating matters concerning the whole world, must first of all inquire into what elements the whole world can be reduced."[167] Plutarch confirms this view at length:

> Investigation must begin from the hearth, so to speak, from the substance of the universe. More than anything the philosopher should expect to differ in this respect from a doctor and a farmer and a flutist. It is enough for the latter to examine the most recent causes. For as soon as the cause

165. Galen, *On the Uses of the Parts* 6.12 (= M.T. May 1968: 308).

166. See Aristotle, *Physics* 194b-195a and 198a-b, *Metaphysics* 983a-b. The material cause refers to the material something is made of and how that affects what happens; the formal cause is the shape or structure into which that material is formed and how *that* effects what happens; while the efficient cause is what we today more commonly mean by a cause—an event preceding the effect without which the effect would not occur; Meanwhile the instrumental cause is the particular instrument or means by which en effect is brought about (e.g. an efficient cause of an object's motion may be a blow; the instrumental cause would be which specific tool or object delivered the blow); and a final cause is the end goal or reason for something is brought about (e.g. the motivation of an agent; the reason they brought about the effects they did; in modern evolution science, and in ancient natural selection theory, it would include adaptive functions, not just the intentional goals of agents).

167. Sextus Empiricus, *Against the Professors* 10.250.

> of an effect is duly noted, for instance that exertion or transfusion cause fever, or blazing sunny days right after a thunderstorm cause the rusting of grain, or bending flutes and connecting them to each other causes a deep sound, that is enough for the expert to manage his job. But for the natural philosopher pursuing the truth for theory's sake, knowledge of the most recent causes is not the end, but the beginning of his journey to the first and highest causes. This is really why Plato and Democritus, when they were investigating the cause of heat and weight, did not stop their account at earth and fire but traced the perceivable back to its rational origins until they arrived, so to speak, at the simplest seeds.[168]

Such, clearly, was the ultimate aim of the *physicus*. "Nevertheless," Plutarch immediately cautions, "it is better to first examine perceptible things."

Hence following from all this, the *physicus* was apt to reinterpret religious beliefs in naturalistic terms (a fact that will become relevant in chapter five). For instance, the *physicus* was regarded as someone who did not take myths or sacred stories seriously, but instead as figurative metaphors for higher or abstract truths. Philo describes this methodology, of not taking a text "literally" but instead as "allegory," as a "way" of doing things that "natural philosophers love."[169] Philo goes on to explain why this is an acceptable recourse even for a pious Jew, at least when some contradiction in the Bible appears unresolvable. But what is notable is that he saw this as an approach especially common among the *physici* in general, a view also shared by Heliodorus.[170] Even without the allegorical approach, replacing divine with natural causation was a defining element of the natural philosopher's method, which was specifically aimed at explaining the world by replacing the gods (as much they could) with natural laws, principles, and processes.

Which is one reason why, rightly or wrongly, the *physici* were often accused of excluding any role for a Creator. Eusebius laments sarcastically

---

168. Plutarch, *On the Principle of Cold* 8 (= *Moralia* 948c).

169. Philo of Alexandria, *On the Descendants of Cain* 7 (see also *On Abraham* 99). For an example of *physici* speaking allegorically, see Plutarch, *Lovetalk* 24 (= *Moralia* 770a) and Servius, *On Virgil's 'Aeneid'* 1.47.1. See *OCD* 1115 (s.v. "personification").

170. Heliodorus, *Aethiopica* 9.9.5 (see also Strabo, *Geography* 1.2.8). For some discussion and sources on this philosophical use of allegory see J. Stern 2003: esp. 52–53, 57–62, 67. And for a summary and bibliography on the practice and its popularity see Carrier 2014a: 114–24.

and polemically of "the all-wise Greeks who were particularly called natural philosophers, whose opinion concerning the composition of everything and its original cosmogony" introduced "no Creator, no Maker of everything" but just "irrational impulse and spontaneous motion."[171] Though he has in mind the Presocratics, he goes on to extend this criticism to several Hellenistic philosophers as well. But "the first natural philosophers" especially set the bar, since, Eusebius claims, "their doctrine about first principles was such as to include no God, no Maker, no Demiurge, nor any cause of the universe, nor even gods, nor incorporeal powers, no intelligent natures, no rational beings, nor anything at all beyond the senses."[172] This does not necessarily mean he thought they were atheists in a modern sense, only that any gods they admitted were subordinate, both in cause and material, to the universe and its principles—and to Christians such beings were hardly gods at all.

There were exceptions. Eusebius admits that some *physikoi* placed 'Mind and God' over everything, including some of the earliest and most famous, "Pythagoras, Anaxagoras, Plato, and Socrates."[173] And while "some of the natural philosophers brought everything down to the senses," still "others went in the opposite direction, like Xenophanes of Colophon and Parmenides the Eleatic," who "questioned the senses, asserting there was no comprehension of things in sensation, therefore we must trust reason alone."[174] So even Eusebius saw a good deal of diversity regarding attitudes toward the divine among the *physici*. Though many of them resembled more closely modern metaphysical naturalists, others maintained a place for souls, spirits, and gods. Galen, for example, explains throughout his treatise *On the Uses of the Parts* that the evidence from human anatomy clearly supported the theory of intelligent design.[175]

Nevertheless, natural philosophers sometimes found themselves at odds even with devout pagans, especially the less educated. In the first century B.C. Vitruvius distinguished the *physici* from priests, and philosophers in general, as the three groups who offer different explanations of the nature

---

171. Eusebius, *Preparation for the Gospel* 1.8.13.

172. Eusebius, *Preparation for the Gospel* 14.14.7.

173. Eusebius, *Preparation for the Gospel* 14.16.11.

174. Eusebius, *Preparation for the Gospel* 14.16.13.

175. See chapter 3.8.IV.

of the world, which produced some friction.[176] For example, in the same century Diodorus said "natural philosophers try to trace back the causes" of catastrophic events "not to the divine, but to certain natural and inevitable circumstances," as opposed to "those who are piously disposed to the divine" who "assign certain plausible reasons for the event" (his example being earthquakes), "alleging the disaster happened because of the gods' wrath at those who committed sacrilege against the divine."[177] Diodorus also reports that some *physikoi* argued that comets are natural objects appearing in regular cycles, rather than divine omens.[178] Likewise, Alexander observed in the early 3rd century A.D. that the *physikoi* explain ocean phenomena by appealing to natural causes and elements, in contrast with theologians who appeal to divine action.[179] Hence in the late 1st century A.D. Plutarch lamented the fact that the common people "despise philosophers who argue that God's dignity lies only in his goodness" rather than in any wrathful or providential meddling with nature.[180] Already by the 3rd century B.C., Strato "the Natural Philosopher" had carried this to its logical conclusion and eliminated god altogether, arguing that "all divine power rests in nature, which alone contains the causes of birth, growth, and decay, yet is devoid of all sensation and form."[181] Likewise, in his book *The Physicus*, Antisthenes is said to have "abolished the power and nature of the gods" by declaring there

176. Vitruvius, *On Architecture* 8.pr.4.

177. Diodorus Siculus, *Historical Library* 15.48.4 (see also 16.61–64).

178. Diodorus Siculus, *Historical Library* 15.50.3. See discussion in chapter 3.3.

179. Alexander of Aphrodisias, *Problems and Solutions in Scholastic Physics* 98. Taub 2003: 125–68 provides a detailed examination of how natural philosophers replaced divine with natural causation in meteorology. And see related discussion in chapter 3.7.III.

180. Plutarch, *On Superstition* 6 (= *Moralia* 167e). Seneca argued the same point in his own treatise *On Superstition*, which is lost but quoted in Augustine, *City of God* 6.10. It appears that disdain for natural philosophy was more evident in early Athens and declined substantially after that (see note in chapter 1.3), though even in the Roman era scientific activities could sometimes be perceived and attacked by an ignorant public as malevolent magic (cf. Apuleius, *Apology* 16.7, 29, 38, etc., with commentary in S.J. Harrison 2000) or as dangerously impious (see the example concluding chapter 4.5 and discussion in chapter 3.5).

181. Cicero, *On the Nature of the Gods* 1.35 and *Prior Academics* (= *Lucullus*) 2.38.121.

is only one Deity, evidently also equating God with nature.[182]

For the same reason, as we have seen, the *physicus* was often accused of atheism by later Christian authors. Even though not all *physici* went as far as Strato or Antisthenes in eliminating the divine from their natural philosophy, a suspicion remained that they all really did, or that their doctrines ended up entailing such a conclusion anyway, and though most *physici* recognized the existence of some god or other, many Christians still did not approve of what most of them meant by 'god'.[183] The ultimate culmination of this criticism can be seen in the late 4th century A.D. treatise *On Natural Philosophers* by Marius Victorinus Afer, who rebelled against the apostasy of the emperor Julian and wrote a book in Latin devoted solely to systematically deriding all *physici* as ignorant atheists. But the same skepticism and commitment to naturalism could anger pagans as well.[184] In the 2nd century A.D. Vettius Valens lamented that some "natural philosophers and picky people, either through envy or crookedness, dismiss or attack the method" of astrologers. Though not an overt accusation of atheism, it still reveals his discontent with the skeptical criticism of natural philosophers.[185] Likewise, Cicero reveals that the *physici* were typically skeptical of popular miracles; and Strabo remarks how the common people did not care much for the *physici*'s naturalistic explanations of cherished myths, and for this very reason the *physici* might avoid mentioning them in public.[186]

---

182. Cicero, *On the Nature of the Gods* 1.32 (probably referring to the Antisthenes who was a student of Socrates and wrote between the late 5th and early 4th century B.C.).

183. A detailed survey of the various beliefs of natural philosophers with respect to the nature and existence of God or gods is provided in Sextus Empiricus, *Against the Professors* 9.49–194.

184. See remarks in S.J. Harrison 2000: 64. For more on the suspected atheism of natural philosophers (and pagan discontent with it) see French 1994: 8–10, 17–18. Popular anxieties of this kind are also reflected in the defensive remarks of Lucretius, *On the Nature of Things* 5.110–125 and the fears voiced in Plato, *Laws* 10.886d-e.

185. Vettius Valens, *Anthologies* 250. This does not mean all natural philosophers rejected astrology—many embraced it and attempted to give it a scientific explanation, most notably Ptolemy, as exemplified in his astrological *Tetrabiblos*, and possibly Posidonius two centuries before him. See Carrier 2016 (index, "astrology").

186. Cicero, *On Divination* 2.27 and Strabo, *Geography* 1.2.8. The strategy of

2.7 Mathematics & Causation

A major question concerning the methods of the *physicus* pertains to the roles of mathematics and aetiology (or causal explanation). Some scholars have argued that there arose in antiquity a sharp division between the 'mathematicians' and the 'natural philosophers', with the former ignoring questions of causation, and the latter ignoring questions of mathematical description, and that a fusion of their methods would not arise until the Scientific Revolution. The evidence does not support this conclusion, but it remains a leading controversy that any discussion of the methods and interests of the *physicus* must address.

It is true that the main interest of the *physicus* was to discover the causes of things.[187] As a result, mathematics was not regarded as an essential part of the activity of the *physicus* except when it could help, such as in mechanics, optics, harmonics—and astronomy (and in some respects even medicine). Just as the best doctors were also expected to be natural philosophers, it was expected that the best astronomers would be as well. But their roles and activities were still distinguished, just as were the roles and activities of doctors and physiologists. Doctors were distinguished from natural philosophers according to their objectives: doctors aim to heal, philosophers aim to know. Astronomers came to be distinguished from philosophers according to their means of analysis: astronomers use mathematics, philosophers do not. But this really amounted to the same thing: mathematics aims to measure and describe, while philosophy aims to explain what is thus measured and described. In effect, mathematics could only improve the accuracy with which observed facts are described and understood. To go any further than that, you had to rely on natural philosophy.

Diogenes Laertius offers a somewhat vague account of the accepted Stoic view on this point, which was that research into the nature of the universe "shared in common with those who use mathematics" various

---

keeping quiet is suggested by Heliodorus, *Aethiopica* 9.9.5.

187. As I've argued in the previous section. Lloyd & Sivin 2002: 140–87 survey this and other objectives of the natural philosophers, which included, also, categorizing and cataloguing the universe, and quantifying and modeling it.

questions relating to measuring astronomical events and phenomena, or studies relating to optics, while most other inquiries are the concern of "natural philosophers alone."[188] His many and various examples imply a distinction between quantitative and qualitative inquiries, but the natural philosopher is not explicitly excluded from either one. To describe those who make mathematical inquiries Diogenes uses the exact phrase *kai tous apo tôn mathêmatôn* twice (first in connection with astronomy, and then optics and meteorology), which means "also those who come from mathematics" or "from among the mathematicians," which is to say, "those who study mathematics" or "those who use mathematics." Since this would also include any natural philosophers who employ mathematics, there is no real contrast here between mathematicians and natural philosophers as such—except on the other side of the equation: for all the more qualitative studies belong *monois tois physikois*, "to natural philosophers alone." This is the only phrase that actually excludes anyone—in this case, all those using mathematics alone. Mathematics is thus subordinated to philosophy.

This may hint at a more specific Stoic view that came to be accepted among Roman scientists generally, which was most influentially stated by Posidonius in the early 1st century B.C. Though his works were not preserved, several extant passages transmit his distinction between the *physikos* and the astronomer, asserting that mathematical analysis was the particular occupation of the latter—although as we have seen, the same person was expected to take both roles. The fact that this ideal was echoed so often in the early Roman era is a significant indication of its influence.

According to the latest surviving account (through a filter of several intervening sources), we are told Posidonius pinned his distinction on the use of mathematics like this:

> It belongs to natural theory to look into the substance of heaven and stars, as well as their power and quality, and their birth and death, and through these one can reveal things about their size, shape, and arrangement. But astronomy attempts to discuss nothing like that. It reveals the arrangement of the heavens, on the premise that heaven is actually an ordered cosmos. It talks about the shapes, sizes, and distances of earth, sun, and moon, about eclipses and conjunctions of the stars, and about the quality and extent of their movements.

188. Diogenes Laertius, *Lives and Opinions of Eminent Philosophers* 7.132–33.

For this reason, since astronomy touches on the theory of how much, of what magnitude, and according to what sort of arrangement, it is reasonable that it needs arithmetic and geometry. Its strength lies in drawing conclusions about these things, the only things of which it promises to give an account, through both arithmetic and geometry. Accordingly, both the astronomer and natural philosopher will often try establishing the same point, such as that the sun is large, or the earth is spherical, but they will not exactly proceed down the same paths. For one will prove each point from the substance, the power, or the possession of a greater attribute, or from its origin and transformation. But the other will do this from the properties of the arrangements or magnitudes or from the extent of the movement and the time coinciding with it.

Also, the natural philosopher will often fix on the cause, paying attention to the productive force, while the astronomer, whenever making a demonstration from the exterior properties, does not become an adequate observer of the cause—as when he shows that the earth and stars are spherical. Sometimes he does not even try to grasp the cause, like when he discusses eclipses. Other times he investigates by hypothesis, demonstrating some of the ways the phenomena will be saved under the circumstances. For instance, why do the sun, moon, and planets appear to move in an irregular way? If we hypothesize that their orbits are eccentric, or that the they revolve in epicycles, the apparent anomalies are preserved. Then it will be necessary to fully examine how many ways these appearances can be brought about. *Then* the treatment of the phenomena of the stars is just like the causal inquiry, which is concerned with possible means.

And that is why a Heraclides of Pontus can come along [in the 4th century B.C.] and say that even if we grant the earth is moving somehow and the sun stands still somehow, the apparent anomaly concerning the sun can be preserved. For it is really not up to the astronomer to know what is still by nature or what kinds of things move, but to propose hypotheses about what stays still and what moves and then see which hypotheses fit the actual phenomena of the heavens. He must take as principles from the natural philosopher that the motions of the stars are simple, uniform, and orderly. Then, through these principles he will show that the dance of all these things is circular, some winding in parallel, others in oblique circles.

So that is how Geminus, or rather Posidonius through Geminus, teaches the difference between natural science [*physiologia*] and astronomy [*astrologia*], taking his departure from Aristotle.[189]

---

189. Text from Simplicius, *Commentary on Aristotle's 'Physics'* 2.2.291–92 (commenting on Aristotle, *Physics* 2.2.193b). Though it is not important here, some scholars suggest the "Heraclides of Pontus" is an interpolation and that the text originally said just

This comes from Simplicius who wrote in the 6th century A.D., here quoting Alexander of Aphrodisias, who wrote in the early 3rd century A.D. and who was in turn quoting Geminus, who wrote in the late 1st century B.C. or the early 1st century A.D. and was himself either summarizing or commenting on Posidonius' *Meteorology*, written in the early 1st century B.C.[190] Therefore, even though Simplicius is generally reliable in his quotations, a lot of what Posidonius really meant may have been lost in all these layers of transmission, especially since we are missing its original context, and can't be sure polemics against him from subsequent authors hasn't altered the presentation over time. But as we'll see, the general idea aligns well with other accounts.

This passage has received a great deal of fairly recent attention among scholars, and commentaries on it abound, not all of them entirely sound.[191] For example, I.G. Kidd unfathomably sees the mention of Heraclides of Pontus here as a criticism, rather than a positive example of the point being made, which is that forming such hypotheses is exactly what natural philosophers do, and seeing if they fit observations is exactly what astronomers do.[192] Posidonius probably did not agree with Heraclides' natural philosophy, but that is not the point at issue here, which is about method, not results. Likewise, Ian Mueller gives the impression that this paraphrase of Posidonius argues for a complete split between *physici* and *mathematici*, but there is no evidence of these ever being separate people.[193] To the contrary, we have already seen evidence they were expected to be the

---

"someone" (cf. Kidd 1988: 133 and *DSB* 15.204, in s.v. "Heraclides Ponticus").

190. On Geminus' date see references in *OCD* 6207, correcting *DSB* 5.344–47, but corrected in turn by Evans & Berggren 2006: 17–22. See also *EANS* 344.

191. The gist and focus of these commentaries vary considerably, and they are worth comparing against each other. For example: Evans & Berggren 2006: 49–58, 250–55; Bowen & Todd 2004: 193–204; Mueller 2004; Russo 2003: 191–94; Kidd 1988: 129–36; Edwards 1984: 155–57; and (in connection with a related passage in Seneca discussed below): Stückelberger 1965: 55–68.

192. Kidd 1988: 134.

193. Mueller 2004. I say 'gives the impression' since what Mueller means to say on this point is unclear (and he seems rather to agree with me elsewhere), but if this was not his intent, then I am responding to anyone who takes it to be, or who embraces such a conclusion on their own (e.g. M.R. Wright 1995: 159–61).

same person.[194] And though a *physicus* and a *mathematicus* were thought to work differently and to focus on different things, in the area of independent research every known *mathematicus* on record also conducted himself as a *physicus*—especially Posidonius himself, who indisputably engaged in both activities quite extensively, working as both *physicus* and *mathematicus*.[195] So clearly *he* cannot have intended these to be different people.

The astronomer Ptolemy also wore both hats, and explicitly said "there is need for a somewhat more mathematical conception in natural theory, and for a more natural one in mathematical theory," and that, in fact, natural philosophy is often hopeless without the aid of mathematics, hence bringing to fruition Aristotle's expectation that to proceed, some sciences must combine both natural philosophy and mathematics.[196] And yet Ptolemy correctly keeps these two aspects as distinct as possible in all his works, since their aims and methods differ, exactly as Posidonius said.[197] We can even see something similar in the works of the engineer Hero, who also separates

---

194. See relevant evidence and discussion in sections 2.2 and 2.5 above.

195. As we saw in section 2.6 above, though more examples will be given in chapter 3.3.

196. Ptolemy, *Analemma* 1 (following Edwards 1984: 79). That natural philosophy benefits from a mathematical method is articulated at length in Ptolemy, *Almagest* 1.1 (see Toomer 1984: 35–37 and discussion in Taub 1993: 19–37). Nevertheless, Ptolemy thought progress could still be made even in non-mathematical natural knowledge, and held nonmathematical sciences in esteem (e.g. he frequently compares "astrology" as one such science, with "medicine" as another successful albeit unmathematized science, in *Tetrabiblos* 1.2–3).

197. We will discuss Ptolemy's work in more detail in chapters 3.3 and 3.4, but illustrating the present point is the fact that he puts his natural philosophy in the first (and possibly last) books of his *Optics*, thus keeping it more or less distinct from his mathematical (and considerably empirical) work in the middle books, and yet these remain thoroughly interdependent. Likewise, he sets up all his essential hypotheses in natural philosophy in the first chapters of the *Almagest* (literally the *Mathematical Treatise*, cf. ibid. 1.4–8), reserving the rest of that book for his mathematical treatment (derived from those initial hypotheses in conjunction with observations), while placing the remainder of his celestial natural philosophy in a completely separate book, the aptly-named *Planetary Hypotheses* (though even there switching between natural philosophy and mathematical argument as the subject requires). On Ptolemy's opinions and arguments in natural philosophy in general see Taub 1993.

his natural philosophy from his mathematical and practical mechanics, yet all the while maintaining their interdependence.[198] This is surely what Posidonius really meant. Just as Geminus himself said elsewhere, though astronomy is a division of mathematics that "considers the cosmic motions, the sizes and shapes of the heavenly bodies, their illuminations and distances from the earth" and everything like that, it still "relies extensively on sense-perception and coincides a lot with natural theory."[199]

Hence in practice astronomy and natural philosophy were inseparable. And this is just what we hear from one of Posidonius' contemporaries, the astronomer Diodorus (according to Achilles a few centuries later), who said "mathematics differs from natural philosophy in that mathematics investigates what follows from the essential nature of things, but it is natural philosophy that actually investigates the essential nature of things," giving as an example the fact that mathematics tells us how an eclipse happens, but natural philosophy tells us what the sun (and, of course, the moon) are made of. But, he says, "though they differ in research goals, one is necessarily linked with the other."[200] Around the same time or shortly after, the Platonist Dercyllides discussed the same point, as quoted a century later by Theon of Smyrna:

> He said, "the same way it is impossible in geometry and music to develop theories from basic principles without first setting down hypotheses, so also in astronomy it is useful to start by granting hypotheses that lead to a theory of planetary movement. But above all else," he said, "everyone

---

198. See analyses in Tybjerg 2004 and 2005.

199. From Geminus as paraphrased in Proclus, *Commentary on the First Book of Euclid's 'Elements'* pr.1.13.41 (early 5th century A.D.). See Evans & Berggren 2006: 43–48, 243–49 for translation and commentary. Geminus' remarks here come from an extensive division he made in a comprehensive treatise on mathematics, which is now lost but summarized in Proclus, *Commentary on the First Book of Euclid's 'Elements'* pr.1.13.38–42. Geminus first divided mathematics into pure and applied (or in our parlance "mathematics" as such and "mathematical sciences"), and then put astronomy under applied mathematics (along with mechanics, optics, and harmonics, as well as surveying and logistics).

200. Diodorus of Alexandria (of the early 1st century B.C., on whom see chapter 3.3), probably from his (lost) commentary on the *Phenomena* of Aratus, as quoted by Achilles (Tatius?), *Introduction to the 'Phenomena' of Aratus* 2 (= Maass 1958: 30), which was written in the 3rd or 4th century A.D.

agrees we should more or less accept basic principles that have a solid support from mathematical studies."[201]

He then gives several examples of such principles, which all derive from natural philosophy but conform to a plausible mathematical model. One of his examples illustrates our point well: astronomers take as a premise from natural philosophy that when stars rise and set they are not blinking in and out of existence, but completing a continuous movement around the earth—obviously a hypothesis that cannot be "proved" mathematically.

All these remarks echo the point made by Posidonius. The distinction they are all making is that mathematics cannot tell us what the underlying facts are, it can only analytically describe what we observe. Hence not only must astronomers rely on natural philosophy for pre-mathematical assumptions about the underlying nature of the universe (like the fact that stars actually rise and set and aren't, instead, blinking in and out of existence), but once astronomers have completed their mathematical analyses they must return to natural philosophy to figure out what's actually going on, since no math can help them sort between mathematical models that otherwise explain observations equally well.

Following the example Posidonius used, the eccentric and epicyclic models of planetary motion were both mathematically identical and yet entailed different physical facts. The epicyclic model involved a system of 'circles on circles', with planets tracing a circular orbit around the earth while at the same time tracing smaller orbits around their own epicenters. But the eccentric model could produce exactly the same motions using an off-center circle for each planet. Though the addition of yet another epicycle was eventually found to be unavoidable, there still remained an equivalence between the single-epicycle-on-eccenter model, and the double-epicycle model. Mathematically there is no way to tell the difference between these two models from the observations available. So only further arguments derived from natural philosophy could resolve the matter, if at all. Hence Theon of Smyrna reports that Hipparchus recommended the epicyclic model, but couldn't muster enough evidence from natural philosophy to prove it. "Because he wasn't supplied with enough resources from natural

201. Quoted in Theon of Smyrna, *Aspects of Mathematics Useful for Reading Plato* 3.41.199 (cf. 3.39–41.198–200), from the lost work of Dercyllides, *The Spindles with which the 'Republic' of Plato is Concerned* (see EANS 241–42).

theory," explains Theon, Hipparchus "himself didn't know for sure which was the true movement of the planets according to nature and the facts, and which matched observations only by chance. And so he hypothesized the epicyclic" model.[202] In other words, Hipparchus "himself" (*autos*) didn't "know for sure" (*sunoiden akribôs*) so he only "hypothesized" (*hupotithetai*) that the epicyclic was "more plausible" (*pithanôteron*) based on what little he could muster from natural philosophy.[203] Hipparchus thus agreed that greater certainty could only be achieved by drawing more arguments and evidence from natural philosophy, since *mathematically* he had done all he could do.

This interpretation runs contrary to that of I.G. Kidd, who again reads Theon as criticizing Hipparchus, as if Theon is accusing him of being a 'mere' astronomer and thus unqualified to discuss physics, even though Theon never says anything like that, nor did any ancient author.[204] All I

---

202. Theon of Smyrna, *Aspects of Mathematics Useful for Reading Plato* 3.34.188 (cf. 3.32.166), translating: *oude autos mentoi, dia to mê ephôdiasthai apo phusiologias, sunoiden akribôs, tis hê kata phusin kai kata tauta alêthês phora tôn planômenôn kai tis hê kata sumbebêkos kai phainomenê: hupotithetai de kai houtos ton men epikuklon....* (the passive infinitive *ephôdiasthai* is from the verb *ephodiazô*, *LSG* 746: "to furnish with supplies for a journey").

203. Hipparchus is said to have credited it as "more probable" in the sentence immediately preceding the above: *hoper kai sunidôn ho Hipparchos epainei tên kat' epikuklon hupothesin hôs ousan heautou, pithanôteron einai legôn pros to tou kosmou meson panta ta ourania isorropôs keisthai kai homoiôs sunarêrota* ("Being aware of this fact, Hipparchus approved the epicyclic hypothesis as his own, saying it was more plausible that all the heavens are laid down evenly balanced against the middle of the cosmos and joined together in the same way.").

204. Kidd 1978: 11. Kidd cites an unrelated passage from Plutarch in defense of his reading of Theon (Plutarch, *On the Face that Appears in the Orb of the Moon* 4 = *Moralia* 921d-e), but this nowhere says Hipparchus was unqualified to discuss physics or that it was inappropriate of him to do so. Cherniss 1957: 45 adds an unjustified interpretive note to that effect. But all the Greek text actually says is that his physics of vision is not generally accepted (*pollois ouk areskei physiologôn peri tês opseôs*), and that the present discussion is not an occasion to debate it. In particular, the speaker says "it is the task" (*ergon*) of someone who believes in the visual ray theory to address questions based on that theory (like the lengthy question the speaker had just asked before this), but "it is not now" (*ouketi*) our task to investigate the visual ray theory itself (though it would be appropriate on some other occasion), or "it is no longer" our task to debate it (because

can see here is a plain statement of the facts. Theon is using Hipparchus as a positive example of his point, not a contrast to it. Far from criticizing Hipparchus, Theon sees him as behaving correctly: he adopted a position only as a hypothesis when he could not adduce enough evidence from natural philosophy to resolve the matter more conclusively. This conforms to the Posidonian argument that hypotheses must be constructed from principles established by natural philosophy, and any decisions we have to make between alternative hypotheses must likewise be guided by natural philosophy, but determining which hypotheses fit the observable facts is the distinct task of mathematical astronomy. I thus do not see any criticism of Hipparchus in Theon. In fact, I see no evidence of any sort of "contemporary debate among philosophers and scientists" that Kidd proposes.[205] To the contrary, all the sources we have present the Posidonian position as undisputed among working scientists—not only do all the sources we have mention no opposition, but there are no scientists on record challenging it, and there is no evidence any did.[206]

This becomes clearer the more we examine the same idea as voiced in other sources. Closer in time to Geminus is Seneca, who summarized in the 1st century A.D. something similar, possibly drawing again on Posidonius, a fellow Stoic—now explicitly defining mathematics as a tool both assisting and depending on natural philosophy:

> Philosophy is one part natural, one part moral, and another part logical ... and when one arrives at natural questions, the matter rests on the testimony of a geometer, so geometry is a part of natural philosophy, because it helps it. But many things help us and are not as a result parts of us. Rather, if they were parts, they would not help. Food is an aid to the

---

we and most others already accept it). There is nothing here about the different "provinces" of mathematicians and philosophers as Cherniss and Kidd claim. On 'visual ray' theory (and the alternative embraced by Hipparchus) see chapter 3.5.

205. Hence Kidd also incorrectly reads Posidonius as criticizing Heraclides (as noted above). It is with these and other errors of interpretation that Kidd 'discovers' a non-existent dispute between 'philosophers' and 'scientists', as if these were ever different people (see Kidd 1988: 134–36, especially in contrast with the *actual* text and context of Theon's other citations of Hipparchus, which do not conform to Kidd's reconstruction).

206. For example, on Ptolemy's embrace of this principle (that hypotheses in astronomy derive from natural philosophy) see Taub 1993: 39–45.

body, yet it is not a part of it.

The service of geometry provides some benefit for us. It is necessary to philosophy in the same way the artisan is necessary to geometry. But he is not a part of geometry, nor is geometry a part of philosophy. Moreover, each has its own objectives. For the sage investigates and learns the causes of natural phenomena, then the geometer follows up and calculates their sums and dimensions. The sage knows the rationale behind celestial phenomena, their power or nature, while the mathematician computes their courses and retrogressions, and certain observations in which they rise and set and produce the appearance of standing still from time to time (even though celestial bodies cannot stand still). The sage will know what causes images to appear in a mirror, while the geometer can tell you how far away a body must be from its image and what shape of mirror reflects what kind of images. The philosopher will prove the sun is large, while the mathematician will prove how large it is, proceeding by some measure of experience and practice. But in order to proceed certain principles must be obtained for him. Yet nothing is an art in its own right if it has a borrowed foundation. Philosophy takes nothing from any other art. It builds up its whole edifice by itself. But mathematics, as I might put it, is a rented abode. It builds on someone else's soil. It takes up first principles and, benefitting from these, arrives at what follows. If it were to venture on to the truth all by itself, if it could grasp the nature of the whole universe, then I would say it would contribute greatly to our minds, which grows by considering celestial things, drawing something from on high.

However, the soul is perfected by one thing alone: the unalterable knowledge of good and evil. Yet nothing, not any other art, investigates good and evil [except philosophy].[207]

Seneca often writes polemically and hyperbolically and is not always faithful to his sources. But there is a more sober author who voices essentially the same point. Strabo, who wrote half a century before Seneca and was a boy when Posidonius was still alive, discusses the same contrast between these two ways of studying celestial phenomena, as a natural philosopher or as an astronomer. Once again Strabo appears to be drawing on the same Posidonian argument:

---

207. Seneca, *Moral Epistles* 88.24–28. For extended commentary see Kidd 1988: 359–65 and Stückelberger 1965: 55–68. The context is an extended argument that philosophy is more important than the 'liberal arts', which included mathematics and astronomy (see chapter 4.6.1 and chapter five of Carrier 2016).

> Regarding the collection of first principles that guide him, he who writes geography must trust the geometers who have measured out the whole earth, and they in turn must trust the astronomers, who in turn must trust the natural philosophers. Natural philosophy is an art, and the 'arts', they say, are those without separate foundation, depending upon themselves, and containing within themselves their own first principles and the proofs thereof.
>
> Now, the sorts of things that are demonstrated by the natural philosophers are that the cosmos and the heavens are spherical, and that the pull of weighty things is toward the center, around which the earth has come together as a sphere, resting homocentric with the heavens, as does the axis through it, which also extends through the center of the heavens. The heavens revolve around both the earth and its axis from east to west, and with the heavens the fixed stars move at the same speed as the celestial sphere. So the fixed stars move along parallel circles. And the best known parallels are the equator, the two tropics, and the arctic circles. But the planets and sun and moon follow certain oblique circles positioned on the zodiac.
>
> The astronomers, trusting these points either entirely or in part, then work out what comes next: the movements, periods, eclipses, sizes, distances, and lots of other things. In the same way the geometers who measure out the whole earth hold to the doctrines of the natural philosophers and the astronomers, and the geographers in turn hold to the doctrines of the geometers.[208]

Nevertheless, such distinctions should not mislead us into thinking different people are necessarily being referred to. Ptolemy, for example, was all these things: geographer, geometer, astronomer, and natural philosopher. But as Strabo would have expected, Ptolemy engaged in each endeavor independently, recognizing their differences in subject and method, and their mutual place in the hierarchy of argument.

That all this talk of different roles for *physici* and *mathematici* in astronomy is really only about different roles assumed by the same person is further confirmed by a direct parallel in another mathematical science: harmonics. This is how Ptolemaïs in the early 1st century A.D. (or late 1st century B.C.) articulated the same distinction in the study of music:

> What comprises the theory that uses [the scientific instrument called] the monochord? The things postulated by musicologists [*mousikoi*] and those

208. Strabo, *Geography* 2.5.2–4.

> adopted by the mathematicians [*mathêmatikoi*]. The things postulated by the musicologists are all those adopted by [harmonic scientists] on the basis of perceptions, for instance that there are concordant and discordant intervals, [etc.].... Those adopted by the mathematicians are all those that [harmonic scientists] study theoretically in their own special way, only *beginning* from the starting points given by perception, for instance that the intervals are in ratios of numbers, [etc.].... Hence one might define the postulates of [harmonic scientists] as lying both within the science concerned with music, and within that concerned with numbers and geometry.[209]

Here we have the same idea of a distinction between what mathematicians contribute to harmonics and what empirical observers (*mousikoi*, here essentially *physikoi* concerned with the study of music and sound) contribute to harmonics, as if these were different people. And yet from her concluding remarks there is no doubt that she means the mathematical and empirical elements of this science are carried out by the same person. Once again this conclusion is confirmed by actual harmonic scientists like Ptolemy, whose *Harmonics* synthesizes both activities, even while keeping them distinct wherever necessary.

There is evidence the same kind of distinction was being made in mechanics. Jaap Mansfeld claims, to the contrary, that mathematicians like Hero and Pappus set themselves in opposition to philosophers, and thus natural philosophers and mathematical scientists were often different people.[210] But again there is no evidence of this. Serafina Cuomo and Karin Tybjerg have already demonstrated the opposite for Hero, showing how he did not regard himself as operating outside or against philosophy, but was establishing himself as a superior philosopher, replacing armchair reasoning with empirical demonstration, thus doing "philosophy with

209. Ptolemaïs, *Pythagorean Elements of Music*, frg. 1, quoted in Porphyry, *Commentary on Ptolemy's Harmonics* 22.22–23.22. For sources on Ptolemaïs see Carrier 2016 (index). The word translated as 'harmonic scientists' is *kanonikos*, those who study the 'canon' of harmonics. On the monochord as an ancient scientific instrument: Creese 2010.

210. J. Mansfeld 1998: 94–95. Cuomo 2000: 81–88 implies a similar conclusion for Pappus, but is rightly challenged by Mueller 2000 (though in fairness even Cuomo concedes there was overlap between the categories of philosopher and mathematician).

machines" (*dia tôn organôn philosophiâ*), which happens to parallel exactly what Galen did, using empirical and experimental anatomy, for example, to refute an old Stoic doctrine of the soul, while repeatedly chastising its advocates for relying on armchair philosophy instead.[211] And yet Galen, like Hero, considered himself a philosopher, wrote philosophy, and defended philosophical positions. Indeed, Galen outright insisted that any doctor worthy of the name must also be an accomplished philosopher, and he regarded almost anyone who wasn't to be a mere hack. We know this is also what the engineer Vitruvius thought, and he appears to represent the general attitude of his profession.[212] No doubt Hero would have said the same (and as we'll see, it appears he did). Jacqueline Feke has made this point before, and proved the same of Ptolemy as of Hero, showing that this contest between "mathematics" and "philosophy" was really a contest, for all these authors, between mathematical empiricism and *armchair* philosophy. In other words, they are not arguing against philosophy, they are arguing that a philosophy reliant on mathematics is superior to any other philosophy.[213]

We have already seen ample evidence of the fact that in the Roman period natural philosophers and mathematical scientists were, and were expected to be, the same people, and that it was the generally accepted view that mathematical science and natural philosophy were inseparable pursuits mutually dependent on each other. But even Jaap Mansfeld's own treatment of the evidence is sufficient to condemn his conclusion. He cites only three passages in his defense: two from the 4th century mathematical writer Pappus, and one from Hero of Alexandria centuries earlier, none of which have Pappus or Hero refusing the label 'philosopher'. Mansfeld only imagines that the introduction to Hero's *Siegecraft* constitutes a condemnation of philosophy, yet it only says philosophers had failed to make progress on a

---

211. On Hero as philosopher see Tybjerg 2003, 2004, 2005; Cuomo 2002; and Fake 2014. In *Siegecraft* 1, Hero says tranquility is achieved not "by the investigation of arguments" (*dia tôn logôn tên...zêtêsin*) but "by a philosophy of machines" (*dia tôn organôn philosophiâ*). Galen's comparable assault against Chrysippus on the anatomical location of the soul is in Galen, *On the Doctrines of Hippocrates and Plato* (for relevant analysis see Tieleman 1996). This is the same Galen who wrote *That the Best Doctor is also a Philosopher*, whose thesis is self-explanatory.

212. For example, Vitruvius, *On Architecture* 1.1.1–2, 1.1.7, and 1.1.11, with 6.pr.6–7. See related discussions in chapters five and seven of Carrier 2016.

213. Feke 2014 (with Feke 2011).

single philosophical question: how to achieve tranquility. Mansfeld somehow telescopes that into an imagined condemnation of all philosophers and the whole of philosophy, a conclusion in no way warranted by this or any other passage in Hero's extant work. That Hero thought little of armchair reasoning does not warrant such a sweeping conclusion, since many *bona fide* philosophers, from Cicero to Galen, said essentially the same thing, a fact that Mansfeld himself is later forced to admit.[214] Like them, Hero was arguing for better philosophy, not its abandonment. Just as Ptolemy and Galen did.

Mansfeld also mistreats a passage from Pappus in much the same way.[215] Pappus says philosophers claim the sphere is the greatest and most beautiful of all forms but provide no demonstration of this claim. This is no polemic against philosophy or philosophers as a whole. It is a simple statement of fact: no philosopher has demonstrated that the sphere is the greatest of all shapes. Again, Mansfeld somehow telescopes this into a sweeping condemnation of philosophy. In the only other passage Mansfeld cites, he claims Pappus "distinguishes philosophers from mathematicians," but Pappus does no such thing there. All he does is distinguish philosophers in general from those, *including philosophers*, who approach the world mathematically. And in neither case are any philosophers depicted negatively. Indeed, the whole of this passage establishes exactly the opposite of Mansfeld's point, since it argues that the mathematical field of mechanics is thoroughly dependent on natural philosophy and therefore inseparable from it. And since part of this must derive (directly or indirectly) from some lost work by Hero, it entails that Hero himself argued that very point.

The relevant passage is therefore worth quoting in full:

> Mechanical theory, Hermodorus my son, being useful for many and important things in life, reasonably deserves the greatest acceptance from philosophers, and is much pursued by all those who study mathematics, since it was practically the first to deal with the natural theory of the material elements of the universe. For a theoretical study of the static and motive tendencies of bodies ... not only investigates the causes of what

214. In J. Mansfeld 1998: 104 n. 355 he presents evidence against himself, noting that, e.g., Plutarch can say much the same thing as Hero and Pappus do "without implying that he prefers not to be called a philosopher himself."

215. Pappus, *Mathematical Collection* 5.19.350.

> moves naturally, but also effects change, forcing things to move against their nature in directions away from their proper places. And this is done by devising mechanisms, using theorems derived from the very same study of matter.
>
> The engineers who follow Hero say mechanics has two parts, one theoretical and the other practical, and that the theoretical part is assembled from geometry, arithmetic, astronomy, and natural theory, and the practical part from metalworking, building, carpentry, painting, and training in these subjects by hand. And so, they say, one who develops in these skills and sciences from childhood, and has a natural talent for them, will be the greatest engineer and inventor of mechanical devices.[216]

The gist of the first paragraph above is that because mechanics has dealt with physics almost from the start, philosophers ought to approve of it (*apodochês êxiôtai pros tôn philosophôn*), and in fact all those philosophers "coming from mathematics" (*pasi tois apo tôn mathêmatôn*) actively pursue it (*perispoudastos*).[217] Hence Pappus does *not* distinguish "mathematicians" from philosophers here, but clearly allows some philosophers among those who approach the world mathematically, as in fact we know was the case. Pappus is thus *associating* mechanics with natural philosophy (*physiologia*), not separating them. The second paragraph not only confirms this, but is even more important, because it evidently derives in some way from the writings or lectures of Hero himself, and Pappus reports, with approval, Hero's view that engineers are expected to study not only practical crafts as well as mathematics, but *also* natural philosophy (*tôn phusikôn logôn*).[218]

Therefore, there is no distinction in Pappus between mathematicians and philosophers as individual persons. To the contrary, these subjects are clearly expected to be pursued by one and the same person. Hero

---

216. Pappus, *Mathematical Collection* 8.1.1022–24.

217. The use of the phrase *kai pasi tois apo tôn mathêmatôn* here is essentially identical to that in Diogenes Laertius, and carries the same connotations when following (again) the lone plural of *philosophos* (see note above).

218. The word used for the "practical" part of mechanics is *cheirourgikon*, literally "hands-on work" or "work done by hand." The word used for the "theoretical" part is *logikon*, which is said to consist not only of physics but also the mathematical study of *geômetria*, *arithmêtikê*, and *astronomia* (and we can assume Hero would also have included harmonics: see discussion of the 'quadrivium' in chapter five, and discussion of engineering education in chapter seven, of Carrier 2016).

and his followers said exactly that, and Pappus raises no objection to it. Instead, the distinction Hero made is identical to that between doctors and physiologists, hence between theoretical knowledge and practical craft. And yet just as in medicine, here in engineering the same person is expected to master *both*. Pappus, meanwhile, implies the same distinction between theoretical and mathematical mechanics as we saw had been made between astronomy and astrophysics: mechanics involves the mathematical analysis of observed phenomena, but relies for this on hypotheses derived from natural philosophy. We can easily presume that if ever an engineer had to decide between mechanical models that were mathematically identical though physically different, he would admit that the decision could only be made on an argument from natural philosophy, exactly as was the case in astronomy. We can see an example of this in the introductory chapters of Hero's *Pneumatics*, where it is not mathematics, but natural philosophy that chooses between physical models of the structure of gases and liquids.

A related distinction that was drawn between mathematics and natural philosophy is that mathematical or logical *precision* is not the particular province of the *physicus* as such, since his conclusions can only aspire to probability. Alexander of Aphrodisias provides the fullest exposition of this point:

> Aristotle now says [in *Metaphysics* 994a14–15] the same thing he said in the first book of the *Nichomachean Ethics* [1094b27] when he was discussing how these arguments ought to be received: "it is nearly the same mistake to accept probable arguments from a mathematician as to demand deductive proofs from a rhetorician." Then he says, "for one must not demand mathematical precision in everything, but only in the case of immaterial things, such as mathematical objects and objects derived from abstraction."
>
> Perhaps he is pointing out to us that such precision is also needed for the present subject [i.e. metaphysics or 'first philosophy']. For his treatise on first principles concerns immaterial things and not ordinary things. "Hence this is not the theory (or method) of natural philosophy" (the text reads either 'method' or 'theory' as being natural).[219] Perhaps he says this because it is precise, for he said this previously. Or maybe

219. This textual note is Alexander's, which means there were two textual variants known to him: the phrase *to de dioper ou phusikos* ended either with *ho logos* or *ho tropos*. Alexander sided with *ho logos*, which we find in the received text of Aristotle. The difference does not matter here.

> his statement means something like this: all natural things seem to be physical, but mathematical things are nonphysical, and for that reason they welcome precise argument, while natural things do not, at least not in the same way. Hence arguments about mathematical objects, since they are about immaterial objects, are not the arguments of natural philosophy, for natural objects do not make room for the same degree of precise argument—because, after all, they are material.
>
> He could also be saying, about the arguments we are looking at now, that they are not the arguments of natural philosophy because we intend to talk about immaterial things, and for that we need more precise arguments than we do in physics.[220]

Alexander does not mean by 'precision' here anything like increasing accuracy of measurement, but decisiveness of argument—such as the difference between deductive and inductive logic. In other words, the *physici* dealt in inductive reasoning, establishing the probable facts of the physical universe, which was not the same as the activity of a mathematician or logician, even though we know the same men often engaged in both activities, and Alexander and Posidonius certainly understood this (no doubt following Aristotle, who had likewise argued the two activities were ultimately inseparable, as we saw in section 2.2). The difference is that when acting in the role of a mathematician, one is expected to deal in deductive reasoning, establishing necessary truths about all being, or about the inevitable consequences of the hypotheses that one developed when acting in the role of a *physicus*. In contrast, in the words of Cicero, "What ought less to be said by natural philosophers than that anything certain is indicated by uncertain things?"[221] Elsewhere Cicero hints at the same distinction Aristotle did, that *physici* discuss "obscure subjects" but *mathematici* discuss "fictions."[222] In other words, the *physicus* discusses real things, which are always messy and imprecise and difficult to figure out, while the *mathematicus* discusses ideal situations that never happen so perfectly in reality.

Thus, a good *physicus* was expected to master and employ both mathematical analysis and empirical observation, but to accept that his

220. Alexander of Aphrodisias, *Commentary on Aristotle's 'Metaphysics'* 169.

221. Cicero, *On Divination* 2.43.

222. Cicero, *On the Boundaries of Good and Evil* 2.5.15.

conclusions can never be as certain as mathematical deductions. This is because in studying real things, either from speculation or observation, a natural philosopher's ultimate interest came to causes and correlations, which can only be arrived at inductively. As an example, Cicero lists several achievements of the Stoic *physici*, revealing the range of their interests. He says there are several exemplary cases where a natural causal connection between phenomena has been (inductively) proven where it would otherwise not be expected. He gives the association of tides with the orbit of the moon as his crowning example.[223] Already this reveals that *physici* were interested in discovering hidden correlations in nature, and had been successful at it. And apart from the role of astronomy and meteorology in the science of tides, Cicero reveals that their investigations touched on subjects as diverse as botany, zoology, animal and plant physiology, and harmonics.

For instance, Cicero says Stoic naturalists had shown:

> That the tiny livers of small mice enlarge in the dead of winter; that on the very day of the Winter Solstice the dry pennyroyal begins to bloom and once its seedpods inflate they burst, and the seeds of this fruit, which were confined within, tumble in various directions; further, that in some strings which have been struck there is a force that makes other strings vibrate; that oysters and all shellfish happen to grow and diminish along with the moon; and that the right time to cut down trees is thought to be in wintertime when the moon is waning, because then they are dry [of sap].[224]

223. Cicero, *On Divination* 2.34. On the ancient discovery of lunisolar tide theory see discussion and notes in chapter 3.3.

224. Cicero, *On Divination* 2.33–34. It is perhaps worth asking if these claims are even true. The last three claims appear to be fact: the phenomenon of 'sympathetic vibration' is well-documented (Rossing & Fletcher 1995); shellfish do grow and shrink with lunar phase as a result of their circalunal rhythm (Cloudsley-Thompson 1980); and though there are elements of superstition in timbering (Meiggs 1982: 331–32), according to some woodcutting experts (e.g. "Primavera" 1994) tree sap tends to be "down" or "low" in conjunction with the drop in temperature and light conditions (both solar and lunar), hence sap is at its lowest on a winter night during a new moon, and since less sap means less food to attract and feed bugs, cutting at the low increases timber quality by reducing discoloration or degradation from insect infestation (Theophrastus was aware of the reason: *Inquiry on Plants* 5.1.1–4; cf. Pliny the Elder, *Natural History* 16.74.188–92; Vitruvius, *On Architecture* 2.9.2–4; Cato, *On Agriculture* 37.3–4). However, what actually 'drops' in these conditions

Cicero concedes that though they might not always know the real causes behind these influences, they were keen to prove such correlations existed, especially for predicting the future.[225]

The place of 'cause' in the research program of the *physicus* is even more obvious in the case of medical subjects. Of course, mathematics was considered less applicable there, but as we've already seen a similar distinction was made between the task of a healer and that of a physiologist, and yet again the two roles were considered inseparably linked. Physiology (the systemic behavior of the internal organs of living beings) was certainly a subject a *physicus* studied.[226] For instance, Galen says a *physikos* was the sort of person who would study the causes of the effects of castration.[227] We have already seen how medical subjects fell within the interests of the *physicus*, and this is further confirmed by Alexander, who repeats Aristotle's point "that it is the first task of the natural philosopher 'to see the first principles concerning health and disease', that is, to figure out from what first principles health and disease come about, and what first principles they are dependent on," in other words, what the ultimate causes were.[228]

Accordingly, in defending Aristotle's argument that a *iatros* (doctor) must also be a *physikos* (natural philosopher), Galen makes this point about the importance of studying causes in human physiology:

---

is still debated, whether it is the actual water weight, or flow pressure, or sugar content of the sap (Edlin 1976: 239–45; Thomas 2000: 48, 57). In contrast, however, the first two claims are dubious: the livers of many animals might grow as they store fat for winter, but I could find no evidence that any species of mouse was more prone to this than any other wintering mammal; and Cicero's description of a plant that starts flowering in winter, and later sheds bursting seed pods, could fit several plants of the region, but not the dry pennyroyal (i.e. fleabane, which is what his phrase *puleium aridum* usually denotes), and even if this denotes a different plant, "on the very day of the Solstice" is certainly hyperbole.

225. Cicero, *On Divination* 1.110–13, 1.126–27. These findings were also used as a 'proof of concept' in support of developing astrology as a science (and likewise other arts of divination).

226. On this point in general: Cicero, *On Divination* 2.37.

227. Galen, *On Semen* Kühn 4.580 (and 4.569).

228. Alexander of Aphrodisias, *Commentary on Aristotle's 'On Sensation'* 6; and see Alexander of Aphrodisias, *Problems* 1.99.

> I should like to ask the Erasistrateans why it is that the stomach contracts upon the food and why the veins generate blood. There is no use in recognizing the mere fact of contraction, without also knowing the cause. For if we know that, we shall also be able to rectify the failures of function. "This is no concern of ours," they say, "we do not occupy ourselves with such causes as these. They are outside the sphere of the *iatros* and belong to that of the *physikos*" .... But how are you going to be successful in treatment, if you do not understand the real foundation of each disease?[229]

On another occasion Galen says:

> It is not enough for you to learn the method of the art whose achievements amaze you. You must also attempt to comprehend the principles, and not only as a doctor, but as a natural philosopher as well. ... Even if you do not have time to learn everything as a natural philosopher does, you ought to put your mind to considering at least as much as you can, because when it comes to improving your ability to predict what will happen, there is no better way than this.[230]

Even Cicero assumed a doctor should know natural philosophy as well as a pilot knows the stars or a magistrate the law. But he also expected that a doctor should not let research take him away from his duties as a doctor, using the same analogy that a pilot should not let his study of the stars take him away from his duties at the helm, or a magistrate let his study of the law take him away from his duties in office, thus reflecting a similar distinction between the different aims of research and practice that Galen and Aristotle express.[231]

---

229. Galen, *On the Natural Faculties* 2.9 (= Kühn 2.126). Though this implies the Erasistrateans separated themselves as doctors from natural philosophers, Galen's representation of his opponents' views is (as here) polemical and frequently dubious. As Erasistratus was famously involved in the study of physiology (cf. sources in Appendix B) and Erasistrateans relied extensively on physiological dogmas (two facts even Galen confirms throughout his writings) clearly Galen's depiction of their opinion on the subject is fundamentally inaccurate.

230. Galen, *On Critical Moments* Kühn 9.738–739. Celsus, *On Medicine* pr.47 echoes the same idea, in effect that a *medicus* performs better when he is also a *physicus*.

231. Cicero, *On the Republic* 5.5.14. See also related point in chapter five of Carrier 2016.

Hence just as Ptolemy and other astronomers kept their mathematical and philosophical reasoning distinct, while still acknowledging their mutual dependency, so Galen could still distinguish medical practice from natural philosophy. This can already be seen in his most monumental treatises on anatomy, *On the Uses of the Parts* and *On Conducting Anatomical Investigations*: the former engaged anatomical research to support Galen's natural philosophy; the latter, to advance medical practice. In maintaining these distinctions, Galen lists some of the roles a *physicus* could take in medical research:

> One use of anatomical theory for the natural philosopher is love of this knowledge for its own sake. Another is not love of knowledge for itself, but to prove that nothing arises in nature without a purpose. Another is to gain knowledge concerning the functions of something, whether physical or mental, by obtaining premises from anatomy. And another use besides these is for one who intends to successfully extract shafts or arrowheads, or to properly cut things out, like correctly performing surgery on ulcerous flesh, deep wounds, or abscesses—these, as I have said, are most essential, and the best doctor needs to practice on them more than anything. Next are the functions of the inner organs deep within. Then, after that, knowledge of what is useful, as much as makes a difference for doctors in the diagnosis of diseases. Some of these aims are more useful for natural philosophers than doctors, especially the first two, as was said: either for the sake of mere theory, or to teach how nature's skill is correctly carried out in every part.[232]

232. Galen, *On Conducting Anatomical Investigations* 2.2 (= Kühn 2.286–87, cf. Kühn 283–86 for context). Galen means it is the business of a natural philosopher to demonstrate the intentions of the Creator in designing particular organs, thus a *physicus* could still be a creationist. But what is important here is that *physici* are interested in anatomical research and the discovery of organ function, e.g. Galen, *On the Natural Faculties* 3.8 and 3.12 (= Kühn 2.174 and 2.185). On all these points see the elegant summary in Galen, *On the Uses of the Parts* 17.1–2 (= M.T. May 1968: 730–33), and see Hankinson 1994b on the diverse uses Galen had for empirical dissection and his advocacy of it as a method, and Hankinson 1988: 142–45 on elements of this passage specifically. Galen also elaborates on the serious need for detailed anatomical knowledge for performing successful surgeries in *On Conducting Anatomical Investigations* 1.3, 2.2, 3.1, 3.9, 4.1, 7.13 (= Kühn 2.229, 2.283–84, 2.340–46, 2.393–97, 2.416–20, 2.632–34), and for developing effective treatments (cf. Ormos 1993: 172). This stood in contrast to doctors of the Empiricist or Methodist sects, like Soranus, who claimed dissection was *only*

In agreement with this, Philo of Alexandria observes that natural philosophers are primarily concerned with theory, but:

> The best doctors *also* carry out investigations of the construction of man and examine in detail what is visible—and also what is hidden from sight, through the careful use of anatomy—so if medical treatment is ever required, nothing that could cause serious danger would be missed out of ignorance.[233]

Similarly, Alexander reiterates Aristotle's reason for closely associating doctors and natural philosophers:

> Aristotle establishes that it is the job of the naturalist and philosopher to inquire into what the first principles of health and disease are, from the fact that the majority of natural philosophers have made arguments about them, and from the fact that natural theory ends in these principles and the most accomplished of doctors make a beginning from them in their theory of medical matters—as they are natural philosophers, too. At the same time he shows us how medical inquiry is united with natural inquiry, and that it falls under the natural, taking its first principles from it, just as optics from geometry, harmonics from arithmetic, and navigation from astronomy.[234]

Consequently, we find the *physici* involved in almost every aspect of medical science. The specific medical subjects that we find mentioned as interests of the *physici* include the study of sense perception,[235] human gestation and fetal development,[236] the anatomy and physiology of a woman's

---

useful for natural philosophy, though even he conceded this should "nevertheless be studied for the sake of profound learning" (Soranus, *Gynecology* 1.2, 1.5).

233. Philo of Alexandria, *On the Special Laws* 3.117.

234. Alexander of Aphrodisias, *Commentary on the Book 'On Sensation'* 6.27–7.6. See also: Alexander of Aphrodisias, *On Fevers* 25.13; Aristotle, *On Sense and Sensibles* 436a17-b1.

235. Galen, *On Irregular Intemperance* Kühn 7.743; Alexander of Aphrodisias, *Commentary on the Book 'On Sensation'* 92; Sextus Empiricus, *Against the Professors* 8.13; Vitruvius, *On Architecture* 6.2.3.2; etc.

236. Philo of Alexandria, *On the Special Laws* 3.117, *On the Creation of the World* 132, *Allegorical Interpretation* 2.6, *Questions on Exodus* 1.frg.7a (knowledge attributed to *physici* in these passages also concerns such diverse subjects as menses, sexual

breasts,[237] the contents and function of blood,[238] the study of the skeletal anatomy of the human jaw,[239] even the nature and causes of heat retention during sleep.[240] In fact, Galen says to truly understand the function of all bones and organs, "you must become a *physikos* and an anatomist at the same time."[241]

Beyond physiology, the interests of the *physici* ranged over just about every other scientific subject as well, everything that fell under the category of *physis* or 'nature'. Though technical distinctions were apparently drawn between natural philosophers and astronomers, since the same person was almost always both the *physici* as a class are specifically cited as experts on celestial bodies, as advancing astronomical theories and making observations of the stars,[242] as calculating the size of the earth and the sun,[243] discovering the moon is lit by the sun,[244] knowing all about the precession of the equinoxes,[245] debating whether the earth spins on its axis,[246] and also, of course, contemplating whether the universe is infinitely old or large

---

dimorphism, heart physiology, and the decay of corpses); Sextus Empiricus, *Against the Professors* 5.59.

237. Tertullian, *On the Flesh of Christ* 39.

238. Galen, *On the Elements according to Hippocrates* Kühn 1.506.

239. Galen, *On the Uses of the Parts* 11.8 (= Kühn 3.922 = M.T. May 1968: 516–20).

240. Galen, *Commentary on the Aphorisms of Hippocrates* Kühn 17.455.

241. Galen, *On the Uses of the Parts* 12.7 (= M.T. May 1968: 560); for example, Cicero cites both *physici* and *medici* as experts on the physiology of human sensory organs (Cicero, *Tusculan Disputations* 1.20.46). For more examples of natural philosophers examining medical questions see Nutton 2013: 145–49.

242. Diodorus Siculus, *Historical Library* 1.28.2; Philo of Alexandria, *On the Life of Moses* 2.103; Varro, *On the Latin Language* 5.69.2; M. Verrius Flaccus, *On the Meaning of Words*, p. 339, §44 (1st century B.C./A.D., via an epitome of Festus in the late 2nd century A.D.); Alexander of Aphrodisias, *Problems* 2.74; Eusebius, *Preparation for the Gospel* 10.14.10 and 15.p.1.

243. Cleomedes, *On the Heavens* 90, 152.

244. Cleomedes, *On the Heavens* 201.

245. Cicero, *On the Boundaries of Good and Evil* 2.31.102.

246. Alexander of Aphrodisias, *Commentary on Aristotle's Metaphysics* 421.

and what the fate of the universe might be.[247] The *physici* are also cited as authorities on such matters as geography and climatology,[248] including the causes of lightning and earthquakes,[249] the source of the Nile and the cause of its annual flooding;[250] the nature of mind and soul;[251] the sense of smell,[252] the nature of sound,[253] the nature of vision,[254] and the habits and characteristics of animals.[255] The *physici* were also known for studying even deeper subjects like the nature of space and time,[256] the nature of heat and fire,[257] and the nature of motion.[258] Hence Seneca divided all research on "the nature of things" into three parts: celestial, atmospheric, and terrestrial, encompassing the entire contents of the universe, from astronomy and meteorology to botany and geology.[259]

---

247. Sextus Empiricus, *Against the Professors* 8.146, 10.169; Pseudo-Plutarch, *Tenets of the Philosophers* 887a; Cleomedes, *On the Heavens* 6.

248. Diodorus Siculus, *Historical Library* 2.37.5, 3.51.1; Cleomedes, *On the Heavens* 22, 60; Lucian, *Demonax* 22; Pomponius Mela, *Description of the Lands* 3.45.1 (1st century A.D.); Apuleius, *On the World* 8.2 (2nd century A.D.).

249. Seneca the Younger, *Natural Questions* 6.12.1.2 (1st century A.D.); Cicero, *On Divination* 2.30, 2.43.8.

250. Strabo, *Geography* 17.1.5; Diodorus Siculus, *Historical Library* 1.38.4; Lactantius, *Divine Institutes* 3.8.

251. Eusebius, *Preparation for the Gospel* 11.28.9, 13.13.30, 15.20.2; Plutarch, *Against Colotes* 21 (= *Moralia* 1119b); Porphyry, *On Abstinence* 3.21.8 (late 3rd century A.D.).

252. Alexander of Aphrodisias, *Commentary on the Book 'On Sensation'* 92.

253. Sextus Empiricus, *Against the Professors* 8.13.

254. Vitruvius, *On Architecture* 6.2.3.2.

255. Claudius Aelianus, *On the Characteristics of Animals* 16.29 and *Miscellaneous History* 13.35 (late 2nd to early 3rd century A.D.); Plutarch, *Alexander* 44.2; Pliny the Elder, *Natural History* 1.8c.3, 8.21.59.

256. Sextus Empiricus, *Against the Professors* 9.331, 10.155, 10.169, 10.177, 10.181; Alexander of Aphrodisias, *Commentary on Aristotle's Metaphysics* 660; Cicero, *On Fate* 24.6.

257. Galen, *Against Lycus* Kühn 18.224, *On the Combinations and Effects of Individual Drugs* Kühn 11.475 and 11.513.

258. Sextus Empiricus, *Against the Professors* 10.42, 10.45; Galen, *On the Combinations and Effects of Individual Drugs* Kühn 11.585; Plutarch, *Agesilaus* 5.3.

259. Seneca, *Natural Questions* 2.1.1–2.2.1 (divisions of natural philosophy

Across all these subjects of study, as noted earlier, the most common aspect of the method of the *physici* was to attempt to explain all these things in terms of fundamental elements, because these had to be the root causes of all things. This involved them in research and speculation on the composition of things as well as their causes, ending in a great deal of diverse opinion and debate.[260] Vitruvius expresses this aspect of their methods and interests quite succinctly:

> For no kinds of materials, nor bodies, nor things can arise or be understood without the coming together of things more basic. Nor otherwise does nature allow things to have true explanations in the teachings of natural philosophers, unless the causes which are present in these things have precise arguments how and why they are so.[261]

Galen, for instance, believes that the "good natural philosopher" will pay closest attention to explaining all phenomena in terms of the role of heat and cold.[262] For example, Galen observes that skin and leather start soft, yet they boil down to something hard, the reverse of expectation since heat usually softens and cold hardens, and he says solving this class of problems is up to the natural philosopher.[263] Thus, in every conceivable aspect of the natural world, causation lay at the center of the natural philosopher's interest. And mathematics was employed as an aid to studying questions of causation where such an approach was observed to be fruitful, hence it saw the most use in fields where it was most effective in narrowing theories down by more precisely describing observations (as in astronomy, optics, or mechanics).

---

match divisions of the universe: 2.1.1 and 2.2.1; astronomy and astrophysics: 2.1.1; meteorology: 2.1.2; geology and botany, etc.: 2.1.2). This division is as old as Aristotle, *Meteorology* 1.1.338a-339a.

260. See: Cicero, *Prior Academics* (= *Lucullus*) 2.36.117; Vitruvius, *On Architecture* 8.pr.1.13, 8.pr.4.7; Galen, *On the Elements according to Hippocrates* Kühn 1.439, *On the Uses of the Parts* 14.7 (= Kühn 4.165 = M.T. May 1968: 632), *Commentary on Hippocrates' 'On the Nature of Man'* Kühn 15.7; Eusebius, *Preparation for the Gospel* 1.pr.1, 1.8.1; Sextus Empiricus, *Against the Professors* 9.365, 10.1, 10.248; Alexander of Aphrodisias, *Commentary on Aristotle's Metaphysics* 180, 202, 224, 262, 602. etc.

261. Vitruvius, *On Architecture* 2.1.9.10.

262. Galen, *On the Uses of the Parts* 14.7 (= Kühn 4.165 = M.T. May 1968: 632).

263. Galen, *On the Combinations and Effects of Individual Drugs* Kühn 11.475.

2.8 Summary & Conclusion

We've already seen the *physicus* featured in the jokes of Lucian and Plutarch, in each case revealing something about who they were thought to be and do. To those we can add another. Nicarchus, an Alexandrian poet from the time of Nero, once wrote:

> Your mouth and your ass, Theodorus, smell the same. So it would be a famous task for natural philosophers to distinguish them. You really ought to write on a label which is your mouth and which your ass, for now when you speak I think you're farting.[264]

Here the joke exemplifies the fact that the *physici* were *the* authorities on natural phenomena, and in that respect they hold a direct parallel with modern scientists. They were the ones looking and examining and thinking about all the phenomena of nature, and everyone knew it. In this sense, as all the evidence surveyed above confirms, the *physicus* was the ancient scientist. They generally shared in common with modern (and early modern) scientists the same fundamental naturalistic assumptions about the nature of the world and the prospects for understanding and explaining it, they were largely interested in all the same subjects, and they were equally averse to trusting traditional beliefs, at least during the period of the early Roman Empire.

Still, it was only occasionally said the *physicus* should not declare more than he can prove, or that he should shun armchair speculation and get down to devising experiments and making careful observations to resolve questions about the nature of the world and the causes of natural phenomena. Some authors, like Cicero and Galen, did insist on such empirical research and demanded humility in declaring opinions without it. But even they did not always follow their own advice (though neither did Galileo or Newton), and these ideals do not seem to have become a universal element of the culture of the *physicus*. It did not hold pride of place as a primary dogma, as it would for the "experimental philosopher" of Bacon and Galileo, Boyle and Newton, and their ideological peers and progeny (however much even they fell short of their own ideal). Natural science and natural philosophy

264. *Anthologia Graeca* 11.241.

in the Roman period remained one and the same thing. It would take a true revolution in thinking to put the soundest views of Galen and Cicero at the forefront of method in all natural inquiry. As we saw in the previous chapter, what prevented that revolution in antiquity is one of the great unanswered questions in history.

Hence the general idea of a 'scientist' as someone who engages a modern scientific method and only gives scientific authority to carefully-controlled observation, is truly a novel product of the Scientific Revolution. The *physicus* was never regarded in such terms, nor did they engage themselves in that way. They occasionally did modern scientific things, but only as an adjunct to their overall method and plan, and rarely with any consciousness of a difference between 'philosophy' and 'science' in the current sense. Although, contrary to legend, scientists would not uniformly distinguish themselves from philosophers or what they did from philosophy until the 20th century. Even Maxwell and Darwin published their ideas, theories, and discoveries as natural philosophers in philosophy journals. And they weren't always right about everything. So the distinction is not as stark as sometimes thought. Ancient natural philosophers were very much like them, differing only in degree. And the only people in antiquity who, as a group, came anywhere near our modern ideal were the *physici*, and they are without doubt the historical and ideological forebears of the modern scientist. The *physici* did somewhat correspond to what we often today call 'naturalists', as those who look around and gather facts about the natural world. But with their primary interest in causes and explanations, even at the deepest level and for everything, the *physici* more closely corresponded to modern day 'philosophical naturalists', who advocate essentially the same worldview as their ancient forebears—though now building upon the findings of a thoroughly modernized science, rather than bottom-up speculation supplemented by the more limited findings of ancient science. Fact-finding and speculation are now held distinct, in both merit and method, but less so then.

This difference is revealed in another humorous story involving the *physici* that comes from the oldest extant text of the *Life of Aesop*, which dates to the 1st century A.D. In the story, told in various ways in all the recensions of this pseudobiography, Aesop at one time is sold as a slave to Xanthus "an eminent philosopher of Samos, accompanied by a retinue of his scholars." Then one day Xanthus had these scholars over for dinner, and:

> While the drink progressed, Xanthus kept farting, so the natural philosophers kept telling him to leave. Aesop left, too, and stood by with a towel and a pint of water. And Xanthus said to him, "Can you tell me why, after we take a crap, we often look at our own shit?"

At which Aesop wittily insults Xanthus with a story about how men do this because they worry their guts will fall out, but Xanthus had no guts so he need not worry. Aesop uses the word *phrên* for 'guts', but also had the Homeric meaning of "soul" or "sense," so Aesop had covertly told his master he had no soul or sense. Xanthus misses the pun and goes on to discuss various natural problems and questions with his guests. It is significant that in all later recensions of this story the guests are only ever called *philosophoi*, but in this, the earliest known recension, belonging squarely within our period of interest (the early Roman Empire), they are here called *physikoi*—more correctly, in fact, given the context of their discussions and the joke.[265] But for us one moral of this story is that the *physikoi* were known as a group not so much for tinkering in laboratories or fiddling with instruments or carefully recording observations (though many did do those things), but were more commonly perceived as the sort who would gather and discuss the nature of things at dinner parties.[266] Therein lies a notable difference between ancient and modern 'science', although as we shall see (for example in chapter 4.6) many *physici* had plenty to do with labs, instruments, and the recording of observations.

---

265. *Lives of Aesop*, vita g § 67 (= e cod. 397 bibliothecae pierponti morgan, recensio 3), on which see Perry 1936.

266. This was a trope of the time, e.g. Lucian, *Hermotimus* 11–12; Plutarch, *Questions at a Party* (= *Moralia* 612c-748d); and see M. Clarke 1971: 92–93 and S.J. Harrison 2000: 30–31. But it was not a fiction: Plutarch, *Advice about Keeping Well* 20 (= *Moralia* 133; discussing natural philosophy: 133e; discussing mathematics: 133a) and *That Following Epicurus is Unpleasant* 13 (= *Moralia* 1095c-1096c: discussing harmonics and music theory); hence academic societies (Museums) were often dinner clubs (see chapter eight of Carrier 2016). However, such talks were not always over dinner, and they often brought up sound empirical and mathematical science: Galen, *On the Affections and Errors of the Soul* 2.5 (= Kühn 5.80–93; esp. 5.92–93; this is sometimes identified as two books, *Affections* and *Errors*, so 2.5 means *Errors* 5); Plutarch, *On the Face that Appears in the Orb of the Moon* 1–23 (= *Moralia* 920b-937c).

3. The Roman Idea of Scientific Progress

From the relative paucity of science and natural philosophy in the educational system of the ancient world (although that science content was not negligible, as I have surveyed in *Science Education in the Early Roman Empire*), one might expect that such a society would have produced little or no scientific work. Yet quite the opposite was the case. The Greeks and Romans generated more scientific advances, and a far wider array of hypotheses in natural philosophy for a broader range of subjects, than any other civilization before the Scientific Revolution. Their achievements in the sciences are astounding, and continue to astound right through the early Roman empire. This fact alone challenges any conclusion that the Romans held science in low esteem. Accordingly, whether there was any belief that natural science could, does, and should make progress is an important element of social attitudes toward the activity of the natural philosopher.

Before we can examine whether there was any idea of 'scientific progress' in the Roman world, we first need to understand the actual scientific progress Romans had been witness to. The second half of this chapter will then present evidence against the claim that the Romans had no idea of the existence or value of scientific progress. But since this requires understanding what the Romans really knew of scientific progress, the first half of this chapter will survey the history of ancient science up to the 3rd century A.D. (after which all notable progress ceased for at least a thousand years). We must begin with a brief comment on the problem of source material.

Obviously to have scientific progress, a society not only needs to be actively engaged in scientific research, but it also needs to preserve the body of knowledge to be built upon or improved by it. And yet it was not the ancients, but the scribes and scholars of the middle ages who did not take a very great interest in preserving works by or about ancient scientists, especially when compared to their inordinate interest in preserving religious literature of every description. Hence, by far, most of what was written on the sciences was not saved at all, and most of what does survive, does so only in quotations, fragments, or translations of inconsistent reliability. Even of whole treatises preserved, many remained completely unknown through most of the middle ages to all but a scant few, and several were preserved only by Muslims and not Christians, and most by far in Eastern Christendom, not Western. Several histories of the sciences and biographies of scientists were also written in antiquity, yet hardly any of those were preserved either. We have the names of far more ancient scientists than we have writings preserved from them, and there was clearly a lot more scientific research in antiquity than survives or than we even know about. Consequently, any survey of the achievements of ancient scientists is only a survey of what we *know* about those achievements.[267]

This appears to have left gaps mostly in regard to competing theories, rather than what achieved anything close to a consensus at the time. But this means we can rarely argue the ancients 'never proposed' or 'never thought' of something, since they may well have, and extant sources just did not preserve mention of it. Likewise, most of the gaps in our knowledge about ancient science relate to how certain scientific discoveries were achieved or settled. Though we have many examples of that, the scholars and scribes of the middle ages (especially Christians, and most especially Western Christians) were far more interested in simply transmitting claims and conclusions than records of methods, experiments, and observations, and yet we know there was a great deal of writing about the latter that no one bothered to keep.

Compounding the problem is the fact that both Christian and Muslim scholars acted with considerable bias in selecting even the few scientific works they bothered to save. The writings of respectably religious creationists (like Galen and Ptolemy) received favored treatment, while

267. See the pointed comments on this problem in Lloyd 1981: 256–60 and Nutton 2013: 1–17.

the works of atheists (like Strato and Erasistratus) were left to rot. This has skewed modern impressions of ancient science into the false belief that all of it was obsessively Aristotelian and highly Platonic, when in fact there were many Stoic and even quasi-Epicurean scientists who often embraced different interests or took a different approach (Posidonius, Asclepiades, and Erasistratus being the most prominent examples), while even the Aristotelians were far more diverse and innovative than their medieval heirs believed. In fact, most Roman scientists were so philosophically eclectic that they lacked significant sectarian allegiances or left those allegiances indiscernible to us. As long as all these defects in the surviving source-record are kept in mind, the history of what we know of the most important developments in ancient science can be summarized as follows.[268]

3.1 The Growth of Ancient Science

As always, the story begins with Aristotle. Although ancient science has an important history preceding Aristotle, starting with Thales in the 6th century B.C. and involving dozens of known theorists and investigators, all subsequent science built on Aristotle, either directly or indirectly.[269] The *Dictionary of Scientific Biography* accurately describes him as "the most influential ancient exponent of the methodology and division of sciences," having contributed himself to "physics, physical astronomy, meteorology, psychology, biology" and several other fields in the middle of 4th century B.C.[270] He is the first philosopher to properly systematize the

---

268. Rather than delve into hundreds of questions and controversies, what follows is simply a summary of established scholarship from standard references (the *EANS*, *DSB*, *NDSB*, and *OCD*; and James & Thorpe 1994) and what is agreed among expert scholars (including Breidbach 2015, Russo 2003, Lloyd 1973, and Sarton 1959, as well as others to be named). For a brief but useful survey of the modern historiography of ancient science see Rihll 2002. Many more scientists are known than I will name (a more complete list is in *EANS*).

269. See Lloyd 1970: 16–98. Aristotle in turn had built on the work of his numerous and divergent predecessors. For a near-comprehensive list of pre-Aristotelian scientists and natural philosophers, see Appendix B.

270. *DSB* 1.250; everything else that follows summarizes *EANS* 141–45 (with 145–52), *DSB* 1.250–81, *NDSB* 1.99–107, *OCD* 159–63, Lloyd 1970: 99–124, and

study of nature and develop a rigorous and success-oriented method for it, making him the most important starting point for our survey. Many of Aristotle's writings were not preserved, including a collection of anatomical drawings and diagrams that he composed and refers to, while some works surviving under his name are actually the work of his students (or even their students), and some are essentially unpolished notes that Aristotle never formally completed, and all his surviving works were edited or corrected by subsequent students and scholars. But it can still be fairly said that Aristotle put every ancient scientific field that existed in his day on a more formal footing than ever before.

Hence Aristotle is the grandfather of scientific method. Method was an issue of conscious debate for centuries before and after him, but Aristotle finally combined the key role of empirical observation with the axiomatic, logical, and mathematical methodologies promoted by Plato, to produce a protoscientific method half-way between the unscientific and the scientifically modern, which established the framework for all further advances and modifications by later scientists.[271] In fact, much if not all of Aristotle's work in the sciences seems only to start the argument or to propose hypotheses from rough or 'commonplace' observations that Aristotle thought could still be confirmed or overthrown by more rigorous use of his own recommended methods. He also appears to have become more empirical in later life. Aristotle had started applying the practice of first-hand research as universally as he could before his death—even in the area of history, for example, he had collected the written constitutions of

---

Shields 2007. For the philosophical function and context of Aristotle's work in biology see French 1994: 6–82 and Lennox 2005. On the method and practice of Aristotle's scientific research see, for example, Boylan 1983 (biology), Taub 2003: 77–115 (meteorology), and Lloyd 1996b (general). For a good discussion of the motives and empirical nature of all of Aristotle's scientific work see Hankinson 1995a and 1995b.

271. Aristotle's systematization of scientific methodology is laid out principally in the combination of the *Posterior Analytics* and the *Topics*, although important digressions add to the subject in the *Physics* and *Metaphysics* (relevantly discussed in Bolton 1991, Lloyd 1992a, and Crombie 1994: 1.229–76). On Hellenistic improvements: Russo 2003: 171–202, Crombie 1994: 131–228, Lloyd 1982, and relevant discussions here in section 3.7. Breidbach 2015 and Lloyd 1979 and 1987 further discuss the origin and expansion of scientific methods and ideals throughout antiquity.

over a hundred Greek city-states and amassed a sizable collection of maps, while his students began writing histories and biographies—even of the sciences, which already demonstrates a recognition of scientific progress, by both documenting and evaluating it.[272]

Biology became Aristotle's greatest area of expertise. He appears to be the first to record reasonably cautious and often accurate eye-witness records of scientific data on an extensive range of animals (over five hundred species), relying on both observation and dissection, in notable cases even vivisection, though often relying on second-hand reports as well. He employed this data to develop or support hypotheses for a wide range of biological questions, and sought to build this work into ever wider generalizations, such as attempting to study animal respiration to explain human respiration and all respiration generally, or dissecting chicken eggs at different stages of development in order to learn things about generation and physiology. Aristotle often arrived at incorrect conclusions from this process, but he also successfully refuted or corrected many previous or popular claims through his own observations. His rate of error was high, but offset by so many accurate conclusions that Aristotle's methods were clearly superior to any that had gone before. Overall, his methods and aims were on the right track, and because of him, experiment and observation became as important as math and logic in the progress of the sciences in subsequent centuries.

More importantly, contrary to the way Aristotle was received by scholastic authors in the middle ages, he was aware of the self-correcting nature of science and had his epistemological priorities in the proper order, for instance concluding his discussion of bee reproduction with the qualification that:

---

272. See previous notes and *OCD* 232–33, 449 (s.v. "biography, Greek," "*didaskalia*") and Zhmud 2003 and 2006. Aristoxenus wrote numerous biographies (*DSB* 1.281–83 and *OCD* 163–64); Dicaearchus of Messana, various histories and biographies (*OCD* 447, s.v. "Dicaearchus"); Eudemus of Rhodes, histories of the sciences of astronomy, arithmetic, and geometry (*DSB* 4.460–65, s.v. "Eudemus of Rhodes" and *OCD* 545, s.v. "Eudemus") and Meno, of medicine (*OCD* 933, s.v. "Meno"; *DSB* 6.421, in s.v. "Hippocrates"). But none of these works survive (except possibly a papyrus fragment of Meno's history of medicine). These Aristotelian historical interests continued into the Roman period.

> This appears to be the method of reproduction of bees, according to theory in connection with the apparent facts. But the facts have not been satisfactorily ascertained, and if ever they are, then credence must be given to observation rather than to theory, and to theory only insofar as it agrees with what is observed.[273]

Many passages like this have led G.E.R. Lloyd to conclude that "Aristotle is not the dogmatist he was sometimes later made out to be," as "statements of remaining difficulties, and of the need for more research, are common" throughout his writings, "especially though not exclusively in his zoology."[274]

In fact, empirical observations were very important to Aristotle, most obviously again in his zoology. D.M. Balme draws an excellent picture of this reality, summarizing the best from Aristotle's biological works and revealing the nature of Aristotle's scientific achievements in this area, which far surpassed any other field he touched, even despite many remaining errors:

> Some of his data clearly come from deliberate dissection, while others come as clearly from casual observations in the kitchen or at augury. One of the best is a full-scale vivisection of a chameleon; and the internal organs of crabs, lobsters, cephalopods, and several fishes and birds are described from direct observation. Many of the exterior observations also presuppose a prolonged study. He speaks of lengthy investigation into the pairing of insects. He satisfies himself that birds produce wind eggs entirely in the absence of the cock. There are graphic accounts of courtship behavior, nest-building, and brood care. He records tests for sense perception in scallops, razor fish, and sponges. He watches the cuttlefish anchor itself to a rock by its two long arms when it is stormy. The detailing of structures in some crustaceans and shellfishes vividly suggests that the author is looking at the animal as he dictates. The sea urchin's mouth parts are still known as "Aristotle's lantern" from his description, and his statement that its eggs are larger at the full moon has only recently been confirmed for the Red Sea urchin. He is able to assert that two kinds of Serranidae are "always female" (they are in fact hermaphrodite). All such data require deliberate and patient observation. How much Aristotle himself did is not known, but it is clear enough that he caused reports to be collected and screened with great care.[275]

---

273. Aristotle, *On the Generation of Animals* 760b.

274. Lloyd 1981: 289.

275. *DSB* 1.264 (in s.v. "Aristotle"). Examples of Aristotle's use of vivisection:

Hence the gold standard was set. Once Aristotle had defined the rules of the game and set the standard to beat or follow in any field, ancient science continued to advance, though more in some areas than in others. We will examine the three most important lines of development: medicine, astronomy, and (in the modern sense) physics.[276] As we shall see, the full extent of scientific progress is difficult to track because we have almost none of what was written by working scientists of the time. Nevertheless, evidence of continual progress stands out clearly.

Before I get to that, however, I must first say something of its general historical context. Just as both the influence and precedent of Aristotle are crucial to understanding subsequent progress, the overall political history of the time also relates to scientific advancement in all fields, insofar as general conditions were conducive or a hindrance to scientific work. Aristotle's school at the Lyceum, as well as Plato's Academy, both in Athens, and several museums and medical schools that arose all around the Greek world, contributed some private support for scientists to meet and collaborate, even if not to as great an extent as some have imagined. But the biological works of Aristotle and his student Theophrastus, for example, show signs of collaboration and subsequent addition. But a far more significant institution for the advancement of science was royal patronage, which had its most prestigious example in Egypt, where the Ptolemaic kings provided substantial support for scientists and scholars in all fields at the Alexandrian Museum for about a century and a half (roughly 296 to 146 B.C.). This play for cultural 'prestige' was then emulated by competing kingdoms, insofar as they had means, and as a result scientists could find support in the royal houses of the Seleucids (centered in Babylon), the Attalids (centered in Pergamum), and, especially after the decline of the Ptolemies, in the court of Mithridates (centered in Pontus). Even the Sicilian royal house had Archimedes under its patronage. As a result, this

---

Aristotle, *On Respiration* 9.3.471b, *History of Animals* 3.12.519a, *On the Movement of Animals* 8.708b. Aristotle's illustrated eight-volume treatise *On Dissection*, however, was not preserved (see French 1994: 40–43).

276. Though tracking a similar path of progress was the increasingly scientific study of logic and language: see Russo 2003: 218–24; J. Barnes 1997; and *OCD* 839–40 and 855 (s.v. "linguistics, ancient" and "logic"). On the entry level logic taught in ancient schools: Huby 2004.

is widely regarded as the heyday of ancient science. The most progress was made then, precisely because scientists could find so much state support for their careers and interests.[277]

But this golden age was very brief, lasting no more than two centuries overall, and barely more than a century in any one place. A rising tide of wars and social and military upheavals in the 2nd and early 1st centuries B.C. disrupted scientific education and activity throughout the Greek world, putting an end to the patronage of scientists in the various royal courts.[278] But little knowledge was lost, and a recovery soon began under the Pax Romana, from the late 1st century B.C. to the early 3rd century A.D., when scientists and scientific literature began once again to rise in quantity and accomplishment.[279] This *new* golden age produced some of the greatest scientists of the ancient world, most famously Galen and Ptolemy in the 2nd century A.D. The sciences could well have continued along that encouraging path, and may even have been on the brink of its own scientific revolution, but history once again intervened.[280] The vast political, military, and economic chaos of the 3rd century A.D. interrupted this course of events and led instead to the subsequent triumph of supernaturalism, not only in the dominance of a rather anti-scientific brand of Christianity, but also in the form of Neoplatonism and other varieties of pagan mysticism, which were also not very supportive of scientific values. This cultural shift

277. The link between state support and scientific progress during this period is most effectively illustrated in Schürmann 1991 (technological sciences) and von Staden 1989 (medical sciences).

278. See, for example, Rawson 1985: 11, 13–18. In the most direct case, scholars and scientists were forcibly expelled from Alexandria for political reasons, by the hostile (and ironically named) Ptolemy the Dogooder [Euergetes II] in 145 B.C. (e.g. Athenaeus, *The Dinnersages* 4.184b-c, who suggests this actually spread science *education* more widely, as fleeing scholars set up schools elsewhere), but this policy did not continue beyond his death in 116 B.C., and scholarship subsequently returned there.

279. Including a substantial revival of Aristotelian studies, spearheaded by Andronicus of Rhodes, who at Athens in the middle of the 1st century B.C. edited, collated, and systematized Aristotle's works, producing a definitive edition that was widely influential. See Gottschalk 1987 (with *OCD* 86 and 238, s.v. "Andronicus" and "Boethus (4)").

280. As argued in Edelstein 1952: 602–04. Lloyd concurs (see following note).

put a decisive end to any recovery in the sciences that was underway, and eventually resulted in a considerable loss of knowledge and understanding of what had been achieved.[281]

Our interest, however, is in the period of Roman revival. Though the Roman emperors never patronized the sciences as many Hellenistic kings had done, the conditions for scientific work under the Pax Romana were good, and scientists did not go unappreciated.[282] The physical sciences (especially mechanics, optics, harmonics, and astronomy) were often pursued by engineers who, if talented, had no difficulty securing well-paying careers in a prosperous age of building, and could often expect even greater rewards from generous patrons or emperors, and retire well. Meanwhile, the life sciences (especially medicine, physiology, zoology, botany, and pharmacology) were often pursued by doctors, who could actually receive some state support in the form of legal privileges and even public salaries,[283] and could also earn very good money in private practice, or find themselves the recipients of a patron's largesse. And though most of the doctors, philosophers, and professors of the liberal arts who commonly entered into the lavish employment of rich patrons were anything but research scientists, such opportunities were plentiful enough for aspiring scientists to benefit from them. Lucian describes educated men of various vocations taking paid positions as attachés to the rich, including philosophers, teachers, and other occupations "more serious" (*spoudaiotera*) than these, which must have meant professionals like doctors and engineers.[284] And we

---

281. See Lloyd 1973: 154–78. For the decline of scientific medicine beginning in the 3rd century A.D. see Kudlien 1968, Nutton 2013: 299–317, Heinz 2009, and Mazzini 2012. For a similar decline in astronomy: Eastwood 1997. For the chaos of the 3rd century see my relevant note in chapter 1.1.

282. Edelstein 1952: 596–602 surveys the lack of direct institutional support for scientific research in antiquity, but he slights the considerable admiration and appreciation it received (an oversight we will remedy in this and the following chapter).

283. On this fact see chapter eight of Carrier 2016.

284. Lucian, *On Attachés for Hire* 4 (usually known by the more contrived title *On Salaried Posts in Great Houses*). Though Lucian seems to ridicule those who take such work, his satire is more a complaint about how much the job sucks than an attack on those who take it. Hence when he himself took a salaried position for the state he had to write an *Apology* for his previous satire, arguing his original

know doctors and engineers could amass considerable wealth from these and other sources.[285]

Whether retired or working, inquisitive doctors, engineers, philosophers, and professors could always find time for scientific study. Galen, for example, worked widely and regularly as both a physician and a teacher, yet was still able to conduct extensive scientific writing and research at the same time.[286] Though he was fortunate enough to be living on a modest inheritance, this was notably earned by his father as an engineer. So it is clear that regular income and benefactions from a professional practice would have provided the same level of support for other scientists of the period. In other words, though Roman science was not supported by the whims of kings or emperors, it had ample support from the general prosperity of the age. Unlike the previous century, scientists of the Roman era enjoyed wealth and peace, and unlike subsequent centuries, they enjoyed a significant level of intellectual freedom and respect. Conditions would not be so favorable again for at least a thousand years.[287]

---

intent was to warn others like him against taking a bad arrangement (cf. *Apology* 3), because the private rich were often ridiculous whereas the government always offers respectable employment (cf. *Apology* 11–12).

285. That doctors could get rich through public and private practice is attested in, e.g., Pliny the Elder, *Natural History* 29.5.6–8 (and see, again, chapter eight of Carrier 2016). That engineers could receive generous pensions for their service is attested in, e.g., Vitruvius, *On Architecture* 1.pr.2–3 (note also the financial success of Galen's father who was an engineer, as discussed in chapter seven of Carrier 2016).

286. Though most professionals probably got most of their original work done in their retirement, as suggested by, e.g., Vitruvius, *On Architecture* 1.pr.1–3; Seneca, *Natural Questions* 3.pr.1–4 and *On Leisure* 4.1–5.7 (= *Dialogues* 8.4.1–8.5.7); Pliny the Elder, *Natural History* pr.18; Quintilian, *Education in Oratory* 1.pr.1 and 1.12.12; and Galen, *On My Own Books* 2 (= Kühn 19.17–18) and *On Exercising with the Small Ball* 2 (= Kühn 5.900–01).

287. For the following sections on the history of ancient science and technology, required reading on all subjects and fields includes Irby-Massie 2016, Russo 2003, and Rihll 1999. Valuable references adding to those include: Oleson 2008; Irby-Massie & Keyser 2002; Lloyd 1973; Sarton 1959; and Cohen & Drabkin 1948. More specific references will be cited below.

3.2 Scientific Medicine up to the Roman Era

Though medicine eventually grew into a wide array of competing sects, there remained a lot of independence and eclecticism among medical scientists, and even the sects most often named (the Empiricists, Methodists, and Dogmatists) were not really 'sects' so much as categories, often encompassing a wide array of ideologies.[288] Despite this diversity, ultimately there were only two dominant paradigms for theoretical advances in biology: atomism and humoralism.[289] The atomist physicians sought to explain all physiology in purely mechanical terms, attempting to reduce all systems and causes to the physical interactions of molecules, atoms, suction, and collision, though sometimes adding some irreducible 'forces' and 'qualities' to the mix. The humoralist physicians, in contrast, gave pride of place to qualitative properties, not only in the form of the causal interaction of imagined 'humors' (theorized biological substances with innate natural powers), but also in a simplified physics of hot, cold, wet, and dry, attributes that were regarded as the natural and irreducible properties of things.[290] Humoralism represented the lasting influence of Empedocles and Hippocrates, while

---

288. A good survey of medical sectarianism is provided in Nutton 2013: 149–53 (Empiricists); 191–206 (Methodists), 170–73 & 207–21 (Dogmatists), and 149 and 191 (various other sects). Nutton traces many lesser known medical writers in the historical development of their respective sects, whereas I will largely ignore these and the history of the sects and focus on the most notable contributions to medical science as a whole.

289. For general context and scholarship on ancient medical science generally see: Nutton 2013; Littman 1996; Scarborough 1993 (which supplements and updates Scarborough 1969); Lloyd 1973: 75–90; and *OCD* 79–82, 441–42, 444, 451, 468, 501–02, 638–39, 712–13, 919–23, 952–53, 1040–41, 1089–90, 1122–23, 1414–15, 1562 (s.v. "anatomy and physiology," "dentistry," "diagnosis," "dietetics," "disease," "embryology," "gynaecology," "humours," "medicine," "midwives," "ophthalmology," "pathology," "pharmacology," "surgery," and "vivisection"). On ancient pharmacology: Everett 2012, Scarborough 2010, Schmitz & Kuhlen 1998, and Riddle 1986. On "veterinary medicine" see *OCD* 1545–46 and summary and sources in Rihll 1999: 132–36. On psychology: *OCD* 502–03 and 881–82 (s.v. "emotions" and "madness"), with P.N. Singer 2013, Roccatagliata 1986, and Siegel 1973.

290. Nutton 2013: 72–86 and Siegel 1968: 196–359 discuss ancient humoral theory in detail.

atomism represented the lasting influence of Leucippus and Democritus, all starting in the 5th century B.C. and both continuing to find adherents among medical scientists well into the Roman period. Meanwhile, the Empiricists refrained from affirming whether atomic or humoralist explanations undergirded observed phenomena, and only documented what could be verified, but this limited theoretical advances.

As an example, toward the end of the 2nd century A.D. Galen's treatise *On the Natural Faculties* aimed at proving there are in fact 'natural powers' that are not reducible to the mechanical action of atoms and molecules, in the process revealing how popular such a reduction still was in his day. To combat his opponents, Galen uses examples outside of medicine, including magnetic and electrostatic phenomena and water absorption in grain, while most of his medical examples have been traced in modern science to known phenomena of biochemistry, demonstrating that both patterns of thought, the atomist and the humoralist, were on to something, though neither was quite on the money. Galen was correct, for example, that the atomists' mechanical explanations of the kidneys' 'intelligently' selective powers in what minerals to eject into the bladder did not match careful observations. What he didn't know was the role played by genetic computers in the cells of the kidney, evolved over millions of years, to effect that selectiveness through the machinery of interacting molecules, which are made of atoms after all. The atomists were right, it turns out, but not even remotely capable of acquiring knowledge of how. In fact no one was until the 20th century, the first time evidence was acquired of the computational role of genes in renal biochemistry.

Hence the ancient disagreement between the atomists and the humorists was simply not one they could have resolved at so early a stage of the science. What all ancient biologists shared, however, was a belief that all phenomena are explicable as the predictable and lawlike outcome of the interaction of natural objects and forces, leaving little (and sometimes no) room for the action of supernatural agents. This drove them all to investigate the nature and causes of medical and biological facts, in an effort to establish a theoretical understanding of health and disease based on observation and reason. Thus medicine started to become increasingly scientific. Pursuit of medical knowledge then sustained the study of botany, zoology, and even mineralogy, at least as related to pharmacology, and, of course, the study of anatomy and physiology.

The most important medical advances in history—the germ theory of disease, the development of anesthesia, and the discovery of antibiotics (most famously, penicillin)—would not be made until the 19th century, well *after* the Scientific Revolution. But at some point before the Roman era ancient doctors discovered primitive anesthetics and antiseptics, and toyed with various 'contagion' theories of disease that approached the modern view.[291] Even so, ancient anesthetics were limited to sedatives and painkillers (like opium) which, though reasonably effective, did not have quite the same duration or effect as modern anesthetics.[292] And though ancient doctors understood the importance of using clean instruments and working in clean environments, they did not yet know how meticulous it was necessary to be. Nevertheless, they *were* cleaning wounds with salves incorporating effective antiseptics such as pitch, vinegar, and turpentine, and they did know certain diseases were infectious.[293] And contagion theories were not limited to the 'bad air' hypothesis, but included a debated 'seed' theory (in which some diseases were thought to be caused by microscopic 'seeds' that could be spread through contact) and even a primitive 'germ' theory, recorded by Varro in the 1st century B.C.[294] But the origin and popularity of these ideas is lost in time. Likewise, for example, numerous forms of chemical birth

---

291. See Jackson 1988: 68, 80, 112–13, 172–73; Nutton 1983 and 2000b; and note below on Varro.

292. See James & Thorpe 1994: 38–41. Pliny the Elder, *Natural History* 25.94.150, for example, discusses painkillers and sedatives.

293. For example, see the extensive list of recognized antiseptic agents in Celsus, *On Medicine* 5.19.1–28. Galen recommended pitch and thick wine (according to extant passages from his *Commentary on the 'Medical Practice' of Hippocrates*, cf. Lyons 1963: 107, 111). The Romans also knew that water is purified by boiling: Pliny the Elder, *Natural History* 31.23.40.

294. Varro, *On Agricultural Matters* 1.12.2–3, who says "certain animals grow" in swamps "that are too small to be seen and float in the air, entering the body through the mouth or nose, causing serious diseases." For discussion: Sarton 1959: 409–10. These 'seeds' could also be imagined as mutating chemicals in air, water or food, a theory articulated in Lucretius, *On the Nature of Things* 6.1090–1144, and debated in Plutarch, *Tabletalk* 8.9 (= *Moralia* 731e and surrounding, where the speaker Diogenianus rejects the theory, but then Plutarch and his medical friend Philo defend it). Breath and bodily fluids were also known to be contagious for some diseases and not others (Pseudo-Aristotle, *Problems* 1.7.859b, 7.8.887a). On these various ideas see Nutton 1983.

control and chemical and surgical abortion were developed throughout Greco-Roman antiquity, but it remains unclear which discoveries were made when.[295] Beyond questions like these, the development of ancient medical and biological science progressed after Aristotle as follows.[296]

Hippocrates began a more scientific treatment of medicine around the time of Plato, but the first 'Aristotelian' doctor was Diocles of Carystus, hailed even in antiquity as the 'Second Hippocrates', and noted as the inventor of a specialized cranial bandage and arrow-extracting spoon.[297] Diocles studied under Aristotle and wrote in the late fourth and early third centuries B.C. At least seventeen of his books are known, but none survive apart from fragments. Diocles was probably the first to write books specifically on the subject of human anatomy, along with commentaries and critiques of the works of Hippocrates. He improved some of Aristotle's work in human and animal anatomy and physiology, even dissecting miscarried human fetuses. He expanded Aristotelian interests to the fields of botany, mineralogy, and pharmacology, writing several books on these subjects, including the first known herbal, which scientifically documented the appearance, origins, and nutritional and medicinal value of various plants.[298] The Diocles herbal became one of the leading texts on the subject until the Roman scientist Dioscorides supplanted it (discussed below). Diocles' contemporary, Praxagoras, also wrote extensively in natural philosophy and medicine, especially on anatomy and the study and treatment of diseases, originating the diagnostic study of the pulse and possibly discovering the difference between veins and arteries. Though his work remained influential even up to Galen, nothing he wrote survives.[299] Alexias was another medical

---

295. *OCD* 1 and 370–71 (s.v. "abortion" and "contraception").

296. I will only survey the best known. For a more complete list of ancient medical writers, see *EANS* 1006–11 (s.v. "medicine") and 1013–19 (s.v. "pharmacy").

297. *DSB* 4.105–07; *EANS* 255–57; *OCD* 453 (s.v. "Diocles (3)"). For references on Hippocrates see Appendix B. For Diocles, Praxagoras, and the development of the life sciences between Hippocrates and the early Aristotelians see van der Eijk 2005.

298. Other pupils of Aristotle also contributed to botany and mineralogy (see section 3.5).

299. *DSB* 11.127–28; *EANS* 694–95; and *OCD* 1205. For Praxagoras and Diocles and other medical writers of the same period see Nutton 2013: 116–29. Ancient

scientist of this same period who was widely renowned but whose writings were forgotten.[300]

After them came Herophilus, who was born in the late 4th century B.C. in Bithynia (near what is now the northern coast of Turkey), though his subsequent pursuit of medicine, including studies under Praxagoras, led him to Alexandria, where in the early 3rd century he launched the systematic investigation of human anatomy and physiology, becoming the most famous medical scientist of his own day, and one of the most renowned in the whole of antiquity.[301] We know he wrote more than ten books on medical subjects, though nothing he wrote survives. It is reported that the kings of Egypt extended him the unique privilege of dissecting human cadavers (and possibly live criminals, though that may be mere legend), which allowed him to advance and correct the scientific understanding of human physiology to an extraordinary level, focusing especially on the brain, heart and eyes, and the nervous, reproductive, and vascular systems. He established conclusively that the brain, not the heart, is the seat of perception and intellect, and analyzed the structure of the brain and central nervous system in detail, as well as the liver and digestive tract, fathering numerous anatomical terms still in use. He traced the path and function of all the major nerves, veins, and arteries, and was the first to distinguish sensory and motor nerves, and to study the specific timing properties of the pulse, contributing to a more widespread use of cardiac rhythm as a diagnostic tool. He also formed the first scientific theory of respiration that came near to being correct, and launched the science of gynecology with his own detailed anatomical investigations of the female reproductive

---

diagnostic use of the pulse was always more divination than science, but it nevertheless increased in sophistication over subsequent centuries, as summarized in an extant textbook on the subject by Marcellinus in the 2nd century A.D. See: *EANS* 526–27; *OCD* 896 (s.v. "Marcellinus (1)"); and Christ 1974; with discussions of the least scientific aspects of this 'science' in Barton 1994a: 133–68 and Kuriyama 1999.

300. *EANS* 59.

301. For the following see: *DSB* 6.316–19, *EANS* 387–90, and *OCD* 677–78; and most comprehensively von Staden 1989. Also, a good survey of the scientific accomplishments of Herophilus and his pupil Erasistratus (discussed next) can be found in Longrigg 1981. And on their legacy up to Galen on the study of the nervous system: von Staden 2000. And gynecology: Bliquez 2010.

system, observational studies of menstruation, and research into the causes of complications in labor.

Most of Herophilus's pupils and adherents set aside anatomical study to advance pharmacology, pathology, symptomatology, and therapeutics, and to produce scientific commentaries and lexicons.[302] But not his most famous student, Erasistratus, who went on to advance the science of anatomy and physiology even further.[303] In the early 3rd century B.C. Erasistratus continued his teacher's work at Alexandria, Antioch, or Pergamum (or any combination thereof, the evidence is debated), having studied medicine himself in several cities, including Athens under an Aristotelian doctor, where he was probably influenced by the teachings of Strato, adopting his highly atomistic revision of Aristotelian physics. Hence Erasistratus originated the long-standing effort to explain all physiology through mechanical principles, rejecting the explanatory value of non-mechanical forces and powers, including humoral theory and intelligent design.[304] Erasistratus wrote a large number of books, especially on anatomy and pathology, yet none were preserved, even though they included crucial advances. Like Herophilus, he is reputed to have performed scientific vivisections on condemned criminals with royal permission, and though scholars are divided on whether that is true, all agree he conducted scientific autopsies on human cadavers (often specifically to study the cause of death), as well as numerous metabolic experiments and vivisections on animals. Erasistratus

302. These Herophileans and their research continued in every century up to the mid-1st A.D.: see von Staden 1989: 445–578 (which includes scientists not listed in the *OCD*). Among them: Andreas (*OCD* 85, *EANS* 77–78); Antonius Musa (*OCD* 113, *EANS* 101); Apollonius Mus (*EANS* 111–12); Bacchius of Tanagra (*OCD* 220, *EANS* 187–88); Callimachus (*OCD* 267, fourth entry, and *EANS* 462); Chrysermus (*OCD* 315–16, *EANS* 473); Heraclides (*OCD* 665, fourth entry, and *EANS* 367); Mantias (*OCD* 894, *EANS* 525–26); Philinus (*OCD* 1127, *DSB* 10.581, *EANS* 645–46); and Zeno (*OCD* 1588, fourth entry, and *EANS* 846). Pretty much all their vast work over the centuries was not preserved through the middle ages.

303. *DSB* 4.382–88; *EANS* 294–96; *OCD* 532–33; von Staden 1997b; and Longrigg 1981: 155–64, 177–85. On the innovative work of Herophilus and Erasistratus (and Eudemus) see Nutton 2013: 130–41. Nutton 2013: 142–59 also surveys other lesser known medical writers before the Roman period that I do not discuss (see also von Staden 1996: 91ff. for Erasistratus and some of his known successors).

304. For examples of his use of mechanical models see Vegetti 1995 and von Staden 1996: 91–98.

improved the Herophilean theory of respiration by hypothesizing that the lungs take in air and distribute it as a vital element throughout the body, and he discovered that the number of folds and cavities in the cortex of an animal's brain increases in proportion to its intellectual capability, that the stomach compresses food with muscular contraction, that the heart operates like a pump, and that different areas of the brain control different faculties and parts of the body, improving on the neurological findings of Herophilus, and further tracing the origin of every kind of sensory and motor nerve to their separate locations in the brain, even severing nerves or resecting the brain of live animals and then observing what faculties were lost as a result. Much of the anatomical terminology still used today originates with Herophilus and Erasistratus.

Though both men made errors and advanced some incorrect physiological theories, together they got a great deal right. Moreover, many of their errors were corrected by the experimental work of Galen and others centuries later. Plus they weren't alone. From the 3rd century B.C. we know the most about Herophilus and Erasistratus, but another renowned anatomist at Alexandria, Eudemus, was publishing at the same time. None of his work survives, but it was highly influential on later writers, especially noted for his advances in the study of the anatomy of bones, nerves, and blood vessels, and their embryonic development.[305]

Other anatomists followed these, continuing their work, such as Antigenes in the later 3rd century B.C., who we know wrote on anatomy, fevers, inflammation, and child care, although none of his works survive; and Apollonius of Memphis wrote on anatomy, pathology, and pharmacology, likewise all lost.[306] Even after the end of the Ptolemaic 'golden age' at Alexandria, anatomical research may have continued among some scientists, but experienced a revival in the early Roman empire, possibly beginning with Hegetor in the 1st century B.C., but certainly others soon after him (see below).[307] Otherwise, early in the 1st century B.C. the Herophilean scientist Demetrius of Apamea advanced gynecology and the study of diseases

305. *EANS* 308.

306. *EANS* 92 and 113–14.

307. *OCD* 652–53 and *EANS* 359, with von Staden 1989: 445–45 (n. 1), 512–14 and P. Fraser 1972: 1.363–64 (with 2.536–39). None of his work survives.

and disorders, building a systematic observation-based catalogue.[308] This research was expanded by Dioscurides Phacas, probably during Cleopatra's reign in the mid-1st century B.C.[309]

These and other facts have even led some scholars to detect a significant revival of scientific research under Cleopatra.[310] But work was also proceeding in pharmacology even before her time.[311] In the 3rd century B.C. Apollodorus of Alexandria had written widely-respected studies *On Poisonous Animals* and *On Poisonous Drugs*, neither of which survive, even though they were considered invaluable sources in the field for centuries.[312] But Alexandria was not the only place where such research was in vogue. During the 1st century B.C. were two scientists named Apollonius, from Citium and Alexandria, one who wrote on joints and surgery, another on drugs and medical issues specifically relating to the care of slaves (possibly the first ever treatise on occupational medicine).[313] In Tlos in the 1st century

---

308. *OCD* 434 (s.v. "Demetrius (21)") and *EANS* 232 with von Staden 1989: 506–11. None of his many books survive.

309. *OCD* 466–67 (s.v. Dioscurides (2)") and *EANS* 270, with von Staden 1989: 519–22 (and not to be confused with the later Dioscorides, discussed below). None of his many books survive. Scarborough 2012 discusses Phacas as well as Philotas and Olympus, two other medical scientists in Cleopatra's court, who also wrote on scientific subjects and none of whose writings survive.

310. Most thoroughly argued in Marasco 1998 (who discusses several other medical researchers of the period, as well as parallel activity in astronomy, geography, philology, and other fields; see also Scarborough 2012). Fraser also argued for a revival of medical research in Alexandria under Cleopatra and suggested this may have been the legacy of the highly-revered work of Heraclides the Herophilean (see above) in the early 1st century B.C. (P. Fraser 1972: 1.361–63, with 2.536–38).

311. Involving several scientists we know very little about, from the late 3rd century B.C. to the 2nd century A.D. See note above on the Herophileans; and *OCD* 288, 896, 1352 (s.v. "Cassius (1)," "Marcellus" [of Side], "Serapion (1)") with corresponding entries in *EANS* 207–08, 530, 733.

312. *OCD* 120 (s.v. "Apollodorus (4)"); *EANS* 106.

313. *OCD* 123, 124 (s.v. "Apollonius (8)" and "Apollonius (10)") and *EANS* 111–12, 113; with Potter 1993 and von Staden 1989: 455–56, 540–54. I am omitting doctors only known to have written commentaries, e.g. *OCD* 1591 (s.v. "Zeuxis (2)"; cf. *EANS* 848), or to have abandoned scientific for magical thinking, e.g. *OCD* 1580 (s.v. "Xenocrates (2)"; cf. *EANS* 836–37), or when we know too little of their contributions to science.

B.C., the woman Antiochis was honored with a statue for her medical science, and we know she wrote books on her research, we just don't have any that survive.[314]

It should also be noted that kings taking a personal interest in the sciences was a trend around this time. Similarly to Cleopatra's interest in the sciences, in the 2nd century B.C., the last king of Pergamum, Attalus III, lost interest in politics, according to R.M. Errington, "devoting himself rather to scientific study, especially botany and pharmacology."[315] Another king, Mithradates VI of Pontus, wrote a treatise on experimental pharmacology in the early 1st century B.C., which may have been of mixed quality, but was valued enough to be translated into Latin by one of Pompey's freedmen.[316] There were also more fanciful botanical and geographical writings by King Juba in the reign of Augustus, whose scientific value is apparently questionable, though the fact that he wrote them still reflects similar interests.[317]

But the most famous doctor of that century was Asclepiades of Bithynia, who made an enormous impact in Rome and Italy in the early 1st century B.C. His fame at Rome became legendary, and he had followers who embraced his principles for centuries.[318] We have seen signs of his

---

314. *EANS* 94.

315. *OCD* 202 (s.v. "Attalus III"; cf. *EANS* 179–80). He eventually granted his kingdom to Rome in his will, one of the few occasions of ostensibly peaceful annexation. Both this Attalus and the Mithradates mentioned next were also said to have tested poisons on condemned criminals (Galen, *On Antidotes* 1.1), which is also alleged of queen Cleopatra (Plutarch, *Antony* 71), who is also reported as having a working knowledge of poisons and chemical tricks (cf. Cassius Dio, *Roman History* 51.11; Pliny the Elder, *Natural History* 9.58.119–121 and 21.9.12). Though the stories of her experiments on humans are doubted (e.g. Marasco 1998: 49), they may embellish reports of genuine scientific activity by doctors in her court (and the same might be said for Attalus and Mithradates).

316. See *OCD* 1179 (s.v. "Pompeius Lenaeus"; cf. *EANS* 684) with 963–64 (s.v. "Mithradates"; cf. *EANS* 557–58), and Pliny the Elder, *Natural History* 25.3.5–7. For more sources and discussion of the scientific activities of Attalus and Mithradates (and other kings) see Marasco 1998: 52 (on Cleopatra: 50–53). On Mithradates, see also Mayor 2011b.

317. *OCD* 777 (s.v. "Juba (2) II"); *EANS* 441–42.

318. R.M. Green 1955, Rawson 1985: 84–85 and 171–78, Scarborough 1993:

influence in chapter 2.3 and will see more in chapter 4.5. Of course, none of his writings were preserved. He advocated the most minimal of treatments possible in any given case (not excluding drugs or surgery, but treating them as a last resort), based on an ethic of compassion for his patients, whose comfort he saw as paramount. He also rejected humoral theory and adopted a mechanical, atomistic physiology that rejected teleological explanations. Both trends in his thinking suggest Epicureanism as his primary influence, very likely explaining the failure of Christians to preserve any of his writings or those of any member of his sect after him.[319] Nevertheless, by radically challenging the dominant Stoic and Aristotelian medical thinking of the time (though not completely rejecting them either), he forced future medical scientists to respond to him, even Galen two centuries later, contributing to an overall advance in medical theory generally, while his subsequent adherents made contributions of their own.[320]

A lot else was going on over the same course of time. Heraclides of Tarentum was advancing Empiricist medical theory in the early first century

---

41–42, and Nutton 2013: 170–73, 190; *DSB* 1.314–15, *EANS* 170–71, and *OCD* 180 (s.v. "Asclepiades (3)"); and (for a reconstruction of his medical theories) Vallance 1990 and 1993.

319. Galen equated the Asclepiads with the Epicureans as advancing similar theories opposed to his own, e.g. Galen, *On the Uses of the Parts* 1.21 (= M.T. May 1968: 104–05). However, modern research has found that much of what Pliny the Elder says about Asclepiades (e.g. *Natural History* 26.7.12–26.9.20) is untrustworthy or demonstrably false, and though what Galen reports (e.g. Galen, *On the Natural Faculties* 1.14 = Kühn 2.45) is more reliable, it cannot be completely trusted either.

320. The Asclepiads were particularly interested in pharmacological research and several of them within a century of their founder had produced books on the subject that were well-regarded by Dioscorides (*DSB* 4.120, in s.v. "Dioscorides"). It is likely the Herophilean physician Alexander Philalethes was a pupil of Asclepiades, and combined his teachings with Herophilean principles and interests toward the end of the first century B.C. (cf. *OCD* 60, s.v. "Alexander (15) Philalethes" with *EANS* 56 and von Staden 1989: 532–39). Another pupil (?) of Asclepiades wrote on chronic diseases (*OCD* 1454, s.v. "Themison"; cf. *EANS* 782–83), and his pupil (?) in turn claimed to have established the Methodist sect *OCD* 1467 (s.v. "Thessalus (2)"; cf. *EANS* 804–05). Nutton 2013: 191–206 treats extensively of the rise of the Methodist sect, and the many medical writers associated with it, which I omit for lack of concrete examples of scientific contributions (until Soranus, discussed below).

B.C. in ways that made him renowned to later scientific writers even of other sects, though none of his books survive.[321] Hicesius wrote respected works in pharmacology and established an Erasistratean medical school in Smyrna (in Greece), sometime in the late second or early first century B.C.[322] And then Zeuxis Philalethes established a Herophilean medical school near Laodicea (midland Turkey) in the 1st century B.C.[323] Though this school lasted only a century (possibly leveled by an earthquake), its last head, Demosthenes Philalethes, wrote a widely influential and comprehensive treatise on ophthalmology in the mid-1st century A.D. which included discussion of the anatomy of the eye. As usual most of this has been lost.[324]

Likewise around the turn of the era some of the most influential research in surgery and biology was published by Sostratus of Alexandria, though none of his books was preserved and his work is known only in scattered quotations.[325] A certain Alexander of Laodicea wrote treatises on gynecology and reproductive science around the same time, but none of his works survive either.[326] Shortly after, Meges of Sidon composed renowned treatises on surgery that revealed detailed anatomical study of the human body, but none of his works survive either.[327] And around the same period Athenaeus of Attaleia founded the 'Pneumatic' sect of medicine, which was distinctively 'Stoic' in character, considerably influenced by the philosophy of Posidonius (whom I'll discuss in the following section), being a student

321. *EANS* 370–71.

322. *EANS* 396.

323. Strabo, *Geography* 12.8.20, says it was a huge school. This benefaction was possibly commemorated on a Laodicean coin series at the time: *OCD* 1591 (s.v. "Zeuxis (3)"; cf. *EANS* 849) with von Staden 1989: 459–62, 529–31 and Benedum 1974. Among the related coins: *Sylloge Nummorum Graecorum* 9 (1964), pl. 125, no. 3855 and 3836/7 (which date between 27 and 7 B.C.).

324. See von Staden 1989: 570–78.

325. *OCD* 1386; *EANS* 754. He wrote on surgery, gynecology, and animals (and on the latter, see the coming discussion in section 3.5). And he is one of several scientists associated with the reign of Cleopatra (P. Fraser 1972: 1.363, with 2.537; Cleopatra's support of science will be discussed in the next section).

326. *EANS* 56.

327. *EANS* 538.

of either Posidonius himself or his system of philosophy.[328] As usual, none of his books survive, even though he wrote extensively on physiology, pathology, embryology, therapeutics, dietetics, and the medical aspects of climate and geography, all from a largely Stoic point of view. His writings had a significant influence on Galen.[329] Other influential members of the Pneumatists included Aretaeus of Cappadocia, who was known for writing books summarizing data and theories on the causes, courses, and treatments of acute and chronic diseases.[330]

In the late 1st century A.D. came Rufus of Ephesus, whose extant books according to Fridolf Kudlien are still "notable for the exceptional richness of their clinical observations" and "the care with which he evaluated" those observations.[331] Rufus became widely renowned as one of the greatest physicians in antiquity, as revered as Hippocrates and Galen. Yet almost nothing he wrote was preserved, except in scattered quotations and a few complete or partial treatises (some known only in translation), even though we can identify nearly a hundred of his books by name. It is clear he sought to write on diseases and anatomy from a perspective of extensive personal observation, cautious theorizing, and careful collection of case notes. Among his most important contributions is what may be the first attempt at standardizing anatomical nomenclature, collecting and sifting the terminology of his predecessors into a single handbook *On the Naming of the Parts of the Human Body*. He is also another of the first doctors known

---

328. Since sources and scholars disagree whether Athenaeus dates to the late 1st century B.C. or mid-1st century A.D.

329. Nutton 2013: 207–08 with *DSB* 1.324–25, *EANS* 176–77, and *OCD* 195 (s.v. "Athenaeus (3)").

330. *DSB* 1.234–35; *EANS* 129–30; *OCD* 147; and Scarborough 1993: 43–44 and Nutton 2013: 210–11. Parts of this latter work survive. Aretaeus was either a colleague of Nero's personal physician Andromachus (who wrote on pharmacology), or a contemporary of Galen (again, our sources are so poor that scholars cannot agree). Other Pneumatists worked in the 1st century A.D. about whom we know little and whose works are lost (e.g. *OCD* 676 and 1167, s.v. "Herodotus (2)" and "Pneumatists"; cf. *EANS* 383–84). For sources on the Pneumatist sect see von Staden 1989: 541, n. 22, and Oberhelman 1994.

331. For the following: *DSB* 11.601–03; *NDSB* 6.290–92; *EANS* 720–21; *OCD* 1298; also M.T. May 1968: 29–30; Scarborough 1993: 44–46; Sideras 1994; Thomssen 1994; Nutton 2013: 214–16.

to have written on occupational medicine, composing a medical study on *Living at Sea* and exhibiting an interest again in the medical needs of slaves.[332]

Meanwhile, working at Rome in the 1st century were Paccius Antiochus, Scribonius Largus, and Claudius Agathinus, among others. Paccius, who eventually published his research in pharmacology (now lost), appears to have enjoyed the patronage of emperor Tiberius.[333] The patronage of either emperor Claudius or one of his staff appears to have subsidized the pharmacological research of Largus.[334] And Agathinus was one of the most important medical theorists in the 1st century A.D. He was another Stoic, who this time established his own eclectic medical sect called the 'Episynthetics', which specifically rejected the splitting of medical theory into sects. This sectarianism had become excessive over the preceding century (reminiscent of the sectarian divisions within early 20th century psychology), and it is notable that efforts were beginning under the Romans to end this. Agathinus sought instead to unify all medical knowledge, an effort that would later be championed to great effect by Galen. Agathinus wrote on numerous subjects, including an empirical treatise on the dosage requirements of hellebore that reported his own experiments performed on animals.[335] Of course nothing he wrote was preserved. His student Archigenes of Apamea advanced the Episynthetic sect in Rome in the late 1st and early 2nd century A.D. Almost everything he wrote is also lost, but we know his books included detailed studies of cancers of the breast and uterus, and treatises on surgical amputation that emphasized the importance of an anatomical investigation of nerves and tendons for successful surgical operations.[336]

332. Similar indications of a growing interest in occupational medicine are indicated by Pliny the Elder's concern for the respiratory health of metalworkers in *Natural History* 34.50.167 (other examples in Nutton 2013: 27). Even without such examples, the claim that "the working man" and "the occupational disease" were "ignored in medical science" until the 18th century (Farrington 1946: 29) is unfounded.

333. *EANS* 95.

334. Nutton 2013: 175–78, *OCD* 1331, and *EANS* 728–29. One of his treatises on *Prescriptions* is extant (see Hamilton 1986). Nutton 2013: 181–82 and 250 discusses a few other imperial medical writers about whose work we know much less.

335. Nutton 2013: 208–09 with *DSB* 1.74–75, *OCD* 35–36, *EANS* 42–43.

336. Nutton 2013: 209–10 with *DSB* 1.212–13, *EANS* 160–61, and *OCD* 140.

Another great scientist of the era, in the middle of the 1st century A.D., was Pedanius Dioscorides, originally from what is now southern Turkey. He is a leading representative of a Roman-era revival of botanical and mineralogical research, in the service of pharmacology. Although his teacher, Arius of Tarsus, was also a noted botanist and mineralogist in his own right (also in the service of pharmacology); just nothing he wrote was preserved.[337] Dioscorides apparently served in the military and says he made good use of that fact to study first-hand the identification, preparation, and use of a wide variety of medicines, including substances extracted from a variety of animals, plants, and minerals. His most renowned book on this subject, *On Medical Materials*, was admired especially by Galen and many others after him, and survives more or less intact.[338] Some manuscripts even preserve attempted reproductions of the meticulous color drawings that originally adorned the text. Skillful drawings had become a part of scientific botanical treatises at least since the early 1st century B.C., when this was most famously a feature of the botanical writings of the Mithridatic physician Crateuas, and two of his contemporaries, Dionysius and Metrodorus, otherwise unknown.[339] In fact, drawings, illustrations, and diagrams had become a standard component of ancient scientific literature in all fields.[340]

---

337. *EANS* 128–29.

338. Riddle 1993: 103–13 and Nutton 2013: 178–81; *OCD* 465–67 (s.v. "Dioscorides (2)"); *EANS* 271–73; and *DSB* 4.119–23, which adds that "numerous treatises in Greek and Latin are falsely attributed to Dioscorides" (4.119), and although numerous later interpolations also entered his authentic text, these can usually be identified through comparison of widely divergent manuscript traditions. Dioscorides did write other books on pharmacology besides *On Medical Materials*, but none survive. He should *not* be confused with Dioscurides Phacas, the medical writer under Cleopatra (as perhaps in Marasco 1998: 43–47).

339. Pliny the Elder, *Natural History* 25.4.8; see *DSB* 4.120 (in s.v. "Dioscorides"). Crateuas: *OCD* 391 and *EANS* 491; Dionysius: *EANS* 264; Metrodorus: *EANS* 553.

340. *OCD* 444 (s.v. "diagrams") and Netz 2010. For examples in mathematics, metaphysics, and astronomy see Obrist 2004. Books on mechanics (e.g. those of Hero, Vitruvius, Philo) frequently refer to accompanying drawings and diagrams (you can see analysis of extant examples in Lefevre 2002 and Leeuwen 2014), as do some medical books (e.g. Apollonius of Citium included instructional diagrams of his procedures for treating dislocations, cf. Potter 1993: 117 and Nutton 2013: 145). Likewise engineering (e.g. Meissner 1999: 247–48; Heisel 1993). Ptolemy, *Harmonics* 3.94 implies all the sciences relied on such artwork, since "what is

Although Pliny observes that copyists could rarely reproduce them faithfully enough to maintain their scientific value, the fact that meticulous empirical drawings were and remained an interest at the time indicates the strength of scientific values.[341] Even apart from this, Dioscorides saw himself as improving on his predecessors, and he was right: his methodology was explicitly empirical and cautious, advancing and revitalizing the field to its most advanced stage. As John Riddle says, "Dioscorides was largely responsible for determining modern plant nomenclature, both popular and scientific" and "so many editions and translations were made from Mattioli's" 16th century critical edition of Dioscorides "that it is said that this printing is the basic work for modern botany."[342] As we shall see again, it often appears that Roman scientists brought their fields to the most advanced levels ever achieved until the dawn of the Scientific Revolution, which took up where the Romans left off.

This is evidenced again in Soranus of Ephesus, who in the early 2nd century A.D. brought the science of gynecology to its acme.[343] Soranus was another biologist who abandoned humoral theory in favor of a more atomistic physiology. He was also among the most famous medical scientists of antiquity, earning the respect of even his philosophical opponents Galen and Tertullian. In fact, he was generally regarded as the equal of Galen—representing in his work some the greatest medical advances of antiquity, yet of a more atomistic and less Hippocratic character (which is

---

given by reason becomes both more teachable and better remembered by us with diagrams and figures." One can find many other examples in all fields (from medicine to geography to engineering). That even geometry texts included diagrams is confirmed by mathematical papyri recovered from Herculaneum (cf. De Falco 1923: 101–03). Aristotle had included drawings and diagrams in some of his works (e.g. Taub 2003: 103–14).

341. On the difficulty of faithfully copying and thus disseminating such visual data: Pliny the Elder, *Natural History* 25.4.8. Ptolemy, *Geography* 1.18 reports the same problem for copying maps, developing a system of map construction in response. Hero recognized the problem for engineering schematics and invented a pantograph as another remedy (see below).

342. *DSB* 4.120, 4.122 (in s.v. "Dioscorides").

343. *DSB* 12.538–42, *EANS* 749–51, and *OCD* 1358; with Lloyd 1983: 168–200; Jackson 1988: 88–90, Scarborough 1993: 46–47, Hanson & Green 1994, Nutton 2013: 199–206; Bliquez 2010.

one reason most of Galen was preserved by medieval Christians, but very little of Soranus). According to Markwart Michler, Soranus sought "a more comprehensive biological view by using vivid comparisons from zoology and agriculture" and he mastered nearly every aspect of medical science that he touched, most notably gynecology, in which he brought together all the best work of his predecessors, improving it with more accurate observations and analysis, and composing a near-definitive treatise on the subject, which has survived—unlike most of his many other works, which (apart from a handful of exceptions) survive only in fragments or translations of uncertain reliability, or only as mere titles. But we know he wrote admired treatises on disease, treatment, bandaging, pharmacology, and surgery, and brought the empirical study of bones and fractures to its most advanced state in antiquity. He also wrote an extensive history of medicine that is entirely lost, except for a chapter on Hippocrates. He also updated anatomical nomenclature with his own treatise on the subject that became definitive for centuries, though it is also no longer extant.

Around the same time Heliodorus wrote advanced works on surgery, wound care, and joint repair; Trajan's personal physician Statilius wrote *On the Composition of Drugs*, among other things; and the leading empiricist Menodotus wrote extensively on medical science.[344] Later in that century a deliberately eclectic Philumenus of Alexandria wrote lost works *On Poisonous Animals*, *On Gynecology*, *On Bowl Disorders*, and more.[345] Aelius Promotus wrote a book on cures called *Potency*, sections of which are extant, but little else is known of him.[346] Antyllus, wrote important works on surgery and other subjects, but nothing survives except scattered quotations.[347] Near the dawn of the 3rd century A.D., the medical philosopher Sextus Empiricus wrote two monumental treatises on epistemological skepticism

---

344. *OCD* 654, 933, 1396 (s.v. "Heliodorus (3)," "Menodotus (3)," "Statilius Crito"); cf. *EANS* 363, 549–50, 494–95. Nutton 2013: 262 discusses Crito's career. As usual, most of what these men wrote was preserved only in fragmentary quotations by later authors. Heliodorus also wrote a treatise *On Weights and Measures*, suggesting a rising medical interest in a subject usually treated by engineers. For more on Heliodorus see P. Fraser 1972: 1.363 (with 2.538).

345. *OCD* 1138; *EANS* 661–62. Only fragments of the named books survive.

346. *OCD* 19; *EANS* 35.

347. *OCD* 114; *EANS* 101–02. See also P. Fraser 1972: 1.363 (with 2.537–38).

that displayed vast erudition and careful study of the sciences—despite his rejection of natural philosophy as ultimately unknowable. Nevertheless, he was himself a medical writer, though none of his scientific works survive. Of these, we know the title of at least one, his *Medical Notes*, which in the tradition of the empiricist sect probably included records of case studies, emphasizing observed correlations between symptoms and successful and unsuccessful treatments.[348]

But historically more important than all these was Marinus in the early 2nd century A.D., who 'revived' anatomical research at Alexandria (according to Galen) by composing the first truly comprehensive anatomical study from personal observations since Erasistratus, which Marinus recorded in his *Anatomy* in twenty books. None of this was preserved, of course, except a summary and table of contents from Galen, but it was especially noted for its detailed study of the human skeleton, which appears to have gone much farther than any before it. This anatomical research was continued by his pupil Quintus and *his* pupils Lycus, Satyrus and Numisianus, as well as the latter's pupil Pelops. Satyrus and Pelops were among Galen's teachers.[349]

It is clear that detailed scientific research in anatomy and physiology was considerably advancing under the Romans. And yet we would not know that any of this was going on, or that there were so many scientists pursuing anatomical research in the Roman period, had Galen not been

348. *DSB* 12.340–41, *EANS* 739–40 and *OCD* 1358–59. There is no basis for Peter Green's assertion (P. Green 1990: 470) that the meticulous Hippocratic method of assembling case histories fell into disuse immediately after Hippocrates invented it. The empiricist sect relied almost exclusively on the method of carefully analyzing case histories, and other doctors employed them as well (Nutton 2013: 150–51; Mattern 2008: 27–47). We have no reason to believe medieval scribes would have preserved anyone's medical case notes, when they did not even deign to preserve a single empiricist medical book. Even some of the case histories of Rufus survive only in Arabic (Nutton 2013: 214). Galen's medical notes had already been lost in a fire and thus were not transmitted to us (refs. in Carrier 2016: 55 n. 135).

349. Marinus: M.T. May 1968: 31–34; *EANS* 532 and *OCD* 899. On the others: M.T. May 1968: 34–38; with: Quintus (*OCD* 1252; *EANS* 717), Satyrus (*OCD* 1323, third entry, there misdated as B.C. instead of A.D., an obvious typo; *EANS* 728), Pelops (*EANS* 634), Lycus (*EANS* 514), and Numisianus (*EANS* 584). See also *OCD* 79–82 (s.v. "anatomy and physiology"). On these and other scientists in the Roman revival of anatomical studies see Nutton 1993b: 15–19.

particularly chatty about it. And he's hardly a comprehensive source.[350] One Roman epitaph from the first century honors an imperial physician Claudius Menecrates as the founder of his own medical sect and author of 156 books in medical science, which earned him public honors from several major cities.[351] Yet we know almost nothing else about him or his sect or any of his hundred books. Another first century inscription honors an equally-unknown Hermogenes of Smyrna for having written a few histories as well as 77 books in medical science.[352] A second century inscription honors another otherwise-unknown Heraclitus of Rhodiapolis who wrote several award-winning treatises in medicine and philosophy.[353] We also know of a woman named Aspasia, probably of this period, who was a revered figure in gynecological science, but none of her books were preserved, nor quoted for many more centuries.[354] Thus it is reasonable to assume there was a lot more science going on in the Roman period than we know.[355]

We do not have any comparably chatty author on the sciences of astronomy or engineering, and few lucky epitaphs.[356] Ptolemy and Hero say

350. Nutton 2010.

351. Gourevitch 1970: 44; *EANS* 544. On the Menecrateans as a possible medical family spanning many generations see chapter 2.3. Galen complains that some medical quacks wrote "hundred volume works" (*On the Therapeutic Method* 1.4.12, cf. also Iskandar 1988: 175, §P.134,4–5) so we can't be sure of the scientific quality of the medical books by Claudius Menecrates.

352. Nutton 2013: 216; *EANS* 379. The content of these books is unknown (see preceding note).

353. *EANS* 373. Of his books we don't even know the titles.

354. *EANS* 172. Her precise date is unknown. But the quality of her work in quotation rivals Soranus, yet neither Soranus nor Galen mention her; and conversely, the content of her work suggests her science predated the 3rd century crisis and subsequent Christianization. So most likely she dates to the early 3rd century.

355. Nutton 2013: 216–21 surveys several other likely medical writers of this period about whom we know almost nothing.

356. Analogous to the mysterious Menecrates is the equally mysterious author of the "Keskinto Inscription," name unknown but clearly an astronomer of considerable skill who surely must have written books (see *EANS* 469 and note at the end of chapter eight in Carrier 2016). Likewise the inscription of the otherwise-unknown engineer Nonius Datus, which is much too wordy and

very little about their teachers or contemporaries or ongoing activities or debates in their own communities, which has more to do with their style and approach as writers (and the choices of medieval bookmen) than with any real absence of considerable scientific activity in their respective fields. Yet we have hints there was a lot. As we shall see, it is reasonable to suspect that theories and research were as diverse in astronomy and engineering as Galen reveals for medical science.

Progress in the life sciences in antiquity ends with Galen, who was by all accounts one of the greatest medical scientists of the age, widely renowned even in his own lifetime, and the last to make any significant advances in the life sciences until the Scientific Revolution.[357] His place in the history of medicine became as central as Aristotle's place in the history of science as a whole. Many of his books became the backbone of medical curricula for centuries, and for more than a thousand years no doctor would be considered educated who had not studied him. Galen flourished in the late 2nd century A.D. and his work represents a perfection and improvement of many elements of medical science up to his time. But his most important contributions to science were his articulation and defense of an increasingly sound scientific method and his advanced empirical research in human and animal anatomy and physiology, which confirmed, corrected, or updated previous work in the field.[358] He even came close to discovering the correct theories of circulation and respiration, and worked out a largely correct

---

exciting to have come from a man who never wrote books (discussed in chapter 4.5). It is also unlikely the three attested "*physici*" in Roman inscriptions wrote nothing on scientific subjects (see chapter 2.3).

357. On Galen's fame: Nutton 1984b. For the rest: *DSB* 5.227–37; *NDSB* 3.91–96; *EANS* 335–39 (cf. 339–42); *OCD* 600–01; Nutton 2013: 222–35; Hankinson 2008; Mattern 2008 and 2013; Riddle 1993: 113–17; Lloyd 1973: 136–53; Scarborough 1970: 303–05; Bowersock 1969: 59–75; Siegel 1968: 4–26. See also: Whitmarsh et al. 2009. Galen also wrote valuable works in language, logic, and scientific method, few of which survive (cf. Nutton 2013: 228). Galen also wrote a treatise on augury, omens, astrology, and dream interpretation (cf. Galen, *On the Natural Faculties* 1.12 = Kühn 2.29) which is lost, but appears to have presented them positively (much as Ptolemy did for astrology). On Galen's medical theories: Nutton 2013: 236–53.

358. Central works in this category include Galen's monumental multi-volume sets *On the Uses of the Parts* and *On Conducting Anatomical Investigations*.

account of the renal and digestive systems.[359]

Galen's methodological improvements include the beginning of a demarcation between science and philosophy, and a conscious effort to develop a correct empirical method in biology by eclectically adapting the best epistemological ideas of all the philosophical schools, and emulating the indisputably successful fields of astronomy and engineering.[360] Though he did not always follow his own advice, he routinely emphasized the need for testing and verification, and for limiting claims to what can actually be proved from observation.[361] He also sought to unify the medical sects by resolving their differences into a common methodology that came strikingly close to modern scientific method. Meanwhile, his extensive anatomical and physiological research indicates a continuation of the scientific traditions championed by Herophilus and Erasistratus under the patronage of the Ptolemies centuries before. And given his numerous mentions of public debates and demonstrations on the subject, it is reasonable to expect medical science in the hands of those embracing Galen's aims and methods would have continued to advance and improve upon his work, and could well have surpassed Harvey, had the events of the subsequent century not reversed the course of ancient society. Instead, Galen's treatises came to be regarded as little more than unsurpassable gospel, and his methodological injunctions to check and improve on his work were largely ignored. Aristotle suffered much the same fate. Many continued to comment and argue with their works, but actually following their own declared methodologies in order to make real improvements in the sciences, as they themselves had done, was not of much interest to anyone until the Renaissance.

Although Roman scientists were apparently prevented from (at least routinely) dissecting human cadavers (see section 3.8.III), they employed monkeys, apes, and many other animals as substitutes, and checked their

---

359. Nutton 2013: 238 surveys some of Galen's scientific discoveries. On his failure to develop correct theories of circulation and respiration see discussion and notes in section 3.9.II.

360. See discussions of his epistemology in section 3.7 and theory of respiration in 3.9.II.

361. As demonstrated throughout Siegel 1968, Galen expressed doubts about some of his own theories and often distinguished proofs from plausible speculations in a way later overlooked.

findings on humans when they could. But even with this limitation upon him, Galen accomplished a great deal. In the words of Ludwig Edelstein and Vivian Nutton, "dissecting animals, especially monkeys, pigs, sheep, and goats, carefully and often," Galen "collected and corrected the results of earlier generations by experiment, superior factual information, and logic" and in fact "his physiological research was at times masterly, particularly in his series of experiments ligating or cutting the spinal cord" to test the attributes of the nervous system.[362] As further examples, Galen ended an ongoing debate about the function and physiological properties of the kidneys and bladder with an extensive and comprehensive system of experiments involving the vivisection of animals, and he specifically uses this research to argue the general point that speculation without solid empirical evidence is vain.[363] He performed detailed experiments on digestion in pigs, including observations during vivisection, which corrected and expanded knowledge of the complex processes involved.[364] He confirmed by observation that saliva had digestive properties.[365] He greatly advanced anatomical understanding of the hands, forearm, upper eyelid, and other areas, consciously filling gaps in the knowledge left by his predecessors, and when fellow intellectuals did not believe him, he gave public anatomical demonstrations to prove his new discoveries.[366] Moreover, he went beyond the practical interests of medicine and extensively dissected animals and studied animal physiology and anatomy for its own sake. He had planned a companion volume to his *On the Uses of the Parts*, which used detailed anatomical study to prove the human body was intelligently designed. His next treatise was to use such detailed anatomical study to prove the intelligent design of animals, including careful observations of animal behavior in the Aristotelian tradition, but his death prevented its

---

362. *OCD* 600 (in s.v. "Galen"). For an exhaustive survey of the wide array of anatomical and physiological experiments conducted by Galen see Debru 1994, along with Tieleman 2002, Rocca 2003, and Siegel 1968, 1970, and 1973.

363. Galen, *On the Natural Faculties* 1.13 (= Kühn 2.30–40), analyzed in Siegel 1968: 126–34.

364. Galen, *On the Natural Faculties* 3.4 (= Kühn 2.155–57).

365. Galen, *On the Natural Faculties* 3.7 (= Kühn 2.162–63).

366. Galen, *On My Own Books* 2 (= Kühn 19.20–22).

completion.[367] Already in his extant work he shows he was observing and documenting animal behavior, just as Aristotle had.[368] His many completed books also document his extensive use, interest, and knowledge of animals of all kinds, including "dissections and vivisections" of "mice, birds, snakes, pigs, goats, oxen, horses," and various monkeys and apes, and many other species, from cats to fish.[369] One of the most telling examples is his study of the elephant, which began when he saw one killed in the arena and he and other doctors eagerly seized the opportunity for dissecting and examining a rare scientific specimen.[370]

Galen's writings reveal a more general revival of scientific interest in the Roman period, beyond medical science. He reports experiments and observations in magnetism and electrostatics that he or others used to 'refute' atomist explanations of magnetic phenomena, and from his comments it is easy to see how close the Romans were getting to the work of Gilbert 1400 years later.[371] Galen also reports his own experiments and observations confirming the property of dry grain to gain weight by absorbing water through terra cotta. What's even more significant about this is that he learned of the phenomenon from peasants who used it as a trick to steal grain, and his disbelief in their report led him to conduct tests, and he was surprised to find that his experiments confirmed the phenomenon.[372] Interacting with the working class to learn and study natural phenomena,

---

367. See Galen, *On the Uses of the Parts* 14.4 (= Kühn 4.153 = M.T. May 1968: 626) and *On Conducting Anatomical Investigations* 11.12.

368. For example, see Galen, *On the Uses of the Parts* 1.3 (= M.T. May 1968: 70–71) and *On Conducting Anatomical Investigations* 6.1 (= Kühn 2.538).

369. Quote and sample of sources in von Staden 1995: 47. Galen discusses a partial list of the animals he had systematically dissected in *On Conducting Anatomical Investigations* 6.1 (= Kühn 2.532–40), specifically adding that he had not dissected insects. More animals he dissected are mentioned in Galen, *On Conducting Anatomical Investigations* 7.11 (= Kühn 2.623–24).

370. Galen, *On the Uses of the Parts* 17.1 (= M.T. May 1968: 724–25) and *On Conducting Anatomical Investigations* 7.10 (= Kühn 2.619–23). See discussion in French 1994: 190–91 and Hankinson 1988.

371. Galen, *On the Natural Faculties* 1.14 (= Kühn 2.45–51).

372. Galen, *On the Natural Faculties* 1.14 (= Kühn 2.55–56). Grain storage manuals now recommend plastic liners over concrete floors to prevent this phenomenon (e.g. Hellevang 1998).

conducting tests to confirm what is claimed, and all simply for the sake of knowledge, are attributes often claimed to be absent in antiquity. Clearly they were not.

We will discuss these issues more later. But so far all of the above demonstrates that Roman medical scientists were conscious of the fact that progress had been made, and were consciously building on past achievements to contribute even more to that progress, and this was all thought to be worthwhile. Now we will see the same in astronomy and physics.

3.3 Scientific Astronomy up to the Roman Era

Astronomy witnessed considerable scientific advancement in antiquity.[373] Under the rubric of 'astronomy' we also include here scientific geography (the study of the size, shape, and nature of the earth), cartography (the study of accurate mapmaking), gnomonics and calendrics (developing and perfecting sundials and calendars), and meteorology (which in antiquity meant the study of both celestial and atmospheric phenomena apart from the study of the sun, moon, planets, and fixed stars), because these fields were all closely related at the time—largely because astronomical theories and data became crucial to mapmaking, while meteorology included the very effort to demarcate astronomical from atmospheric phenomena.[374] Although astrophysics and cosmology were also subjects of speculative interest among natural philosophers, they were never placed on a scientific footing until more modern times.[375]

373. For context and scholarship see Neugebauer 1975, Lloyd 1973: 53–74 (with 33–52); van der Waerden 1963; *OCD* 188–90 (s.v. "astronomical instruments" and "astronomy"), and 910–11 and 1483 (s.v. "mathematics" and "time-reckoning"). For the mathematical background (across all fields of inquiry, not just astronomy): Cuomo 2001.

374. On ancient geography see *OCD* 611–12 (s.v. "geography"), with Dueck 2012, Hubner 2000, Rihll 1999: 82–105, French 1994: 114–48, and Gorrie 1970. Ancient meteorology also included geological phenomena such as earthquakes: Taub 2003 with *OCD* 482 and 941–42 (s.v. "earthquakes" and "meteorology"); cf. Aristotle, *Meteorology* 1.1 (338a-339a); Seneca, *Natural Questions* 2.1.2–5.

375. Russo attempts to argue that Hellenistic astronomers had actually achieved an astrophysical dynamics rivaling Newton's (Russo 2003: 231–42, 282–320), but

Apart from its use in timekeeping and navigation, the most central challenge in ancient astronomical science was to explain the startling observation that the planets do not cross the sky at constant velocities—they even appear to stop and reverse course for brief periods. This singular problem inspired some of the most scientific aspects of Greco-Roman astronomy: attention to detailed observations—which led to the discovery of the unusual planetary motion (and continued disconfirming attempts to explain it)—and the quest for explanatory models of those observations, which led to a sophisticated theory of the solar system that became increasingly more accurate over the centuries. Other problems that occupied a place of central concern in antiquity included the long-sought ability to predict lunar and solar eclipses, and (eventually) an astrological interest in computing the course and position of planets and constellations for any given month and year.

Immediately after Aristotle, and bridging the 4th and 3rd centuries B.C., our story begins with Autolycus of Pitane, another scientist from the west coast of what is now Turkey.[376] Two of his treatises survive, one in spherical geometry and the other recording the times of rising and setting for stars throughout the year, demonstrating the combined interests of observation and mathematical explanation. These were related projects, since the geometry of spheres was essential to determining the geometrical properties of the star field as it changed throughout the year over a curved path. But most important is the record he preserves of astronomers' empirical objections to the simple planetary model of Eudoxus, Callippus, and Aristotle, especially the observation that Venus and Mars change in brightness throughout the year (which suggested they were not always the same distance from earth) and the fact that some solar eclipses are annular while others are total, which all but proved that the moon varies its distance from the earth. Astronomers were thus paying attention to the facts and criticizing theories that did not fit

---

though this maverick effort is clever, it is ultimately flawed and unconvincing. It's not impossible. But it is very unlikely.

376. *DSB* 1.338–39, *EANS* 183, and *OCD* 214 (s.v. "Autolycus (2)"), whose treatise *On Risings and Settings* is the earliest known scientific star catalogue. For pre-Aristotelian astronomy see Appendix B (esp. Eudoxus and Callippus). After Aristotle I'll only remark on the best known astronomers; for a more thorough list of all known astronomers in antiquity, see *EANS* 995–96 (s.v. "astronomy").

them, in an ongoing process of research and debate.[377]

Contemporary with Autolycus was Dicaearchus of Messana, a pupil of Aristotle and possibly the first scientific geographer. Although none of his books were preserved, his known achievements entail considerable mathematical and astronomical skill. According to C.B.R. Pelling, he composed world maps and "established with some accuracy a main parallel of latitude from the straits of Gibraltar to the Himalayas," which would not have been possible without astronomical observations and calculations.[378] And though Pelling says he "overestimated" the heights of mountains, he was nevertheless remarkably accurate: the elevation Dicaearchus reported for the highest mountain known to him, Mount Pelion, was just over 6000 feet—an estimate only 700 feet too high.[379] Such a close value could only be the product of geometric survey. In fact, evidence suggests ancient surveyors subsequently produced increasingly accurate measurements of mountain heights over time.[380]

A more renowned astronomer working around the same time was Euclid, who is most famous for establishing the basic principles of geometry so thoroughly and successfully that his *Elements* remained the standard geometry textbook in schools well into the modern age, and was the foundation upon which all subsequent mathematicians built. Euclid is believed to have taught at Alexandria in the early years of the Museum, but nothing else is known about his life or what his research interests were, though we know he wrote on optics and astronomical phenomena, while his geometrical work systematized and improved upon that of previous astronomers who wrote on geometry, and laid the foundations for the future of both astronomy and engineering, and influenced methodology even in the life sciences. Euclid applied his geometrical findings to basic astronomical problems in the *Phenomena* and wrote the first known treatise on the theory of perspective in the *Optics*, both of which survive in edited versions. He also wrote an *Elements of Music* on harmonic theory, and some

377. Empirical observations of varying distances for the planets and moon continued into the Roman period (cf. Cohen & Drabkin 1948: 103–05 and 142, quoting Ptolemy and his contemporary Sosigenes).

378. *OCD* 447; *EANS* 246.

379. Pliny the Elder, *Natural History* 2.65.162.

380. M.J. Lewis 2001b: 157–66, 335–39 (on Dicaearchus: 158–62).

Arabic sources suggest he wrote something on mechanics, but the original text of these is lost.[381]

Also from that time is the most renowned scientist in antiquity, Strato of Lampsacus, whom Diogenes says was "held in the highest regard" even in the Roman era, and widely "nicknamed 'The Natural Philosopher' because he took the greatest care and consumed himself with natural theory more than any other," studying a wide variety of subjects, including medicine, husbandry, meteorology, psychology, physiology, zoology, and mechanics.[382] Lampsacus lay on the Eastern side of the Hellespont and was a noted center for teaching by both Anaxagoras and Epicurus. So when Strato came to study Aristotelianism under Aristotle's successor Theophrastus, he already came with a sympathy for atomist philosophies. He then served as royal tutor for several years in Alexandria, likely in close connection with the Museum, and eventually became the third head of Aristotle's school at the Lyceum in Athens. Philosophically, Strato is most noted for having combined Aristotelian and atomist natural philosophy and reinforcing this prescient mix with a strong empirical and experimental scientific spirit. He thus became the father of an entire tradition in the history of ancient science that was nearly erased by the deliberate neglect of medieval Christian scribes and scholars, who preferred to save works that agreed with the less atomistic (and thus less disturbingly atheistic) tradition, which instead came closer to merging Aristotelian and Platonic natural philosophy and favored Hippocrates over Erasistratus in medicine, and Hipparchus over Aristarchus in astronomy. Consequently, not a single book Strato wrote was

---

381. *DSB* 4.414–59, *EANS* 304–06 and *OCD* 544; and see discussion in *DSB* 13.321–25 (s.v. "Theon of Alexandria"). Besides the *Elements* Euclid also wrote several other books on geometry, only one of which survives intact (the *Data*, which includes theorems relevant to algebra), although a few others survive as fragments in Arabic translation. Regardless of whether other extant works in optics, catoptrics (i.e. reflection), and harmonics are his, he probably did write on those subjects.

382. Diogenes Laertius, *Lives and Opinions of Eminent Philosophers* 5.58 (as noted in chapter 2.5, the honor of this epithet was awarded him throughout ancient literature). He also wrote on logic, ethics, and technology, among many other subjects. For the rest of the present discussion, see: *EANS* 765–66; *NDSB* 6.540; *DSB* 13.91–95; *OCD* 1406 (s.v. "Straton (1)"); Lloyd 1973: 15–20; and sources and discussion in Berryman 1996 and Desclos & Fortenbaugh 2011.

preserved, despite including some of the most important scientific content in antiquity. Nevertheless, due to his unquenchable fame, we know a great deal about him and his work from quotes, discussions, and comments in later authors.

Though Strato also contributed to physics (hence we will discuss him again in the next section), this led him to theories of considerable relevance to the history of astronomy. First, he rejected providence, creationism, and intelligent design, and instead sought a system that would explain all phenomena in terms of natural weights, movements, and powers, which led him to reject several Aristotelian dogmas, adopting in their place some of what we now know to be scientifically correct theories. For example, in his lost books *On Lightness and Heaviness* and *On Motion*, Strato abandoned the doctrine of 'natural places' in exchange for a more mechanical view of why some objects rise and others fall, which happened to be nearly correct (all objects are drawn to earth by a force but lighter objects are squeezed upwards by heavier ones). He also abandoned Aristotle's astrophysics, arguing in his lost treatise *On the Heavens* that the same principles, elements, and physics operate in the heavens as on earth, even insisting the stars and planets are subject to the same pull towards earth as everything else—which is incorrect in its geocentricity, but remained in antiquity the only answer for what causes the movements of the moon and planets that was close to being correct.

Strato based both his dynamics and his cosmology on a primitive theory of inertia. This he borrowed from the atomists, particularly the Epicureans, who held that everything falls at the same rate regardless of mass, and changes direction or speed only when struck, whether by a blow or a medium.[383] He then combined this with the Aristotelian conclusion that falling bodies accelerate, which Strato proved empirically by observing falling stones and streams of water.[384] In the same way Strato refuted the Aristotelian belief that objects gain weight as they fall, observing instead, for example, that stones make a greater impact the farther they fall solely

383. A prototype of inertial theory in atomism can be seen in Lucretius, *On the Nature of Things* 2.62–166 and 2.184–332; and in Seneca, *Natural Questions* 7.14.3–5.

384. Reported (though perhaps without fully comprehending the original argument or context) in Simplicius, *Commentary on Aristotle's 'Physics'* 5.6.916; and Simplicius, *Commentary on Aristotle's 'On the Heavens'* 1.8.267.29 and 1.8.269.4.

because of their increased speed, not their increased mass. It seems he also observed the fact that heavy drops of water do not fall faster than light ones, yet all fall faster the farther they have fallen, which would suggest a nearly modern view of gravity, but since we do not have a full or clear account of Strato's physics we can say nothing certain on this point.

All this led to Strato's student, Aristarchus of Samos, who in the early 3rd century B.C. was the known scientist in history known to propose a heliocentric theory of planetary motion, possibly building on partially heliocentric theories proposed by others before him.[385] And Seleucus (whom will discuss shortly) is known to have embraced the Aristarchan model a century later.[386] There were probably others unknown to us, and we cannot assume it was ever abandoned—any rival astronomical work even in Ptolemy's day may simply have been relegated to oblivion, just like the atomistic medicine of Galen's contemporaries—but we will return to this subject later. None of Aristarchus' works survive, except an early treatise *On the Sizes and Distances of the Sun and Moon*, which predates his heliocentric theory. Vitruvius reports of Aristarchus that he advanced the field of sundial construction and was also competent in a full range of physical sciences, not just astronomy, which suggests Aristarchus wrote books on many other subjects. Modern scholars agree he was a brilliant mathematician, inventing the first procedures for determining the distances of the sun and moon, though with only crude observational data, which later astronomers would improve upon. But to accomplish this feat Aristarchus began work in trigonometry that would be greatly expanded in subsequent centuries.

Eratosthenes of Cyrene in the late 3rd century B.C. was already employing better methods and data to produce a surprisingly good estimate of the size of the earth.[387] Eratosthenes conducted his research as head of the Alexandrian library for much of that century, after completing an education in several

---

385. For the following: *DSB* 1.246–50; *EANS* 131–33; *OCD* 153; Heath 1913.

386. Plutarch says Aristarchus proposed heliocentrism as "only a hypothesis" but that Seleucus "demonstrated it" (*Platonic Questions* 8.1 = *Moralia* 1006c). He does not say how. Though heliocentrism never dominated, it was not ignored, e.g. Panchenko 2000 argues its challenges continually led to improvements in geocentric models. See also discussion in section 3.5.

387. For the following: *DSB* 4.388–93, *EANS* 297–300, and *OCD* 533–34. Eratosthenes was also a published poet and philosopher and frequently combined science and literary scholarship (see Pfeiffer 1968: 152–70).

philosophies at Athens. His work was primarily in the field of history and literary studies, but he also produced the first scientific geography based on the best mathematics and astronomy of his day, even making some original scientific observations for the purpose, and with this he also launched the science of cartography.[388] Although both fields would be greatly improved by later scientists, Eratosthenes' estimate for the circumference of the earth (roughly 29,000 miles) was based on sound methodology and very close to the true value (about 25,000 miles).[389] He also solved various mathematical problems in arithmetic, mechanics, astronomy, and harmonics, and was the first to attempt a scientific chronology of historical events.[390] Once again, nothing he wrote survives, except in scattered quotations, references, or paraphrases in later authors.

In the late 3rd and early 2nd century, the astronomer Apollonius of Perga perfected the study of conics (the geometrical properties and laws governing parabolas, hyperbolas, and ellipses), as an aid to astronomy, especially sundial construction.[391] His *Conics* made him famous, winning him the appellation "The Great Geometer."[392] This became the standard textbook on the subject throughout antiquity, and would not be improved upon until the 16th century, yet only portions of it have survived (some in Greek, some in Arabic). He also wrote *To Those Who Study Mirrors* and possibly *On Burning Mirrors*, and numerous treatises in geometry and

388. On ancient explorers and exploration in relation to scientific geography see Appendix A.

389. When converting measures in ancient stades (or "stadium lengths," similar to Americans measuring distances in "football fields") to modern miles I follow the critical conclusions of Engels 1985 and Pothecary 1995 that the stade used by scientific authors measured somewhere between 600 to 610 feet (or roughly 8.75 stades per modern mile).

390. It was possibly in connection with his chronographic work that Eratosthenes wrote on mathematical and calendrical problems in astronomy (cf. Geminus, *Introduction to Astronomy* 8.24). His chronographic work was subsequently extended and improved by Apollodorus of Athens in the following century (*OCD* 120, s.v. "Apollodorus (5)").

391. For the following: Fried & Unguru 2001; *EANS* 114–15, *DSB* 1.179–93, *NDSB* 1.83–85, and *OCD* 122–23.

392. Eutocius, *Commentary on the 'Conics' of Apollonius* 2.170 (early 6th century A.D.).

astronomy, none of which were preserved, though one, *On the Cylindrical Helix*, suggests an interest in mechanics, since the figure described has no astronomical application but would correspond to Archimedes' Screw and other mechanical screws.[393] He may also have written a treatise on the principles and construction of a robotic flute player, if an Arabic fragment is authentic.[394] Likewise a treatise on gears.[395] But in astronomy Apollonius was especially renowned for his studies of the moon, including attempts to calculate the lunar distance, and for his work on planetary motion, which included proving the mathematical equivalence of epicyclic and eccentric models, a fact accepted by all later astronomers.[396]

Dionysodorus, a colleague of Apollonius, adapted his principles of conics to the geometry of the torus and may have used this advance to invent an improved conical sundial.[397] Also dating sometime after Apollonius in the 2nd century B.C. is another writer on the mathematics of the torus, the astronomer Perseus, about whom we know next to nothing.[398] And his contemporary, the noted astronomer Zenodorus, also remains in obscurity, though we know he wrote on (and probably established) the geometry of isoperimetry, which would have been of use not only to astronomers (Ptolemy relies on it in that capacity), but also to geographers and engineers.[399] Likewise, the astronomers Dositheus and Diocles, in the 3rd and 2nd centuries B.C. respectively, employed the same principles to write treatises on (and actually construct) parabolic burning mirrors, and further discussing the relevance of conics and optics to engineering and

393. Russo 2003: 98, 120. For Archimedes see discussion in section 3.4.

394. M.J. Lewis 2000: 352–54 (and 1997: 49–57 & 86–88).

395. M.J. Lewis 1997: 24, 50.

396. These alternative models of planetary motion were discussed in chapter 2.7.

397. *DSB* 4.108–10; *EANS* 266.

398. *DSB* 10.529–30, *EANS* 636, and *OCD* 1111 (s.v. "Perseus (3)").

399. *DSB* 14.603–05, *NDSB* 1.83–85, *EANS* 845, *OCD* 1588. Even Quintilian shows a sound grasp of the uses and principles of isoperimetry and gives several examples of why generals, historians, surveyors, and lawyers need to learn it (*Education in Oratory* 1.10.39–45). Examples of its application and discussion are found in extant surveying manuals from the early Roman empire (e.g. B. Campbell 2000: 12–13) and it found use even in biology (e.g. Cuomo 2000: 57–90 for its use in apiology; Aristotle, *Posterior Analytics* 1.13.79a for its use in medical physiology).

astronomy, especially in the construction of sundials, although these fields were useful also for scientific geography and cartography and in solving various engineering problems.[400] Fittingly, in the *Conics* Apollonius had already expressed his own explicit consciousness of progress in mathematics, describing how he was drawing on the work of multiple predecessors and advancing it with his own discoveries.[401] Hence later scientists were likewise following his own example and advice. Also at work in the middle of the 2nd century B.C. was the heliocentric astronomer Seleucus, the student of Aristarchus already discussed above, who also discovered the combined lunar-solar effect on the tides.[402] And around the same time an astronomical geographer named Crates constructed the first known scientifically-based cartographic globe of the earth, painting a map of the known world on a sphere, probably relying on the work of Eratosthenes.[403]

All this led to Hipparchus, who began his own research in his native

---

400. *DSB* 1.187 (in s.v. "Apollonius of Perga"). Dositheus: *DSB* 4.171–72; *EANS* 277; *OCD* 477. Diocles: *DSB* 4.105 (updated in *DSB* 15.115–18); *EANS* 255; and *OCD* 453 (s.v. "Diocles (4)"). For Diocles' extant treatise on burning mirrors see Toomer 1976; and for a broader history of the scientific study of them in antiquity see Acerbi 2011. On ancient sundial technology see Evans & Berggren 2006: 34–38, Evans 1999: 243–51, and Gibbs 1976 (plus other sources on sundial technology mentioned in following notes).

401. Apollonius of Perga, *Conics* 1.2.4. His predecessors included the 4th century founders of the study of conics: first Menaechmus (*OCD* 929, second entry; *EANS* 542–43), then Aristaeus (*DSB* 1.245–46; *EANS* 130–31) and Euclid (above); and from the early 3rd century, Nicomedes, who wrote on the mathematical uses and properties of conchoids (i.e. three-dimensional spirals, cf. *DSB* 10.114–16; *EANS* 580; *OCD* 1015, s.v. "Nicomedes (5)").

402. *OCD* 1342 (s.v. "Seleucus (5)") and *EANS* 730. On ancient lunisolar tidal theory (which was studied and developed further after Seleucus) see: Pliny the Elder, *Natural History* 2.99.212–218 and 2.102.221, with: Cicero, *On Divination* 2.34 and *On the Nature of the Gods* 2.7.15–16; Seneca, *On Providence* 1.4; Cleomedes, *On the Heavens* 156; and Ptolemy, *Tetrabiblos* 1.2.3–6; as well as Strabo, *Geography* 3.5.8 and 1.1.8–12, who confirms that the role of the moon had already been established by Eratosthenes shortly before Seleucus, who probably discovered the role of the sun (see Kidd 1988: 522–25, 759–65, 772–92). On ancient tide theory and its significance see Russo 2003: 305–15 and 360–65, though some of his conjectures exceed the evidence.

403. Strabo, *Geography* 2.5.10. See *EANS* 490 and *OCD* 390–91 (s.v. "Crates (3)").

town of Nicea, in what is now northern Turkey, but then spent the rest of his life, and completed most of his work, on the island nation of Rhodes in the middle of the 2nd century B.C.[404] He became one of the most famous astronomers in antiquity, the only one known to have been honored on ancient coins.[405] He is also the most important astronomer before Ptolemy, and a notable contributor to geography as well, yet none of his many books were preserved, other than a brief commentary on the *Phenomena* of Aratus.[406] The most alarming losses are his works in combinatorial arithmetic, and his scientific study of falling objects, which appears to have expanded on Strato's work on the same topic.[407] This is especially tragic since we know Hipparchus rejected the Aristotelian physics of motion and followed Strato in embracing an early impetus theory in advance of Galileo.[408] Hipparchus

---

404. For the following: Neugebauer 1975: 1.274–343 with *DSB* 15.207–24, *EANS* 397–99, and *OCD* 685–86 (s.v. "Hipparchus (3)").

405. These coins appeared only in the Roman era, not during his life, a fact we will discuss in chapter 4.3. For scholarship see Schefold 1997: 418–19, 543 (Abb. 302) and in *DSB* 15.207–08 and 15.222 (in s.v. "Hipparchus").

406. On this poem by Aratus see discussion and notes in chapter four of Carrier 2016.

407. *DSB* 15.220 (in s.v. "Hipparchus") with Russo 2003: 281–82 and Netz 2003: 283–84. Combinatorial arithmetic involves factorials and permutations (the ability to calculate accumulating products and sums and determine the number of possible ways to arrange a collection), but we do not know how much of this Hipparchus studied or to what end (see Plutarch, *Tabletalk* 8.9 = *Moralia* 732f-733a; and the bibliography in *DSB* 15.223–24). We at least know the title of his treatise on what we now call gravity: *On Objects Carried Down by their Weight* [*Peri tôn dia Barutêta katô Pheromenôn*], cf. Simplicius, *Commentary on Aristotle's 'On the Heavens'* 1.8.264.25 (see Desclos & Fortenbaugh 2011: 313–52).

408. *DSB* 7.136 (in s.v. "John Philoponus") and Simplicius, *Commentary on Aristotle's 'On the Heavens'* 1.8.264.25–265.6. However, Russo's attempt to argue that heliocentrism was embraced by both Hipparchus and Archimedes (Russo 2003: 78–89, 282–319) is flawed and unconvincing, e.g. Strato's theory of motion is more compatible with geocentrism than Russo allows, and many subsequent sources knew the work of Hipparchus and Archimedes yet never list them among the heliocentrists, despite the fact that they were far more famous than either Aristarchus or Seleucus, so their endorsement of the theory would have been too notable not to mention. I also do not believe Hipparchus has been correctly interpreted when Simplicius quotes him (indirectly from his lost *On Objects Carried Down by their Weight*) that bodies are heavier the higher they are (e.g. Wolff

also wrote lost works on optics, developing an atomist theory of light that was close to correct, though not universally adopted.[409] Most importantly, we know Hipparchus collected a large body of observational data, from others as well as more accurate observations of his own, and used this to greatly improve both Eratosthenes' geography and Apollonius' theory of the solar system. He also introduced crucial innovations, including the first full formal development of trigonometry, and the practice of using extensive eclipse records (mostly accumulated in Babylonia over many centuries) to test models of solar and lunar movement, and likewise testing theories of planetary movement against Babylonian records of planetary positions across recorded constellations.

Although evidence suggests he might not have completed this project, it did lead him to the most astounding discovery of the precession of the equinoxes, a finding so profound some believe it transformed Hellenistic religion.[410] This discovery was also made possible by detailed records of

---

1987: 100–05 and Wolff 1988: 489 n. 19, in reference to Simplicius, *Commentary on Aristotle's 'On the Heavens'* 1.8.265.10)—I suspect (for reasons too numerous to list here) that Hipparchus was actually speaking of the impact weight (we would say 'force') of dropped objects, not their static weight at elevation, hence following Strato's discovery that falling objects accelerate regardless of mass (thus Alexander of Aphrodisias, as paraphrased in Simplicius, *Commentary on Aristotle's 'On the Heavens'* 1.8.265.29–266.4, clearly did not understand the Hipparchean explanation of the acceleration of falling objects, ibid. 1.8.264.25–265.6, nor did Simplicius, who had clearly never read Hipparchus himself). Though Wolff 1988 attempts to defend the interpretation of Alexander and Simplicius, I believe there are flaws in Wolff's argument as well, which I may explore in future.

409. Plutarch, *On the Face that Appears in the Orb of the Moon* 4 (= *Moralia* 921d-e). See discussion in section 3.5 (and related note in chapter 2.7).

410. Influencing Mithraism: Ulansey 1989; influencing messianic Judaism and thus Christianity: Charlesworth 1978. On the science of precession see Russo 2003: 315–16. "Precession" is the result of the earth's slow wobble (like a rotating top, though Hipparchus might have assumed it was the sphere of the stars that wobbled), which results in a regular shift in the observed positions of stars. As a result, the pole of the sky rotates in a circular arc over a period of roughly 26,000 years, with the effect that a solar year will begin with the rising of a different constellation roughly every 2200 years. This meant astrological signs shift on the calendar, an appalling yet revolutionary fact for astrologers, hence influencing all astral religions (and Pliny the Elder, *Natural History* 2.24.95, suggests Hipparchus wrote works on astrology).

star positions (and their times of rising and setting) that had been made by earlier colleagues in the same century, including Aristyllus and Timocharis (about whom we know little and none of whose works were preserved) and Conon, an elder friend and correspondent of Archimedes, who also collated records of eclipse reports from Egypt—though again nothing he wrote survives, even though we know he wrote extensively on astronomical theories and observations and was regarded in antiquity as one of greatest astronomers of all time.[411]

All these data and innovations made it possible for Hipparchus to develop the first method of predicting lunar and solar eclipses reliably and accurately.[412] He was also able to use precise observations of eclipses made simultaneously from different locations on earth to recalculate the size and distance of the moon, and the distance of the sun. With this method he estimated the moon must lie between 59 and 67 earth radii (Ptolemy would later confirm it must be nearer 59). Hence Hipparchus, later improved by Ptolemy, had "arrived at a value for the lunar mean distance that was not only greatly superior to earlier estimates but was also stated in terms of limits that include the true value (about 60 earth radii)."[413] He also calculated the diameter of the moon as roughly a third the earth's, which is in the right ballpark (it is closer to a fourth), and though his estimates for the size and

---

411. Aristyllus: *DSB* 1.283; *EANS* 155–56. Timocharis: *OCD* 1483; *EANS* 812–13. Conon of Samos: *DSB* 3.391, *EANS* 486, and *OCD* 361 (s.v. "Conon (2)"); cf. Seneca, *Natural Questions* 7.3.3. The achievements of Hipparchus all but eclipsed his own contemporaries, hence we know very little about them, like Leptines (unknown but for a papyrus fragment of his introduction to astronomy, with illustrations, written around 165 B.C., cf. *NDSB* 4.271–72, *EANS* 505, and Evans & Berggren 2006: 10–12, 79); or Hypsicles of Alexandria: *DSB* 6.616–17 (with *DSB* 15.210, in s.v. "Hipparchus"), *EANS* 425, *OCD* 718, and Evans & Berggren 2006: 74, 79–80.

412. Either Hipparchus or subsequent astronomers before the 1st century A.D. could make eclipse predictions down to the hour according to Pliny the Elder, *Natural History* 25.5.10, although Ptolemy's system (perfected a century after Pliny) made prediction easier and more accurate.

413. G.J. Toomer (in *DSB* 15.215, s.v. "Hipparchus"). The ratio holds regardless of the true earth radius (about 4000 miles), but Hipparchus was working from a slightly high value (roughly 4600 miles according to Eratosthenes), so 59 to 67 radii translated then to an absolute value for the earth-moon distance of 271,000 to 309,000 miles. The actual distance varies from 221,000 to 252,000. Remarkably close given the instruments available.

distance of the sun were incorrect, he was aware of their inaccuracy, and was at least correct that the sun was many times larger than the earth and far more distant than the moon.[414] Hipparchus also recorded data on the rising times of stars and produced the first scientific star chart, and with it the first mathematically arranged star globe, and developed a system for calculating the time on any given night of the year using precise observations of star positions. He may also have invented (and certainly used) a simple diopter, something comparable to a plane astrolabe, and methods of stereographic projection.

Overall, the evidence is clear that Hipparchus embraced an "attitude toward astronomy as an evolving science that would require observations over a much longer period before it could be securely established," not only by assembling the work of previous astronomers over many centuries, but by recording his own observations and thus "assemble observational material for the use of posterity."[415] In this he set a standard that subsequent astronomers would follow. Theodosius of Bithynia in the late 2nd century B.C. wrote practical treatises based on the findings of Hipparchus and an important book on the geometry of the sphere, which laid major groundwork for Menelaus centuries later (see below). Theodosius also applied the findings of Hipparchus to develop the first portable sundial.[416]

---

414. Though it is worth noting that his estimate of solar distance (2500 earth radii) and size (1880 earth volumes) are still impressive. This equated to over a million miles distant (actual is about 93 million) and over a hundred thousand miles in diameter (actual is about 870,000). These figures are given in Theon of Smyrna, *Aspects of Mathematics Useful for Reading Plato* 3.39.197 (though Theon does not state the Hipparchean value for the earth-sun distance, we can deduce it from the method and figures Theon records; Cleomedes, *On the Heavens* 2.1, gives a different amount, but Theon's more detailed report is more credible than this passing remark by Cleomedes, which was more vulnerable to error or textual corruption).

415. G.J. Toomer (in *DSB* 15.220, s.v. "Hipparchus"). See relevant example in section 3.9.II.

416. *DSB* 13.319–21, *EANS* 789–90, and *OCD* 1459 (s.v. "Theodosius (4)"). The technology of portable sundials would be greatly advanced under the Romans, who developed versions the size of a human thumb that could determine the hour of the day, at any time of year, for a variety of latitudes (the Roman invention of all-latitude sundials is attributed to "Andrias," a name possibly garbled in Arabic translation: *EANS* 77). See Arnaldi & Schaldach 1997, which includes a historical

Then in the early 1st century B.C. Diodorus of Alexandria built on the work of Theodosius and Hipparchus by writing an influential treatise on the construction of the plane astrolabe.[417] And by then, possibly after examining Babylonian records, astronomers like Apollonius of Myndus had begun arguing systematically that comets are not an atmospheric phenomenon as Aristotle had concluded, but actual stellar bodies with very wide, eccentric orbits, a view that gained currency into the Roman period.[418]

But the most famous astronomer around this time was the Stoic philosopher Posidonius, who lived from the late 2nd up to the mid-1st century B.C., befriending both Cicero and Pompey.[419] Strabo regarded

---

discussion and a recovered example. For other kinds of portable sundial see Dilke 1971: 70–73. Several geared sundial calendars have also been recovered of Byzantine date (M.T. Wright 1990) whose design could long predate extant finds. Their technology is similar to the Antikythera computer (Evans 1999: 267–70), and some are mechanically adjustable for latitude, so some ancient references we have to advances made in portable sundials could refer to devices like these.

417. *OCD* 455 (s.v. "Diodorus (4)"), *NDSB* 2.304–05, and *EANS* 247, not to be confused with the historian from Sicily. Only a fragment of the Alexandrian's work was preserved and only in Latin and Arabic, plus scattered quotations (see discussion in Edwards 1984: 152–82).

418. The idea that comets could be planetary bodies precedes even Aristotle, who argued against it (cf. Aristotle, *Meteorology* 1.4–7), but subsequent defenses of it became more sophisticated, most notably, around this time, in the lost works of Apollonius of Myndus (*EANS* 114; Seneca, *Natural Questions* 7.4 and 7.17, reporting in 7.19 that some Stoics agreed). Whatever its origin, a nearly correct theory of comets evolved and continued into the Roman period: see Diodorus Siculus, *Historical Library* 15.50.3; Manilius, *Astronomy* 1.867–75; Pliny the Elder, *Natural History* 2.23.91 and 94; and Ammianus Marcellinus, *Deeds of the Divine Caesars* 25.10.2. It was verified and defended by Seneca (cf. *Natural Questions* 7.22), who combined past records with his own observations of comets in 54 and 60 A.D. (cf. e.g. *Natural Questions* 7.17, 7.21.3–4, 7.23.1, 7.26.2, 7.28.3–7.29.3, etc.) and described the most advanced cometary theory of his time, very close to the modern view, in books 2 and 7 of the *Natural Questions*. See Heidarzadeh 2004; Keyser 1994; and Kidd 1988: 490–96 (with 1999: 184–88), as well as the background provided in *DSB* 12.309–10 (s.v. "Seneca, Lucius Annaeus").

419. *DSB* 11.103–06, *EANS* 691–92, and *OCD* 1195–96 (s.v. "Posidonius(2)"). See also Edelstein & Kidd 1989, Kidd 1988, and Kidd 1999. In the *DSB*, Warmington's cynical conclusion that Posidonius was not influential is wholly untenable in light of copious evidence of his broad influence in literature throughout the early Roman

him as the greatest philosopher of his time.[420] Galen called him "the most scientific of the Stoics" because of his use and mastery of mathematics.[421] And Seneca said he should be counted "among those who have contributed the most to philosophy."[422] Originally a Syrian Greek from Apamea on the Orontes, Posidonius studied and later taught in Athens, lectured in Rome, and became a prominent citizen of Rhodes. Some scholars now argue he was "determined to bring Stoicism back to the empirical sciences" and "cut it free from overdoctrinaire scholasticism," hence engaging in hands-on scientific research, including making maps, observing the tides, measuring the size of the earth, and building astronomical computers. Though Peter Green thinks "he was on his own" in his passionate scientific investigation of natural causes, this is unlikely.[423] Though men who shared his methods and interests were certainly rare, and most philosophers were indeed more concerned with moral theory and armchair reasoning, we have already seen why we cannot make an argument from silence when it comes to the existence and activities of ancient scientists—our sources have disproportionately preserved the discussions of moralists almost to the exclusion of mentions of the activity, methods, interests, and sometimes even the findings of ancient scientists.

But however unique Posidonius may have been, he could be called the next Aristotle, a star example of the revival of a scientific spirit in the Roman era, which makes the loss of all his writings such a tragedy. We know of at least thirty books by him, and there were no doubt more, but none were

---

empire, from Strabo, Livy and Diodorus, to Seneca, Plutarch, Pliny the Elder and Galen—and many others, as Warmington's own notes ironically demonstrate. Kidd's more moderate conclusion in the *OCD* is more reasonable, and well supported by evidence in Kidd 1999.

420. Strabo, *Geography* 16.2.10.

421. Galen, *On the Doctrines of Hippocrates and Plato* 8.1.14, using the word *epistêmonikôtatos*, the superlative of *epistêmonikos*, "capable of knowing, scientific" (*LSG* 660).

422. Seneca, *Moral Epistles* 90.20.

423. P. Green 1990: 644 (for Green's overly cynical picture of Posidonius in general: 642–46, 596–97). For more general aspects of the relationship of Stoicism to science see references in *DSB* 14.605–07 (s.v. "Zeno of Citium"), *EANS* 846–47, and *OCD* 1403–04 and 1587–88 (s.v. "Stoicism" and "Zeno (2)"). See also Carrier 2016 (index, "Stoicism").

preserved—all we have are quotations, paraphrases, or references in later authors. The subjects he covered ranged across the whole gamut of logical, moral and natural philosophy, including books on astronomy, meteorology and climatology, earthquakes and lightning, seismology and volcanology, mathematics, geography, oceanography, zoology, botany, psychology, anthropology, ethnology and history. He might also have contributed to medical theory or even mechanics.[424] He had some knowledge of lenses and magnification and may have begun research on the subject.[425] He also wrote

---

424. Medicine: Kudlien 1970: 16 and Rawson 1985: 178, though this is disputed by Marasco 1998: 44–46 (and others cited there). Mechanics: Cuomo 2001: 164, though this is questioned by Kidd 1988: 714–16 (= F199b).

425. Strabo, *Geography* 3.1.5 (with Kidd 1988: 464 and Edelstein & Kidd. 1989: 115), which contains a reference to knowledge of lenses that magnify through refraction, attributed to Posidonius—in a discussion of atmospheric refraction (cf. also Cleomedes, *On the Heavens* 2.6 and Sextus Empiricus, *Against the Professors* 5.82) that bears comparison with later research by Ptolemy on exactly the same subject (see notes below). A century later Seneca mentions in passing lenses that magnify well enough to assist reading (*Natural Questions* 1.6.5–7). No scientific treatise on the subject survives from antiquity, although missing sections of Ptolemy's *Optics* may actually have included it (cf. Russo 2003: 331, with A.M. Smith 1996: 47–49), and there is archaeological and literary evidence that Romans may have started to experiment with lenses and magnification. See Dillon 1970 (with Kisa 1908: 355–59, Trowbridge 1930: 182–83, and Healy 1999: 147–50), Bastomsky 1972, Sines & Sakellarakis 1987, Enoch 1998 (with James & Thorpe 1994: 157–61), and Draycott 2013. Though skepticism is maintained by Plantzos 1997 and Krug 1987 (who correctly rebut, among other things, the notion that Nero had spectacles, though he may have used monocular or binocular sunshades; cf. also Disney et al. 1928: 43–65, though much of that is obsolete), and magnifying glasses may have been unknown to Galen, though he had seen microscopic art (*On the Uses of the Parts* 17.1 = M.T. May 1968: 731), which was not uncommon (e.g. Pliny the Elder, *Natural History* 7.21.85), as also microscopic texts (ibid. and Millard 2000: 169–70). On early references to using lenses to start a fire (e.g. Aristophanes, *Clouds* 768–75; Aristotle, *Posterior Analytics* 1.31.88a14–17; Theophrastus, *On Fire* 73) see Trowbridge 1930: 178–80, a property employed in Roman medicine to cauterize wounds (Pliny the Elder, *Natural History* 37.10.28). For a possible optical cauterizer recovered from antiquity see Plantzos 1997: 460. Likewise, magnifying *mirrors* were certainly well known and in use, both to magnify and burn, and their principles were scientifically understood (e.g. Plutarch, *On the Face that Appears in the Orb of the Moon* 17 and 23 = *Moralia* 930b and 937a; Seneca, *Natural Questions* 1.15.7–1.16.8; Pliny the Elder, *Natural History* 33.45.128–129; and sources cited in

histories and on scientific and descriptive geography, and a treatise on *The Art of War* that was so advanced the Roman tactician Arrian would later complain it was only useful to experts.[426] With all this scientific, historical and philosophical work Posidonius truly exemplified the Peripatetic tradition in Stoic form.

Posidonius was also an accomplished astronomer. He empirically confirmed the previously-developed theory that the tides were caused by the movement of the moon and sun, and further discussed estimates of the sizes and distances of earth, moon, and sun.[427] His own measurement of the circumference of the earth, about 20,500 miles, though less accurate than Eratosthenes' result, was adopted by subsequent experts, and was still not very far off.[428] Posidonius also built a machine that replicated the movement of the seven known planets.[429] Cicero's description of this device certifies

---

previous notes).

426. Arrian, *Art of War* 1.1.

427. See discussion of Seleucus above (tide theory), including subsequent note (size of the earth), and earlier note (sizes and distances of sun and moon). Posidonius speculated several different estimates for the size and distance of the sun and moon, but his best were: 57 million miles for the distance of the sun (actual: 93 million), 344,000 miles for the diameter of the sun (actual: 870,000), 229,000 miles for the distance of the moon (essentially correct, which casts doubt on the claim that Posidonius found the moon's diameter to be 4500 rather than its actual 2200 miles: see Cleomedes, *On the Heavens* 1.7, 2.1, and 2.3, and Pliny the Elder, *Natural History* 2.21.85).

428. Besides 20,500 miles, Posidonius also said the earth could be as much as 27,500 miles in circumference, a result actually *closer* to the truth than Eratosthenes'. For the best account of what happened to lead subsequent experts to prefer the lower value see M.J. Lewis 2001b: 143–56 and 332–34 (substantially correcting Taisbak 1974), who also demonstrates that no one in antiquity believed these figures were anything more than approximate and hypothetical. Ptolemy, for example, explicitly said he invented the system of latitude and longitude (essentially the same one we use today, although with the mean line now moved from Alexandria to Greenwich) precisely to bypass the problem of not having an accurate measure of the earth's diameter (and he called upon future scientists to therefore develop better measures of it).

429. Cicero, *On the Nature of the Gods* 2.34. Novara 1996 provides a literary-historical analysis of this passage. The seven known planets were: earth, moon, sun, Venus, Mars, Jupiter, and Saturn.

it was a proper orrery (a luniplanetary armillary sphere)—a machine that represents the solar system in three dimensions, in rings that can be rotated to reproduce the actual relative motion and position of the seven planets over time.[430] This was probably a significant improvement on a similar machine Archimedes had built over a century before; Posidonius would have known of important corrections and improvements to planetary theory developed after him.[431]

An armillary sphere is essentially an analog computer, since it allows analogous mechanical motion to compute and thus 'predict' astronomical

430. Various kinds of armillary sphere were constructed during and after the Renaissance—and no doubt in antiquity, as various texts and depictions on ancient coins, reliefs, and mosaics attest. On ancient astroglobes and armillaries see Evans & Berggren 2006: 27–34 and 47, Beck 2006: 120–28, Evans 1999: 237–43, Murschel 1995, and Aujac 1993: 157–78. Arnaud 1984 extensively surveys the iconography of globes in ancient art (on armillary and other astronomical spheres specifically: 59–77). For the best recovered depiction of an armillary sphere, from an early Roman villa in Solunto, Sicily (late 2nd or early 1st century B.C.): von Boeselager 1983 and Evans & Berggren 2006: 32. Most armillary spheres did not include the planets, but only the position of ecliptics and other astronomical lines with respect to an observer's position on earth, adjustable by season. Few added the movement of the sun or moon. Very few did include all the planets (thus luniplanetary). Though we have no visual representations of these from antiquity, we have literary descriptions. Some of these machines were automated by water power, as attested in Galen, *On the Uses of the Parts* 14.5 (= M.T. May 1968: 627) and Pappus, *Mathematical Collection* 8.2.1026.

431. Mentions or descriptions of these and similar orreries and armillary spheres include: Cicero, *On the Republic* 1.14.21–23, *Tusculan Disputations* 1.25.63, *On the Nature of the Gods* 2.34.87–88; Ovid, *Fasti* 6.270–77; Geminus, *Introduction to Astronomy* 5.62–63, 6.21, 12.23, 12.27, 16.10–12; Aulus Gellius, *Attic Nights* 3.10.3; Galen, *On the Uses of the Parts* 14.5 (= M.T. May 1968: 627); Sextus Empiricus, *Against the Professors* 9.115; Lactantius, *Divine Institutes* 2.5.18. See also Rawson 1985: 163 and Simms 1995: 53–55. In late antiquity: Claudian, *Epigrams* 51.68; Macrobius, *Commentary on the Dream of Scipio* 2.15; Martianus Capella, *The Marriage of Philology and Mercury* 8.815; Julius Firmicus Maternus, *Eight Books of Astrology* 1.pr.5 and 6.30.26. On the most advanced armillary spheres in antiquity: Ptolemy, *Almagest* 1.6, 5.1, 13.2. See also Nicolaus Copernicus, *De Revolutionibus* 2.14. It is not known what the "Billarus Sphere" captured in Sinope by Lucullus in the early 1st century B.C. was or when it was made (Strabo, *Geography* 12.3.11; *EANS* 192), but it probably would not have been singled out as remarkable unless it was some kind of armillary sphere.

events. It is also possible Posidonius constructed a dial computer, a kind of astronomical clock, which indicates planetary positions (and even lunar phases and other data) two-dimensionally, through a gear-driven dial readout. We have actually recovered one of these devices, built shortly before 100 B.C., from a ship that sank in the 80's B.C. near the island of Antikythera (ancient Aigila), near the sea routes from Rhodes to Athens and Athens to Rome, exactly where and when Posidonius is known to have traveled and had many contacts. The astrological terms and function of the device accord very well with Posidonius' known penchant for astrology, and if he did not design or build it, then there was clearly a lot of scientific work going on around him that we have otherwise heard nothing about. From abundant circumstantial evidence some scholars have plausibly concluded this ship contained a cache of Sulla's loot en route to Rome from his sack of Athens in 86 B.C. But regardless of whether this was built or owned by Posidonius, it is definitely an astronomical computer for calculating the positions and conjunctions of the sun and moon and planets, with brilliantly crafted gearing, confirming that such computers existed, and had achieved remarkable precision and sophistication for the age, another example of how arguments from silence about what ancient scientists accomplished are not so tenable.[432]

Posidonius is an important example of the growing revival of science in the Roman era. As Dobson observes, "it is important to note that Posidonius, whom Galen calls the most scientific of the Stoics, aimed, so far as he could, at accuracy in material observations" and "from very numerous quotations in Strabo and others, we gather that he was at any rate a scientist of very considerable repute."[433] According to Ian Gray Kidd, his methodology adapted deductive methods from the geometry of Euclid to develop a complete natural philosophy, using the empirical sciences as a source of verified premises. Kidd concludes that "Posidonius' position in intellectual history is remarkable." He very much followed the ideals of Aristotle, making "an audacious aetiological attempt to survey and explain the complete field

---

432. On this now-famous Antikythera mechanism see Marchant 2010 (for a popular account) and (for expert analysis) A. Jones 2016 and 2017, Sticks 2014, and Hannah 2009: 59–67. Additional specialist literature: Edmunds 2011; M.T. Wright 2007; Freeth et al. 2006; Freeth 2002a and 2002b; Economou 2000.

433. Dobson 1918: 189–90.

of the human intellect and the universe in which it finds itself as an organic part," pursuing a thoroughgoing "analysis of detail and the synthesis of the whole, in the conviction that all knowledge is interrelated."[434]

Soon after Posidonius, in the middle of the 1st century B.C., there were several scientists whose writings are completely lost. The Alexandrian astronomer Sosigenes was recruited to reform the Roman calendar on a scientific basis, and we know he was making his own planetary observations, yet none of his books survive.[435] Some pupils of Posidonius were also busy with scientific research. Asclepiodotus wrote a lost work on *Investigations of the Causes of Natural Phenomena*, probably on seismology and volcanology (if not other subjects as well), and Athenodorus wrote a lost work *On the Oceans*, which included the most sophisticated account of lunisolar tide theory then known.[436] The astronomer Serapion of Antioch attempted his own calculations of the size of the sun and wrote scientific critiques of Eratosthenes' *Geography*, although none of his books survive either.[437] Then around the turn of the era Strabo produced his own cartographic geography by integrating and selecting from the works of previous geographers, from Eratosthenes to Hipparchus and Posidonius—and Artemidorus of Ephesus, who also composed an advanced but now-lost geography of the world a century before Strabo.[438] Though Strabo more heavily emphasized literary, descriptive and historical geography, he may have seen himself as merging

434. *OCD* 1196 (in s.v. "Posidonius (2)").

435. Rawson 1985: 112–13, *EANS* 752, and *OCD* 1385, with Feeney 2007 and Pliny the Elder, *Natural History* 2.6.39 and 18.57.211–212.

436. Asclepiodotus is known for an extant (and rather sophisticated) treatise in military tactics (*OCD* 180; *EANS* 172). His lost work on natural causes provided scientific material on earthquakes and volcanoes for Seneca's *Natural Questions* (cf. 2.26.6, 2.30.1, 5.15.1, 6.17.3, 6.21.2; and discussion in Kidd 1988: 30–33), although it may have treated other subjects, and there is no telling what else he wrote books on—we are lucky even to know of these. On Athenodorus see Strabo, *Geography* 1.1.12; with *OCD* 195, and *EANS* 179.

437. *OCD* 1352 (s.v. "Serapion (2)"); *EANS* 733 (s.v. "Serapion of Antioch"). Hipparchus also wrote such a critique, and Cicero had read all three (Eratosthenes, Hipparchus, and Serapion: Cicero, *Letters to Atticus* 2.6.1). Serapion's size of the sun was a sixth of the correct value.

438. Artemidorus: *EANS* 165. For the others, see previous discussions and references.

a total field of human intellect, acknowledging scientific detail as no less important than history and ethnology. Strabo relied on the reports of several explorers before him and conducted his own explorations in eastern regions, from Armenia to the Nile. He also included much of what could be called scientific geology, if we mean 'scientific' in the ancient sense of empirical, observational, and rational, though still pre-modern and as often wrong as right.[439] Around the same time, Geminus wrote popularizing summaries of astronomy, meteorology, optics, and theoretical mathematics, probably drawing a good deal on the teachings of Posidonius, but frequently introducing his own judgment and criticism.[440]

The next most important contribution to astronomy came from Menelaus of Alexandria in the late 1st century A.D., who developed a completed spherical trigonometry, and was the first to put plane trigonometry on its own theoretical footing.[441] This was a crucial advance on Hipparchus, which Ptolemy would make good use of.[442] Menelaus was improving on his predecessor, Theodosius of Bithynia, who had written important works in observational astronomy around 100 B.C. that helped him develop an advanced spherical geometry.[443] Menelaus combined this

---

439. *DSB* 13.83–86, *EANS* 763–64, and *OCD* 1404–05. Theophrastus began this research (see section 3.5) and Strabo confirms it was continued by others like Eratosthenes and Posidonius.

440. Of these, only the *Introduction to Astronomy* survives. See *DSB* 5.344–47, *EANS* 344–45, *OCD* 607, and most importantly: Evans & Berggren 2006. On dating Geminus see note in chapter 2.7. We also have some less technical astronomical textbooks from (or shortly before) the 2nd century A.D. Besides Cleomedes (see note below) we have similar textbooks from Hyginus (*OCD* 714, first or third entry; *EANS* 454), Theon (*OCD* 1460, second entry; *EANS* 796), and Achilles (*OCD* 7–8, first or second entry; *EANS* 51–52).

441. On Menelaus: *DSB* 9.296–302 and 15.420–21, *EANS* 546, and *OCD* 932 (s.v. "Menelaus (3)") as well as *OCD* 1507 (s.v. "trigonometry") and Russo 2003: 52–55 and Van Brummelen 2009. Menelaus also wrote a handbook on geometry, and probably others unknown to us. A papyrus fragment of Menelaus' work on astronomical theory probably survives (cf. A. Jones 1999).

442. *DSB* 15.209 (in s.v. "Hipparchus").

443. See Evans & Berggren 2006: 7 (and references on Theodosius cited earlier). In turn, Autolycus and Euclid wrote treatises on spherical geometry that Theodosius improved upon.

with the trigonometric foundations of Hipparchus. Menelaus may have studied at Alexandria, but we know he made astronomical observations at Rome that confirmed Hipparchus' discovery of the precession of the equinoxes. He was famous enough for Plutarch to use him as a paradigmatic example of a contemporary accomplished astronomer.[444] None of his works were preserved in Greek, and only a few in Arabic, though we know he wrote several treatises on geometry, and an observational study on the setting times of major stars and constellations. Menelaus also wrote on mechanics (see next section).

Next came the crucial work of Marinus of Tyre at the end of the 1st and start of the 2nd century A.D. This Marinus was a key intermediary in the progress of geography and cartography, developing improved methods of cartographic projection and further employing astronomical and other data to construct accurate maps of the Roman empire.[445] Roger Batty rightly identifies him as "one of antiquity's least known but most influential authors," concluding that "his works were innovative and widely utilized—and in the history of ancient geography, they proved to be quite decisive." Ptolemy built on but considerably improved his work. Marinus and Ptolemy both stand in contrast to hack geographers like the early 1st century writer Pomponius Mela, who composed (in Latin) little more than a tourist's handbook of mixed quality for the known ports and coastal lands of the time.[446] The fact that medieval Christian scribes preserved Mela's work but hardly even a mention of the more scientific work of Marinus is symptomatic of the problem facing the study of Roman science. This medieval preference for simple, fabulous, amusing, or entertaining work, over the boring but otherwise technically superior scientific books on the same subjects (exemplified perfectly here by the very different fates of Marinus and Mela) leaves a skewed picture of the state of ancient science. Imagine if countless books by hacks, quacks, fabulists, and spiritualists of the 19th century had been preserved, but almost nothing by scientists of that era. Any picture of modern science we based on such a state of evidence would not be accurate.

All this accumulated to produce the pinnacle of geography,

444. Plutarch, *On the Face that Appears in the Orb of the Moon* 17 (= *Moralia* 930a).

445. On Marinus of Tyre see Batty 2002, Berggren & Jones 2000: 23–25, Dilke 1985: 72–75; and *NDSB* 5.27 and *EANS* 533.

446. See Appendix A for a brief history of ancient exploration.

cartography, and astronomy in the ancient world, accomplished by Claudius Ptolemaeus.[447] Working at Alexandria in the mid-2nd century A.D., Ptolemy is best known for his *Mathematical Syntaxis*—also known from the Arabic as the *Almagest*—even though this was among his earliest achievements. The *Almagest* presents a complete description of the most accurate, and at the time most defensible, model of the planetary system, including how to establish its parameters from observations, and how to use all this to make good predictions of astronomical phenomena—especially regarding the size, time, and duration of solar and lunar eclipses for any given location on earth, which Ptolemy's system could do fairly well. The *Almagest* represents not only the first real systematization of the work of Apollonius and Hipparchus, but also a considerable improvement on them in key details. Based on new observations, including many made by Ptolemy himself, problems had been found with Hipparchus' work, many of which Ptolemy solved (although his own work was not free of error either). One of his most novel innovations was the introduction of a system of double eccentricities that entailed inconstant velocities for the planets, with an equal-angles-in-equal-times law of planetary motion, coming remarkably close to Kepler's elliptical model.[448] G.J. Toomer rightly concludes "the *Almagest* is a masterpiece of clarity and method, superior to any ancient scientific textbook and with few peers from any period."[449] Thus, again, science reaches its acme under the Romans.

Ptolemy also produced a more detailed star catalogue than Hipparchus had begun, listing the longitude, latitude, and magnitude of over a thousand stars in nearly 50 constellations, many confirmed by his own observations—though most were merely remapped by calculations from preceding work, based on his confirmed sample. Ptolemy also composed the *Astrological Influences* (or *Tetrabible*, describing his quasi-scientific system of astrology), the *Phases of the Fixed Stars* (collecting previous data on the rising and

447. For the following: *DSB* 11.186–206; *NDSB* 6.173–78; *EANS* 706–09; *OCD* 1236–38 (s.v. "Ptolemy (4)"); A.M. Smith 1996: 1–5; Riley 1995; Lloyd 1973: 113–35.

448. The only missing element was the ellipse (and the heliocenter), which may have been avoided for more practical reasons than is usually claimed (including the need to simplify calculation and build computer models of the solar system in the form of armillary spheres): Russo 2003: 89–93.

449. *DSB* 11.196 (in s.v. "Ptolemy").

setting times of stars, and combining this—as many astronomers did—with a weather almanac), the *Handy Tables* (an updated and more user-friendly collection of mathematical tables from the *Almagest*), the *Planetary Hypotheses* (a more speculative work of astrophysics, though including an updated discussion of planetary theory and methods for calculating planetary positions, sizes, and distances, and procedures for making corrections for any observation point on earth), *On the Criterion* (a brief treatise on scientific epistemology), some lost works in mechanics, as well as *Optics*, *Harmonics*, and his monumental cartographic work, the *Geography*. We will only discuss his achievements in astronomy and geography here, covering physics later.

Ptolemy's *Geography* is a phenomenal *tour de force* that eclipsed every previous effort at the task.[450] Not only does it reproduce an updated scientific geography, but it includes detailed geographic data and instructions for how to construct an accurate world map, in twenty-six interconnecting sections, while accounting for the challenges of projecting a spherical earth onto a flat surface, along with additional instructions for constructing cartographic globes.[451] It is clear Ptolemy greatly improved on the work of his recent predecessor Marinus of Tyre, especially in correcting errors and organizing information in ways that are more practical for the reader, and providing several different (and two better) methods of projection, explaining the disadvantages of each. Ptolemy himself invented systems of conic projection and (what would later be called) Bonne projection. Toomer concludes that Ptolemy "took a giant step in the science of mapmaking"

450. Well discussed in Berggren & Jones 2000 and A. Jones 2012. Though Ptolemy ultimately preferred the lower (and less accurate) value for the circumference of the earth (which ultimately derives from Posidonius—rather than Eratosthenes, whose value Ptolemy appears to have relied on in his earlier *Almagest*), he explicitly argues that this value needed revision through more accurate observations (which was true even for Eratosthenes' measurement) and thus should be accepted only provisionally (hence criticism of Ptolemy on this point, e.g. Russo 2003: 69–70, 273–77, is often unjustly excessive). And yet this value would not be improved upon until Muslim scientists followed Ptolemy's advice in the 9th century (M.J. Lewis 2001b: 156).

451. That Ptolemy actually made maps and globes and had considerable skill and experience with this is shown in Berggren & Jones 2000: 46–48.

that "had no successor for nearly 1,400 years."[452] This despite the fact that Ptolemy knew his data (and thus his maps) could be greatly improved with the accumulation of astronomical observations from a wider variety of additional locations, since he was forced to rely on sparse data from only a very few places (since at the time the major historical centers of serious astronomical research could probably be counted on one hand).

Ptolemy was clearly aware of the fact that he was improving on past work.[453] And not only did he make progress on the work of his predecessors, he made progress on his own work. For instance in the *Almagest* he attributed the apparent enlargement of the moon, sun and planets when near the horizon to physical causes, but in the *Optics* to psychological causes, apparently having confirmed in the interim that the observed enlargement is only apparent and not actual.[454] And though, according to Lloyd, "Ptolemy himself was well aware that problems, some of them serious, remain[ed] unsolved," and according to Toomer, "Ptolemy himself regarded his work as provisional," and explicitly said some of his findings or data were much in need of improvement by his successors, these statements of his were subsequently ignored.[455] Just like Galen's work in biology, Ptolemy's work in astronomy (and physics) "was treated as definitive" by later astronomers, despite errors in certain fundamental variables that would have been obvious to almost anyone who made real observations as checks against his, indicating how poor astronomical science subsequently became, until Muslims sometime between the 9th and 14th centuries caught and corrected the most obvious of these errors. But even they made no major changes to his planetary model, and then dropped the whole project, leaving it to

452. Quote and other details: *DSB* 11.200 (in s.v. "Ptolemy").

453. See, for example, Ptolemy, *Almagest* 9.2.

454. Ptolemy also discusses the optical illusion created by horizon observations in *Planetary Hypotheses* 1.2.7. See Goldstein 1967: 5, 9 (with *OCD* 190, in s.v. "astronomy"), Lloyd 1982: 134–35, and A.M. Smith 1996: 2–3. Ptolemy first assumed it was an enlargement caused by atmospheric refraction, a phenomenon studied by Posidonius and later by Ptolemy himself (see related notes above and below; note this also means they knew refracting lenses could achieve magnification). More examples of Ptolemy revising his own theories in light of new evidence are surveyed in Lloyd 1982: 139–40 and Hamilton et al. 1987: 57 and 68.

455. Quote from G.J. Toomer and other points in this paragraph: *OCD* 190 (in s.v. "astronomy"). Quote from G.E.R. Lloyd: Lloyd 1981: 279.

Christian scientists of the 16th century to finally pick up where Ptolemy left off. But even without these corrections, with Ptolemy's work Romans could predict the courses, positions, eclipses and conjunctions of the moon, the sun, and every known star and planet, from any observation point on earth. Some inaccuracies remained, but the general model was remarkably good and unparalleled in history before the Scientific Revolution. Even the Copernican model was inferior in predictive success—it would take the innovations of Brahe and Kepler to produce an actual improvement in predictive power.

There was certainly other astronomical research going on around and near Ptolemy's day, probably as much as we hear from Galen in anatomical and medical research. Yet not much survives to tell us about it. Ptolemy was never as chatty as Galen when it came to describing contemporary affairs in his field. But we know, for example, of an older contemporary of Ptolemy, Theon of Smyrna, who wrote a Platonic handbook on the quadrivium, showing how arithmetic, music, astronomy, and geometry (including stereometry) are interrelated. He also made numerous astronomical observations that Ptolemy employed, leading scholars to suggest Theon may have been one of Ptolemy's teachers.[456] Like Theon, Cleomedes also appears to have been a full-fledged astronomer who made his own astronomical observations, although none of his original scientific work survives.[457] Meanwhile, another Sosigenes, the Aristotelian tutor of Alexander of Aphrodisias and a younger contemporary of Ptolemy, wrote lost works on optics and astronomy based on his own observations.[458] We

456. See *DSB* 13.325–26 (s.v. "Theon of Smyrna") and 11.187 (in s.v. "Ptolemy"), *EANS* 793 and 796, and *OCD* 1460 (s.v. "Theon (2)"). See also Evans 1999: 296–97. (That these two Theons were the same man is not certain; but their dates and interests do align.)

457. Cleomedes: *OCD* 331, expanded and corrected by *DSB* 3.318–20 and Bowen & Todd 2004: 1–4. Though his date has long been uncertain, the most recent and careful analyses place him somewhere between the 1st century B.C. and the early 2nd century A.D. Later dates have been suggested but are very improbable. A basic astronomical textbook aimed at laypeople is his only extant work (various titles are given but *On the Heavens* is most credible).

458. *OCD* 1385 (s.v. "Sosigenes (2)"); *EANS* 753. His extensive treatise on vision is mentioned in Alexander of Aphrodisias, *Commentary on Aristotle's Meteorology* 143.13. For a fragment of his astronomical work see Cohen & Drabkin 1948:

also have mentions of the astronomers Dioscorides Salvius and Apollinarius of Aizani, both probably of the late 1st century or early 2nd century A.D. Apollinarius, evidently a very renowned astronomer, compiled tables of astronomical data and wrote on eclipses, but we know little else about him or his work; and even less about Salvius.[459] Ptolemy employs a technical scientific observation recorded by the astronomer Agrippa of Bithynia in 96 A.D., yet except for this one mention of him we would not even know he existed—even though he must have written books for Ptolemy to have his data.[460] And Plutarch describes debates in astronomy around the same time that confirm there was much more going on.[461]

From this survey it is clear, once again, that progress in the astronomical sciences was constant, acknowledged, and evidently considered worthwhile, well into the Roman period.

3.4 Scientific Physics up to the Roman Era

For this section we mean by "physics" something closer to the modern sense of the word, as a study of the fundamental mechanical principles of the universe. In antiquity, "physics" in this sense saw significant scientific progress in the fields of harmonics and acoustics (the behavior of 'sound'), optics (the behavior of 'light'), and mechanics (which included the behavior of simple and complex machines, the principles governing architectural structures, and subjects like pneumatics, hydraulics, statics and hydrostatics).[462] Other important fields, such as dynamics (the study of

---

103–05.

459. Toomer 1985: 203–04. For Apollinarius: *OCD* 118; *NDSB* 1.82–83; *EANS* 105.

460. Ptolemy, *Almagest* 7.3; *EANS* 47. A. Jones 1999 discusses Agrippa and other 1st century astronomers.

461. Plutarch, *On the Face that Appears in the Orb of the Moon* 1–23 (= *Moralia* 920b-937c). See Russo 2003: 286–93, though he sometimes goes beyond what the evidence actually supports.

462. For general context and scholarship see Lloyd 1973: 91–112, Schürmann 1991: 33–59 and 2005, Vitrac 2009, and *OCD* 8–9, 291, 714, 917–18, 975–84, 1042, 1145–48, 1166–67, 1396 (s.v. "acoustics," "catoptrics," "hydrostatics," "mechanics," "music" esp. §5, "optics," "physics," "pneumatics," "statics"). Scenography (the

movement, especially in terms of velocity, energy, gravity, or force) did not experience as much advancement, and more obscure fields like magnetism and chemistry were never placed on a scientific footing at all (a question we will examine in the next section). Here we shall focus on the progress that *was* made.

As before, I will survey only known authors of scientific works that conveyed original discoveries. So, for example, I leave out of account the many authors in mechanics and engineering whose scientific contributions (if any) remain unknown, like the many known only from inscriptions, or mentioned by authors like Pliny or Vitruvius, or elsewhere. For example, we can't say anything useful about the otherwise unknown Abdaraxos "who built the machines at Alexandria," who appears in a fragmentary list of 'famous engineers' on a recovered scrap of papyrus.[463] Even though his "machines" were evidently so famous they needed no explanation, we know nothing at all about him or them, an example of how incomplete our understanding of the history of ancient science is. But even from what we know, much can be said.[464]

The most important physicist after Aristotle, born in the late 4th and working in the early 3rd century B.C., was the renowned Strato of Lampsacus, whose fame and role in astronomy and astrophysics I already discussed earlier.[465] As with his maverick physics of gravity and inertia, by merging Aristotelian with a more atomist physics Strato developed a theory of void and air pressure that, with some further developments, became central to

---

science of visual representation in the arts) and other practical applications were a formal part of optics, e.g. Camerota 2002; Russo 2003: 58–65; Evans & Berggren 2006: 45–46, 244, 248. For the connection between ancient acoustics and ancient music: Landels 1999: 130–47, 190–95. Ancient optics: A.M. Smith 2014: 25–129.

463. See Diels 1920: 29–31; M.J. Lewis 1997: 60–61; *EANS* 29. The papyrus here reads "*ho Abdaraxôs ho ta en Alexandreiai mêchanika suntellôn*" and includes other architects and military engineers, known and unknown.

464. I will only survey the best known. For a list of *all* known physicists and engineers in antiquity see *EANS* 993 (s.v. "architecture"), 1002–03 (s.v. "harmonics"), 1005–06 (s.v. "mechanics"), and 1012 (s.v. "optics"). For writers making advances purely in mathematics (advances which continued after Archimedes well into the Roman era, but almost all of which was not preserved; even the advances of Archimedes barely survived): *EANS* 1003–05.

465. For the following, see references cited for Strato in section 3.3.

engineering for the remainder of antiquity, even though his crucial treatise *On Void* is lost. This also led him to anticipate many developments in modern physics, such as his explanation of wind as caused by differences in air pressure produced by differences in air temperature, which he described in his lost treatise *On Wind*. His theories of light and sound, presented in his lost treatises *On Sound* and *On Vision*, expanded on atomist explanations, coming nearer the truth than any others in antiquity—though only his theory of sound was widely adopted. Strato was also the first philosopher, as Gottschalk says, "to use experiments systematically to establish" elements of his natural philosophy, and his methodology was nearly modern, for his "experiments are not isolated, but form a progressive series in which each is based on the result of the previous one." In fact, "characteristic of Strato are the care taken to define the conditions in which the experiment takes place and to eliminate all possible alternative explanations of the result" as well as "the practice of pairing controlled experiments with observations of similar phenomena occurring under natural conditions." By emphasizing the methodological standard of physical experimentation far more than Aristotle had done, Strato set the gold standard for all subsequent physicists. He also began scientific interest in technology, writing *On Mining Machinery* and *Examination of Inventions*.[466] Of course, nothing he wrote was preserved by medieval Christians, except in the quotations or paraphrases of later authors. Yet his experimental methods were picked up and used and promoted by Hero and other scientists of the Roman era.[467]

Surviving sources are too poor to be certain who discovered what, but several basic conclusions in physics had been empirically established before or during Strato's time, which he at least built and expanded upon. It was known, for example, that sound is a vibration transmitted through the physical medium of the air or other material, and that a sound varies according to its wavelength, both facts having been confirmed experimentally through the phenomenon of sympathetic vibration. Likewise, Strato may or may not

466. According to Diogenes Laertius, *Lives and Opinions of Eminent Philosophers* 5.59–60 (the latter is cited as simply *On Inventions* in Clement, *Stromata* 1.14.61 and 1.16.77). Others were writing similar books in the 3rd century B.C., e.g. *OCD* 1136 (s.v. "Philostephanus of Cyrene"; cf. *EANS* 660) and 1144–45 (s.v. "Phylarchus"). See Rihll & Tucker 2002: 287–89. Likewise, Oleson 2004, Dunsch 2012, and Greene 1992 argue there were a lot of technical manuals like these, all now lost.

467. Papadopetrakis & Argyrakis 2010; Boutot 2012.

have been the first to discover or prove that hot air expands and cool air contracts, but either way the fact itself became well-known—it was much relied on by the Roman engineer Hero, for example—though perhaps the strangest example of this knowledge is a Roman-era child's game described by Galen, in which pig bladders were filled with air that was then heated with friction to make the bladders expand into a balloon.[468] Likewise, research on the principles of mechanics was ongoing around the same time, but here, too, it is hard to discern what was discovered when. We would know more, for example, if we had the *Mechanics* written by Callistratus, though the extant pseudo-Aristotelian *Mechanical Problems*, by someone of the same era, gives us some clues as to the early stages of research on the physics of the lever.[469]

We know more about Strato's pupil Ctesibius, the son of a barber and himself a practical mechanic, who nevertheless received a good education and served in the Alexandrian Museum in the early 3rd century B.C.[470] Ctesibius wrote several treatises in physical science, himself launching the fields of pneumatics (the study of the compression and behavior of air and steam) and hydraulics (the study of the mechanical uses of water). Like Strato, he employed physical experiments to prove air is a body and that it is compressible, inventing the air-powered catapult (involving the first

---

468. Galen, *On the Natural Faculties* 1.7 (= Kühn 2.16–17). For examples of Hero's use of heat to cause air to expand and thus drive machinery see Hero, *Pneumatics* 1.12, 1.38–39, 2.3, 2.34–35.

469. Callistratus (*EANS* 466) was no idle philosopher but a working engineer, so his textbook on the subject would have been valuable (cf. Athenaeus the Mechanic, *On War Machines* 28.7–8, with Whitehead & Blyth 2004: 140). The Pseudo-Aristotelian *Mechanical Problems* survives in extant collections of Aristotle's works. For other engineers who may have written scientific books prior to the Roman era, see lists in Whitehead & Blyth 2004: 24–25 and Cuomo 2007: 61–62, though others are mentioned throughout *On Architecture* by Vitruvius (e.g. 1.1.17, 7.pr.14, 9.8.1–2, etc.; cf. M.J. Lewis 1997: 45–46), or in scattered sources elsewhere. Athenaeus, *On War Machines* 5.3, identifies Archytas of Tarentum (see Appendix B) and Hestiaeus of Perinthus (a pupil of Plato; *EANS* 391) as having written on mechanical theory, although nothing from them survives (see Whitehead & Blyth 2004: 68–69), unless Winter 2007 is correct that Archytas wrote the *Mechanical Problems* mistakenly attributed to Aristotle.

470. For the following: *DSB* 3.491–92, *EANS* 496, and *OCD* 396, as well as Vitruvius, *On Architecture* 9.8.2–7 and 10.7–8.

known air pump), the water organ (also operating on compressed air), the cylinder-and-plunger pump, and the reciprocating double-force pump that produces a continuous stream of water.[471] He wrote a scientific treatise on the construction and testing of artillery and other siege equipment that included his own advancements on existing technology. Neither his air-spring nor bronze-spring catapults were adopted, probably because they were impractical in battlefield conditions or not actually competitive with existing technology, but they were nevertheless ingenious.[472] He also wrote a treatise on his various inventions in other fields, including theatrical machinery. He also invented the continuous water clock, as well as a parastatic clock face that indicated the hours of different lengths according to the time of year, a design that required knowledge of mathematical astronomy. His other inventions included the cuckoo clock and various other mechanical automata. All this we know from later writers, for none of the writings of Ctesibius were preserved. But he clearly embraced hands-on experimentation and the application of mathematical theory to explain observations, and he established new areas of scientific research that became standard subjects of study for all future engineers.

A younger contemporary of Ctesibius was the engineer Biton, who wrote an extant booklet on the *Construction of War Machines and Catapults*, as well as a lost work on optics, and possibly other subjects, around the middle of the 3rd century B.C.[473] And Apollonius of Perga was writing on mechanical subjects around this time, but too little of his work in this field survives to assess.[474] When it comes to engineering we know more about Philo of Byzantium, who advanced the work of Ctesibius a generation after

---

471. On Hero's emulation of Strato's experimental methods, e.g. in demonstrating the corporeality of air: Papadopetrakis & Argyrakis 2010.

472. For the history of ancient artillery technology and why some devices won out and not others: Rihll 2007; Landels 2000: 99–132; DeVoto 1996; Marsden 1969 and 1971. See also D. Campbell 2011 and *OCD* 178 (s.v. "artillery") and 1364–65 (s.v. "siegecraft, Greek" and "siegecraft, Roman"). Cuomo 2007: 41–76 is also worthwhile, though she overdraws some conclusions and might err in some basic physics.

473. *OCD* 235; *EANS* 193–94. Marsden 1971: 5–6 makes a fair case that Biton wrote in the mid-3rd century B.C. rather than (as some have thought) mid-2nd century B.C.

474. See mention and note in section 3.3.

Biton.[475] Philo was a native of Byzantium (now Istanbul) and traveled to Rhodes and Alexandria, two leading centers of engineering, specifically to study catapult design, learning of the work of Ctesibius in Alexandria and of a certain Dionysius in Rhodes, who had invented the automatic catapult, which Philo rightly noted could not compete tactically, but its design demonstrated remarkable ingenuity. Toward the end of the 3rd century B.C. Philo wrote a comprehensive treatise on mechanics, of which only a few chapters survive, and half of those only in Arabic translation, but we know this book discussed the practical matters of using levers, constructing ports and fortresses, building torsion catapults, "besieging and defending towns," and other strategic material, including the use of poisons, cryptography, and his own invention of an optical telegraph. But the same treatise also discussed the science of pneumatics and the building of robotic theaters, and could have contained other material unrelated to warfare. He may have been the first to use geometric theory to develop the equivalent of scaling laws in engineering; these were certainly known to later engineers.

Then came Archimedes, a Sicilian engineer also of the late 3rd century B.C.[476] The son of an astronomer and a friend or relative of the Sicilian royal house, he is believed to have studied in Alexandria, but completed most of his work in Sicily, while still corresponding with several leading scientists in Alexandria, including Eratosthenes. Archimedes was the first to put the science of mechanics on a solid mathematical footing, discovering and demonstrating fundamental laws of mechanics, and establishing the fields of statics and hydrostatics, which consist of the study of equilibrium in systems of weight distribution and the scientific principles of floating and immersed bodies. The latter led to his invention of a procedure for calculating the density of different materials by immersing them in water to determine their volume and then comparing their weight to their volume. The law he developed in this procedure is even now called "Archimedes' principle," and density is still measured relative to water, the original medium he chose to employ.

475. For the following: *DSB* 10.586–89, *EANS* 654–56, and *OCD* 1133 (s.v. "Philon (2)"). On dating and contributions of Ctesibius and Philo, see Marsden 1971: 6–9.

476. For the following: *DSB* 1.213–31, *NDSB* 1.85–91, *EANS* 125–28, *OCD* 141–42, and Simms 1995 and 2005. Also M.J. Lewis 1997: 37–41 (and 137 n. 86); and Russo 2003: 25–27 and 70–75, who shows how Archimedes put mechanics on a scientific footing by refuting Aristotle, an example of the actual tendency in antiquity *not* to treat Aristotle as gospel, exactly opposite the behavior of medieval Christians.

Archimedes further advanced the geometrical mathematics of calculating volumes and surfaces, and wrote a treatise on method that linked empirical discovery to mathematical proof, specifically aimed at helping future mathematicians discover new theorems. He also wrote *On Spheremaking*, a manual for the construction of armillary spheres—the mechanical orreries that reproduce the motions of the sun, moon, planets and stars (as I mentioned in the previous section), for which Archimedes became famous, though like most of his work, this book does not survive.[477] He invented several other machines, including (at least by ancient accounts) Archimedes' screw (a rotating helical water pump) and a compound pulley, which eventually became standard equipment for Roman engineers, and he may have written on these in his lost mechanical writings. He might also have written on the water organ and on odometer and waterclock design.[478] More famously, he built an arsenal of mechanical artillery for the defense of Syracuse against the Romans that gave their legions a rare black eye.[479]

Several of Archimedes' mathematical treatises were preserved, though many only in Arabic (including his treatise on the construction of waterclocks) and many of them only in fragments, while even those that survive in Greek were altered, mutilated, or badly edited over time, and some have only recently been recovered from palimpsests (including Archimedes' treatise on method).[480] Among his books on mechanics are *Equilibrium of Planes* (on principles of weight and leverage) and *On Floating Bodies* (on hydrostatics). But many of his works were not preserved at all, including his treatises on optics and catoptrics, and most of his mechanics. And though we know he must have written books on astronomy (he touches competently on astronomical subjects in *The Sand Reckoner*), we do not know even the

477. Simms 1995: 53–55. See also related note in section 3.3 above (under Posidonius).

478. Tertullian, *A Treatise on the Soul* 14, credits a water organ to Archimedes. Archimedes, *On the Construction of Waterclocks* appears to have survived in Arabic (cf. Hill 1976 and 1984: 230–32). That Archimedes may also have written on odometer design is only conjectured (see note in chapter 3.6.IV).

479. Simms 1995: 60–71. Discussed in chapter 4.6.I.

480. On palimpsests (documents that were erased and written over but then forensically recovered with modern technology) see *OCD* 1069 (s.v. "palimpsest") and examples to follow (Ptolemy's *Analemma*; and in section 3.8.IV, Archimedes' *On the Method of Mechanical Theorems*).

title of a single work (except his lost treatise *On Spheremaking*). But it is clear from what survives that he consciously built on the achievements of several predecessors and sought to provide a firm mathematical foundation for engineering. As Archimedes himself says, he wrote his treatise on method "because I believe it will be of no little service to mathematics," for "I believe some of my contemporaries or successors will be able to use the method thus established to discover yet more theorems that have not occurred to me."[481]

After Archimedes the history of physics gets fuzzy. Apellis wrote a lost treatise on his advances in winch and pulley design sometime around the dawn of the 2nd century B.C.[482] Hermogenes of Alabanda wrote a lost treatise on the design principles he employed in the construction of various temples in the 2nd century B.C.[483] Later that century (or early in the next) Agesistratus substantially improved torsion catapult design and wrote renowned works in engineering that also no longer survive.[484] Posidonius may have made some few improvements in the mechanics and machinery of Archimedes near 100 B.C., though it is unclear what.[485] A certain Apollonius (of Athens and Rhodes) was earning a reputation for mechanical genius around this time, but if he wrote anything none of it survives.[486] At the same time the engineer Andronicus of Cyrrhus was building impressive monumental clocks and sundials, but anything he may have written on these or other subjects doesn't survive either.[487] And Carpus of Antioch was renowned as a

---

481. Archimedes, *On the Method of Mechanical Theorems*, pr. (addressed to Eratosthenes).

482. *EANS* 103.

483. *OCD* 670–71; *EANS* 379.

484. *EANS* 45; the works of Agesistratus were consulted by Vitruvius and his much less competent contemporary, Athenaeus the Mechanic (author of an extant *On War Machines*, cf. *EANS* 176 and *OCD* 195): cf. Marsden 1971: 4–5 and Whitehead & Blyth 2004: 172–74. That Athenaeus was pretty much a hack pretending to engineering ability is convincingly argued in Whitehead & Blyth 2004: 34–39, 187-92. I therefore do not include him beyond this note, although his book attests to the achievements of his sources.

485. See mention and note in section 3.3 above.

486. *EANS* 113. See Whitehead & Blyth 2004: 18, 26, 47, 137–38.

487. *EANS* 81. See my later discussion of the Tower of the Winds, and sources

writer in engineering, mathematics, and astronomy in the late first century B.C. or early A.D., but none of his works survive, not even their titles, other than a lost *Mechanics* that included advances on Archimedean machines.[488]

Likewise, outside the field of engineering, also in the early 1st century A.D. or late B.C., we know there was an important but lost work of the only known female research scientist in our period of interest, Ptolemaïs of Cyrene.[489] What we know of her is that she composed an important treatise in harmonics that sought to bring disparate doctrines into a single unified science, and that her achievement in this regard was well regarded by subsequent scientists in the field. She also wrote a treatise on combining empirical with rational methodology, which was also not preserved. Again there is no telling what else she may have done, but these works suggest a trend seen also in Galen, of seeking to unify a scientific field and establish the correct methods for pursuing it. We know of other important empirical scientists in the field of harmonics in the Roman era, such as Heraclides of Heraclia, who wrote in the first century A.D.[490] But we know nothing else about them, either, apart from vague mentions or brief quotations in later authors.

Our only significant body of evidence lies in the extant works of the Roman engineer Hero of Alexandria, who wrote books in mechanics and other physical sciences in the mid-1st century A.D.[491] His synthesizing and improvement of past work may indicate a Roman-era revitalization of the study of mechanics after the decline of Ptolemaic sponsorship. Hero was clearly interested in solving mechanical and technological problems by combining theoretical analysis with hands-on experimentation. Even Hero's mathematical works display wide erudition yet skillfully blend the formal with the practical—so much so, demonstrates Karin Tybjerg, that "it is not possible to maintain the notion that Euclidean-Archimedean geometry

---

cited there.

488. *EANS* 468.

489. See quote in chapter 2.7, and discussion in section 3.7; with Levin 2009: 230–93; Plant 2004: 87–89; Irby-Massie & Keyser 2002: 344–45; Barker 1989: 239–42; *OCD* 1234, *NDSB* 6.172–73, and *EANS* 705–06.

490. *EANS* 369–70.

491. For the following see *DSB* 6.310–15, *EANS* 384–87, and *OCD* 676–77; cf. also Keyser 1988 and Russo 2003: 130–37.

was sealed off from traditions of professional problems and calculation techniques."[492] Hero clearly understood that the achievements of scientists, past and present, are indispensable to technology and engineering, and theory was as important as practice.[493]

Although the preponderance of attention paid to "playthings, puppet shows, or apparatuses for parlor magic" in his treatise on *Pneumatics* has led some scholars to imagine a decline in the technological aspirations of the Romans, this is not a sound judgment. These interests occupied Philo and Ctesibius as well, under the Ptolemies centuries before, so there is no discernible difference between eras. Hero built on their work to add even more, in both practical as well as theatrical machinery. And those were not mutually exclusive categories. Hero's supposedly "trivial" technologies were actually nothing of the kind.

Religion and entertainment were big business in the ancient world, and catering to the market demand of well-paying clients by developing 'marvels' for local temples and theatres to draw crowds (who presumably would make donations or buy various religious services), was no mere idle pastime, but a serious enterprise.[494] Robotic entertainments were also a means for the rich to advertise their wealth, serving the same function as modern 'useless' technologies like yachts and sportscars, or designer shoes and handbags.[495] Even beyond that, Hero certainly did not see his work in theatrical robotics as frivolous, but as "worthy of approval, because of the complexity of the craftsmanship involved and the striking nature of the spectacle produced," and because "every facet of mechanics is encompassed

---

492. Tybjerg 2004: 34–35. The whole of Tybjerg 2004 establishes her point.

493. In addition to the analysis of Tybjerg 2004 see that of Tybjerg 2003 and Cuomo 2002.

494. Indeed, machinery for temple marvels were so standard that it was simply assumed any new temple built under the empire would be so equipped: Cassius Dio, *Roman History* 69.4.1–5.

495. See, for example, the analysis of automata in Schürmann 2002 and Schneider 1992: 201–07. For a different perspective, note how Russo 2003: 140–41 compares Hero's automatic theatre with a related modern invention: the cinema. Though Hero's theatrical scenes could not have moved that quickly (despite Russo's effort to argue they did), the function and value is similar.

in the building of robots."[496] Even the technologies that seem to be little more than toys (like a steam turbine that does nothing but spin) would have had obvious uses as tools for demonstrating scientific principles and as test apparatuses for further experimentation and development.[497] They certainly were not intended for children. Moreover, as Karin Tybjerg points out, many of his simple pneumatic devices may seem trivial by themselves, yet Hero explicitly introduces his *Pneumatics* by announcing that the simple things he describes can be combined into more elaborate devices. Hence his supposed 'toys' play the same role as the five basic machines of ancient mechanics, another example "where simple machines are combined systematically to produce more complex ones."[498] Modern scholars are quite wrong if they imagine Hero intended all his pneumatic devices to be completed products.

As usual, Hero's works were not preserved very well. In fact, his *Mechanics* was mostly saved only in Arabic translation and his *Optics* only in Latin translation, and his other extant works have not always been transmitted accurately (many have been heavily edited). But altogether we know Hero wrote on mechanics, pneumatics, optics, and surveying and was himself a competent astronomer and mechanic. He thus mastered all the expected fields of an engineer enumerated by Vitruvius except harmonics, although there is no reason to suppose he did not write on that subject as well. Scholars generally agree that Hero invented an improved screw-

---

496. Hero, *On Constructing Automata* 1.1.1 (cf. Murphy 1995: 11). On Hero's elevated and quite serious appreciation and use of mechanical marvels see Tybjerg 2003.

497. See, for example, Russo 2003: 75–78 and Landels 2000: 192–93. Tybjerg 2003 demonstrates how the ancient idea of 'wonder' (which Hero embraced and employed) included scientific demonstration of counter-intuitive principles (often with a mechanical apparatus). Hero, *Pneumatics* 2.11 describes his steam turbine, which combines rotary motion with a previous scientific demonstration device known to Vitruvius, *On Architecture* 1.6.2, which dates back to Philo, *Pneumatics* 57, and is related to another device in Hero's collection: a steam levitator (Hero, *Pneumatics* 2.6). That Hero's steam turbine was a scientific demonstration device, evolved from previous devices of the same general purpose, and deployed to challenge Aristotelian physics, is convincingly argued in Keyser 1992.

498. Tybjerg 2004: 50–51. The five basic machines are (still) the wheel, lever, wedge/ramp, pulley, and screw. Though the gear was also developed, which made six in all, this was rightly considered a combination of lever and wheel (as I'll discuss later).

cutter, more advanced odometers for use on land and water, a coin-operated vending machine, an automatic door, a simple steam turbine, and many other devices, not just entertainments but also practical or scientific equipment.[499] He also passed on the principles and instructions for the inventions of his predecessors, such as the water organ, the continuous-stream fire pump, and various mechanical clocks. And he used or developed several mechanical instruments to carry out complex measuring operations.[500]

Hero's extant writings include *On Constructing Automata* (a book about the principles and construction of robotic theaters and devices), *The Baroulkos* (a treatise on the design principles underlying a massive lifting machine of Hero's invention), *The Dioptra* (on the construction and use of several sophisticated surveyor's instruments that Hero improved upon), *Siegecraft* (a treatise summarizing the early history of artillery development), *Pneumatics* (a notebook on machinery powered by water, air, or steam), and treatises on the practical applications of geometry (including the use of stereometry for land surveyance and procedures for measuring the volume and surface area of objects of complex shape), a commentary on Euclid, and extensive treatises on *Mechanics* and *Mirrors*, in the latter proposing a principle of least action to explain the laws of reflection.[501] Hero may have written a great deal more, including treatises on waterclocks, crossbows, and architectural subjects. He is credited with a treatise *On Vaults* (or *On Domes*) for which the 6th century engineer Isidore of Miletus wrote

---

499. His robotic doors (Hero, *Pneumatics* 1.38–39) were not steam powered (as I have cited some claiming in the past) but pneumo-hydraulic, and operated on a small-scale replica (not an actual temple), although such building of models often preceded full scale implementation (see Di Pasquale 2002). And though Hero's vending machine was based on earlier water-dispensing technologies (Philo, *Pneumatics* 28–34), the idea of dropped-coin activators first appears in Hero, *Pneumatics* 1.21 (cf. James & Thorpe 1994: 128–29). On the development of ancient odometers see M.J. Lewis 2001b: 134–42, 329–31 (and the note below on mechanized carriages).

500. Tybjerg 2004: 40–48.

501. On Hero's principle of least action, anticipating but not yet developing the modern equivalent, see A.M. Smith 1999: 81 (cf. also 134, and 145 nn. 9 and 10) and Boutot 2012. On Hero's *Mechanics* and *Baroulkos* see Vitrac 2009 and Drachmann 1963: 19–140.

a commentary, but neither text nor commentary survives.[502] There are several extant references to his treatise *On Waterclocks*, and there are extant fragments of a treatise on constructing a portable metal-frame torsion catapult (*Cheiroballista*, i.e. *Handgun*) that most scholars believe is his (and possibly his invention).[503]

Shortly after Hero came Menelaus, the astronomer I already mentioned in section 3.3.[504] He was also an engineer and may have written a great deal on that subject, including a broader treatise on mechanics. However, the only specific work we know for certain is his treatise *On Knowing the Weights and Distributions of Different Bodies*, apparently written and researched at the request of emperor Domitian. This was an extensive study of the geometric analysis of weight and density, including new findings for the specific gravity of different metals and alloys, and describing a new kind of balance that Menelaus invented for the purpose. An Arabic translation of at least part of this treatise has recently been discovered, which gives its title as "the book of Menelaus to Emperor Domitian on the mechanism by which one can know the amount of each of a number of substances that are mixed together," and its content updates and improves upon the statics and hydrostatics of Archimedes.

The astronomer Ptolemy was also an engineer.[505] Nearly everything we know he wrote survives in some form (though sometimes only in parts, and sometimes only in Arabic), *except* his treatises in mechanics (including one with the title *On Balances*), which is an unfortunate loss.[506] Considering

502. *DSB* 7.29 (in s.v. "Isidorus of Miletus").

503. Hero's 'handgun', which became standard equipment in the Roman legions, incorporated numerous significant advances on previous catapult design. See Marsden 1971: 206–33 and Landels 2000: 99–132. The historical development of waterclocks is hard to track, but by Hero's time they had become quite sophisticated: Russo 2003: 101–05; M.J. Lewis 2000; Evans 1999: 251–56. Doctors even had special adjustable pulse-timing clocks (originating with Herophilus): Russo 2003: 145–46 and von Staden 1996: 89.

504. See references there.

505. See references in section 3.3.

506. Also missing are his works in geometry—one in which he claimed to have proved Euclid's parallel postulate, another (*On Dimension*) in which he claimed to have proved there can only be three dimensions, and another (*On the Elements*), which appears to have been a more comprehensive treatise in geometry in relation

the advanced state of his astronomy, geography, optics, and harmonics, his mechanical work may have included or mentioned some of the most advanced scientific studies of that subject—and, also given the content of his other work, would likely have included descriptions of experiments and various experimental apparatuses.[507] Likewise, the fact that Ptolemy wrote on all the fields that Vitruvius identified as essential for an engineer—not only astronomy and geography, but also optics and harmonics, as well as mechanics—strongly supports the conclusion that Ptolemy himself was an engineer, though with a strong personal interest in the astronomical side of his field.

Engineering interests also underlie the second book of the *Almagest* and his *Handy Tables*, which had explicit uses for the construction of accurate sundials for any latitude at any time of year. He explains how to build several precision instruments in the *Optics* and *Harmonics* and even in the *Almagest*, clearly showing his knowledge not only of theoretical mechanics but even practical shopwork, as we also see in his advice regarding several cartographic instruments in his *Geography*.[508] In fact, he gives the impression not only of having built many of them himself, but of working closely with craftsmen in commissioning their construction. Ptolemy also wrote a treatise on sundial construction called the *Analemma* (a Latin translation of which has recently been recovered from a palimpsest), which shows he made improvements in this field, too,

---

to theories of matter. An excerpt from Ptolemy's *On Balances*, with discussion of a problematic experiment it contained, survives in Simplicius, *Commentary on Aristotle's 'On the Heavens'* 4.4.710–11 (cf. translation and note in Cohen & Drabkin 1948: 247–48). According to the *Suda* (a 10th century Byzantine encyclopedia) Ptolemy even wrote a whole three volume *Mechanics*.

507. Though they might also have contained some erroneous theories and conclusions (just as his other works do), at least judging from the one reference to Ptolemy's *On Balances* in Simplicius (see previous note).

508. In the *Almagest* this includes construction and use of an armillary sphere in book 5, as well as a meridian ring and plinth (1.12), a parallactometer (5.12), a specialized diopter (5.14), and a practical star globe (1.22–23, 7.6–8, 8.2–3), just to name a few. Several other instruments are described in the *Optics* and *Harmonics*, and even in the *Geography* (e.g. 1.22). On these see A.M. Smith 1996, 1999, and 2014: 25–129 (*Optics*), Solomon 2000 (*Harmonics*), Berggren & Jones 2000 (*Geography*), Toomer 1984 (*Almagest*), and in general Lloyd 1982: 136–44 and Evans 1999.

drawing on a new construction procedure invented in the early 1st century B.C. by Diodorus of Alexandria.[509] Moreover, his *Handy Tables* specifically declare the purpose of assisting the construction of precision armillary spheres, and he includes astute technical advice on that task, producing the most accurate mechanisms in antiquity.[510] Ptolemy also wrote a treatise on the construction of the *Planisphere* (extant in Arabic translation), which describes procedures for constructing a plane astrolabe, and a lost treatise on constructing a new instrument, possibly of his own invention: the *Meteoroscope*.[511] And in conjunction with his geography he composed his own projectional world map and a series of specialized sectional maps that appear to have survived at least into the 4th century.[512]

Ptolemy's *Optics* and *Harmonics*, like his *Almagest*, represent the greatest scientific achievements of their respective fields, preserving, systematizing, and experimentally confirming prior work, and advancing both subjects with his own experiments, observations, and theories. His *Optics* included discussions of stereometry (the differences between monocular and binocular vision), and a detailed experimental study of the laws of catoptrics (the science of mirrors and reflection). Ptolemy appears to have been the first to experimentally calculate the size of the human visual field and to observe and measure the index of refraction in different materials; he also made the first stab at finding a law of refraction, and proposed a principle of least action to explain it.[513] He also covered questions of perspective and illusion in considerable detail, and the geometrical difference between the actual and apparent size, distance,

509. *DSB* 15.519 (in s.v. "Vitruvius Pollio"). See discussion and sources in section 3.3.

510. *DSB* 15.219 (in s.v. "Hipparchus"). See discussion and sources in section 3.3.

511. On Ptolemy's meteoroscope see Ptolemy, *Geography* 1.3 with Evans & Berggren 2006: 48 and Berggren & Jones 2000: 61 n. 12.

512. *DSB* 10.300 (in s.v. "Pappus of Alexandria").

513. Ptolemy proposes his principle of least action at the end of the surviving fragment of *Optics* 5 (see Cohen & Drabkin 1948: 281), where his discussion breaks off and is now lost. On Ptolemy's anticipation of the (now correct) theory that refraction is caused by a slowing of the rays passing through an object see A.M. Smith 1996: 42–43. On how he fell short of discovering the correct law of refraction but came close: Wilk 2004.

motion, and shape of objects, with a special discussion of how refraction can affect astronomical observation.[514]

Ptolemy shows a keen interest in developing special instruments for testing several questions in both optics and harmonics. For example, Toomer describes how Ptolemy "determin[ed] the relationships between the images seen by the left and right eyes and the composite image seen by both, using an ingenious experimental apparatus with lines of different colors," something never before attempted.[515] The *Harmonics* is similarly devoted to resolving remaining disputes in order to establish a unified science and produce a definitive text in the field, by carefully combining mathematical analysis and physical experiments, including observations of how actual musicians tuned their instruments. Though Porphyry near the end of the third century claimed that Ptolemy's second century work in this field was largely dependent on the work of a little-known Didymus in the first century, this might only refer to its theoretical content, and either way this still places a Roman date on the achievement of the most advanced developments in the science of harmonics in antiquity.[516]

Once again, we know there were other engineers doing original work around his time, but next to nothing survives to tell us about it. In the early first century a certain Mantias, for example, was writing important treatises

---

514. On Ptolemy's discussion of atmospheric refraction affecting astronomy see A.M. Smith 1999: 134–37 and 1996: 46, with Ptolemy, *Optics* 5.23–31. On atmospheric refraction in general, see relevant notes above, and quotations of relevant passages in Cohen & Drabkin 1948: 281–85.

515. *DSB* 11.200 (in s.v. "Ptolemy").

516. *DSB* 11.200 (in s.v. "Ptolemy") citing Porphyry, *Commentary on the Harmonics* 5. See Barker 1994: 62–73 and *OCD* 451 (s.v. "Didymus (3)"), *NDSB* 2.284–86, *EANS* 244–45. Scientific harmonics (theoretical and empirical acoustics) had already begun before Aristotle and continued since, but it is difficult to reconstruct its progress from extant sources: see *OCD* 8–9 (s.v. "acoustics"). Barker 1994 argues scientific harmonics stagnated after Aristotle and then underwent a major revival in the early Roman empire. Adding to the evidence of Ptolemy are writers on musical scales (Alypius: *OCD* 67; *EANS* 62) and the philosophy of music (Quintilian Aristides: *OCD* 155; *EANS* 134) from the 3rd or 4th century A.D., which contain evidence Ptolemy himself wrote a similar treatise specifically on *Music*. There were other writers on harmonics during the 2nd and 3rd centuries A.D. (e.g. Cleonides: *OCD* 332; *EANS* 481) and in the 1st century A.D. (e.g. Ptolemaïs of Cyrene, discussed earlier). See Barker 1994: 54 n. 2 for a longer list of known examples.

on hydrostatics, but we know so little about him, we are not even certain of his name.[517] And flourishing near the dawn of the 2nd century A.D. was the Roman military surveyor Balbus, who wrote a handbook in Latin on practical geometry dedicated to a fellow engineer named Celsus—who is otherwise unknown, yet in his introduction Balbus says Celsus had recently invented some kind of instrument that Balbus found very useful. This handbook has survived only in corrupted and incomplete form, but clearly draws on the writings of Hero of Alexandria. An extant Latin catalogue of Roman cities, evidently part of another surveyor's treatise, probably also derives from a lost work by Balbus, and other writings by him are known to have provided material for later compilations on geometry and surveying.[518] Likewise, in the late 1st century A.D. Sextus Julius Frontinus, though a layman, wrote impressively on surveying, military science, and aqueduct management.[519] In the early 2nd century A.D. Apollodorus of Damascus became a renowned and accomplished architect and engineer whose writings included an extant textbook on military machinery, though we have no idea what else he may have written.[520] Before them, sometime in the 1st or 2nd century B.C., Dorion and Epicrates wrote treatises on

---

517. *EANS* 525.

518. *DSB* 1.418–19, *EANS* 189, and *OCD* 222; see also *OCD* 636–37 (s.v. "*gromatici*") and B. Campbell 2000: xxxix-xlii, and the relevant sections of Cuomo 2001. There were other engineers who wrote on surveying and military camp construction and other subjects in the same era, about whom we know even less (e.g. Hyginus: *OCD* 714–15, second or fourth entry, and *EANS* 426–28; Marcus Iunius Nipsus: *OCD* 766, *EANS* 457; and Siculus Flaccus: *OCD* 1363, *EANS* 740; cf. also B. Campbell 2000: xxxv-xxxix).

519. *OCD* 762–63 and *EANS* 453, with Rodgers 2004: 1–20. Two of his treatises survive (*On the Aqueducts of Rome* and *Stratagems*), though his more detailed *Art of War* does not, and his works on surveying survive only in fragments (see B. Campbell 2000: xxvii-xxxi).

520. *OCD* 120 (s.v. "Apollodorus (7)") and *EANS* 107–08, with Cuomo 2007: 131–32 and La Regina 1999. We know this Apollodorus designed and built Trajan's famous bridge across the Danube (Serban 2009). Architects and engineers are relatively poorly represented in the sources (*OCD* 142–44, s.v. "architects" and "architecture"), yet we know about a lot of them (including far more names than I mention).

machinery, but we know nothing more about them.[521] Yet others we don't even know the names of.[522]

Finally, relevant to engineering in regard to calculation and planning is the advancement of algebra, which had been developing since pre-Roman times, though hardly any information about its use or advancement was preserved. Our only surviving discussion comes from Diophantus of Alexandria, author of a detailed textbook on algebra, the *Arithmetic*, which originally covered 'arithmetic' from the basics all the way up to what is now called algebra, as far as quadratic equations, discussion of negative numbers, and squares and roots up to the sixth power. All this survives only in edited and mutilated sections. Diophantus is usually placed in the middle of the 3rd century A.D., by connecting his name with the Christians Dionysius and Anatolius, but the evidence for this is highly conjectural, and there are good arguments placing him two centuries earlier. Also highly conjectural is the claim that Diophantus actually invented symbolic (as an improvement on merely propositional) algebra, rather than only codifying methods of symbolic notation already developed, though either way such notation is well featured in his work.[523] However, we know the basic idea of algebra—solving for unknown quantities—had already begun as early as the 4th century B.C. and there is no telling how it did or did not develop over the many centuries before Diophantus.[524] But we can clearly see that in this, as in the several areas of physics just surveyed, and again just as in the astronomical and medical sciences, progress was seen, valued, and made.

521. *EANS* 275, 292.

522. Such as a 1st or 2nd century B.C. author of a work on spherical and parabolic mirrors, whose name has become corrupted beyond recognition ("Dtrums," *EANS* 278).

523. Derbyshire 2006: 31–42, *DSB* 4.110–19 and 15.118–22, *EANS* 267–68, and *OCD* 465, which correctly dates him "between 150 BC and AD 280," hence probably Roman-era. A good case for dating Diophantus to the 1st century A.D. is presented in Knorr 1993 and Russo 2003: 322–23 (esp. n. 230).

524. *DSB* 13.399–400 (s.v. "Thymaridas"; cf. *EANS* 808–09). See also the debate on the status of pre-Diophantean algebra between Unguru 1975 and 1979 and van der Waerden 1976 and Freudenthal 1977; discussed in Fried & Unguru 2001.

3.5 Other Sciences?

My survey of scientific progress up to the Roman period has emphasized what we can now confirm were genuine advances toward a correct understanding of the natural world. This is no mere anachronism, since that progress is objectively real, a fact that was even noticed by ancient scientists (like Galen, Ptolemy, and Hero) who frequently called attention to the evidence of the senses in confirmation of the fact: the accumulation of new knowledge was proved by its increasing success in practice, both in a practical sense (e.g. medicine, technology, predictive astronomy) and in a straightforwardly empirical sense (e.g. one could see for oneself the discoveries of anatomists or the operation of a physical principle in a workshop apparatus). And though even the greatest scientists in antiquity could not see so clearly as we can the difference between genuine and illusory progress, we can nevertheless confirm their methods and efforts were indeed working to sort between the two in favor of the genuine. Slowly (and not always consistently), gradually, and overall, correct ideas were defeating incorrect ones.

But even if we expanded our survey to include the illusory elements of ancient scientific progress—the 'advances' in the sciences that were only regarded as such *then*, yet are now known to be false (which includes everything from inaccurate observations and untrue beliefs to incorrect theoretical models)—we would still see many ancient scientists regarding their own history as one of progress. Though not every development was regarded as progress, and though which developments were, or were not, so regarded did vary with the experience and ideological assumptions of the scientist, and though we still sometimes hear calls from them to go back to the 'correct' views of various long-dead predecessors, all the most prominent experts of the time, and even their lay admirers, believed that some sort of progress had been made—Ptolemy regarded his astronomy as better than that of Hipparchus, and that of Hipparchus as better than any before him; Galen regarded his medicine as better than that of Herophilus, and that of Herophilus as better than any before him; and so on. We'll see evidence of this in later sections. But if we accept that fact, then whether we define 'progress' in terms of their genuine successes or their imagined successes, we still see the same thing: ancient science had continually progressed and improved through the centuries well into Roman times, and they knew it.

In fact, in both respects, there were so many advances made in medicine, astronomy, and physics, advances that were clearly continuing even under the Romans, that it is incredible anyone would think Greek and Roman scientists had no idea of either the possibility or value of scientific progress. Clearly such an idea was alive and sustained for centuries. Nevertheless, scholars still wonder why only some fields saw such development and not others. Such wonder is probably often misplaced, since an active interest in expanding the methods of these fields into other areas was more an evolving product of the Scientific Revolution than a cause of it. But the question remains: Why did such an idea not come to the Romans, or even the Greeks before them—despite watching and pursuing, over six centuries, scientific progress in the medical, astronomical, and engineering sciences?

In some cases our expectations are probably unreasonable. For example, consider the lack of sound scientific progress in chemistry and magnetism. Peter Green claims a "fear of inaccuracy and a lack of proper controls" prevented the development of a scientific chemistry.[525] But neither a lack of 'proper controls' nor this (unevidenced) 'fear' stopped progress and experimentation in other ancient sciences. So Green's theory has no explanatory value. Although Greek and Roman alchemists *were* experimenting in chemistry, and experiments in magnetism *were* being made by natural philosophers (even, as we have seen, by Galen), neither chemistry nor magnetism were organized as distinct sciences or explored with any systematic empirical method.[526] Yet such advances took place

525. P. Green 1990: 457.

526. Empirical study of magnetism and chemistry was not neglected, it just did not rise to the highest status of 'science' the way astronomy, physiology, or mechanics did. Ancient theories and discussions of magnetism are surveyed in Lindsay 1974: 245–72 (with examples in Cohen & Drabkin 1948: 310–14). Though no books on the subject have been preserved, we know some existed (e.g. see Theophrastus, *On Stones* 28). For an example of the state of 'theoretical' chemistry under the Romans see Alexander of Aphrodisias, *On Blending and Growth* (with commentary in Todd 1976). For discussions of practical and empirical chemistry in antiquity: Martelli 2011, Russo 2003: 165–70, Wilson 2002, Healy 1999: 115–41, Cohen & Drabkin 1948: 352–73. For a selection of literary evidence for a wide array of ancient chemical technologies see Humphrey et al. 1998: 205–34, 354–78, 880–90. The degree of scientific chemistry in the practice of alchemy in antiquity can be gleaned from the extant 3rd century writings of Zosimus (*NDSB* 7.405–08; *EANS* 852–53); and *OCD* 51–52 (s.v. "alchemy"). See *EANS* 992–93 for a list of ancient alchemical writers.

for other sciences. So no broad social causes can be at fault here. There must have been something different about magnetism and chemistry that impeded their development.

As we will see shortly, it appears that only sciences with practical applications experienced considerable development, and usually where theoretical models could most easily be tested against observations—hence anatomy, astronomy, and mechanics experienced the most genuine scientific advancement. In contrast, though magnetism was just as accessible to empirical and theoretical study as any of these, the most obvious difference is that, unlike optics, harmonics, mechanics, astronomy, or medicine, magnets as yet had no practical use, certainly none that was sufficient to attract the attention of scientific engineers.[527] The compass would not be discovered for centuries yet—and even once in wide use, it would not be studied scientifically for many centuries more. Only when warring empires had been sailing the Atlantic and competing over the New World and the Atlantic slave trade for nearly a century, did a man like Gilbert finally get the idea that a proper scientific study of the compass might prove useful.[528]

Conversely, though chemistry is obviously useful, the underlying causes of chemical phenomena are very difficult to observe, much less correctly understand. How exactly does one go about studying the composition of water, for example? You cannot cut it, smash it, or burn it, and boiling or freezing it gets you no nearer the answer. It is not at all obvious that one should instead be asking, "Which gases pyrochemically combine to make water?" Hence chemistry did not become a proper science until the 18th century, well after the Scientific Revolution had already transformed the world by providing its own model and motive to study chemistry scientifically—something not even the startling discovery of gunpowder had managed, despite centuries of its widespread use and manufacture. If even the moderns needed a scientific revolution to finally put chemistry on a scientific footing, then we should not expect the ancients to have done so.

But what of other sciences? From the late 4th to the early 3rd century B.C., Theophrastus, Aristotle's successor at the Lyceum in Athens, wrote

---

527. Even at best, magnetism was only rarely employed to produce public tricks in temple displays (cf. Pliny the Elder, *Natural History* 34.42.148 and Claudian, *The Magnet* = *Minor Poems* 29, with discussion in Schürmann 1991: 234).

528. See brief discussion and notes in chapter 1.1.

over two hundred books on almost every conceivable subject, scientific and otherwise, although little of this work survives. Most of what remains consists of two treatises on botany, *Inquiry on Plants* and *On the Generation of Plants* (which essentially expand to botany what Aristotle did for zoology), plus a collection of his briefest works, and quotations, that even combined barely equal half the length of his extant botanical books.[529] But judging from what we have, all his work is significant not only for the fact that, like Aristotle's zoology, it seeks the study and understanding of their subject for its own sake, and contains a remarkable amount of empirical and observational detail to that end, but also because it ranges so widely, observing and noting the industrial applications of natural resources, and recording the observations of the craftsmen who use them. For example, the fifth book of his *Inquiry on Plants* is mostly devoted to recording which types of wood are used for which industries and applications, while (among a great deal else) he researched ore processing for *On Stones*, studied the perfume industry for *On Smells*, and wrote *On Fishing*.[530]

Sometimes Theophrastus seems almost unique in his interests, although this may again be owing to the destruction of books on the same subjects by later scientists. His extant treatise *On Fire* is the only surviving observational study of the nature of heat and flame. His lost treatise *On Fossils* was the first (and only known) scientific treatise on paleontology.[531] Overall, much of his work consisted of recording data that could be used to produce generalizations that would advance natural philosophy, or criticizing and correcting or clarifying Aristotle, and it is believed even some of Aristotle's

529. For the following: Lloyd 1973: 8–15; Rihll 1999: 106–18; *DSB* 13.328–34 (s.v. "Theophrastus"; cf. *EANS* 798–801 and *OCD* 1461) and 245–46 (s.v. "botany"). Theophrastus also wrote numerous zoological and other works that were not preserved (and his *Meteorology* survives only in Arabic: Taub 2003: 115–24), and a few that survive only in mangled or abridged versions (e.g. *On Weather Signs*, cf. Sider & Brunschön 2007). The philosophical function and context of his more scientific work is discussed in French 1994: 83–113. But the premiere authority on all things Theophrastean is now Fortenbaugh et al. 1992–2007.

530. Russo 2003: 165–66, 210–12.

531. See Mayor 2011a (index). The whole of Mayor 2011a documents a lot of interest in this subject spanning antiquity. Though no one's books on it were preserved, we know such study influenced ancient geological theories: Russo 2003: 161–63 (e.g. Strabo, *Geography* 1.3.4).

extant works are really editions reworked by Theophrastus. But whatever the case, Theophrastus carried on Aristotle's passion for understanding the natural world, and in broad scope. Additionally, as J.B. McDiarmid observes, his works are filled with "frequent reminders to himself and his readers that there must be further investigation" than what he had completed, since he mainly gathered data collected by others, and knew it was of mixed reliability, and that his conclusions would need confirmation or even revision.[532]

And yet "science for its own sake" *almost* appears to have ended with Theophrastus (and some of his colleagues).[533] Only a little scientific botany, zoology, or mineralogy followed in antiquity that wasn't just pursued for medical or commercial purposes; and the study of 'pyrology' appears to have been abandoned completely.[534] Though we can't be sure of this. Theophrastus also wrote a lost work *On Metals*, for example, but there were probably other writers on the same subject. J.T. Vallance observes that "archaeological evidence for ancient mining and metallurgy" actually "suggests degrees of technical sophistication and understanding which are not equally evident in the surviving literary sources," so there might have been a lot more written on the subject of mineralogy than we know.[535]

---

532. *DSB* 13.333 (in s.v. "Theophrastus").

533. See *OCD* 87–90 and 483 (s.v. "animals, knowledge about" and "ecology (Greek and Roman)"). Clearchus, another of Aristotle's pupils, also wrote lost works on zoology and mathematics (*OCD* 329–30, s.v. "Clearchus (3)"; cf. *EANS* 477), and Phanias did the same on botany (*EANS* 641; see note in chapter 2.5). At some point in the 2nd century B.C. a certain Damigeron also studied and wrote on mineralogy, and large fragments of his work *On Stones* survive (*EANS* 225; cf. *DSB* 4.121, in s.v. "Dioscorides"), but he did not quite rise to the standard of Theophrastus. We also know someone named Democritus wrote on magnets and other stones sometime in the last three centuries B.C. (*EANS* 236). Lennox 1994 discusses the vanishing interest in this theoretical zoology and botany championed by Aristotle and his pupils, and advances several fanciful theories relating to this, but he ignores or hastily dismisses most of the Roman evidence I am about to present.

534. Although we shouldn't cite the practical sciences; e.g. that king Ptolemy's physician Apollodorus (*EANS* 105 and 106) was writing scientific treatises on wine, perfumes, and poisons and their antidotes, is still a significant continuance of scientific work.

535. *OCD* 957 (s.v. "mineralogy"). See also *OCD* 938–39 and 957–58 (s.v. "metallurgy" and "mines and mining").

The same could probably be said of any other subject. But in the surviving record, as we suggested above, after Theophrastus scientific progress appears to have been tied to practical applications, particularly where observations could be most directly linked to a workable theory. Hence the history of progress in ancient biology is really the history of progress in medicine and pharmacology. Likewise, progress was made in physics largely insofar as it informed engineering, and astronomy was pursued insofar as it, too, had practical applications—particularly in timekeeping, celestial prediction, cartography, and even 'astrology'. The pursuit of optics and harmonics were notably related to their uses for engineers—and in the case of optics, for astronomers as well. Although in these, and to a lesser extent other subjects, research often went beyond known applications, they were already being pursued for practical reasons and thus were already attracting attention, so their most easily observable structural or mathematical behaviors could retain extended interest for centuries no matter what their use.

This might have been changing. By the turn of the era, Nicolaus of Damascus, the personal friend and court historian of Herod the Great, was an accomplished Aristotelian and wrote a treatise *On Plants* in the Aristotelian tradition (so much so that it was falsely attributed to Aristotle).[536] A certain Oppius had written detailed treatises on trees and insects, including anatomical observations (all now lost, but later quoted).[537] Antonius Castor (possibly Mark Antony's freedman) cultivated his own botanical garden and wrote scientific treatises on botany for which he was regarded as the foremost authority in the field, yet none of his works were preserved.[538] Cicero reports that his polymathic friend Publius Nigidius

---

536. *DSB* 10.111–12, *EANS* 577–78, and *OCD* 1014; see also *DSB* 1.268 (in s.v. "Aristotle") and Gottschalk 1987: 1122–23. We know he wrote a great deal more, including histories, ethnographies, several widely respected commentaries on Aristotle, and works in many other genres (including an autobiography), none of which was preserved. Even the extant Greek text of *On Plants* that had long been attributed to Aristotle (and is still included in some collections of his works) is a Renaissance back-translation into Greek of a Latin translation from an Arabic translation of a Syriac translation of Nicolaus' original Greek, an absurdity typical of the middle ages.

537. *EANS* 594.

538. *OCD* 112 and *EANS* 100 (maybe the same as Antonius Rootcutter: *EANS* 101); Pliny the Elder, *Natural History* 25.5.9 (with quotations or paraphrases in

Figulus, better known as a Pythagorean astrologer, was also a "sharp and diligent investigator" of natural phenomena, and around that time or shortly afterward Papirius Fabianus became renowned for his work in the natural philosophy of stones, animals, plants, and more.[539] In the mid-1st century A.D. a certain Pamphilus wrote a work *On Botany* whose contents remain unknown but were notable, despite some elements of fancy.[540] And Apion of Oasis wrote books on mineralogy, and other subjects, that were cited by later science writers.[541]

We know many others were engaged in botanical, zoological, mineralogical, and other kinds of research in the Roman era.[542] Posidonius, for example, was reviving scientific geology and mineralogy as part of his observational research on volcanology in the early 1st century B.C.[543] In the

---

20.66, 20.89, 20.98, 23.83, 26.33, etc.).

539. On Figulus see Rawson 1985: 94–95, 180–83, 288, 291–92, Griffin 1994: 707–10, Horsfall 1979: 81; with *OCD* 1016 and *EANS* 572–73. Whatever research he published has not survived, so its merit cannot be assessed. Cicero's praise of him appears in the fragmentary preface to his translation of the *Timaeus*. On Fabianus see Capitani 1991: 98–101 and Griffin 1976: 37–42.

540. *OCD* 1071 (s.v. "Pamphilus (2)") and *EANS* 606. Although Galen complained that he included digressions on local Egyptian magic (Galen, *On the Combinations and Effects of Simple Drugs* 6.pr = Kühn 11.792–98), it is unclear whether this was merely literary digression or a real defect. Pamphilus also wrote a *Physics* and a comprehensive dictionary of the Greek language, neither preserved.

541. *EANS* 104.

542. Including Sextius Niger (*EANS* 738–39), Julius Bassus (*EANS* 451), Gaius Valgius (*EANS* 822–23), Niceratus (*EANS* 575–76), Petronius Musa (*EANS* 639), Diodotus, etc., although still of varying scientific merit (cf. *OCD* 245–46, s.v. "botany"). Pliny the Elder names several other Roman-era botanical writers otherwise unknown (20.100, 20.109, 23.83, 24.120, 25.3, 25.110, 26.93, 27.120, etc.). Healy 1999 surveys the botanical, zoological, mineralogical and other data accumulated in Pliny's *Natural History* and finds that a great deal more knowledge was available to him than could have been derived from the era of Theophrastus, which entails a lot more had been written in the interim that we have simply lost, a conclusion supported by Hardy & Totelin 2016 and the diverse contributions in French & Greenaway 1986. On the nascence of anthropology in antiquity: G. Campbell 2006; Sassi 2001.

543. We know Posidonius made good first-hand observations of the properties of bitumen, naphtha, petroleum, pumice, and asphalt, and was possibly the first to

late 1st century B.C. the historian Pompeius Trogus wrote treatises on botany as well as zoology, and in the early 1st century A.D. observational research on fish and birds was conducted by the Roman consul Lucius Lucullus and written up by his colleague Trebius Niger.[544] We know Apuleius wrote scientific books in the early 2nd century A.D., including *On Medicines*, *On Trees*, and works in astronomy, mathematics, agriculture, music, and zoology, involving him in astronomical and pharmacological observations and even the scientific dissection of fish, yet none of these books were preserved.[545] In the 2nd century A.D. the poet Oppian wrote *Fishing*, an epic poem on fish and fishing—though ranging across all creatures of the sea, not just fish—which frequently displays enough scientific accuracy that modern ichthyologists regard it as "the most accurate and comprehensive ichthyological treatise up to" the 16th century.[546] It almost certainly was

---

do so in such detail: Kidd 1988: 826–36, 951–53. He was also a renowned authority on volcanology and appears to have initiated the field as an observational science: Kidd 1988: 809–16, 824–26.

544. Trogus: *OCD* 1181, *EANS* 685. Trebius & Lucullus: *OCD* 1503, *EANS* 815. For examples: Pliny the Elder, *Natural History* 9.41.80, 9.48.89–93, 10.20.40–41, 32.6.15 for Trebius; and 10.51.101, 11.94.229, 17.9.58, 31.47.131 for Trogus. Both Trogus and Trebius mixed in apocryphal and legendary material, but Pliny reveals enough explicit references to occasions of careful observation to suggest their work was not frivolous. Even Aristotle and Theophrastus were not immune to the flawed and fanciful, so without these Roman books we cannot assess their overall scientific quality. But they still demonstrate a renewed interest in the subject. Similar interest (in both the scientific and the fantastic) is shown by the lost but oft-quoted books on animals and plants by Alexander of Myndus, written in the early 1st century A.D., who appears to have been a lay compiler of others' work and not an original researcher: cf. *DSB* 1.120–21, *EANS* 57, and Irby-Massie & Keyser 2002: 271–72. The same can be said of Aelian, a late 2nd or early 3rd century compiler of animal lore: cf. *EANS* 32–34 and *OCD* 18.

545. Apuleius, *Apology* 29; Servius, *Commentary on the* Georgics *of Virgil* 2.126. See *OCD* 127–28, *EANS* 119–20, and S.J. Harrison 2000: 29–32, 65–69. Apuleius made astronomical observations to verify theories: *Florida* 18.32. He also shows empirical interest in medicine and pharmacology: *Apology* 40, 41, 48; in the anatomy and physiology of fish: *Apology* 38; and in the scientific study of the laws of reflection: *Apology* 16. He also translated into Latin some Platonic works in math and philosophy, though whether extant translations are his is disputed.

546. Kellaway 1946: 120; *OCD* 1041; *EANS* 593–94 (not to be confused with the Oppian who wrote *On Hunting*: *EANS* 594). Oppian was certainly not writing

relying on now-lost scientific treatises, including lost treatises on fish and fishing by Metrodorus of Alexandria (late 2nd century B.C.) and Leonidas of Byzantium (early 1st century B.C.); and Demostratus.[547]

That Demostratus (or Damostratus) was both a statesman and a scientist, and wrote sometime in the early Roman empire a twenty-volume treatise *On Fishing*, plus books on rivers and other subjects, quotations of which reveal he was making careful observations, performing dissections, and conducting simple experiments.[548] Even Pliny the Elder's *Natural History* reflects a range of interest and quality of research comparable to the minor works of Theophrastus.[549] At the same time, Galen appears to have written a lost book on the physiology of plants, while his extant experimental and anatomical studies on animals rival Aristotle's, and he had more work

---

an original scientific treatise, hence he must have had access to advanced zoological works now unknown to us. Since his descriptions suggest discoveries and observations apparently unknown before his time, he probably had at hand research produced within a century of his own writing. It is worth noting that around the same time the musician Mesomedes (*OCD* 936) was writing lyrics on sundials and glassmaking, and versifications of scientific astronomy, geography, zoology, and mineralogy were also known from this time (see discussion and notes on Aratus, Dionysius, and scientific poetry in general, in Carrier 2016: 49–51).

547. *EANS* 554 & 503.

548. *EANS* 228. See discussion at the end of chapter 4.5. All his works are now lost, though scattered quotations survive (e.g. his *On Fishing* is quoted or cited in Aelian, *On the Characteristics of Animals* 13.21, 15.4, 15.9, 15.19; his *On Rivers* is quoted in Pseudo-Plutarch, *On Rivers* 13; he is cited on mineralogy in Pliny the Elder, *Natural History* 37.11.34; etc.). Collectively citations of him indicate a 1st century Roman official of significant status, although Pliny lists him among his 'foreign' sources (*externis*) in 1.37c (possibly because he wrote in Greek or was a native of Greece; he certainly employed Roman sources, e.g. Pliny the Elder, *Natural History* 37.23.85–86).

549. *EANS* 671–72; *NDSB* 6.116–21. This is often disguised by hyperbolic claims about the nature and quality of Theophrastus' minor works. Hence Stahl 1962 (and 1971) unjustly disparages the scientific content of Pliny's *Natural History* (as if comparable errors were never heard even from the greatest of ancient scientists), but a sober corrective is provided by Healy 1999, French 1994, Beagon 1992, and French & Greenaway 1986, who find Pliny more reliable than has been assumed. See also the brief account of his faults and virtues (and legitimate excuses) in Lloyd 1983: 135–49.

planned.[550] Similarly remarkable were the lost works of Sostratus, a surgeon and zoologist at Alexandria whose early 1st century work on animals was extensive, and remains impressive enough in quotations by later authors for William David Ross to conclude that "in zoology he perhaps ranks next after Aristotle." His known titles in this area include *On Animals, On the Nature of Animals*, and *On Striking and Biting Animals*, suggesting scientific interests beyond mere medical applications.[551] There were others writing on the subject in Galen's time, too. For instance, Galen admired a Lucius Calpurnius Piso who wrote a treatise *On Animals* in the late second century. Of which only a single quotation survives.[552] And Pelops (one of Galen's teachers), wrote scientific treatises on the dissection and anatomy of animals; also lost.[553] Countless authors and treatises on horticulture and agriculture and veterinary science also existed, as well as writers on other subjects, in every century from Aristotle's to Galen's, which conveyed scientific botanical and zoological content (and occasionally meteorology, geology, and mineralogy as well).[554]

Might work like this have grown and continued under the Pax Romana had everything not fallen apart a few centuries after it began? Even the encyclopedias of science and nature launched by Varro, Celsus, Seneca, and Pliny the Elder resemble very much the sort of thing natural philosophers were doing just before and during the time of Aristotle.[555] Seneca's friend

---

550. On Galen's completed and planned studies on animals see discussion in section 3.2 above. Galen mentions and describes his lost book on plant physiology in *On My Own Opinions* 3.5–6. Nutton incorrectly interprets this as a reference to his (extant) *On the Natural Faculties* (Nutton 1999: 148, §P.62,4): the latter is almost entirely devoted to human, not plant physiology, and Galen is quite clear when he says he wrote three volumes proving the physiological faculties of *plants*; moreover, in *On My Own Opinions* 3.6 he specifically distinguishes his book on plants from *On the Natural Faculties*, so they cannot have been the same.

551. See references in section 3.3.

552. *EANS* 204.

553. *EANS* 634.

554. *EANS* has extensive lists of the known authors and works; almost none preserved (991–92, s.v. "agriculture/agronomy"; 996–98, s.v. "biology"; 1003, s.v. "lithika"; 1011–12, s.v. "meteorology"; 1020, s.v. "veterinary medicine"). For some we don't even know the author's name: see Carrier 2016: 52.

555. The philosophical functions and context of these products of the Roman

Lucilius and Pliny the Younger's friend Sura were also interested and engaged in such studies, indicating more than an isolated fad.[556] Could this work have spawned more careful pursuit of the same subjects by others, which we have not been told of, or that was cut short by the chaos of the third century, when the scientific spirit that might have progressed was muffled instead by the rise of mysticism and supernaturalism? It is impossible to know for sure.

For example, Galen was showing interest in the study of magnetism and was beginning to make empirical observations relating to heat and fire.[557] Though he had more pressing scientific interests and thus only made a start on these subjects and never completed a proper scientific study of them, there is no telling who else may have shared Galen's interests and methods and been more engaged on these subjects than he. Even those who would reject this idea of a Roman revival must accept that these avenues of careful scientific interest had already been largely abandoned by the end of the 3rd century B.C., well before Rome ruled the Mediterranean—indeed, even before the peak of the golden age of science under the Alexandrian kings. Nevertheless, the unique achievements of Theophrastus and Aristotle continued to be preserved and admired throughout antiquity, and a rising interest in the same subjects in the Roman era is clearly evident. There had to have been many who thought progress in these subjects was possible and worthwhile. How long would it have been, honestly, before

---

period are briefly surveyed in French 1994: 149–95 (Pliny specifically: 196–255) and discussed further here in chapter 4.3.Aspects of the decline in proto-scientific natural history from the 3rd century on are briefly surveyed in French 1994: 256–303. But French generally does not discuss any of the non-extant works in the early Roman period, even though almost all first-hand research in natural history from that period is not extant.

556. *OCD* 835 and 862 (s.v. "Licinius Sura, Lucius" and "Lucilius (2) (Iunior), Gaius"). Seneca had Lucilius investigate first-hand various natural phenomena in and near Sicily (*Moral Letters* 79); Pliny the Younger requests help from 'the most learned' Sura in explaining a strange spring in Italy, which Pliny had investigated himself first-hand (*Letters* 4.30; Pliny later requests Sura's opinion on the reality of ghosts, again including his own first-hand experiences, in *Letters* 7.27). These examples are discussed further in chapter 4.3 and 4.4. Sura's immense fame as a scholar is attested in Martial, *Epigrams* 1.49 and 7.47. Nothing he wrote survives.

557. Heat and fire: Galen, *On the Causes of Disease* (cf. Mark Grant 2000: 47–51). Magnetism: Galen, *On the Natural Faculties* 1.14 (= Kühn 2.45–51).

the methodologies championed by scientists like Galen and Ptolemy were adapted to other fields of inquiry?

A more notorious problem is the lack of a scientific study of dynamics and ballistics. Though, as we will see, arguments from silence here don't hold up. The treatises that treated dynamics empirically and mathematically are lost, along with any evidence of their influence. So the absence of its development, though usually blamed on some sort of mental block, is not wholly supported by the evidence—a point we will examine later.[558] But even were it known to be the case, we could just as easily say developments in dynamics required the unrelated happenstance of developments in gunpowder artillery, which extended gun range to such a degree that calculating the course of projectiles only then became a military necessity.[559]

However, the evidence seems to point in a different direction. Ancient artillery certainly had ranges that would benefit from an improved dynamics. And Hellenistic governments were actually funding scientific research on improving artillery range, power, and accuracy. And we know Hipparchus wrote on the subject in such a connection (we just don't know what his results were). Conversely, Renaissance debates on dynamics and ballistics do not seem as much concerned with artillery as with philosophy, mechanics, and astronomy.[560] In fact, the motivations here resemble those of similar debates in antiquity. Even Galileo's explicit motive for solving fundamental questions in dynamics had little to do with artillery, but was driven instead by his desire to prove heliocentrism, and thus solve a number of astronomical problems left unresolved by both Ptolemy and Copernicus.[561] So we cannot presently explain what happened to dynamics in antiquity, as we do not have the works

558. For example, Vernant 1983: 288–89; P. Green 1990: 472; etc. Against these arguments see Russo 2003: 26–27 (though Russo takes his argument too far in later sections of his book) and Cuomo 2007: 3–4.

559. Suggested in Crombie 1959: 2.141–42 and Russo 2003: 110, and defended in Lindsay 1974: 383–406 (383: "ballistical problems did not come up strongly in the ancient world on account of the relatively short distance to be covered by the missiles," cf. 383–84 and 390–93).

560. See discussion in Crombie 1959: 2.131–226.

561. This is obvious in Galileo's *Dialogue Concerning the Two Chief World Systems*, which articulated his dynamic and ballistic theories in 1638, following a trend he had already begun by attempting to use his own (incorrect) tidal theory to prove heliocentrism in his 1616 treatise *On the Tides* (cf. e.g. Naylor 2007).

of Strato and Hipparchus on the subject (*On Motion* and *On Lightness and Heaviness*, and *On Objects Carried Down by their Weight*, respectively), nor do we have any other references to its status.[562]

Meanwhile, this brings us to one of the most infamous failures of ancient science: the rejection of heliocentrism. Many mythical explanations have been offered for this, though the actual reason was scientific. There were in fact three competing astronomical theories in the ancient world, which still had competing adherents in the Roman period. Static geocentrism held that the earth did not move at all, dynamic geocentrism held that the earth spun on its axis but that the rest of the cosmos still revolved around it, and heliocentrism held that the earth spun on its axis *and* revolved around the sun. Ptolemy gave three reasons for rejecting both heliocentrism and dynamic geocentrism.[563] First, if the earth revolves around a stationary point, then we should be able to observe stellar parallax, but we do not. This observation could also be explained by postulating that the universe is extraordinarily vast, but that conjecture seemed more incredible than that the earth remains at rest, as it does appear to do.[564] Second, if the earth rotates on its axis, the speed of this rotation would be so vast it would produce observable consequences, like torrential winds, toppling towers, or projectiles moving oddly, yet we observe no such thing. And third, if heliocentrism is true, then we are left with no explanation for gravity, since

562. Except possibly one: a sequence of two passages in the Arabic translation of Hero's mechanics appears to repeat obsolete Aristotelian dynamics (in Hero, *Mechanics* 2.33, part of a Q&A section where he appears to answer two questions about differential speed of fall), but their translation into Arabic may have been compromised (if Hero was originally writing of impact force and not time of fall, a problem already noted in interpreting passages from Strato); or their inclusion could be a Muslim interpolation and not in the original Greek (Q&A sections in ancient texts sometimes became expanded by later editors); and even if genuine and correctly translated, that Hero was repeating obsolete science does not entail all Roman physicists agreed with him (just as we have seen in the case of geocentrism and visual ray theory).

563. Ptolemy, *Almagest* 1.5–7.

564. Ptolemy estimated the distance of the star field to be less than 20,000 earth radii (roughly 92 million miles), which happens to be almost exactly the actual distance of the sun (Ptolemy estimated solar distance to be considerably less). Heliocentrism required accepting vastly greater distances for the stars. See comments of G.J. Toomer in *OCD* 190 (§8 in s.v. "astronomy").

the most reasonable explanation at the time was Aristotle's theory of natural places, or similar theories of gravitation to a common point, which entailed the center of the universe was located at the center of the earth. Positing multiple gravitational centers violated their own conception of Occam's Razor. Aristotle's division of terrestrial and celestial spheres as operating on different physics, wherein change does not occur in the celestial realm, also suited the single-center hypothesis.

Notably, none of Ptolemy's objections to heliocentrism are religious. He does not argue the earth 'must' be the center of all Creation or that it would be sacrilegious to suggest otherwise. Instead, his objections are all empirical and derived from scientific theories—which happened to be false, though not obviously so.[565] However, some pagan natural philosophers did object to heliocentrism on religious grounds.[566] And the heliocentric theory of Aristarchus really only made sense in the context of Strato's 'godless' physics, which stood in opposition to the more popular 'god-friendly' system of Aristotle. Yet, just as we saw for medicine, rival schools of thought existed among astronomers and physicists even in the Roman era. In addition to Strato's system, throughout Greek and Roman times there were many fans of Democritean and Epicurean physics who embraced something close to a theory of inertia and assumed as a matter of course that the same physics operated above the moon as below it (and some Aristotelians thought likewise, from Strato to Xenarchus). Atomist sympathizers also argued the universe was not only vast in size, but infinite, and had no common center, and that the stars were actually distant suns. But just as happened to atomistic medicine, its parallels in astronomy and physics were expunged from the historical record by disinterested medieval scribes.

This is a tragic loss. For atomist and Stratonian theories of inertia would have answered Ptolemy's second objection, and their celestial physics, which held that gravity is a universal force or effect and that no difference obtained between the physics of the terrestrial and celestial spheres, answered his

565. Hence B.L. van der Waerden: "in my opinion, the Greeks were quite right... to reject the hypothesis" of heliocentrism (van der Waerden 1963: 57).

566. For example: Dercyllides (cf. Theon of Smyrna, *Aspects of Mathematics Useful for Reading Plato* 3.41.200) and Cleanthes (cf. Plutarch, *On the Face that Appears in the Orb of the Moon* 6 = *Moralia* 923a), but this was not typical among the elite (on pagan hostility to atheism see the end of chapter 2.6).

third objection.[567] As for his first objection, it is likely that Strato, and thus his student Aristarchus, accepted and expanded on the Epicurean theory that the stars are actually other suns, and thus of the same size as our sun, but appearing small due to their great distance.[568] That would entail the distance of the stars was indeed extraordinarily vast, since many astronomers, from Aristarchus to Posidonius, were already confirming vast distances for the sun. That left no remaining objection to heliocentrism. But this meant the heliocentric model only made sense within the context of some form of atomistic physics, which happened to be largely correct and yet was widely regarded as godless. Consequently we cannot rule out religious reasons for the popularity of geocentrism, since its atheism appears to have been a major reason for the rejection of atomism.

However, as we have seen, Roman science was not monolithic. The work of Strato, Aristarchus, and others had not been forgotten. A debate still raged about the validity of Aristotelian physics even in Roman times, and there is clear evidence that not every Roman agreed with Ptolemy. The Aristotelian Xenarchus in the 1st century B.C. wrote a treatise *Against the Fifth Element*, attacking the celestial mechanics of Aristotle—yet another book medieval scribes chose not to preserve.[569] At the dawn of the first century A.D., the poet Ovid described dynamic geocentrism as if it were an obvious fact, which entails he was not alone.[570] Seneca said a few decades later that experts still disagreed about this, and he couldn't decide himself whether the earth moves or not, concluding instead that further research is needed to decide the question.[571] And a few decades after that, Plutarch attests to the existence of Roman philosophers and astronomers who rejected Aristotelian dynamics and were engaging sophisticated debates on the subject, even contemplating theories of inertia and universal gravitation.[572]

---

567. On these points see related discussion and notes in section 3.3.

568. On the Epicurean theory in this regard see Lucretius, *On the Nature ofThings* 1.1052–1113, 2.62–166, 2.184–332.

569. *DSB* 7.134 (in s.v."John Philoponus"). This Xenarchus was the tutor of Strabo and friend of Augustus, and thus no insignificant figure (Strabo, *Geography* 14.5.4).

570. Ovid, *Fasti* 6.269–71. The same is attested in other authors: Cohen & Drabkin 1948: 105–07.

571. Seneca, *Natural Questions* 7.2.3.

572. Plutarch, *On the Face that Appears in the Orb of the Moon* 6–11 (= *Moralia*

Engineers had mixed views, but some appear to have been more sympathetic to an eclectic atomism.[573] Even Galen wrote a treatise attacking Aristotelian dynamics—and yet, once again, medieval scribes did not bother to preserve it.[574] And that's just what we are lucky enough to know about. There is no telling how much else was being written on this subject.[575] But even from what little we know, had things continued in the direction Galen and others were pressing, the issue might have been resolved correctly a thousand years or more before Galileo or Newton. At the very least, this evidence of Roman discontent and disagreement in matters of dynamics and planetary theory suggests the means, motive and opportunity were well enough in place for progress even in these subjects, had circumstances allowed science to continue.

The same can be said of other theoretical questions confounding the ancients, like the physics of vision and the origin of species, where several competing theories were still being debated. Those scientists whose books were preserved argued there was a 'visual ray' emitted from the eye and

---

922f-926b). Sambursky 1962: 234–44 provides an apt analysis of relevant sections of this text.

573. Russo 2003: 279–80. For example, see Philo, *Pneumatics* 7 and Hero, *Pneumatics* 1.pr. Keyser 1992 even argues that Hero developed some of his pneumatic machinery specifically to refute (by demonstration) certain elements of Aristotelian dynamics.

574. We are lucky even to know this: in a treatise that survives only as fragments from an Arabic translation, Alexander of Aphrodisias (in the early 3rd century A.D.) attempted to refute Galen's criticisms (in yet another lost work) of the Aristotelian physics of motion. See: Pines 1961, Nutton 1984b, and Nutton 2013: 235 (with Simplicius, *Commentary on Aristotle's Physics* 6.10.1039.12–33). For background on this Alexander: *DSB* 1.117–20, *EANS* 54–55, and *OCD* 59 ("Alexander (14)"), with Todd 1976: 2–20 and Sharples 1987. There were other Roman commentators on Aristotle whose works are largely lost (e.g. *OCD* 14, s.v. "Adrastus (2)"; cf. *EANS* 31–32).

575. However, the evidence of these debates and witnesses (of the theories and works of Strato and Aristarchus and others) in the Roman period is still sufficient to refute Russo's already-implausible contention that proto-Newtonian models of the solar system were discovered and somehow 'lost' before the Roman era. See note on this point in section 3.3. Nevertheless, Russo does present good evidence that Aristotelian dynamics was not universally accepted (cf. Russo 2003: 293–96, 302–09).

reflected back, and when light (such as from candles or the sun) struck objects it made them apparent to this 'visual ray', a theory that sounds strange to us, though it did have plausible arguments to commend it. The converse theory (which happens to be correct) eliminated the visual ray and had light do all the work.[576] Galen and Ptolemy defended the wrong theory of vision, but in doing so reveal they had opponents defending the correct one—including Strato and Hipparchus, who still had their defenders in the Roman period and thus could not be ignored.[577] However, like heliocentrism, the correct theory was associated with atomism and hence atheism—and nothing its advocates wrote was preserved. Likewise for the ancient theory of natural selection, which was gradually developed by various Presocratic thinkers, and embraced by Epicureans and others sympathetic to atomism well into the Roman period.[578] Their view was attacked and rejected by Aristotle and later creationists like Galen. Once again the theory's association with atomism and atheism seems to have ensured that almost none of the works defending it survive.[579] Yet these books existed and the debates surrounding them had not ended. Hence I suspect it was, ultimately, political and economic catastrophe, not ideology or disinterest, that put an end to progress in these subjects.

---

576. On ancient visual theory see A.M. Smith 1999 and 2014: 25–129, who explains the controversies and why incorrect positions were thought convincing.

577. Galen conducted extensive mathematical and empirical studies of vision (*On the Uses of the Parts* 10.12–15 = M.T. May 1968: 490–503, with 472, esp. n. 19), as did Ptolemy in his *Optics*—though his first chapter discussing his physical theory is lost, A.M. Smith 1999 partly reconstructs it from the surviving books, and it aligns with Galen's on several but not all points. Siegel 1970: 10–126 surveys both visual models and their connections to the related experiments and theories of both Galen and Ptolemy. Lehoux 2007a analyzes the role these studies played in ancient epistemological debates.

578. Evolution, however, was as yet invisible, for lack of good chronological data on speciation; it would be centuries even after the Scientific Revolution before enough paleontological data would exist to see that evolution had occurred. Darwin then combined ancient natural selection theory with the modern observed pattern of evolution to produce his famous theory.

579. See discussion in Sedley 2007 and Russo 2003: 160–65. Descriptions of ancient theories of natural selection (presaging Darwin's) can be found in Aristotle, *Physics* 2.8.198b-199b, and Lucretius, *On the Nature of Things* 2.1150–56, 5.783–877.

3.6 Technological Progress

Such was the state of science in the Roman empire before a general social and economic decline began in the 3rd and 4th centuries A.D. But many scholars also link the history of science to the history of technology, even though most technological progress before the modern era had little or nothing to do with science. There were many technological advances produced by scientists or aided by scientific knowledge, but a great deal more was developed without them. Although the lack of useful records often makes it impossible to determine the connections any given technology may have had with ancient science, even at the dawn of the Scientific Revolution science usually followed technology, not the other way around—gunpowder and the magnetic compass were invented long before they were studied scientifically, while even the telescope was invented by a jeweler a century before Newton's *Optics* properly explained how it worked.[580] Nevertheless, attitudes toward scientific and technological progress are related, and many of the claims scholars have made about ancient attitudes toward scientific progress are challenged as much by the history of technology as by the history of science.

A typical example is the strange assertion by Jean Pierre Vernant that the Greeks "were not technological innovators," that they used only "human or animal force" and never "the forces of nature," and consequently there was a "stagnation of technology" when there should have been "decisive progress in this field" comparable to their progress in the sciences.[581] Yet he goes on to admit that engineers like Philo, Ctesibius, and Hero were "inventors" who "produced theories about various types of machines, how to make them, the way they worked, the rules for their use," and in fact "a series of remarkable inventions were produced by this technological ingenuity combined with research into general principles and mathematical rules." How this can possibly warrant his claim that the Greeks "were not technological innovators" is beyond comprehension.[582] Even Vernant's claim that the Greeks failed to exploit the forces of nature is based on his

580. See discussion and notes in chapter 1.1.

581. Vernant 1983: 280–81.

582. Vernant 1983: 283, where he lists numerous inventions (and more in n. 19 on p. 297).

inexplicable assumption that the watermill was not introduced until the 2nd century A.D., even though it had been described as already a standard technology by Vitruvius in the 1st century B.C.[583]

We hear something similar from Peter Green, who claims "the actual list of known technological advances is minimal" from the 4th to the 1st century B.C.[584] He rattles off a short paragraph of items and concludes "this is not exactly an overwhelming list." But suspicion is warranted. Like Vernant, Green contradicts himself, this time by giving a completely different list of inventions from that period elsewhere in the same book—several times longer than the first.[585] And yet even combining both lists, the sum of them is still much too short if intended to encompass the early Roman period, and is probably too short even for the period he claims to be describing.

Altogether, Green first credits the Greeks with inventing the iron plowshare, the ox-powered wheel-pump and grain mill, an improved saddle quern, the rotary quern, the screw press "for crushing grapes and olives," the compound pulley, the Archimedean screw pump, the Ctesibian piston pump, the lateen sail, domestication of the camel, "a new quick-growing wheat which gave a double harvest and a higher yield," double and triple crop rotation, and (not the invention, but the introduction to the West of) the commercial cultivation of sugar, cotton, peaches, cherries, and apricots. Then a hundred pages later he remembers to add "cogged gears," and "glassblowing, hollow bronze casting, surveying instruments, the torsion catapult," "an odometer and a pantograph," "the water clock and the water organ," "the machine-gun" (meaning the automatic catapult), "the fire-engine pump," "the automatic puppet theater," and various other automated devices for amazing crowds, new "modes of transport for heavy beams and columns," and "celestial spheres for the benefit of astronomers." He also allows the invention of the watermill and steam-powered demonstration devices, and minor improvements to lifting technologies.[586]

---

583. Vernant 1983: 296 n. 9. See below for sources on ancient watermills.

584. P. Green 1990: 367.

585. P. Green 1990: 467–69.

586. Ibid. Though he's wrong about the camel (P. Green 1990: 367). Camels were domesticated in Arabia and the Levant long before. And hollow bronze casting (contrary to P. Green 1990: 467): that technique actually originated in Mesopotamia millennia before its appearance among the Greeks (Dalley & Oleson 2003: 7–11).

Of course, Green *means* to argue that most inventions had little economic impact or were only enjoyed by the rich, which is often true, although I am not sure how he could know that beyond a few obvious cases—neither of his lists fit this argument very well. But even if he's right, this would not be relevant to the question of Greek inventiveness or interest in technological progress. Even if only the rich were buying advanced technologies, and even if much of what they were buying was not put to industrial use, there still had to have been a profitable economic trade in these luxury technologies which continually drove their invention and production.

Whatever the case, the fact remains that science-based technologies were more commonly developed and employed in Greco-Roman society than any other until the early Renaissance. This included an extensive mathematical understanding of the operation of the wheel, windlass, pulley, lever, wedge, ramp, screw, piston, valve, gear, spring, and siphon, all in terms of the mechanical advantages these devices provided, and thus in terms of how much each device multiplied the labor of a man or animal. In other words, ancient scientists actually developed and employed a limited but nevertheless scientific understanding of efficiency. They were specifically working out how to build a crane operated by ten men that could do the work of a hundred, or a pump operated by one man that could move as much water as four. Scientific engineers were also looking for ways to store and release power, introducing to that end the use of torsion, steam, suction, elevated weights, and compressed air. As a result of all this, the building industry was advanced by a variety of crane and lifting technologies, the food industry was advanced by the exploitation of screw presses for crushing grapes and olives, watermills for grinding grains, and aqueducts for irrigation, and even the mining industry was improved by a variety of pump technologies; and so on. And that's judging only from Green's lists.[587]

There was clearly some interest in all this research and development. Attempts to slight this fact as vehemently as Green does are more fiction than history. For example, Green rather infamously argued that "the Greek world, which knew all about the principles of lever and pulley, could not even dream up so simple a device as the wheelbarrow."[588] Only a few years

587. For example, on the economic impact of screw press technology: Lewit 2012.

588. P. Green 1990: 474.

later he was refuted. The wheelbarrow is attested in a Greek construction inventory from the late 5th century B.C.[589] There is no reason to believe it was not in common use thereafter. Just because extant art and literature do not mention it is no reason to assume it was not used. Where would we expect such a trivial piece of construction equipment to be discussed or depicted? And since these "one-wheeled carts" would have been made of wood, to expect archaeological evidence of them would be just as foolhardy. Arguments from silence in the history of technology are thus always worthy of suspicion.[590] And using such an argument to condemn an entire civilization of incompetence is itself incompetent.

So why not attempt what Green failed to do, and produce a more thorough list of the technological inventions that were either made or exploited by the Romans?[591] We already know Green was relying on outdated scholarship, deriving most of his argument (and much of his lists) from the deeply flawed work of Moses Finley.[592] And yet on the specific matter of ancient technology, Finley has already been decisively refuted by a considerable body of more recent scholarship.[593] To be fair, both Finley

---

589. M.J. Lewis 1994, who notes the inscription in question literally says "one-wheeled cart," which can only be a wheelbarrow or its functional equivalent.

590. See Renn 2002 and Marchis & Scalva 2002 on the various problems with 'arguments from silence' in the area of ancient technology. Green commits a similar boner when he claims "the astrolabe" was "restricted to pragmatic arts, such as navigation" (P. Green 1990: 457). As noted in section 3.3, Diodorus and Ptolemy both wrote treatises on the construction and astronomical use of the plane astrolabe, and Ptolemy developed a more complex armillary astrolabe, and also discusses the use of the quadrant.

591. For context see *OCD* 1435 (s.v. "technology").

592. P. Green 1990: 367 and 467–69 essentially repeats the same arguments as P. Green 1986, and in both cases he closely follows Finley 1981 and 1985: 109, 113–14, 145–47. Just as Finley is obsolete on the technological point, so are almost all who preceded him (e.g. almost every conclusion in Reece 1969 is now known to be false, as is much of Pleket 1973), and so also are many still (like Peter Green) who have not caught up with current research (a point made more generally in Greene 1990).

593. There was already a more impressive list and survey of technologies in K.D. White 1984, who includes many more inventions than I will here. But his work has been greatly multiplied and reinforced by others, e.g. Wikander 1990, Schneider 1992, Chevallier 1993, Greene 1994, and Russo 2003: 95–141, and most

and Green attack ancient culture on a double front, on the one side claiming (a non-existent) technological stagnation, while also claiming broader and more fundamental failures in the ancient economy.[594] My only real concern here is with the question of scientific and technological progress, not the more complex question of 'economic' progress, since that can be defined and assessed in so many different ways. Though I will inevitably touch upon some aspects of this, and though I think there are definite signs of economic progress in *some* sense under the Romans (certainly in terms of increased efficiency, production, infrastructure, and trade), I will not be resolving that debate here.[595] I will merely demonstrate that technological progress existed

---

directly by Greene 2000, who puts together a point-by-point refutation of Finley's entire project (see also following note on ancient economics), and Simms 1995: 83–93, who effectively provides a refutation of Green's own "list" (factually and methodologically). On technological progress as a feature of the Roman world see Schneider 1992: 219–23; and a great deal more has been established in the twenty years since.

594. This is the general thrust of P. Green 1990: 366–81 and 467–79, and the context of Finley's every mention of technology. Oleson 1984: 397–408 also offers reasons for the slow pace of ancient technological progress that are just as clichéd and dubious, and not well explored.

595. For ongoing debates regarding the nature of the ancient economy see Derks 2002, Scheidel & von Reden 2002, Manning & Morris 2005, and now Scheidel et al. 2013 and Andreau 2015. I've concluded the evidence in no way supports the Finley camp. It never did. But it certainly doesn't now. On whether or what kind of 'economic rationalism' existed in antiquity see Macve 1985, who refutes several myths about ancient economic attitudes and abilities, as do D'Arms 1981, Andreau 1999 (with Pleket 2001), Meissner 1999: 99–122, Greene 2000, Christesen 2003, Russo 2003: 243–67, and Morley 2007. A more accurate account now is represented in Temin 2006. See also *OCD* 222–23, 276–77, 391 and 899 (s.v. "banks," "capitalism," "credit" and "maritime loans") with support from: *OCD* 484–86 (s.v. "economic theory (Greek)" and "economy, Greek," "economy, Hellenistic," and "economy, Roman"); *OCD* 1490–93 (s.v. "trade, Greek," "trade, Roman," and "traders"); and *OCD* 734–35 and 1526–29 (s.v. "industry" and "urbanism"). Less informative is *OCD* 787–88 (s.v. "labour"). Much better on that topic is Temin 2004 and Brunt 1987, especially in conjunction with Manning 1987, who surveys a massive increase in the size and scale of all manner of industrial operations under the Romans; and Parker 1987, who surveys a correspondingly enormous boom in all manner of trade operations under the Romans. See also Mattingly & Aldrete 2000 (on the commercial implications of the Roman food supply) and DeLaine 2000 and 1997 (on the labor implications of the Roman building industry) and

in antiquity, was recognized, and evidently valued enough to be continued and exploited.

I shall accomplish this by doing what Finley and Green did: listing the examples and observing whether they are many or few. D.L. Simms rightly qualifies such an approach, noting that lists like ours are always too small because they exclude numerous improvements to existing technologies, which have an equal right to be counted. Of course, that only means my list will refute Green's *a fortiori*, by yet excluding a great deal more than could be included. Simms also argues we ought to distinguish among different *types* of innovation. But since progress need not be measured by only one set of values, I will focus on significantly life-changing innovations of whatever type. So with those caveats in mind, let us assess what the Greeks and Romans actually did invent.

I. Techniques as Technologies

If techniques count, then we would have to include countless useful discoveries and inventions in medical practice, in surgery, in pharmacology (including birth control and abortifacients), in cartography, and in all the sciences just surveyed, including physiology, astronomy, optics, harmonics, pneumatics, hydrostatics, and mechanics—and all the discoveries made in mathematics, even besides the invention and development of the entire fields of geometry, algebra, isoperimetry, combinatorics, and both plane and spherical trigonometry. There are also many minor examples, such as the development of liquid mirrors to observe eclipses, portable sandboxes to work problems in geometry, and fully articulated anatomical wooden dolls for medical training.[596] The Romans even developed electroshock therapy, using electric fish as a local anesthetic, and to alleviate gout, migraine

---

Shaw 2013 (on the role of labor in Roman agriculture). And in general: Erdkamp & Verboven 2015.

596. For sandboxes see Carrier 2016: 84; and, e.g., Seneca, *Moral Epistles* 88.39. The use of water, oil, or pitch mirrors is attested by Tertullian, *To the Nations* 2.6 and Seneca, *Natural Questions* 1.12.1 and 1.17.2–3. Anatomically correct dolls with moving joints are mentioned by Galen as the preferred method of teaching the art of bandaging, in a lost work quoted in Arabic (Lyons 1963: 101). These dolls would have been full or nearly-full scale and must have been finely crafted to mimic an actual human range of motion to teach bandaging as Galen recommends.

headache, and hemorrhoids.[597]

We can also add to all the sciences the invention of proto-scientific methods of composing histories and chronographies; and the invention of proto-scientific textual analysis.[598] And outside the sciences there were surely also countless improvements in the techniques of craftsmen in every field, from carpentry to smithing, which are difficult or impossible to document.[599] But in the area of technologies as skills, we know a Latin shorthand was invented by Cicero's scribe Tiro around 63 B.C., and though the evidence is unclear whether Greek shorthand followed or preceded the Latin, both systems were in wide use by the 2nd century A.D., greatly economizing secretarial work and speech reporting.[600] I already mentioned the invention of various systems of cryptography and optical telegraphy, to which we can add the introduction of the carrier pigeon, and all three systems were improved by the Romans, who developed a telegraphy code as efficient as modern Morse code.[601] Both Archimedes and Apollonius of

---

597. Kellaway 1946. Though Kellaway's dating of some authors is obsolete, his citation of sources is thorough and his conclusions indisputable: the use of electroshock therapy was discovered in the reign of Tiberius, was further tested and developed by Scribonius in the reign of Claudius, and its therapeutic value was confirmed experimentally by Dioscorides and Galen.

598. Pfeiffer 1968.

599. Though see Mercer 1975 for an extensive discussion of the wide array of carpentry tools (and related techniques) developed by the Greeks and Romans (and you will see some specific examples in coming pages, like the Roman invention of the carpenter's plane), while Mols 1999 surveys Roman advances in carpentry techniques in the construction of furniture. K.D. White 1967b and 1975a provides a similar survey for agricultural tools. R. Taylor 2003 surveys Roman innovations in construction techniques (all throughout, but esp. pp. 44–48), and Absmeier 2015 does the same for wooden buildings; while O'Connor 1993: 44–62 surveys Roman construction tools and equipment. See also coming references on wooden machinery (cranes, waterwheels, bonesetters, pumps, harvesters, wagons, ships, presses, etc.). Similarly there is a lot to explore in the technology of Greco-Roman sculpting in ceramics, stone, and bronze (Hasaki 2012). Strong & Brown 1976 and Oleson 1986 also treat a small but representative sample of technologies employed in a wide range of Roman industries.

600. *OCD* 1425–26 (s.v. "tachygraphy"); James & Thorpe 1994: 510; Marrou 1964: 448–50 (= Marrou 1956: 312–13).

601. Cryptography: James & Thorpe 1994: 507–12. The carrier-pigeon: James &

Perga invented place-notation systems for representing large numbers, and at some point in the Greek period the decimal place-notation system that we use today became a standard calculating tool—the abacus was based on it, and the decimal abacus was in regular use before and all through Roman times.[602] Agriculturally, even Peter Green acknowledges the Greeks introduced "double or even triple crop rotation," adopted the cultivation of cotton and sugar for the luxury market and the domestication of the peach, cherry, and apricot, and invented a new fast-growing wheat that could produce a double harvest.[603] The cultivation of melons, lemons, mangos, and pineapples also spread under Roman tenure.[604] To all that we should add Julius Caesar's calendar reform (as noted in section 3.3), which became the foundation of our modern calendar, keeping the solar year in proper alignment for centuries.[605] And of course we should not forget the Roman invention of the first fully-developed postal service.[606]

II. Inventions Already Mentioned

That is certainly not a complete list of techniques invented in antiquity

---

Thorpe 1994: 525. Optical telegraphy: James & Thorpe 1994: 531–36 (and see notes on Philo in section 3.4 above, and the discussion of telegraphy's development in Polybius, *Histories* 10.43–47).

602. *OCD* 1 (s.v. "abacus"); Turner 1951; O'Connor 1993: 61–62; Maher & Makowski 2001; Russo 2003: 43; Hermanns 2010. On place notation in Archimedes: Netz 2003 (also discussing abacus: 260–61; and the system of Apollonius: 284–86), which is sufficient on the facts, though some of his added speculations are questionable.

603. P. Green 1990: 367. In fact there were several systems of ancient crop rotation in use, demonstrating an increasing sophistication of options: cf. K.D. White 1970: 110–24 and Pliny the Elder, *Natural History* 18.50.187.

604. Renn 2002: 15–17.

605. See Pliny the Elder, *Natural History* 18.57.211. The Julian calendar was not improved upon until the Gregorian reform of the 16th century: see *ODCC* 705 (s.v. "Gregorian Calendar").

606. Hyland 1990: 250–62 (based on Persian precedents: Humphrey et al. 1998: 425–26). Though the Imperial Post was not (officially) available to private citizens, it was still an extensive and efficiently organized postal system for rapidly and systematically transporting government mail, baggage, and personnel throughout the empire, a remarkable achievement in its own right.

and employed by the Romans. But even if we limit ourselves to physical inventions, we have already mentioned advances in the use of the wheel, windlass, lever, wedge, and ramp, and the invention of the compound pulley, the pressing screw, locking screw, water screw, the piston and valve, and the gear, spring, and siphon.[607] Ancient gearing included cogged and toothed gears, gear trains and transmissions, gear-driven chain-belts and bucket chains, reduction gearing, worm and ratchet gears, and the rack and pinion.[608] The Romans also invented the crank.[609] Likewise, many types of valve were invented, including the beveled spindle valve.[610] And the ancients also understood the principle of the cam.[611] We listed many other inventions in our survey of the sciences: the torsion catapult (and trials of the pneumatic catapult, bronze spring catapult, and automatic catapult), the metal-frame catapult, the pneumatic water organ, the cylinder-and-plunger pump and the reciprocating double-force pump, Archimedes' screw, the continuous parastatic waterclock, the screw-cutter, a variety of terrestrial and aquatic odometers, and advanced sundials (including the conical and the portable), and the cuckoo clock.

The Greeks also invented countless other robotic amusements, a wide

607. On all inventions listed in this and the following paragraph see references provided in previous sections above where each invention is mentioned, and also lists and notes in P. Green 1990: 367, 467–69. On applications of the five 'basic' machines (plus the gear) in antiquity, Drachmann 1963 is still useful, though somewhat out of date. And there is some dispute as to whether the Greeks invented or 'reinvented' the waterscrew: cf. Dalley & Oleson 2003.

608. For the full range of ancient gearing see all the cited sources on ancient machinery, above and below (I have seen each type listed in several ancient sources and artifacts).

609. Originally debated (e.g. Drachmann 1973 vs. Simms 1995: 57, Landels 2000: 10–11, and Di Pasquale 2004: 150–64), a third century inscription now establishes its use in Roman industry (Ritti et al. 2007: 147–48), and one has even been recovered from the excavation of a 2nd century Roman sawmill (Schiøler 2009).

610. Russo 2003: 123 (with diagramatic reconstruction: 124). Hero, *Pneumatics* 1.27–28 describes the use of spindle valves.

611. Cams and camshafts are employed in many of Hero's automata—even his wind-powered organ employed a cam-driven piston (cf. M.J. Lewis 1993: 143–45 and 1997: 84–115; Hero, *Pneumatics* 1.43). There is also evidence they were used in industrial machinery (see discussions of mechanized hammers and sawmills below), and to operate cylinder block force pumps (M.J. Lewis 1997: 111–13).

array of other siege machinery and tools, and a whole galaxy of scientific and surgical instruments, well beyond those already mentioned.[612] Though all of that was invented before Roman times, all of it was widely employed by the Romans. We also saw that the Romans added the automatic door, the coin-operated vending machine, the geared crane (Hero's Baroulkos), the Menelausian balance, and the Celsean surveyor (though the latter's function is unknown). But of greatest significance to scientific progress are the scientific instruments, since new and better instruments entails not only advances in scientific knowledge, but also a persistent appetite for making these advances, and for developing technologies to do this. We've already mentioned parabolic mirrors that were actually used to burn or magnify, star globes, star charts, scientifically designed maps, simple and complex diopters, quadrants, astrolabes and armillary spheres, gear-train computers, and an experimental use of lenses for burning and magnifying. But countless other instruments can be added to this list, far more than I have mentioned, devices built and used not only by astronomers, but also for research in optics, catoptrics, cartography, harmonics, pneumatics, and mechanics, as well as practical applications in surgery and architectural and geographical survey.

III. Hellenistic Inventions

That much I have said already. To that list we can add a number of inventions that were developed between 400 and 100 B.C., though employed and often improved in subsequent centuries.[613] This included the invention

612. Some of which survive only in medieval Arabic translations of ancient Greek treatises (cf. Schomberg 2008).

613. I will leave out entirely trivial inventions, like the bottle rocket, e.g. Archytas is said to have invented a toy jet airplane, described as a wooden dove propelled by "a current of air" from within. Though Gellius' description of how it worked is inconveniently missing due to a lacuna in the manuscript (Aulus Gellius, *Attic Nights* 10.12.8–10), he seems convinced the method he was to describe would work. Some scholars regard the story as a legend, but Gellius' confidence in the face of his own skepticism leads me to conclude it was probably an ordinary soda rocket (employing vinegar and sodium bicarbonate, which were readily available). See Berryman 2003: 354–55 (and sources there) for alternative suggestions, which I find much less plausible.

of parchment and the codex,[614] the gimbal,[615] the universal joint,[616] the butt hinge,[617] the water level,[618] the mesolabe,[619] the anemoscope,[620] the thermoscope,[621] the hydrometer,[622] the volumetric table,[623] the perfumer's press,[624] the sphere lathe,[625] the iron frame-saw,[626] the miner's

---

614. For parchment (and the bound codex, i.e. a proper book as distinct from a scroll): James & Thorpe 1994: 485; Reynolds & Wilson 1991: 3, 34–35; Skeat 1982; Roberts 1954.

615. James & Thorpe 1994: 118.

616. Simms 1995: 63–64; Russo 2003: 110; Grewe 2009; Athenaeus, *On Siege Engines* 35–36.

617. Like the modern door hinge, with two plates attached to abutting surfaces and joined by a rotating pin: British Museum 1908: 160; Hero, *Pneumatics* 1.11.

618. Water level serving the same function as the modern bubble level: Russo 2003: 238–39; M.J. Lewis 2001b: 89–96; O'Connor 1993: 59–60; Dilke 1971: 74–76.

619. A kind of slide-rule for calculating scaling functions for architects and engineers: Russo 2003: 111, Netz 2002: 213–15, Knorr 1989: 131–53, Cohen & Drabkin 1948: 62–66.

620. A sophisticated combination of windvane and windrose for tracking the wind: Taub 2003: 103–07, 148–49, 178–79. Some even had mechanisms for a readout indoors, so an observer could know the wind conditions before going outside (Varro, *On Agricultural Matters* 3.5.17).

621. Essentially the world's first thermometer: Philo of Byzantium, *Pneumatics* 7 and Hero, *Pneumatics* 2.8 (see Keyser 1992: 109–10).

622. An instrument for weighing the density of liquids, described in late antiquity but invented sometime before (probably by Menelaus): Hill 1993: 61–65 and Khanikoff 1860: 40–53, with *DSB* 10.300–01 (in s.v. "Pappus of Alexandria"); and Synesius, *Letters* 15, with *DSB* 13.225 (in s.v. "Synesius of Cyrene") and *OCD* 281 (s.v. "*Carmen de ponderibus et mensuris*").

623. A systematically constructed table of stoppered basins for measuring the volumes of dry and liquid goods for sale: Mau 1908: 88–89 = Mau 1982: 92–93 (discussing an example recovered from Pompeii).

624. A powerful wedge-block press for ultra-fine extraction of liquids: Drachmann 1963: 55–56 and Mattingly 1990.

625. A lathe for turning out balls and spheres: Strabo, *Geography* 1.3.3 and Pseudo-Aristotle, *On the Universe* 391b22 (though it was simply called a lathe, cf. *LSG* 1807, s.v. "*tornos*" §II, in these contexts a sphere-making lathe is clearly meant).

626. Meiggs 1982: 346–49.

lamp,[627] the pile driver,[628] the acoustic resonator,[629] the garden fountain,[630] the snorkel,[631] the diving bell,[632] the multihook fishing line,[633] the folding pocket knife,[634] the whaling harpoon,[635] the heated bath,[636] indoor plumbing, the shower, the toilet sponge, and the most practical public and private toilet facilities known in the West until the Renaissance.[637] At some point a drop-boom fishing spear was introduced, employed in shallows to spear numerous fish simultaneously, while in deeper waters inflated bladders were widely exploited as floats for nets and lines, and then to float rafts, and even military pontoon bridges.[638] The screw press was adapted to fulling

627. A lamp bound to the forehead of miners: according to Agatharchides as reported by Diodorus Siculus, *Historical Library* 3.12.6.

628. Mechanical pile driver: Vitruvius, *On Architecture* 3.4.2 (and something similar used to compact earth is mentioned in Columella, *On Agricultural Matters* 1.6.13), with discussion in O'Connor 1993: 50–51.

629. Mathematically designed metal jars that enhanced theatrical music: Landels 1967.

630. Using pressurized water: Schürmann 2002: 49–53.

631. For divers, compared to an elephant's trunk: Aristotle, *Parts of Animals* 2.16.659a8–12.

632. A small inverted pot for delivering a pocket of air to a diver: Pseudo-Aristotle, *Problems* 32.5.960b31–33.

633. Bekker-Nielsen 2004: 89–90.

634. British Museum 1908: 139 (with fig. 157).

635. Oppian, *Fishing* 5.131–51 (discussed in Rihll 1999: 111–13).

636. Connolly & Dodge 1998: 34–35, 238–47.

637. Shower: James & Thorpe 1994: 460. Toilets, toilet sponge, and indoor plumbing (taken for granted in Seneca, *Moral Epistles* 100.6): Connolly & Dodge 1998: 130–33, 148–49 and Pavlovskis 1973 *passim*. For a thorough study of Roman toilet technology: Jansen et al. 2011.

638. Boom-spike: Oppian, *On Fishing* 4.535–48. Inflated bladders to buoy whaling lines: Oppian, *Fishing* 5.131–51. To buoy fishnets: Bekker-Nielsen 2002: 219. Using inflated bladders as floats (even to float rafts) was a common sight: Plutarch, *On the Face that Appears in the Orb of the Moon* 12, 15 (= *Moralia* 926c, 928b); Aristotle, *Physics* 4.9.217a, 8.4.255b and *On the Heavens* 4.4.311b; Pseudo-Aristotle, *Problems* 25.13.939a. For examples of rafts and pontoons exhibiting float technology in ancient art: Casson 1971: 3–4, 371–72; Munteanu 2013.

and pressing cloth.[639] A technique for mining, processing and weaving asbestos into fire-proof towels, nets, and shawls was developed.[640] More important still was the invention of the treadwheel pump, employing one or more men walking a treadwheel for lifting water, and then the treadwheel windlass, using the same device for powering cranes.[641] The treadwheel crane underwent continuing improvements in pulleys, winches, and hoists, and the introduction of revolving booms, until these arrangements became standard construction machines.[642] Comparable machines, possibly even gear-train cranes like the Baroulkos, were also employed in Roman ports. In fact, port and construction cranes had become so common by the early 2nd century A.D. that Maximus of Tyre could assume any urban audience would be familiar with them, announcing that "you have surely before now seen ships being hauled up out of the sea and stones of enormous bulk being moved by all sorts of twistings and rotations of machinery."[643] The Greeks also invented heavy beam land transports, and a variety of locking pins and loading bolts for hoisting large stone blocks.[644] They even invented

---

639. As shown in a painting recovered from Pompeii: Mau 1908: 414 = Mau 1982: 395; and a physical example recovered from Herculaneum (cf. Feldhaus 1954: 120–21, w. Abb. 77).

640. Unmistakably in Plutarch, *On the Cessation of Oracles* 43 (= *Moralia* 434a-b), who attests that such fire-proof articles were still in common use in the Roman era.

641. On the treadwheel 'pump' (and other water-lifting machinery) see Oleson 1984 and 2000, and Landels 2000: 11–13, 58–83.

642. Conceded even by P. Green 1990: 467–68. For a good sketch of a 1st century A.D. Roman tomb relief of a large crane in use see James & Thorpe 1994: xxi (for a photograph of same: O'Connor 1993: 44 and Di Pasquale 2002: 78). Vitruvius, *On Architecture* 10.2.1–10 describes the various kinds and components of cranes up to his time, including the swivel-and-boom. More detailed and advanced discussion can be found in book 3 of Hero's *Mechanics*. Both these sources on cranes are discussed in Schürmann 1991: 146–57 and Landels 2000: 84–98.

643. Maximus of Tyre, *Orations* 13.4. Vitruvius, *On Architecture* 10.2.10 also notes the use of crane technologies in the shipping industry.

644. Heavy-beam transports: Vitruvius, *On Architecture* 10.2.11–14, with Meiggs 1982: 338–46; P. Green 1990: 467–68; Schürmann 1991: 140–43; Landels 2000: 183–85; M.J. Lewis 2001a: 14. Locking pins and loading bolts: Hero, *Mechanics* 3.5–8, with Drachmann 1963: 103–06, Rosumek 1982: 128–31, Schürmann 1991: 144–46,

the railway.[645] The earliest example is the *diolkos*, built in the 6th century B.C. for hauling ships across the isthmus of Corinth, which remained in use throughout Roman times.[646] Rails for hauling carts were then expanded for use in theatres and temple magic shows, and archaeology has uncovered their use in mines of the Roman period.[647]

There were several improvements in general land transport, including an improved ability to harness teams and trains of animals, an increasing specialization of cart and wagon design, and pivoting axles and suspension systems.[648] But even more significant were improvements in naval transport.

O'Connor 1993: 54–55, Landels 2000: 89–92.

645. Complete survey of evidence in M.J. Lewis 2001a.

646. MacDonald 1986; Werner 1997; M.J. Lewis 2001a: 10–15. See also Pettegrew 2011 and Humphrey et al. 1998: 417–18; and *OCD* 458 (s.v. "*diolkos*").

647. Theatrical and stage railing: Hero, *On the Construction of Automata* 1.2.2; M.J. Lewis 2001a: 9–10. Roman mine railways: Wilson 2002: 21 and M.J. Lewis 2001a: 15–17. That some roads were likely rutted intentionally (and thus were effectively railways) is argued in Landels 2000: 182–83 and Humphrey et al. 1998: 418–19.

648. See *OCD* 1501–02 (s.v. "transport, wheeled"), Burford 1960, Röring 1983, Oleson 1986: 339–54, Schneider 1992: 130–40, Landels 2000: 170–85, and Adams 2012 and 2007, esp. 65–69, 81, 199–205 (wagons) and 74–77, 203–04 (harness). See also G. Mansfeld 2013. McWhirr 1987 offers a broader perspective, tying in widespread road, harbor, canal and lighthouse construction as Roman improvements to a whole ransportation 'system'. Specialized wagons were developed for hauling special cargoes overland, like bulk liquids (Kneissl 1981 and McWhirr 1987: 662). Pivoting front axles were proposed at least as early as the 2nd century A.D. (by Athenaeus the Mechanic, *On War Machinery* 33–37) and actually in use by at least the time of Diocletian (late 3rd century A.D.) and probably earlier (Whitehead & Blyth 2004: 192 n. 19; Landels 2000: 180–81; K.D. White 1984: 133–35). Röring 1983 (and Schneider 1992: 136 & 236) surveys evidence for pivots and suspension systems from the 2nd and 3rd centuries A.D., though more primitive suspensions had seen specialized use centuries before (e.g. Diodorus Siculus, *Historical Library* 18.27.3–4). It should also be noted that in the general field of transport and traction systems much has been made of a supposed Roman failure to invent the modern harness or exploit the horseshoe (e.g. P. Green 1990: 474, and even Landels 2000: 174–79, whose remarks on the economics of animal selection, here and at pp. 13–14, are otherwise correct), however "the unsuitability of ancient harness to equines has frequently been remarked in modern times, but the most recent experiments indicate that this has been exaggerated" (*OCD* 708, s.v. "horses")—in fact, almost wholly fabricated: modern experiments have confirmed that horses

The Greeks not only added the lateen sail to their rigs, but also spritsails, lugsails and topsails, and began building double- and triple-masted ships and lead-plated hulls. Significant improvements were made in side rudders, which were as capable as stern rudders, and in some respects superior.[649] Then there was an explosion of mercantile shipbuilding under the Romans, during which, according to A.J. Parker, "variations in ship construction" were introduced for the transport of "special cargoes such as marble or roof-tiles," while a "greater economy of labour and materials" in shipbuilding was introduced sometime in the 3rd century A.D. Though most recovered merchant vessels have been small, we know much larger merchantmen and warships existed.[650] We know large ships were used in the Alexandrian grain trade, for example, and were also required by the merchant tactic of riding seasonal monsoon winds, introduced at least by the early 1st century B.C. for navigating the trade route between Arabia and India.[651] There were also

---

are unimpaired by the ancient harness system (e.g. Spruytte 1983; Schneider 1992: 136–39). Though galling of the neck was still an occasional problem even for oxen, drivers were expected to take care to prevent it, and ancient harness was even built to help this (e.g. Pliny the Elder, *Natural History* 18.49.177; K.D. White 1967a: 644). See also Burford 1960, who dispels many other myths about ancient harness (though she still clings to some). The importance (and lack) of horseshoes has also been exaggerated (see below).

649. Rudders: Casson 1971: 221–28, Landels 2000: 139–40. On sails and masts see sources in following note. Green incorrectly thinks they only developed the lateen sail (P. Green 1990: 367). For lead-plated hulls: Russo 2003: 115–16 and Casson 1971: 195.

650. *OCD* 1359–60 (s.v. "ships" and "shipwrecks, ancient"). As today, large ships were outnumbered by smaller boats, and less likely to sink, hence the disparity in the archaeological record. On Roman ship technology in general: Casson 1971; K.D. White 1984: 210–13; Meijer 1986; Oleson 1986: 354–95; Basch 1987; Schneider 1992: 140–55; Landels 2000: 133–69 and 219–24, Russo 2003: 112–16, Polzer 2008, Whitewright 2009, and especially Davis 2009 and Harris & Iara 2011; with *OCD* 1002–03 and 1508 (s.v. "navies" and "navigation," and "trireme"). The supposed ox-turned paddle boat in the anonymous 4th century treatise *On Matters of War* will be discussed in section 3.8.IV below.

651. See *OCD* 546, 685 and 967–68 (s.v. "Eudoxus (3)," "Hippalus" and "monsoon"). Though certain details are disputed, monsoon riding was definitely in existence by the time of Posidonius: Kidd 1988: 254–57; Casson 1980 and 1991; Pliny the Elder, *Natural History* 6.26.100–106.

rare cases of single-use cargo ships of fantastic size, exhibiting tremendous engineering abilities, most famously the *Alexandris*, a luxurious triple-masted cargo transport designed by a colleague of Archimedes, which was over 350 feet long, displacing well over 3500 tons.[652] Large merchant ships in regular use were of more modest size, like the 1200 ton, 180-foot merchantman described by Lucian.[653] In contrast, though fishing boats could never be large due to the inability to refrigerate a large catch, nevertheless numerous highly efficient netting techniques were developed and in wide use under the Romans, sometimes involving teams of boats, demonstrating considerable ingenuity and industry.[654]

There were also many important architectural inventions, which were most widely exploited by the Romans, whether originating with them or not. This included not only the roof truss, but most importantly the arch and the barrel vault, which led to greatly improved bridges, buildings, and aqueducts.[655] There were many other developments in construction technologies and techniques.[656] The Greeks should be credited with the idea of the aqueduct itself—as distinct from the canal, which was also developed to an advanced state, complete with locks and sluice-gates and other elements.[657] Aqueduct design was certainly much improved by Greek

652. Athenaeus, *The Dinnersages* 5.206d-209e (5.203c-209f describes several other exceptional superships). Giant cargo ships continued to be built for other special occasions even in the Roman era: see Duncan-Jones 1977 and Casson 1971: 183–99. On the comparable evolution of Greco-Roman warship technology see Foley & Soedel 1981.

653. Lucian, *The Ship* 5, on which see Casson 1950, 1956, 1971: 186–89 and K.D. White 1984: 212 (w. 155).

654. Bekker-Nielsen 2004: 90–93 (on nets specifically: Bekker-Nielsen 2002).

655. See K.D. White 1984: 86–90, 206–07.

656. See *OCD* 140, 142–44, 250, 254–55 (s.v. "arches," "architecture," "bridges," "building materials"); also: Oleson 1986: 183–211, Schneider 1992: 155–70, and G.R.H. Wright 2005: 1.89–109 (Greek) and 1.110–28 (Roman). On Roman innovations in the use of metal reinforcement in their architecture: Loiseau 2012. See also sources in previous note on construction and carpentry.

657. See *OCD* 128–29, 274, 1316, 1571 (s.v. "aqueducts," "canals," "sanitation," "water supply"). See also Schneider 1992: 181–93 and Oleson 1986: 211–29. Seneca gawks at how rapidly Roman canals could be filled and emptied in *Moral Epistles* 90.15. On Roman canal technology: Peacock 2012, Wikander 2000c: 321–

scientific knowledge—most impressively in the use of the inverted siphon, which literally made water run uphill.[658] Norman Smith even begins his study of this development by repeating many of the usual myths about Roman attitudes and achievements, and yet what he documents regarding the Roman use of the siphon confirms exactly the opposite: technological sophistication and ingenuity, and a widespread use of scientifically-based technologies in a deliberate defiance of nature.[659] Among other improvements, the Romans developed a standardized system of duct sizes to improve efficiency of aqueduct design, construction, and repair.

The Romans also developed an increasingly standardized system of brickstamping that indicates a similar mindset.[660] The recovery of the contents of a 1st century Roman nail warehouse, containing over 800,000 iron nails in all (exceeding ten tons in total weight) confirms the Romans had also standardized nail manufacture into six different types of considerable quality and consistency.[661] Indeed, improving on the technologies they inherited (from Greece or beyond) was a signature Roman characteristic. Their mastery of the science of bridge design would produce structures unrivaled for over a thousand years, and in some respects not matched until the 18th century.[662] Large-scale manmade reservoirs were introduced, some lined with concrete and dammed with stonework.[663] High-rise apartments

---

30 and K.D. White 1984: 110–12 (plus, more briefly, McWhirr 1987: 667 and M.J. Lewis 2001b: 167–96, 340–44).

658. See Russo 2003: 118–23, Wikander 2000c: 39–94 and 103–216, Landels 2000: 34–57, Dodge 2000, and O'Connor 1993: 150–62 (plus the whole of O'Connor 1993 pertains to advances and achievements in Roman aqueduct technology).

659. N. Smith 1976.

660. The aqueduct standards may have been invented by Vitruvius: Rowland & Howe 1999: 6 and 277, with Sextus Julius Frontinus, *On the Aqueducts of Rome* 25.1 (with 26–34). Brickstamping: *OCD* 250 (s.v. "brickstamps, Roman").

661. Pitts & St. Joseph 1985: 109–13, 289–99. Buried to hide them from the enemy when a legionary camp was abandoned in Scotland, it is of more than passing economic and industrial significance that a distant Roman military outpost could have nearly a million nails in its storerooms.

662. See Barow 2013 and O'Connor 1993 for complete studies (O'Connor offers comparisons with later achievements: 187–88).

663. Hodge 2000; James & Thorpe 1994: 384–85; Reynolds 1983: 44.

appeared—five story buildings were common in major Roman cities, and up to seven stories were known.[664]

And one can hardly ignore the large-scale introduction of well-designed paved roads. Lynn White would absurdly claim that "the cost of maintenance was out of all proportion to [the] benefits derived" from Roman roads, an unproven remark typical of an arm-chair historian who fails to note the incalculable economic and military advantages of all-weather roads that always drain well, are almost always level, are almost never washed out by seasonal floods, never slow or bog down animals or vehicles in puddles and mud during the wet season, and raise much less dust in the dry season, which impairs the health and vision of military units and betrays their size and movement.[665] Indeed, rather than a wasteful expense, road building and repair even kept idle legions busy.

Also invented by the Greeks but far more widely and systematically exploited by the Romans were various technologies involved in the construction of artificial harbors, dry docks, and the invention of the lighthouse. In fact, Roman harbor design would remain unrivaled until modern times. Avner Raban, a director of excavations at Caesarea, says archaeology has now confirmed the once-unbelievable description of the Roman port at Caesarea by Josephus, remarking that "this Herodian port is an example of a 21st century harbour built two thousand years ago." Indeed, he says, "if the modern harbours of Ashdod and Haifa had employed such systems of design and engineering, they would not have had the problems they face today."[666] Similarly advanced harbor design is witnessed at Cosa, where a fully integrated port, factory, and fishery complex has been excavated, revealing that bucket-chain water-lifting machinery supported "an industrial complex for the raising and catching of fish," and the

664. James & Thorpe 1994: 365–67. Seneca is annoyed at how common highrises had become in *Moral Epistles* 90.7–8.

665. L. White 1963: 274. In contrast see *OCD* 1282–83 (s.v. "roads") with Chevallier 1976, Schneider 1992: 171–80, O'Connor 1993: 4–34, M.J. Lewis 2001b: 217–45, 347–48, and Barow 2013.

666. Raban quoted in James & Thorpe 1994: xx. On the Caesarean harbor (described in Josephus, *Jewish War* 1.5.408–1.7.414): Hohlfelder et al. 1983. On Roman harbor technology in general: *OCD* 645 (s.v. "harbours"); Hohlfelder 1997; Schneider 1992: 178–81; Houston 1988; Oleson 1988; Rickman 1988; K.D. White 1984: 106–10; Casson 1971: 365–70.

processing, packaging, and shipping of the resulting fish products, proving the Romans had "developed the skills of mass production," including a mass-scale amphora factory, and a tidal fish catchery, which exploited the natural forces of the ocean in combination with gate and aqueduct technology to bring the fish to the factory rather than merely sending out boats to find them.[667] And though several lighthouses were built by Greeks before Roman times, the Romans expanded the use of this technology by developing an entire system of lighthouses throughout the Mediterranean, including one at the Cosan industrial complex.[668] There are even hints that some of these may have employed rotating parabolic reflectors, and even foghorns.[669]

Another area of notable progress was mining technology, with advances continuing under the Greeks and the Romans.[670] According to George

667. McCann 1987 and 2002 (quotes from latter: 30, 32; mechanized waterhouse: 35–46, with Oleson's contribution to McCann 1987: 98–128). Though they identify the facility as a tidal catchery, it may have been a tidal fish farm as described by Columella, *On Agricultural Matters* 8.17, though the function would be similar. Notably, extensive evidence of the use of glass jars for pickling and storing products was also recovered at the site. For evidence of nearly industrial-scale fishing under the Romans see Bekker-Nielsen 2004 (and all the contributions to that same volume by other authors) and Marzano 2013.

668. See *OCD* 836 (s.v. "lighthouses"); Hague & Christie 1975; Seidel 2010. Pliny the Elder, *Natural History* 36.18.83 mentions the appearance of new lighthouses, and archaeology confirms that an entire network of them was systematically constructed around the Mediterranean in the first two centuries A.D. On the lighthouse at Cosa, see previous note.

669. Russo 2003: 116–18 makes a reasonable but inconclusive case for parabolics. There is no direct evidence of their use in lighthouses, but such reflectors did exist and books were written about them, and it is hard to imagine what else they were used for (Russo adds additional evidence from accounts of the Lighthouse of Alexandria). Mechanized fog horns are implied if *Aetna* 294–96 reads *ora* rather than *hora* (the mss. disagree, but *ora* is more probable), and if a water-powered horn echoing off the shore (and associated with Triton, the son of Neptune) is indeed a foghorn. Again, it is hard to imagine what else it would be. But if the passage reads *hora*, then it refers to a water-powered horn that blows on the hour. Either way, the use of a mechanized horn as an analogy in this passage entails such a thing was common enough to be familiar to any reader.

670. See *OCD* 938–39, 957, 957–58 (s.v. "metallurgy," "mineralogy," "mines and mining") with: Wilson 2002: 17–29 and 2000: 135–42; James & Thorpe 1994: 410–11; Woods 1987; Oleson 1986: 55–100; K.D. White 1984: 122–24; Rosumek 1982;

Sarton, "they developed new ways of flushing, pitting, driving galleries, sinking shafts, lighting and ventilating, draining, propping, hauling, and surveying," as well as "better methods of crushing ores, washing, roasting, better furnaces of many kinds, better smelting, liquation, cupellation, and so forth."[671] Agriculturally, the Greeks introduced the widespread employment of the iron plowshare and ox-driven waterwheel, and various improvements in grain processing—including the invention and improvement of the hand-rolled saddle quern and the donkey-powered grain mill.[672] The Greeks also invented the watermill, which the Romans exploited far more widely than once believed—though Peter Green thinks waterwheel technology "was not put into general use until about the third century A.D.," abundant evidence confirms it was already widely employed more than a century earlier.[673] And though there is no evidence of ancient windmills, modern windmills are essentially a combination of the Roman-era watermill and a windpump described by Hero, who assumes the windwheel was so common his readers would know it by name and that his using one to power a water organ was a novel application of an existing technology.[674]

The Greeks and Romans were also very creative with theatre technology. Though stage machinery had been in use since Aristotle's time, it continued

Healy 1978: 86–102; Sarton 1959: 376–79; Forbes 1950.

671. Sarton 1959: 377. We already mentioned the Roman use of railways above. Schneider 1992: 71–95 and Rosumek 1982 provide more recent surveys of evidence establishing how much progress the Romans made in nearly every aspect of the mining industry. See additional discussion in section 3.6.IV.

672. P. Green 1990: 367, 467–68; Reynolds 1983: 11, 25.

673. P. Green 1990: 469. For sources and discussion on Roman watermills see section IV below.

674. Hero, *Pneumatics* 1.43 (which instructs the reader to build his organ pump *platas echetô kathaper ta kaloumena anemouria*, "with plates like those things called windles"). The meaning of 'windles', i.e. *anemouria* (*anemos*, "wind" + *ouros*, "favorable or useful wind" + *-ion*, "little") is debated. M.J. Lewis 1993: 143–47 argues persuasively that *anemouria* were mechanically driven wheel-fans designed to blow air (a notable invention in its own right, for which Lewis presents evidence of regular use) and that Hero was the first to reverse their operation. Russo 2003: 125–26 argues less convincingly that they were actual windmills (cf. also Feldhaus 1954: 82–83 and Landels 2000: 26–27). Other possibilities (such as wind-powered irrigation pumps, very common still to this day) cannot be ruled out.

to be employed and developed. In the 1st century A.D. Seneca followed Posidonius in regarding as better than common artisans "the stage-machinists, who invent scaffolding that goes aloft of its own accord, or floors that rise silently into the air, and many other surprising devices, as when objects that fit together then fall apart, or objects which are separate then join together automatically, or objects which stand erect then gradually collapse," a list he obviously intends as only examples of a much wider array of stage technologies, which we know included thunder machines and rigs for making actors and props fly.[675] Seneca's description also confirms the ancients had invented the stage elevator. They had also developed revolving ceilings and stages, aromatic sprinkler systems, and bizarre parade floats, like a gigantic ship that drove seemingly by itself down city roads, or a mechanized snail that left slime as it crawled.[676] Mechanical starting gates were also invented for athletic competitions and horse races, an example

---

675. Seneca, *Moral Epistles* 88.21. See also Vitruvius, *On Architecture* 10.pr.3–4 (with 5.6). Thunder machines: Hero, *On Constructing Automata* 2.20.3–4. On the rest see James & Thorpe 1994: 589–92 and Murphy 1995: 6–7.

676. Seneca, *Moral Epistles* 90.15, who mentions "spraying perfumes to a tremendous height from hidden pipes" (in Seneca, *Natural Questions* 2.9.2 these systems are described as powered by compressed air and are treated as commonplace; waterjets driven by compressed air are discussed in Hero, *Pneumatics* 1.10 and 2.2) and "a dining room with a ceiling of movable panels" that change with the courses of the meal (Suetonius, *Nero* 31 says Nero had one such ceiling installed that revolved "day and night in time with the sky," which Wikander rightly notes was probably water-powered: Wikander 2000b: 409). On revolving theatres (with both stage and audience turning about): Pliny the Elder, *Natural History* 36.24.116–120, whose description is muddled but contains enough incidental detail to confirm a real account. In the 2nd century A.D. Herodes Atticus commissioned for a religious procession a gigantic ship "that wasn't hauled by animals but moved along by machines below deck" (Philostratus, *Lives of the Sophists* 2.1.550, who reports this marvel was later parked and was still on display, as confirmed by Pausanias, *Description of Greece* 1.29.1). On the robotic snail: Polybius, *Histories* 12.13.11. Rehm 1937 plausibly argues that these used concealed human treadwheel propulsion machinery. On these and other ancient robotic vehicles see Schürmann 1991: 235–49 and M.J. Lewis 1997: 84–86 (and Wachsmann 2012). Not all were internally propelled, some only carried robotic displays powered by the float's forward motion (e.g. Athenaeus, *The Dinnersages* 198c-200b and Appian, *Civil Wars* 2.147).

of applying technological innovation to solve even mundane problems.[677] The same innovative spirit, aiming at problem solving and perfecting an outcome, is evident in the technology of Roman racing chariots.[678] New technologies were also developed for Roman amphitheatres, such as sophisticated awning systems for shading audiences, and mechanisms for flooding arenas to simulate naval combats.[679]

We already mentioned the pneumatic water organ (which was similar to the modern pre-electric church organ), but we should not overlook the fact that this included the invention of the keyboard, and as we just noted, by Roman times some water organs were powered by windpumps. Hero also improved water organ design in other ways, apparently leading to a "completely new kind of water organ" that won the attentions of a much-impressed Nero.[680] Peter Green doubtfully asks whether "anyone, I cannot help but wondering, ever really play[ed]" the water organ, but I cannot count the number of them I have personally seen on coins, medals, and seal-rings recovered from the Roman period, which does not sound like an instrument no one played. Contrary to Green's arbitrary skepticism, many mosaics depicting arena orchestras feature a water organ, many writers mention having heard it, and we've even recovered pieces of one. There was even an organ playing competition in Delphi.[681] It was clearly a widely employed machine.[682] On the more practical end, another machine

677. H.A. Harris 1968; Schürmann 1991: 235–36; James & Thorpe 1994: 553; Balabanes 1999.

678. Sándor 2012; Crouwel 2012.

679. Connolly & Dodge 1998: 190–208; Pliny the Elder, *Natural History* 19.6.23–24.

680. Suetonius, *Nero* 41.4 (more ambiguously: Cassius Dio, *Roman History* 63.26.4). See Keyser 1988 for a discussion of the new organ design and Nero's eagerness to introduce it at Rome.

681. M.J. Lewis 1997: 71.

682. P. Green 1990: 478. On the water organ in general: Russo 2003: 228–30; Landels 1999: 202–04, 267–70; James & Thorpe 1994: 602–05; and Apel 1948. For a good example of a mosaic depiction: Connolly & Dodge 1998: 217. That gladiators sometimes fought to the sound of this organ is attested in Petronius, *Satyricon* 36.6. That it was also played in theatres is attested in *Aetna* 297–299. For the recovered pieces of an actual organ (with sketched reconstruction) see sources in Rowland & Howe 1999: 306. I have personally photographed several water organs on coins, seals, and medals on display at the British Museum. I would not be surprised if it

we already mentioned that also saw wider use than often thought, is the fire-engine pump. By the Roman period this combined the continuous-stream, double-force valve-and-piston pump of Ctesibius, with a turret valve, producing a water gun that could be elevated and swiveled, throwing a steady stream of water wherever the operator pointed the nozzle. Some of these Roman fire engines have been recovered, and Pliny the Younger appears to refer to them as common or expected equipment in major cities in the early 2nd century A.D.[683]

IV. Roman Inventions

Already we have far outstripped Peter Green's ridiculously short lists, and yet more was probably invented *after* 100 B.C., even beyond the few examples we have already mentioned. Most famously, the Romans invented hydraulic concrete, a discovery that transformed the construction industry, allowing concrete harbors and other structures to be built underwater.[684] Several advances in artillery design were made in the Roman period, not only

---

was the most widely depicted machine in extant Roman art. Several witnesses report its sound was beautiful (Pliny the Elder, *Natural History* 9.8; Athenaeus, *The Dinnersages* 4.174a-b; Cicero, *Tusculan Disputations* 3.18.43, who also attests to the organ's use at banquets; while Seneca, *Natural Questions* 2.6.5 reports that water organs could be louder than any human-blown horn). J. May 1987 makes a persuasive case that one of Seneca's neighbors was a water-organ tuner, which certainly suggests widespread use of the instrument.

683. James & Thorpe 1994: 368–70; Landels 2000: 79–81; Schiøler 1980; Pliny the Younger, *Letters* 10.33–34. Hero, *Pneumatics* 1.27–28 describes their use and construction (Apollodorus of Damascus, *Siegecraft* 174.1–7 also describes their counter-incendiary use in combat). Tacitus, *Annals* 15.43 says Nero required landlords in Rome to keep *subsidia reprimendis ignibus* ("equipment for suppressing fires") *in propatulo* ("out in the open"), which may have included firefighting pumps. See Oleson 1984: 324–25 (and 396) on the Roman 'pump corps' organized by Augustus. The idea of using these pumps as flamethrowers was realized in the Byzantine era but already imagined in the 1st century A.D. (cf. *Aetna* 294–96).

684. See K.D. White 1984: 85–90, 204–05 and G.R.H. Wright 2005: 2.1.181–217. Additionally, Malinowski 1982 presents scientific evidence confirming the remarkable quality of ancient concretes and mortars and the sophistication of their employment. Courland 2011:71–135 summarizes several Roman advances in concrete, which would not be replicated for a thousand years. Brandon et al. 2014 provides the most thorough history and study of this Roman marvel.

Hero's 'handgun' already mentioned, but other improvements, including the introduction of oval washers allowing more line to be torqued (magnifying the power of torsion catapults), and the invention of the grappling harpoon.[685] The Romans also developed hipposandals, horseshoes, and a new four-horned saddle that provided nearly the same effectiveness as the stirrup.[686] They also improved the use and design of the steelyard scale,

---

685. The invention of the grappling harpoon is described in Appian, *Civil Wars* 5.118–19. Improvements in catapult design have been confirmed archaeologically, and go beyond mere washer design: K.D. White 1984: 217–19 (with Whitehead & Blyth 2004: 21). These Roman developments post-date our last surviving treatises on ancient artillery, though such advances were probably mentioned in contemporary works that do not survive—yet another example of the dangers of arguing from silence in the area of technological progress in antiquity (the surviving part of Hero's *Siegecraft* only discusses the history of catapults up to the 4th century B.C.). There is likewise evidence of continuing Greek and Roman innovation in incendiary combat (well preceding the much later development of Greek Fire): see Partington 1960: 1–41 and Lindsay 1974: 368–77 (with Simms 1991, who argues the myth that Archimedes burned warships with parabolic mirrors likely arose from a more standard innovation in incendiary weaponry; although Rossi & Unich 2013 argue it arose from what was actually the invention of a steam cannon).

686. Hyland 1990: 131–34 and Dixon & Southern 1992: 70–74; also Schneider 1992: 139. Even though the stirrup was a significant improvement on it, the four-horned saddle was still a major advance in riding technology. The importance of the horseshoe has been exaggerated. Many experts now conclude shoeing is unnecessary as long as the hooves are not overworn and are regularly hardened (e.g. www.healthehoof.com and www.thenakedhoof.com.au), which ancient horse care attended to (e.g. Xenophon, *On Horsemanship* 4). Though Thucydides, *Peloponnesian War* 7.27.5, is often cited as evidence of the risks of unshod cavalry, this passage actually describes an exceptional forced action that could have lamed the same proportion of even modern cavalry. The superiority of proper hoof care to shoeing may have been recognized in antiquity, since the Romans actually had both hipposandals (similar to the modern horseboot) and nailed horseshoes, yet chose to use them sparingly. Surviving examples of the latter have been recovered from as early as the 1st century A.D. (e.g. Hyland 1990: 123–24 and 234; C. Green 1966; A.D. Fraser 1934; Ramsay 1918: 142–43). Hipposandals began earlier (possibly even pre-Roman) and are much more numerous in extant finds (see ibid. and Beckmann et al. 1846: 1.442–54). Mules were also shod (e.g. already in the 1st century B.C. Catullus mentions a mule losing its shoe in *Carmina* 17.25–26, probably a hipposandal). Roman hipposandals have also been recovered that bear cleats for ice and turf (Hyland 1990: 123–24).

which employs a sliding counterweight balance like many modern scales.[687] The Romans also introduced the dual beam vertical loom, the horizontal loom, and possibly the foot-treadled loom.[688] Archaeology confirms that a large variety of smartly crafted scissors and shears were manufactured for a variety purposes in the Roman period, and though their invention may be pre-Classical, the Romans improved their design and greatly expanded their use.[689] And though tumbler locks also predate the Roman period, again the Romans improved them, and then invented the first rotary lock or "deadbolt" (like today, in such locks a bolt was extended and retracted by turning a key), as well as the first padlock, and then combined these into the first rotary padlock.[690] The Romans also scaled down the size and thus increased the application of screw fasteners and threaded nuts and bolts, a

---

687. See *OCD* 1572 (s.v. "weighing instruments," though this does not mention the more sophisticated weighing instrument developed by Menelaus, already noted earlier). Simple steelyard scales were in occasional use since Classical times, but more advanced versions first appear (and come into common use) under the Romans (cf. e.g. British Museum 1908: 152–46, with figs. 170–74; and more detailed discussion in Damerow et al. 2002, who also presents evidence that steelyards were in use before Archimedes, contrary to Simms 1995: 52).

688. *OCD* 1446–47 (s.v. "textile production"); Rogers 2001; Schneider 1992: 125–28; Carroll 1985; Wild 1987. In *Moral Epistles* 90.20 Seneca says the Romans had developed a means of weaving shear garments and that a new loom for this was invented after 50 B.C., which may indicate one of the new looms just mentioned or yet another invention. It should also be noted that variety and sophistication were realized even at the level of ordinary needles and hooks for sewing and knitting (cf. e.g. British Museum 1908: 137–38, with figs. 154–56).

689. See Notis & Shugar 2003; K.D. White 1967b: 119–20; British Museum 1908: 137 (with fig. 153); and Nicolson 1891: 51–56. There is some textual but no archaeological evidence of (some form of) shears or scissors in use before Roman times, but they become archaeologically abundant under the Romans, showing design improvements over time and widespread use for many different purposes. Shears were more common than scissors. Shears are any double-bladed spring-levered version of scissors, which are any double-bladed pivot-levered cutting instrument. Analogously, both spring-levered and pivot-levered tongs (and pivoted compasses) were in use in antiquity, and shears and scissors are essentially tongs with blades instead of grips. Metal pivot-levered nut crackers are treated as commonplace in Ps.-Aristotle, *Mechanics* 22.854a-b.

690. *OCD* 784–85 (s.v. "keys and locks") with James & Thorpe 1994: 472–73 and British Museum 1908: 139–46.

marked improvement on these already-novel inventions of the Hellenistic period.[691] They also developed folding chairs, tables, and lampstands, often with adjustable heights.[692] They introduced artificially-heated hothouses, and wheeled cold frames, for growing flowers and vegetables out of season.[693] They also developed water heaters employing coiled-pipe heat exchangers, for a variety of uses.[694] They even invented a bread-kneading machine.[695]

---

691. Large nuts and bolts were employed in screw presses, and screw-cutting machinery was developed to manufacture them shortly before the Roman period (e.g. Russo 2003: 97–98, 151). But smaller-scale nuts, bolts, and screws begin to appear under the Romans. See Mercer 1975: 272–73, with photograph and discussion of a 5cm threaded metal nut recovered from a Roman military site in Germany dating to the late 2nd century A.D. (overlooked by both O'Connor, who discusses the equivalent Roman use of nails and eyebolts in O'Connor 1993: 45–46, and Deppert-Lippitz 1995, which otherwise surveys examples of ancient screws of many types and sizes). For small screws as fasteners in the 1st century A.D. see the relevant sections of Hero's *Dioptra* and sources in Burkert 1997: 40. Threaded bolts as structural elements are described in Josephus, *Jewish Antiquities* 3.3.120–21. Threaded screws were also employed as adjustable valves and stopcocks in Roman pipe systems (cf. e.g. Marchis & Scalva 2002), which were clearly designed to be turned by a wrench, which entails another Roman invention.

692. Folding height-adjustable table: G. Richter 1926: 138 (with fig. 322). Folding chairs and stools date back to Egyptian times, but became increasingly popular in the classical era, and the Romans produced some of the finest examples (cf. G. Richter 1926: 39–43, 126–27). Lampstands that could be folded into themselves and adjusted for different heights: Mau 1908: 395 (= Mau 1982: 367).

693. Seneca, *Moral Epistles* 122.8, describes (with the annoyance of an old codger) hothouses with water heaters for growing spring flowers in winter (along with orchards cultivated atop roofs and walls, which would also present both irrigation and architectural challenges). Cold frames (wheeled trolly gardens, either with transparent mica roofs or able to be parked beneath them as needed) are described in Columella, *On Agricultural Matters* 11.3.52–53 and Pliny the Elder, *Natural History* 19.23.64.

694. Seneca describes a coiled-pipe heat exchanger for heating water in Seneca, *Natural Questions* 3.22.2 (which he says is a commonplace technology available in many forms). Romans innovated in the design of heating systems in a number of ways besides: Schiebold 2010.

695. Examples were recovered from Pompeii, powered by one or two men turning a handlebar, and there is evidence they were in wide use: Mau 1908: 410–11 (with fig. 241) = Mau 1982: 391–92 (with fig. 224).

They also invented the carpenter's plane.[696] And judging from extant artifacts in several industries, there must have been significant advances in lathe technology, especially for metal and glass turning.[697] We already noted that Hero invented the pantograph, or copying instrument—in fact two of them, one for automating the duplication of plane schematics and another for producing schematics of three-dimensional objects.[698] Hero was also developing an automated bellows, and other bellows machinery may already have been in use.[699] Around the same time, other Roman engineers were mounting rotating turrets on conical rollers (a precursor to modern ball bearings).[700] By the end of the 2nd century A.D. someone appears to

---

696. Gaitzsch & Matthäus 1981a: 25–29 and 1981b, with recovered pieces and a reconstruction, demonstrating sophistication of design.

697. Cave 1977 documents what we know about developments in ancient lathe technology (and its use on wood, metal, and stone) and why our knowledge must be incomplete given abundant evidence of advanced products. Pliny the Elder, *Natural History* 36.66.193 attests to the existence of glass turning, which also requires technology not otherwise attested (evidence also in Lierke 1999), as does metal polishing: though most precision pump machinery was manufactured by lost-wax metal casting rather than lathe turning, wax models of such precision must have been turned on a lathe, and precision metal parts were polished on lathes (as reported in Vitruvius, *On Architecture* 10.7.3; and confirmed by Schiøler 1980: 24–25 and Marchis & Scalva 2002: 27–28, 32–33). To generate the necessary capabilities, Cave hypothesizes a hand-powered belt-drive, but with the widespread availability of water-turned millstones, it is hard to imagine no one would think to grind other materials against them, and adapting the same or similar machinery to a fast-turning lathe may then have become obvious.

698. Pantograph: P. Green 1990: 467; Drachmann 1963: 33–43, 159; cf. Hero, *Dioptra* 34.292, *Mechanics* 1.15, 1.18, and 2.30 (in fact the description in 1.18 appears to incorporate a lead pencil, which would be yet another noteworthy invention).

699. Hero, *Pneumatics* 2.34–35 describes two demonstration devices that used a fire to cause expanding air or steam to vent back onto the fuel, becoming a self-powered bellows. The possibility of a water-powered bellows in mining operations has been proposed (see note below), and bellows machinery may be the intended analogy in *Aetna* 555–65, e.g. "what greater engines can art move by hand" than those heating volcanic furnaces (*quae maiora...artem tormenta movere posse manu*) appears to reference a bellows operated by windlass (*tormenta*) and in some manner cleverly constructed (*artem*).

700. Russo 2003: 264–65.

have invented a mechanized carriage seat that automatically turned away from the sun or into the wind.[701] And by the 2nd century A.D. Roman engineers had invented and disseminated something far more useful: the cylinder-block force pump.[702]

Even further afield, soap was imported from Germany, and Roman doctors began recommending it for medical and household hygiene.[703] Likewise, the chimneyed shaft furnace, which had been invented in central

---

701. This was a luxury carriage owned by Commodus and later sold at auction, according to 'Julius Capitolinus', *Life of Pertinax* 8.2–7 (one of the more trustworthy books of the *Augustan History*). M.J. Lewis 1992 argues this passage derives from lost sections of the *Roman History* of Cassius Dio (who would have been an eye witness). Lewis argues this had a seat geared to the wheel-train so it would always face away from the sun (probably north). The seat was also geared to turn into the wind, which implies an overall mechanism of considerable sophistication—even if this meant the seat could be disengaged from the directional train and swiveled manually, though the text implies both functions were automated. Lewis speculates the design went back to Archimedes, but he offers no good case for that; it is explicitly said to have been a "new" design in the reign of Commodus. The same passage also mentions other mechanical carriages among his property, including an odometer and (possibly) a traveling clock. Lewis thinks the latter is a mistake for a static clock, but his reasoning (that such a clock would have no use and would hardly function in transit) is not conclusive: a carriage-mounted clock would not have to function while moving, and would certainly be useful when encamped (even portable sundials existed at the time). Sleeswyk 1981 makes a more convincing (though still inconclusive) case that Archimedes invented the first odometers (described by Vitruvius in the 1st century B.C.), which were more primitive than the versions developed by Hero (a century later), but that would not mean all mechanized carriage equipment originates with Archimedes, or that any such equipment had not been substantially improved (like the odometer) by Roman engineers (e.g. stationary mechanical clocks and sundials even predate Archimedes, yet continued to be improved in the Roman period).

702. Stein 2004, Oleson 2004, M.J. Lewis 1997: 111–13.

703. Mishnah, *Shabbat* 9.5e-f and *Niddah* 9.6; Galen, *On the Composition of Drugs According to Location* 2 (= Kühn 12.589) and *On the Therapeutic Method* 7.4 (= Kühn 10.569); and *LSG* 1583 (s.v. "*sapôn*"), which confirms that the use of soap is attested from the medical writings of Rufus, Asclepiades, Galen, and Aretaeus. Previously the most common detergent had been various forms of sodium carbonate (cf. *LSG* 1177, esp. s.v. "*nitron*" and "*nitroô*"). Partington 1960: 306–09 and Beckmann et al. 1846: 2.92–108 treat extensively the evidence for soap and other detergents in antiquity.

Europe in the 5th century B.C., was finally adopted by the Romans, and was in wide use by the 2nd century A.D.[704] The Romans had also made significant advances in the technology of fish farming in the early 1st century B.C.[705] One of the most important of these developments was the invention of the hypocaust, which was employed to warm fish farms and eventually public baths, and then its use spread to every other kind of building. An early form of central heating, the hypocaust warmed rooms and buildings through their floors and walls, while keeping the furnace (and its exhaust) outside the interior spaces, on both counts superior to the indoor chimney—although chimneys were known and used, especially for larger hypocaust boiler rooms, and in major industries like smelting.[706]

Another major development was the invention of glassblowing, which appeared sometime in the 1st century B.C., originating in Syrian Palestine but spreading very rapidly, with glassblowing operations in Italy already by the next century, where several innovations soon followed.[707] A market soon

---

704. Frere 1987: 287–88, Healy 1978: 188–89 and Blair 1999. Blair's claim that we "know" Roman smelting facilities did not employ another more efficient convection furnace otherwise known outside the empire's borders is actually highly questionable (both methodologically and archaeologically), but neither can a Roman adoption of it be proved, so I leave it out of account.

705. James & Thorpe 1994: 399–404; Marzano & Brizzi 2009; Columella, *On Agricultural Matters* 8.16–17. And previous notes on Roman fish farming.

706. *OCD* 717 (s.v. "hypocaust"); James & Thorpe 1994: 424, 462–63; and for a thorough study: Lehar 2012. That large hypocaust boilers used chimneys: K.D. White 1984: 44. Hypocaust ducting passed under the floor of a house, up the walls, and out the rooftop. Commercial bakeries employed ovens with ceiling vents of similar design (Mau 1908: 273–75, 409–10 = Mau 1982: 266–67, 391), though Roman cooking usually involved ventless front-loaded wood furnaces (like those still employed in traditional pizzerias), open braziers (like modern barbecue grills), and double-boilers (another Roman invention), always in well-ventilated kitchens (*OCD* 649, s.v. "heating").

707. For this and following see *OCD* 618 (s.v. "glass"); G.R.H. Wright 2005: 2.1.279–92; E.M. Stern 1999; Fleming 1999; James & Thorpe 1994: 464–68; Schneider 1992: 108–19; P. Green 1990: 467; K.D. White 1984: 41–42; Grose 1977; and Trowbridge 1930 (esp. 95–137 on the ancient glass industry, and 138–93 on its products). For ancient discussions see Strabo, *Geography* 16.2.25 and Pliny the Elder, *Natural History* 36.66.193–67.199 (with examples in Seneca, *Moral Epistles* 9.31, 86.8, 90.25, etc.).

grew in blown glass cups, jars, vases and other containers. Around the same time, rolled glass was developed, and thus window panes, frosted glass, and swing frame windows appeared on a more significant scale, an often-overlooked advance in heating technology (allowing solar heat to enter a home or building while preventing the warmed air from escaping).[708] Both markets led to the development of substantial glass factories in Roman Germany (and possibly elsewhere) by the 2nd century A.D. Glass mirrors had been introduced in the previous century.[709] Glass lamps and streetlights eventually followed, though it is unclear when.[710] In any case, lanterns with transparent casements of glass, soapstone, skin, parchment, or horn, were certainly in use by the 1st century A.D.[711]

Blown glass also improved scientific instruments, from glass alchemical apparatus to the glass ampules employed in water levels to the suction cup or 'cupping glass', which primarily had dubious medical uses but was also

---

708. In addition to the sources in previous note see Trowbridge 1930: 186–90 and Ring 1996 (who demonstrates the Romans achieved large gains in fuel efficiency with window design).

709. See *OCD* 962 (s.v. "mirrors"); James & Thorpe 1994: 252; Trowbridge 1930: 184–86; Pliny the Elder, *Natural History* 33.45.130 and 36.46.193.

710. See *OCD* 791, 836 (s.v. "lamps," "lighting") and Trowbridge 1930: 190–91. The earliest extant mentions of household glass lamps begin in the 4th century A.D., when archaeology also confirms their existence (along with the expanded use of household glass in general), though the invention and use of a thing can long predate its literary or archaeological appearance. Likewise, cities had often been lit at night on special occasions (e.g. Suetonius, *Caligula* 18), but it is unknown when cities began to engage this as a regular expense, though again the earliest extant mentions of municipal streetlights begin in the 4th century. Beckmann et al. 1846: 2.172–85 is still a useful survey of attestations of occasional and municipal streetlighting in antiquity.

711. British Museum 1908: 108–10 (with fig. 114) discusses a portable bronze lantern frame recovered from Pompeii that contained a cylindrical transparent case (now lost), with evidence of transparent soapstone in other contexts. For a better photograph see Ciarallo & De Carolis 1999: 260, who propose a casing of parchment or gut. But this frame's construction looks very similar to that of a small glass menagerie described by Hero, which he recommended be cased in either glass or transparent horn (*diaphaneis êtoi hualinoi ê keratinoi*: Hero, *Pneumatics* 2.3). At any rate, whether using horn, soapstone, parchment, gut, or glass, the technology of transparent lantern encasement had certainly arrived by the 1st century.

employed as a demonstration device in pneumatic science.[712] But even apart from the use of glass, there were still other Roman additions to the arsenal of available scientific instruments, even beyond the dozen or more we already mentioned in earlier sections. For example, Hero invented the piston syringe.[713] And we already mentioned his development of a complex diopter so advanced it rivals early modern surveying instruments, incorporating fine-cut metal screws and gears, designed for use by astronomers, architects, tacticians, and surveyors for calculating distances, sizes, and angles by manipulating a set of geared sights.[714] There were even a few monumental astronomical instruments, such as: a large set of equinoctial rings erected for astronomical use in Alexandria shortly before the Roman era, still standing in Ptolemy's day centuries later; a monumental sundial commissioned in Rome by Augustus, built by the Roman scientist Facundus Novius in 9 B.C. using a massive obelisk as its gnomon, which fell out of alignment in the 30's or 40's A.D. but was lavishly redesigned under Domitian; and an elaborate public clocktower, the "Tower of the Winds," built by Andronicus of Cyrrhus in the agora of Athens in the early 1st century B.C., sporting a sophisticated eight-pointed wind vane, nine different sundials, and an elaborate astronomical waterclock that indicated both daily and calendrical time.[715] Evidence suggests these were exceptional only in scale, and that

---

712. Glass ampules for the water level: Dilke 1971: 76–79 (and discussed in Hero's *Dioptra*). Pneumatic cupping glass: Hero, *Pneumatics* 2.17 (incorporating two small bronze shaft valves). Glass instruments in alchemy: Irby-Massie 1993: 362–63; Martelli 2011.

713. Hero, *Pneumatics* 2.18 (essentially identical to the modern syringe).

714. Moreno Gallo 2009; Coulton 2002; James & Thorpe 1994: 417–18; P. Green 1990: 467; K.D. White 1984: 170–71. A binocular diopter had already been invented by Democlitus and Kleoxenus in the 2nd century B.C. (*EANS* 234 & 484). The Romans used many other surveying instruments, like the groma, libra, and water level, whose origin is less certain (M.J. Lewis 2012; Grewe 2009; M.J. Lewis 2001b: 109–33, 318–28; Dilke 1971: 66–70). Diopters had undergone a whole series of improvements over time, from the Classical into the Roman period, and simpler models remained in use. On the variety of ancient diopters and the history of their development see Evans & Berggren 2006: 38–42; M.J. Lewis 2001b: 36–108, 305–17; and Dilke 1971: 76–79.

715. Rings: Cuomo 2001: 151–52 and Taub 2002. Augustan dial: Pliny the Elder, *Natural History* 36.15.72–74; Beck 1994: 100–05; Cuomo 2001: 151–153; and references in Swan 2004: 280. Athenian clocktower: Vitruvius, *On Architecture* 1.6.4

other cities had public monuments of comparable function.[716]

Map technologies also improved under the Romans, both in scientific accuracy and practical uses.[717] Anamorphic travel maps were employed at least as early as the 2nd century A.D.[718] Agrippa commissioned a massive wall-mounted map of the world in Rome near the end of the 1st century B.C. and a generation later a similar world map on a tapestry was presented as a gift to the Roman emperor (either Tiberius or Caligula).[719] By the end of the 1st century A.D. the *Forma Urbis* was built, a massive municipal map of the city of Rome that must have provided a more efficient management of numerous city operations (which at Rome were enormous in scale). This was inscribed on 151 slabs of marble fixed to a wall in the Temple of Peace, possibly in the office of the Urban Prefect. Pieces of a version of this from

---

and 9.8, with *DSB* 15.518–19 (in s.v. "Vitruvius Pollio"), *OCD* 336 (s.v. "clocks"), Noble & de Solla Price 1968, Rawson 1985: 163, and Schürmann 1991: 261–70.

716. Cuomo 2001: 151–153 describes how other cities built public sundials of their own, though much less lavish in scale than the Augustan monument; likewise, Schürmann 1991: 258–72 discusses evidence of monumental waterclocks in Samos, Pergamum, Prienne, etc. A large clock face (two feet in diameter) dating from the 1st or 2nd century A.D. was excavated in Austria (cf. Noble & de Solla Price 1968: 352; and Eibner 2013), suggesting public waterclocks were not rare. Cicero knew of mechanical clocks as complex as armillary spheres (*On the Nature of the Gods* 2.38.97) and Lucian says waterclocks were expected at any decent public bath (*Hippias* or *The Bath* 8). Varro expected them even at the best country villas (*On Agricultural Matters* 3.5.17). Public anemoscopes were also commonplace instruments (see earlier note).

717. See *OCD* 895 (s.v. "maps"), Talbert 2012; Talbert & Unger 2008; and Dilke 1985.

718. See *EANS* 640 and *OCD* 1118 (s.v. "Peutinger Table") with Dilke 1985: 112–20 and Talbert 2012: 163–92. The extant Peutinger map was based on a 4th century A.D. modification of a 2nd century (or earlier) design and thus might not represent the quality of the original, but even the Peutinger is a reasonably accurate anamorphic map of roads, cities, and waystations, marked with distances, from Britain to India.

719. Agrippa's map was constructed in the Porticus Vipsaniae. Since Agrippa also wrote a geographical commentary (now lost), his map probably incorporated scientific knowledge. See *OCD* 1554–55 (s.v. "Vipsanius Agrippa, Marcus") with Dilke 1985: 39–53 (which also discusses evidence of other publicly displayed maps) and Talbert 2012: 163–92. For a survey of debate on the existence of this map see Scott 2002: 13–16 (and on the tapestry see the whole of Scott 2002).

the early 3rd century survive, but evidence suggests this was a revision of an earlier map that was actually of superior quality and probably built in the late 1st century A.D.[720]

By Roman times surgical instruments had achieved a level of remarkable versatility and craftsmanship. Ralph Jackson says "the single most striking feature of surviving [medical] instruments is their quality," in fact "almost without exception they are precision tools" so well made that "the quality of Roman medical instruments was not surpassed until recent times."[721] John Healy concurs, noting "the examination of Roman instruments proves that the craftsman took pleasure in producing a tool which was elegant in appearance and highly efficient in use."[722] Archaeological finds confirm that Roman doctors were using numerous sophisticated surgical tools, from finely-crafted retractable needle syringes (such as for cataract removal) to multi-component gear-screw speculums, whose purchase and manufacture would have required considerable consultation and interaction with craftsmen.[723] The dental drill, "a tiny drill to release diseased matter from inside decaying teeth," was reportedly invented by Archigenes around 100 A.D.[724] Galen explains how scientific theory applied to anatomical evidence led to the invention of the catheter, and he uses this as a model example of the need for scientific research to make useful progress in medical treatments and technologies.[725] Tertullian describes an elaborate abortion instrument

720. See *OCD* 585 and 895 (s.v. "Forma urbis" and "maps") and Dilke 1985: 103–10 (who also discusses evidence of other municipal maps like this one in cities throughout the Roman empire).

721. Jackson 1988: 113–14.

722. Healy 1978: 250.

723. See Healy 1978: 246–51; Jackson 2010, 1995, and 1988: 92–94, 113–29; James & Thorpe 1994: 11–17, 19, 29–30; Nutton 2013: 186–88. On the cataract needle syringe specifically, one the most impressive achievements of precision craftsmanship: von Staden 2002: 43; James & Thorpe 1994: 19; Jackson 1988: 123. For a comprehensive survey of archaeologically-recovered medical instruments see Künzl 1996. For a similarly comprehensive survey of discussions and descriptions of medical instruments in ancient literature see Milne 1907. For a general study of both: Bliquez 2015.

724. James & Thorpe 1994: 35. This microdrill saw medical applications beyond dentistry, cf. Milne 1907: 133 (with 21, 25, 126–32).

725. In Galen, *On Medical Experience* = Walzer 1944: 140–41 (this is an English

that combined a geared speculum with a fetal limb amputator and a removal hook, not much unlike some equipment used today, and he regards this as a common item.[726] Tertullian's description is similar to another mechanism that had been invented for breaking up kidney stones *in situ* and extracting the fragments.[727] Several kinds of bonesetting and jointsetting machines had also been invented. Galen used and described these admiringly as a model example of applying the science of mechanics to the medical art. Several models were in use, but all involved an ergonomic box arranged with a system of small winches and pulleys. Though not of Roman invention, these machines were clearly employed in Roman times, and would again have required extensive interaction with craftsmen to build and maintain.[728] In fact, Galen says surgeons use "many" machines in common with engineers, giving the bonesetter as just one example.[729] There were certainly other types of machinery employed by ancient doctors, and many other instruments of varying complexity in components or design, including a variety of simpler machines for joint-setting (already known to Hippocrates and still in use in Roman times), and a windlass for assisting delivery at birth.[730] We should also count a number of alchemical instruments perfected by the Greeks and Romans, including invention of the still.[731]

---

translation from an Arabic translation of a lost Syriac translation of Galen's original Greek). See also James & Thorpe 1994: 15–16.

726. Tertullian, *On the Soul* 25. He follows by describing a simpler instrument that he says was used by older doctors, implying the more elaborate device was a relatively recent invention.

727. Celsus, *On Medicine* 7.26.3b.

728. And to invent (see von Staden 1998). On such medical machinery see von Staden 1989: 453, 474 and Drachmann 1973: 38–42 and 1963: 171–85 (also M.J. Lewis 1997: 54–56), as well as descriptions in: Celsus, *On Medicine* 8.20.4; Galen, *Commentary on Hippocrates' 'On Joints'* 1.18 and 4.47 (= Kühn 18a.338–39 and 18a.747) and *Commentary on Hippocrates 'On Fractures'* 2.64 (= Kühn 18b.502–06); Oribasius, *Medical Collection* 49.4.8–13, 49.4.19–20, 49.4.45–50, 49.5.1–5, 49.6; and in following note.

729. Galen, *On the Uses of the Parts* 7.14 (= M.T. May 1968: 364–66), which also describes a bonesetter and discusses the mechanical principles behind it.

730. Soranus, *Gynecology* 21.68; Apollonius of Citium, *On Joints according to Hippocrates* 2 and Hippocrates, *Joints* 42–47 and 72.

731. Martelli 2011; Wilson 2002.

Science thus benefitted from technological advances. But so did industry and trade. We have already listed several examples. Blown glass, for instance, generated a new area of economic opportunity based on a newly invented industry that was rapidly and extensively exploited. A lesser known example is the rise of an ice vending industry. Roman merchants discovered they could make good money selling snow and ice for cooling liquids, and kept up business by packing snow in insulated pits to store it long past its normal melt date (sometimes compressing it into ice).[732] A better known example is the Roman invention of mass production in the form of a mold-pressed terracotta industry, which became increasingly well-organized, economizing and streamlining production by developing a "workshop system" that simplified manufacture, for example making universal press-molds that could be used simultaneously to produce hundreds of lamps, figurines, lids, or pots; and there is evidence of other kinds of production lines and efficiency adaptations in Roman manufacturing.[733]

We also already mentioned Hero's coin-operated vending machine, but it is worth bringing up again as an example of technological ingenuity

---

732. James & Thorpe 1994: 320–22, with: Plutarch, *On the Principle of Cold* 15 (= *Moralia* 951c), *Questions at a Party* 6.6 (= *Moralia* 691c-692a); Seneca, *Natural Questions* 4b.13.8–11 and *Moral Epistles* 95.21, 95.25; Pliny the Elder, *Natural History* 19.19.52–56; Galen, *On Venesection against the Erasistrateans at Rome* 3 (= Kühn 11.205) and *On the Therapeutic Method* 7.4 (= Kühn 10.467–68); and perhaps: Petronius, *Satyricon* 31; Pliny the Elder, *Natural History* 31.23.40; Suetonius, *Nero* 27.2; Pliny the Younger, *Letters* 1.15.2. Snow storage predates the Roman era, but probably not its use as a business enterprise, cf. Beckmann et al. 1846: 2.142–60 (who also reports from his own observation that the very same methods, still in use in Portugal in the late 18th century, could keep snow through a whole summer).

733. See *OCD* 791, 836 (s.v. "lamps," "lighting"); Grandjouan 1961 (on streamlining of manufacturing methods: 2–3); also W.V. Harris 1980 and Oleson 1986: 335–37 and Simms 1995: 87 (plus examples in British Museum 1908: 172–78). Mold casting was also commonly employed in the metal and glassware industries. Organized 'assembly lines' had long been in use for various industries in antiquity: Humphrey et al. 1998: 390–400. On a related note, Pliny the Elder, *Natural History* 34.6.10–11, discusses what appears to be the development of interchangeable components for bronze chandeliers, crafted separately and assembled elsewhere. On gargantuan ovens for the mass production of ceramics: see Schneider 1992: 104, 234. Efficiency in industrial-scale pitch and charcoal industries under the Romans: Orengo et al. 2013;.

serving the purpose of profit and efficiency. The machine Hero describes was apparently invented to relieve temple custodians of the laborious task of selling fixed quantities of holy water to pilgrims and supplicants, thus permitting a substantial increase in sales through mechanization. This would seem to be a case of a scientific engineer contracted by a business to produce a machine that would reduce labor and increase profits. It hardly matters that the business in this case was a religious operation, rather than industrial or agricultural.

Yet the mechanization of larger industries was beginning. And an inscription of the 2nd century, for example, reports that the city of Beroea in Macedonia was deriving substantial income from a whole array of "watermachines" (*hydromêchanai*), a term that entails far more than just milling grain.[734] We know the technology had disseminated to other industries. For lumber and stonecutting there is clear evidence the Romans invented and widely disseminated the water-powered sawmill, while in the mining industry there is evidence of the use of water power to mechanize ore-crushing, and water power was certainly employed in early forms of hushing and strip mining.[735] Rotating mills, turned by animals or men, were

---

734. Ritti et al. 2007: 146.

735. On ore-crushing see following two notes. On simple hushing and strip mining (involving manmade reservoirs, aqueducts, and steerable piping aimed by an operator): Pliny the Elder, *Natural History* 33.21.74–75. Sawmill: Rosumek 1982: 134–38 and M.J. Lewis 1997: 114–15. Our knowledge of ancient water-powered saw mills used to be based solely on the (albeit sound) conclusions of Simms 1983 and 1985, regarding a description of an automated stone saw in Germany (near Trier) in Ausonius, *Mosella* 361–64, who wrote in the 4th century A.D. (but giving no indication of when the sawmill began operation). Though Lynn White challenged the authenticity of this poem, he is adequately rebutted by Simms; and the evidence has multiplied since. Fragments of mechanically sawed stone from the Roman period were later found on the isle of Thasos and at Trier (Ritti et al. 2007: 156; Neyses 1983: 218–21); and three Byzantine sawmills of the sort described by Ausonius were excavated, in Ephesus, Jerash, and Gerasa (all operating in the 6th century: Ritti et al. 2007: 149–53 and Seigne 2002)—probably not the first of their type, just the first ones we were lucky to find. As proved by subsequently recovering the metal crankshaft of a Roman sawmill in Switzerland dating to the 2nd century (Schiöler 2009). An inscription from the 3rd century even depicts one already operating in Hierapolis, in what is now Turkey (Ritti et al. 2007; Grewe 2010). Another inscription of the same place and century establishes a whole watermiller's guild had arisen to care for that region's waterwheel tech, entailing

also used to grind ores, as well as sand for the glass industry, and there is no particular reason to believe watermills were not thus employed as well.[736] Water-powered mechanical hammers were also in use, in agriculture (for hulling and pounding), and in mining and metalworking, and possibly for fulling and felting as well.[737] And as we noted already, the Romans certainly expanded the use of water power to grind grain. In fact, Kevin Greene confirms "it is now known from archaeological evidence" that watermills "were used extensively in the Roman empire," with remains recovered from widely diverse locations, from Palestine to Rome to the frontiers of Britain

---

a substantial industry; and similar guilds are known as far back as the 2nd century (Ritti et al. 2007: 144–45, Wilson 2002: 11, M.J. Lewis 1997: 71, and Wikander 1990: 73). In the 1st century, Pliny the Elder, in *Natural History* 36.9.51–53, described the sawing procedure employed in Roman stonecutting mills, which matches the evident operation at Ephesus, Gerasa, and Hierapolis—Pliny just doesn't mention the power source (whether water, human, or animal). Two other passages in the 4th century also suggest mechanized sawmills were commonplace: Gregory of Nyssa, *Homilies on Ecclesiastes* 3.321 (= Hall 1993: 63), refers to the clever contrivances (*mêchanêmata*) by which stone is sawed "with water and iron," as if anyone would readily know what he meant, and Ammianus Marcellinus, *Deeds of the Divine Caesars* 23.4.4, assumes his readers would be more familiar with "sawing machines" (*serratoriae machinae*) than onagers (a common siege weapon at the time).

736. On ore mills: Healy 1978: 142–43 (cf. also Spain 2002: 50). Pliny the Elder, *Natural History* 36.66.194 mentions sandgrinding by pestle and millstone, and in the Roman period the latter were typically powered by animals or water. Though we have no specific evidence of watermilled ore or sand, we have no reason to expect such evidence even if it was common, yet such an application would have been as obvious and unremarkable as grinding by ox-mill. It would also have been obvious that millstones used in rotary machines could be manufactured using the very same rotary machines.

737. A fragment of Plautus (fr. 12, from the 2nd century B.C.) uses a simile involving a "Greek hammer" in the sense of a reciprocating machine, as if it were a readily known mechanism (see *OLD* 1380, s.v. "*pilum*[1]"). On evidence for the use of mechanized ore-crushing hammers in Roman mining operations see Wilson 2002: 21–24 and Wikander 2000b: 406–07 (who also mentions evidence of automated hammers in Roman iron works, and possibly a water-powered bellows). For evidence supporting an even wider exploitation of robotic hammers in the Roman period see Spain 1985: 121–23 and M.J. Lewis 1997: 84–115 & 123–24 (including possible use in Roman fulling mills: M.J. Lewis 1997: 89–100 and Wikander 2000b: 406; and in Roman armor manufacture: Fulford, Sim, and Doig 2004: 218–19).

and almost everywhere in between—in fact, the technology was so common that by the 2nd century A.D. there were entire watermiller guilds.[738] And Pliny the Elder reports that by mid-1st century A.D. "most of Italy uses the bare pestle, as well as wheels turned by passing water, and the [ordinary] mill."[739] The most spectacular example recovered so far is the massive

738. *OCD* 955 (s.v. "mills"), plus James & Thorpe 1994: 389–92 and Landels 2000: 16–26 (there is also a "List of ancient watermills" maintained at Wikipedia, collecting examples of watermills for both grinding and sawing). On the watermiller's guilds see previous note. On ancient watermills, still useful is Reynolds 1983: 9–46, despite the fact that he makes many dubious and unwarranted generalizations from the evidence (and overlooks some as well, e.g. compare K.D. White 1984: 193–201). Together, Wikander 1990 and 2000a, M.J. Lewis 1997, Spain 1985 and 2002: 54–55, and Wilson 2002: 9–15, provide a much-needed corrective (though not without their own flaws). See also Spain 2008. But Reynolds' errors range well beyond those demonstrated by recent scholarship, e.g. Reynolds claims Vitruvius describes watermills "in a section dealing with rarely employed machinery" (Reynolds 1983: 17 and 30) and then concludes they were rarely used in the 1st century B.C., ignoring the fact that Vitruvius includes cranes and catapults in the same section, which were actually technologies in frequent use. In fact, Vitruvius does not say these were rarely *employed*, but rarely *known*, i.e. seldom encountered or understood by laypeople. Literally, things *raro veniunt ad manus*, "that seldom come to hand" (or are "seldom understood," per *OLD* 1077 §14, in s.v. "*manus*"), hence Vitruvius aims to make them *nota*, "known" or "understood" (Vitruvius, *On Architecture* 10.1.6; cranes are the first subject immediately following this remark: 10.2; watermills: 10.5; indeed, watermills come immediately after treadmill pumps, and other wheel and bucket pumps, also common technologies: 10.4; catapults and ballistae: 10.10–12). Vitruvius considers these things 'rarely encountered' only relative to "everyday things, ready to hand, like handmills, blacksmith's bellows, passenger wagons, two-wheeled carts, lathes, and other things that find general use in daily life," i.e. things the average Joe would see every day.

739. Pliny the Elder, *Natural History* 18.23.93: *maior pars Italiae nudo utitur pilo rotis etiam quas aqua verset obiter et mola*. Because this passage is obscure and believed to be corrupt, it has been creatively interpreted to mean everything from horizontal watermills to mechanical trip-hammer mills (see Reynolds 1983: 355 n. 51), despite the fact that vertical watermills are far more plausible (e.g. when Pliny wrote there was a vertical watermill in operation just outside Pompeii, not far from where Pliny lived: Reynolds 1983: 36; Roman-era turbine watermills also existed but probably post-date Pliny: Schneider 1992: 48; Wikander 2000a: 377). If the extant text is as Pliny wrote it, then *mola* is in the ablative singular, and three machines are meant (pestle, watermill, and the more common millstone turned by mules, oxen, or slaves). But if the text is corrupt, it may have originally said *et*

industrial millery at Barbegal, France, where a Roman aqueduct powered a binary system of sixteen overshot watermills, a facility now known to date from the early empire.[740] There is some evidence the Romans may also have developed tide mills.[741]

The Roman development of mass-scale agricultural operations in the form of the *latifundia* is well known, and harvesting in many of these operations was mechanized by the invention of an animal driven reaping machine sometime in the early 1st century A.D., while mechanization of threshing had already begun in the previous century with a variety of new animal-powered machines.[742] But the Romans introduced other innovations

---

molit, "wheels which passing water turns and [as a result] grinds [the grain]." This is even more likely if *obiter* refers not to the water chancing by but means "at the same time" or "incidentally, besides, into the bargain" (*OLD* 1213, s.v. "*obiter*"), hence "most of Italy uses the plain pestle, as well as wheels that water turns, also [turning] a millstone into the bargain," which an emendation from *mola* to *molit* (or even *molas* or *molam*) would support.

740. Russo 2003: 255–56 and Sellin 1983. There are many other examples of massive industrial capital investment under the Roman empire: Bowman & Wilson 2013: 107–42. Though the Barbegal flour factory is often cited as an "early medieval" development, it has been redated to the 2nd century A.D. (Leveau 1996 and Bellamy & Hitchner 1996: 172–73). There is ample evidence of other overshot watermills in antiquity (e.g. Neyses 1983), though in all periods of history the undershot variety was cheaper and abundant, and usually more than adequate. Unfortunately, Columella (who would surely have known all the available options) does not discuss grain mills but merely says a farm should construct milling facilities according to its needs (*On Agricultural Matters* 1.6.21).

741. See the extensive report in Spain 2002, which analyzes considerable evidence of a large tide mill in operation on the Thames of London from about 60 to 180 A.D. He also mentions evidence of another tide mill on the coast of Brittany, which is unfortunately undatable but intriguingly close to a known Roman settlement (ibid.: 52).

742. *OCD* 794–95 (s.v. "*latifundia*"). For the harvester: Pliny the Elder, *Natural History* 18.72.296 (1st century A.D.); Palladius, *Agricultural Opus* 7.2.2–4 (4th or 5th century A.D.). This machine was pushed by mules, horses, or oxen and is represented in extant reliefs (suggesting it was widely used). See James & Thorpe 1994: 387–89; L.J. Jones 1979; K.D. White 1966, 1967a, 1967b: 157–73, and 1969; Cüppers 1964; Mertens 1958. For the thresher: Varro, *On Agricultural Matters* 1.52 (and, in passing, Columella, *On Agricultural Matters* 1.6.23); Pliny the Elder, *Natural History* 18.72.298. See K.D. White 1984: 30, 1970:449–50, and 1967b: 152–56.

in agriculture besides machinery, including various improvements in standard agricultural equipment.[743] Even Lynn White concedes the Romans introduced the giant scythe, the hinged flail, "the most advanced form of vineyardist's pruning knife," and the barrel.[744] They also invented elaborate but effective mechanical devices to restrain animals for medical treatment and mating.[745] More significantly, in the 1st century A.D. the Romans invented a heavy wheeled plow drawn by multiple teams that also turned the soil.[746] Many other developments show a more than occasional interest in saving labor.[747] Roman agricultural writers like Columella also argued for the importance of making large capital investments (such as in buildings and irrigation works) that would increase a farm's long-term productivity.[748] They did not always have the right ideas, but there were many like him seeking to promote a productive industry, only a few of whose works survive.[749] And

---

743. *OCD* 42–43 (s.v. "agricultural implements: Roman"). See also Rees 1987 and K.D. White 1967b and 1975a. K.D. White 1970 (esp. 446–54) is a bit over-pessimistic but nevertheless documents advances, while K.D. White 1984: 195–96 provides a brief list of examples of ancient progress in agricultural tools and techniques.

744. L. White 1963: 281. Though he notes these (like the most popular Roman wagon designs: White 1967b: 13 n. 3) were Celtic in origin (just as soap was German), they were still invented more or less within the Roman empire and adopted throughout. See K.D. White 1967b: 71–85, 102–03, 207–10 (development of the scythe) and 93–96 (advanced pruning knife), and 1975b (on the Roman adoption of Celtic technology in general).

745. Columella, *On Agricultural Matters* 6.19 describes the veterinary corral in detail—he calls it a *machina*, and it was more than elaborate enough to justify the label. A modification of the same mechanism to assist the mounting of mares by donkeys is then described in 6.37.10.

746. Pliny the Elder, *Natural History* 18.48.172–173. Pliny says employing two or three pairs of oxen on one large plow was typical. The Romans employed a variety of other plows according to circumstances: ibid. 18.48–49.171–183. See *OCD* 1164 (s.v. "ploughing (Roman)") and K.D. White 1967b: 123–45. There is evidence Eratosthenes had written on plough design in a lost treatise on architecture (M.J. Lewis 1997: 77); so there may have been Roman treatises on the same topic.

747. For example, see Pliny the Elder, *Natural History* 18.49.181 and 18.67.261.

748. Columella, *On Agricultural Matters* 1.1.1–2 and 1.1.18 (for examples of his recommended investments to improve the efficiency and productivity of a farm: 1.4.7–8, 1.6.4–24, 2.2.9–14).

749. *OCD* 43 and 227 (s.v. "agricultural writers" and "bee-keeping"). Besides

they were certainly open to innovation. Pliny the Elder reports with approval several advances in wine press technology made within his own lifetime and only a generation before.[750] Columella was quite proud of his invention of a new grafting drill.[751] And Seneca actually *complains* that agriculturalists "even in the present day are inventing countless new methods of increasing the fertility" of their fields, just as the anonymous *Aetna* complains how

---

Cato, Varro, Columella, and Palladius, from whom agricultural works survive, some of the lost writers include: *OCD* 288 (s.v. "Cassius Dionysius," translated a major Carthaginian treatise on agriculture, and also wrote his own, in the early 1st century B.C.), *OCD* 1320 and 1504 (s.v. "Saserna" and "Tremelius Scrofa, Gnaeus," each wrote agricultural manuals in the early 1st century B.C.), *OCD* 714 (s.v. "Hyginus (1), Gaius Iulius," wrote *On Agriculture* and *On Bees* in the late 1st century B.C.), *OCD* 756 (s.v. "Iulius Atticus," wrote on how to reduce costs and maximize production in the wine industry in the early 1st century A.D.), *OCD* 603 (s.v. "Gargilius Martialis, Quintus," wrote on tree farming and veterinary medicine in the early 3rd century A.D.); many more agricultural and veterinary writers are listed in *EANS*. See also *OCD* 134 and 353 (s.v. "arboriculture" and "Columella, Lucius Iunius Moderatus"). The dates of Aristomachus of Soli and Philiscus of Thasos are uncertain, and their writings are lost, yet they were the most famous and diligent scientific bee researchers in antiquity, cf. Pliny the Elder, *Natural History* 11.9.19 and Russo 2003: 251 (with *EANS* 138–39 and 649). It appears that the most experimental agricultural research occurred in Ptolemaic Egypt in the 3rd century B.C. (*OCD* 123, s.v. "Apollonius (3)") and it is from that time that we hear of the first zoos, whose subsequent fate is unknown (one at Mithridates' palace in Pontus: Strabo, *Geography* 12.3.30; another at Ptolemy's palace in Alexandria: Athenaeus, *The Dinnersages* 14.654b-c).

750. Pliny the Elder, *Natural History* 18.74.317, on which see K.D. White 1984: 184–85. Hero, *Mechanics* 3.13–20 (cf. Drachmann 1963: 110–35) details an even wider variety of innovations in press design, and Hero, *Mechanics* 3.21 (cf. Drachmann 1963: 135–40) discusses his related screw-cutting machine. So the point holds even if Pliny has his history wrong as Russo claims—however, Russo mistakes Pliny as referring to the screw press in general rather than new versions of it (Russo 2003: 150–51).

751. Columella, *On Agricultural Matters* 4.29.15–16, describes the advantages of his new grafting auger. Pliny the Elder, *Natural History* 17.25.116, is equally impressed with it. Such experimentation was not unusual. Galen's father "carried out experiments on his crops and wines to improve their quality" according to Nutton 2013: 222, 247 (cf. e.g. Galen, *On the Powers of Foods* 1.37 = Kühn 6.552–53 = Mark Grant 2000: 107–08). And we have many epigraphic examples of engineers expressing pride at their innovation and problem solving (Kolb 2015).

more time is spent on soils research for customizing and maximizing agricultural yield than in less avaricious branches of geological science.[752]

Although, according to Dominic Rathbone, "by modern standards Roman agriculture was technically simple, average yields were low, transport was difficult and costly, and storage was inefficient," all factors that limited urbanization and thus industrialization, "nevertheless, in the late republic and earlier Principate agriculture and urbanization" actually "developed together to levels probably not again matched until the late 18th" century, and in his opinion, contrary to the assertions of previous scholars:

> Roman estate owners showed considerable interest in technical and technological improvements, such as experimentation with and selection of particular plant varieties and breeds of animal, the development of more efficient presses and of viticulture techniques in general, concern with the productive deployment and control of labour, and, arguably, a generally 'economically rational' attitude of exploitation of their landholdings.[753]

Like the sciences, it is not unreasonable to expect that improvements in agriculture and industry would also have continued, had events of the 3rd century not reversed the course of ancient culture away from its budding scientific and industrial spirit. Until then, Roman industry had matched pace with agriculture, with more major buildings, roads, bridges, aqueducts, harbors, ships, and other construction projects than would ever be seen again until early modern times, and even "ordinary farmers and urban craft-workers possessed more iron tools, architectural stonework, and fine table-ware than ever before," as well as many other manufactured products, "to an extent that would not be matched again until the post-medieval period," and there is still more "evidence for extensive industries and widespread application of technology in the ancient world" which is not at all comparable to the early medieval period.[754]

---

752. Seneca, *Moral Epistles* 90.21 and *Aetna* 265–70.

753. *OCD* 44–45 (s.v. "agriculture, Roman"); with Schneider 1992: 52–71, Shaw 2013, and Bowman & Wilson 2013. Lirb 1993 argues that even small-scale farmers benefitted from equipment and animal sharing cooperatives in antiquity, which would have expanded access to agricultural technologies beyond the elite.

754. Quoting Kevin Greene, *OCD* 1435 (in s.v. "technology"). See also *OCD* 128–29 (s.v. "aqueducts"), and Schneider 1992: 181–93, for some of the peculiar features

V. The Prospect of Steam

Nevertheless, scholars still ask why the Romans did not hit upon the cornerstone of a true industrial revolution: industrial steam power. I suspect circumstances mattered more than ideology. Coal mining was introduced in Roman England only around 100 A.D.[755] Coal was then used to fuel hypocausts and forges, and heavily employed in the army, which is yet another example of a new industry launched and exploited by the Romans. But the late discovery and exploitation of coal may explain why the Romans never developed the steam engine. As even Peter Green reports:

> The Greek inability, despite possessing all its separate parts, to develop an efficient steam engine was long ascribed to a lack of the technique that would enable them to precision-turn and cast close-fitting metal cylinders and pistons... [and yet]...four bronze-cast force pumps found in the wreck of a first-century-A.D. Roman merchantman were tooled to an all-around clearance between piston and cylinder of between 0.1 and 0.35mm, and when greased could operate at over 95 percent efficiency.[756]

The Romans had thus achieved all the skills and component parts for constructing a steam engine—not only had they invented a simple steam

---

of ancient technology that exceeded medieval. Scholarship on the massive decline from Roman prosperity in industry and agriculture experienced in the middle ages is collected in Carrier 2014b (reinforced by Loseby 2012 and Brun 2012).

755. Travis 2008; Frere 1987: 288; James & Thorpe 1994: 409. The less efficient fibrous lignite was already in limited use in early Italy and Greece as an industrial fuel (Theophrastus, *On Stones* 13 and 16; cf. Caley & Richards 1956: 81–82, 85–86 and Healy 1978: 149), but even less efficient charcoal remained the most common industrial fuel across the empire (Healy 1978: 150–52; Oleson 1986: 172–81). On this situation inhibiting development of steam power see Oleson 1984: 402–03, though in the end he still resorts to the 'ancient people were stupid' argument, which is quite out of place in a book he had just filled with unchallengeable evidence of their intelligence and ingenuity (which is amplified abundantly in the present chapter).

756. P. Green 1990: 474. The quality of Roman metal tooling was indeed superb: see Schiøler 1980 and 1989.

turbine but, as Green observes, they had developed precision machinery to a remarkable level, and were using in other functions all the parts needed for a steam engine. Hence many scholars, including Green, blame the failure to take the next step on some sort of mental block (an implausible argument I'll get to in a later section).

There is no room at present to fully examine the question of why the Romans did not invent the steam engine. But since it probably had more to do with historical happenstance than any sort of mental block, I'll pause briefly on it. Significant scientific research in mechanics was primarily taking place in the coastal areas of the Mediterranean, especially Egypt, where there was no readily available cost-efficient fuel that would make steam power economically useful.[757] I suspect it was no accident that a proper steam engine was not invented except as an adaptation from an already-existing steam-powered waterpump, which in turn was invented only when and where a primitive steam pump would be useful: for draining coal mines in England, a circumstance in which the pump's fuel was effectively cost-free, since it allowed the extraction of more fuel than it consumed (a circumstance that could easily inspire the device's invention in the first place).[758] Yet coal exploitation began only just before the decline of the Roman empire, in an area far from major centers of scientific research, and before coal mines needed draining. The steam-powered pump was only developed at the end of the 17th century because only then did it become a practical solution to a new problem, as coal mines were becoming exhausted below the water table after centuries of use. Had the Pax Romana continued another century or two, it is conceivable a steam-powered pump could have been developed for the same purpose, in the same place and way, which may well have gone on

757. To be at all attractive to an ancient investor, the wood (or other fuel) to power a steam engine must cost less per unit of work done than the food to power a man or animal at the same task, which is not easy to do, considering how little men eat, and how cheaply animals can be fed. Landels 2000: 28–33 discusses the prospects for steam power, though is over-optimistic about the relative cost of charcoal (especially in Alexandria).

758. See Hills 1989: 13–30 and Savery 1702. See also Simms 1995: 88 and Russo 2003: 126–28. Over the course of about a century before Savery's invention a variety of steam equipment had been inspired by Hero's precedents, but none of these imaginative devices ever saw industrial application (e.g. Keyser 1992: 114–17; Dickinson 1939: 4–18, 192–93; Galloway 1837: 7–15).

to inspire a steam engine, exactly as happened in early modern England. No intellectual disease need be posited.

VI. Science and Technology

This completes my survey of actual Greek and Roman inventions. Even without a practical steam engine making the list, and even though we have left out countless items, we have identified well over two hundred Greco-Roman inventions employed in Roman times. Peter Green's first list of a mere thirteen or fourteen items looks absurd by comparison. We also identified several general categories of inventions (like scientific instruments and theatrical robotics) from which we could enumerate many dozens more individual items. And we did not even consider the vast array of techniques and equipment invented for chemical processes in antiquity.[759] Although many of the technologies listed on the preceding pages might not have been widely used or were not universally available, this would only indicate, at most, a certain failure of economic progress, not a lack of scientific or technological progress. For as we noted in the very beginning, even the rarely employed technologies were nevertheless invented, and continued to be invented, improved, and employed for centuries. Even if only by the wealthiest families, municipalities, or magnates, there was nevertheless a demand for technological inventions, however limited, and it was met by creative and enterprising inventors.

In many cases these new technologies were aided by progress in the relevant sciences, as for example in crane and fire engine design, or in aqueduct and bathhouse construction, while in most cases ordinary human creativity and enterprise led to the development and exploitation of new industries and numerous improvements in old ones, quite apart from scientific influence.[760] Nevertheless, Green's assertion that "the remarkable scientific advances of the Hellenistic period contributed virtually nothing

759. Whose origins are most difficult to date, but for examples of what was in use see sources on practical and empirical chemistry in the related note in section 3.5.

760. On this distinction in the context of Greco-Roman times see Russo 2003: 209–10. Edelstein 1952: 579–85 surveys a few examples of ancient science influencing technology.

to society's technological or economic betterment" is quite implausible.[761] Architecture, most importantly in port and aqueduct design, improved significantly from the science of mechanics and the technologies built upon its principles, such as cranes, pulleys, pumps, siphons, and gears. Surely all this produced some "technological and economic betterment." Ancient medicine bettered the lives of all who had access to it, and not a trivial number of people did. Army life was especially improved by it, but the availability of subsidized city doctors yielded some benefits to urban civilians as well.[762] Many technologies were specifically developed for employment in medical care, from bonesetters to catheters to cataract syringes, and for use by surveyors, engineers, and astronomers, hence craftsmen were well employed in producing a large array of instruments based on scientific principles and advice. The force pump extinguished urban fires, filled urban baths, primed fountains, irrigated gardens, cracked heated rock in mines, washed vaulted ceilings, and cleared the sumps of buildings and the bilges of ships.[763] Scientifically designed weapons contributed to securing and maintaining peace and prosperity. And were there no social or economic benefits from Roman advances in cartography? Were no economic or social benefits produced by the scientific regulation of the calendar or the production of portable sundials? What of the watermill, which combined Archimedean principles of hydraulics, levers, and wheels, and was adapted widely to multiple industries, automating labor? Quite simply, the idea that science contributed "virtually nothing" to ancient technology or prosperity is not easy to maintain.

Green also claims that "in no case, except possibly that of grain milling, is there any attempt, predictably, to improve industrial efficiency," because "change implies degeneration" and was therefore opposed—but this is not a credible assessment either.[764] Waterpower did not just automate grain milling. It also turned saws on lumber and stone and ground ore for the

---

761. P. Green 1990: 363 (similarly Vernant 1983: 283: "yet this ingenuity did not transform the technology of the ancient world").

762. See chapter eight of Carrier 2016.

763. Oleson 2005: 211–31.

764. P. Green 1990: 468. Russo 2003 refutes almost every aspect of Green's assessment of Hellenistic science, even after subtracting Russo's untenable theories from his overall case.

mining industry (and possibly sand for the glassblowing industry). The gradual adoption of the shaft furnace and of mass production in the brick and lamp industries, the development of bread-kneading machines and oil and fulling presses, and numerous technological improvements in harbor and mining technology, are just a few more examples that refute Green's point. Indeed, Green tries to maintain his case by pretending that weapon and building technologies—and, somehow, hydrological and mining technologies—which he concedes advanced continually, do not count as "industries." Yet he offers no plausible basis for this distinction, nor would it rescue his theory if he did. For if "change implies degeneration," then why did the Romans work so hard to "degenerate" their military and building capabilities by constantly improving them? Why improve their mining industry? Their water industry? Their ports and harbors? Green says "we will look in vain" for "labor-saving devices, servo-mechanisms, inventions designed to promote increased efficiency or to streamline production," but what about cranes, waterwheels, screw presses, harvesters, threshers, bread-kneaders, or pressmold lamp manufacture?[765] As we've already seen, this list could be expanded considerably.

Likewise, Green asserts that "a pressing, but economically viable, public need is often essential to facilitate [industrial] development, and even then it can often be frustrated by innate conservatism or social (most often religious) prejudice," a throwaway comment that bears little merit on close scrutiny.[766] Many of the developments in our survey do not reflect any "pressing need" (what pressing need was there for an ice vending or

765. P. Green 1990: 469. I do not know what exactly Green means by "servo-mechanisms," which are by definition electronic devices (consult any common dictionary). If he means mechanisms that multiply human strength, the crane alone refutes him. If he means mechanisms that automate mechanical processes, the watermill alone refutes him, as do Roman sawmills and evidence of mechanized hammers. If he means industrial robotics that do not require electrical power, there was nothing of that kind before the 18th century, hence *after* the Scientific Revolution. Perhaps he means the adaptation of water power to more diverse industries, which began in the 14th century (see Reynolds 1983). Even supposing none of these applications existed in Roman times (again, arguments from silence are feeble here), there is no reason to assume comparable developments would not have taken place in the 3rd and 4th centuries if the peace, prosperity, and dominant zeitgeist of the 1st and 2nd centuries had continued.

766. P. Green 1990: 467.

glassblowing industry?); there is no discernible religious opposition to either science or technology that had any identifiable effect on progress in either domain before the 4th century A.D.; and there is no evidence of "conservatism" preventing the gradual embrace of a large array of technical or scientific innovations. Though ancient culture was indeed conservative, and conservative enough perhaps to impede the rate at which innovations were thought up or deployed, this is true only relative to modern cultures. In every other respect, a slow rate of progress is still progress.

It is also true that the ancients did not realize the gains in efficiency they could have achieved if they had invested more heavily in technological research. But it cannot be said that the same people who specifically sought out systematic ways to improve the labor and output efficiency of their water system, were somehow uninterested in doing the same to make money, improve their military, or achieve any other desired goal.[767] The happenstance of gradual technological and industrial improvements can be found in all ages of human history, but the idea of *actively* setting out to develop new technologies for the purpose of increasing production "efficiency" is a decidedly modern one, hardly to be found at all before the 17th century.[768] But though maybe falling short of *recognizing* the link between *investment* in scientific research and economic betterment, the Romans were more than open to technological innovation, even for economic benefit, and well aware of the connection between them.

---

767. Interest in efficiency in the water supply is a central and continual concern of Julius Frontinus in his manual *On the Aqueducts of Rome*, as rightly noted by K.D. White 1984: 188.

768. See introduction to section 3.9 below. Ancient dreams of automation, imagining robot servants and laborers that would render slavery obsolete, were never connected to realistic technologies, but remained mere fantasies (like 20th century dreams of bubble cities and flying cars—or, still my favorite, the little pill that becomes a fabulous chicken dinner in *The Fifth Element*). For example: Aristotle, *Politics* 1.4.1253b-1254a (referencing Homer, *Iliad* 18.369–381 and Plato, *Meno* 97d-e) and Athenaeus, *The Dinnersages* 6.267e-270a (discussed cynically in P. Green 1990: 392). Although at the dawn of the 1st century A.D. the poet Antipater of Thessalonica praised the watermill as a labor-saving machine that would bring us back to a golden age without toil (*Palatine Anthology* 9.418), which is a more serious recognition of the value of automation, notably from the Roman era (see Reynolds 1983: 17, M.J. Lewis 1997: 66–69, and Humphrey et al. 1998: 31, with *OCD* 107, s.v. "Antipater (5)").

The Romans were indeed conscious of the fact that science produces practical results and was thus worth pursuing for that very reason. This was already obvious in the area of military technology. In fact, it is the explicit thesis of Hero's *Siegecraft* that scientific research leads to a better society through its contributions to technological advance. It was also clear in architecture and mechanics. Vitruvius makes a particular point of explaining how scientific advances had led to many useful technologies.[769] So he was definitely aware of this fact. And promoting it. The same was also obvious in medicine, which saw continual improvement in applied knowledge, techniques, and equipment.

This was also evident in the mathematical sciences. Galen was even happy to cite the latter in support of the former:

> I had observed the incontrovertible truth manifested—and not just to myself—in predictions of eclipses, in the working of sundials and waterclocks, and in all sorts of other calculations made in the context of architecture, and I decided that this geometrical type of proof would be the best to employ [in medicine as well].[770]

From several passages like this, Serafina Cuomo finds a consistent pattern in Galen's attitude toward science and technology:

> On top of its compelling form of argumentation, and the positive consequences this had in the establishment of shared belief, [Galen argued that] mathematics deserved recognition because of its concrete workings in the world. Galen never lost sight of the fact that the people engaged in mathematical practices ([e.g.] calculators, geometers, architects, astronomers, musicians, gnomon-makers) produced something: predictions of eclipses, buildings, instruments like sundials and waterclocks. ... For Galen, mathematical truth is demonstrated both by its products and by its proofs, and its validity is guaranteed by the role it has in the community, by shared assent and collective persuasion.

---

769. Vitruvius, *On Architecture* 9.pr.3–16 and 10.1.5–6 (e.g. 9.pr.4–8: discoveries in theoretical and applied mathematics; 9.pr.9–12: Archimedes' discovery of the principle of hydrostatics).

770. Galen, *On My Own Books* 11 (= Kühn 19.40). Galen repeats and elaborates this sentiment throughout his works, always including logic and empirical research in his methodological ideal. See related discussion in chapter seven of Carrier 2016 (on mathematics) and discussion here in section 3.7 (on method).

> Assent in mathematical proofs is generated by the experience itself of going through the demonstration or of learning a certain method to solve geometrical problems [and] one can see that it works. Analogously with the embodied mathematics of sundials, waterclocks, predictions of eclipses or architectural calculations: one can see that they work, they too are proofs of the incontrovertible truths of mathematics, and a proof which is often out there in the street, under everybody's eyes.[771]

While Galen was keenly aware of the connection between mathematical science and technology, and between demonstrable results and sound method, modern historians often seem relatively clueless. For example, Vernant claims "the five simple instruments," the lever, wedge, pulley, wheel, and screw, "formed a coherent and self-enclosed system that excluded innovation or progress," after having just described the ancient addition of numerous other "simple instruments" beyond the basic five, such as the piston, gear, valve, spring, and siphon, refuting himself almost in the same breath.[772] But even apart from this, his argument remains that of an arm-chair historian mired in ignorance. For even *modern* engineers agree that all mechanical functions can be reduced to just five basic principles—the very same ones Vernant lists, in fact. According to the most advanced naval fighting force in human history, if we add the gear (as in fact the ancients did), then...

> There are only six simple machines: the lever, the block [i.e. the pulley], the wheel and axle, the inclined plane, the screw, and the gear. Physicists, however, recognize only two basic principles in machines: those of the lever and the inclined plane. The wheel and axle, block and tackle, and gears may be considered levers. The wedge and the screw use the principle of the inclined plane. ... [so all] complex machines are merely the combinations of two or more simple machines.[773]

Though here by adding the gear the original five are increased to six, gearing is nothing more than a combination of wheels and levers and is thus reducible to the other five, which are in turn reducible to only two. *And the Greeks also knew this.*

---

771. Cuomo 2001: 187–88.

772. Vernant 1983: 289, 283 (cf. 287).

773. NAVEDTRA 1994 1–1 and 1–2. This is presently a textbook used by the officer training programs of the United States Navy.

Hence the Greek understanding that all machines are compounded from only a few simple machines is not a limitation in thinking, but is, to the contrary, brilliant. Not only did this "coherent system" of the five basic machines plus the gear *not* prevent the innovation of new "basic machines," it actually led to innovative progress in the way all basic machines were combined and improved, leading to such widely applied inventions as the crane, reciprocating bilge pump, and watermill. Even the waterscrew is an example of combining the wedge and the wheel to produce one of the most revolutionary pump designs in technological history, significantly increasing the efficiency of waterlifting. There was therefore no "exclusion of innovation or progress." To the contrary, when all evidence is considered, modern experts are compelled to conclude that the "ancient Romans reached a technological level that was only regained in the sixteenth century," and "this [later] recovery was, as a matter of fact, a new discovery of procedures which had been completely forgotten."[774]

3.7 Was Roman Science in Decline?

Nevertheless, some scholars have claimed science suffered a stagnating decline during the Roman period. Lucio Russo even claims (absurdly) that "the Romans were not interested in science" and had abandoned the superior aims and methods of their Hellenistic forebears.[775] Such assertions are ultimately baseless. There is no evidence of any difference, much less decline, in scientific aims or methodology between, for example, Hipparchus and Ptolemy, or Herophilus and Galen (Russo's favorite examples). Only by romanticizing Hellenistic scientists, and imagining (implausibly) that they never held or defended any absurd or erroneous beliefs, can Russo contrive any appearance of decline. A more frequent mistake is to compare Hellenistic

774. Marchis & Scalva 2002: 26. On Roman waterlifting machinery: Bowman & Wilson 2013: 273–305 (supplementing references on waterwheel technology in following notes).

775. Russo 2003: 266 (he offers several negative assessments of Roman science, none of which are demonstrated by any adequate evidence: 15, 215, 231–41, 264–70, 282–86, 318; yet ironically he challenges the basis of such assessments from other authors: 197–202). Contrast Russo's assessment with that of Chevallier 1993.

scientists with Roman laypersons, as if *Hellenistic* laypeople would come out any better in comparison with Roman *scientists*. Myths of a 'Roman decline' are thus often based on assessments of lay authors like Pliny the Elder, rather than actual Roman scientists like Dioscorides, Hero, Marinus, Menelaus, Ptolemy, Galen, or Soranus.[776] But the mistakes and flawed or inexact methods of an author like Pliny tell us only about the standards and practices of lay admirers of science, not what actual *scientists* were doing.

Of course, negative assessments of Pliny's merits are also often exaggerated.[777] But more importantly, an individual author does not always represent their whole society—one need only compare Pliny's treatment of medical science with that of his predecessor Celsus to see how superior a treatment the same subject could receive from another lay author of the very same time.[778] Even picking on individual scientists is not always apt. Hero might not always appear as rigorous and brilliant as Archimedes, but that may be the very reason why Hero's works were preserved, and not those of even Archimedes, much less other Roman engineers who may have been similarly rigorous and thus too unintelligible to medieval antiquarians to warrant copying. One can only wonder, after all, what happened to the mechanical writings of Menelaus, Apollodorus, and Ptolemy, much less authors unknown. In the same fashion, one cannot claim Strabo's failings in geography or astronomy were symptomatic of the Roman era, when that same era also produced the superior work of Marinus and Ptolemy in those same fields, and especially when it cannot be established that *none* of Strabo's Hellenistic predecessors were any worse than he was.[779]

Nor does it make any sense to maintain there was a "resurgence of

---

776. An example of this error on a grand scale is Stahl 1962 and 1971.

777. For modern mis-assessments of Pliny (and their recent correction) see earlier note.

778. Pliny discusses medicine in the 29th book of his *Natural History*, Celsus in his extant volumes *On Medicine*, both in Latin. Romans did not all agree with Pliny, e.g. Aulus Gellius (in *Attic Nights* 10.12) takes Pliny's credulity to task, somewhat unfairly according to Beagon 1992: 11 n. 31, but there would have been many laypeople of the day who could correct Pliny on many points. Similarly, Quintilian (in *Education in Oratory* 10.1.128) complains that Seneca was a brilliant man but relied too much on research assistants who sometimes led him into error.

779. For a relatively (and sometimes unfairly) negative assessment of Strabo see Aujac 1966.

religious enthusiasm" in the Hellenistic age that worked against scientific advancement.[780] There is no good case to be made that religiosity and superstition was ever in any state of decline. Skepticism and rationalism remained as present but as uncommon as ever, hardly more than the preoccupation of a rarefied elite, while superstition and irrationality remained the norm against which exceptional men had battled even in Classical Athens.[781] And though ancient scientists in *every* era had embraced bad ideas, and did not follow their own recommended methods as consistently as we would like, the very same could be said of the savants of the Scientific Revolution. Galileo's ideas about tides and visual rays were often wildly wrong, Kepler was obsessed with the harmony of the spheres, and Newton pursued alchemy and worked profusely on biblical theories of history, prophecy, and cosmology, spending considerable time trying to predict the apocalypse.[782] Meanwhile, bloodletting continued as a 'scientific' medical treatment well into the 18th century. The 19th century became an infamous age of medical quackery. So we moderns are in no position to judge.

The first half of this chapter has already shown how claims of both scientific and technological stagnation under the Romans are implausible. Peter Green concedes that "progress of a sort did take place" but then claims there remained "a dead-weight legacy from the past that in many ways made true progress virtually impossible," a judiciously meaningless statement, since he does not explain what we are supposed to count as "true" progress or why.[783] We have already seen that scholars like Peter Green are obsessed

780. P. Green 1990: 481. Farrington 1965 attempts a similar but even more inept argument.

781. See Dodds 1951 and note in chapter 1.3 on religious persecution in Classical Athens.

782. On various absurdities among 17th century scientists see: Russo 2003: 355–59, 363–64, 366–69, 385–88 (likewise, for Newton, Rossi 2001: 203–29); more examples in Zimmermann 2011. Ultimately there was nothing any more boneheaded in ancient scientific treatises than can be found in even the most respected authorities of the Renaissance.

783. P. Green 1990: 480–81. Similarly, Moses Finley concedes "there were improvements of one kind or another," and "technical refinements," but insists these were only "marginal" and not "radical improvements," without defining either 'marginal' or 'radical' (Finley 1985: 109, 114).

with finding fault with what the ancients *did not* invent or discover, while ignoring almost everything they *did* invent and discover, and then accusing them of having invented and discovered nothing. Which they then proceed to explain with one or another fanciful hypothesis. It is a peculiar way of doing history. As an example at the very nexus of science and technology, Green complains that the ancients failed to invent "steam gauges, thermometers, microscopes, telescopes, [and] fine-calibrated lathes," as if these were somehow obvious and easily conceived technologies, while ignoring the countless instruments ancient scientists *did* invent to further their research.[784]

More absurd is Aubrey Gwynn's claim that "the Roman Empire never produced a scientific discovery that has been of permanent use to mankind."[785] Even a lot of obsolete science was still a necessary step toward modern science. For instance, Ptolemy's law of refraction was not entirely correct, but it was close, and his idea and procedures for experimentally discovering a mathematical law of refraction were certainly of permanent use to mankind, and though Hero's experimentation with steam-powered machinery did not lead immediately to a practical steam engine, it was a necessary first step that eventually inspired it, so Hero's discovery that steam could be used to produce mechanical motion was of permanent use to mankind. Meanwhile, many Roman discoveries (such as in pharmacology) were certainly of permanent use to mankind, or may have been yet were lost, while others (like electroshock therapy) remain in use, even if in different applications. Roman discoveries still (more or less) in use include

---

784. P. Green 1990: 481. He also complains of a lack of formal statistics and "advanced technical instruments" in antiquity (P. Green 1990: 457), even though neither existed until *after* the Scientific Revolution. Likewise for Zilsel's complaint that they didn't have periodicals (Zilsel 1945: 327). However, as noted in the previous section, the Romans must have had more advanced lathes than we are otherwise aware. Indeed it is ironic that (as also noted in the previous section) Green cites ancient precision tooling of nested cylinders to within a tenth of a millimeter, and yet he somehow thinks this was achieved without fine-calibrated lathes, which would have been needed for turning the wax molds to such a precise clearance.

785. Gwynn 1926: 146. Such dismissiveness, which can still be found (we opened with an example from Russo), is rightly criticized in Nutton 2013: 13–16 (though only for medicine, his remarks are as relevant for astronomy and physics) and also challenged by Chevallier 1993.

Ptolemy's system of cartographic projection, Hero's principle of least action in reflection, Galen's experimental discoveries relating to kidney function, the spherical trigonometry of Menelaus, and the idea symbolic algebra of Diophantus—we just do not use these same systems of trigonometry and algebra today any more than we speak Latin or ancient Greek. Ptolemy's most crucial innovation in planetary theory, the acceptance of inconstant planetary velocities and proposing a law of planetary motion (equal angles in equal times), turned out to be essential to Kepler's solution for the planetary motions and orbits (updating Ptolemy's law to equal areas in equal times), while the efforts of Ptolemy and Galen to unify their sciences and epistemologies were of even more general benefit to modern science. And then there were useful discoveries we often ignore. For example, one of the areas Galen knew he was making considerable advances in was the physiology of voice and speech, pursuing a comprehensive research program involving extensive physiological and anatomical observations and experiments on every related organ from the lungs and thorax to the nerves and muscles of the throat, larynx, tongue, and more.[786]

Like Gwynn's antiquated nonsense, most of the claims of a Roman decline are so contrary to the facts that they hardly need refutation. The most famous example is a raft of assertions by Samuel Sambursky, all plagued by fanciful and inaccurate conceptions of ancient science, many of which have already been exploded in previous sections of this chapter.[787] Ancient scientists were not isolated from each other, but enjoying frequent communication and interaction, and the sharing and accumulation of results.[788] There was no relevant disdain among them for shopwork and

786. See Galen, *On the Doctrines of Hippocrates and Plato* 2.4.33–39 and *On the Uses of the Parts* 7.14 (= M.T. May 1968: 367).

787. Sambursky 1962: 253–76. The same points (or claims even more ridiculous) are still echoed in more recent scholarship, e.g. Vernant 1983: 294–295 and 366, Reynolds 1983: 32–35, Lewis & Reinhold 1990: 2.210, and Stark 2003: 151–54 (with the same material almost verbatim in Stark 2005: 17–20, though adding more false assertions about ancient science and technology in 2005: 12–17). All of which are adequately refuted by the substance of the present chapter. As also (more succinctly) in Carrier 2010 (supported by Efron 2009).

788. Contrary to Sambursky 1962: 254–55. See examples in chapters three and four here, and throughout Carrier 2016.

technology.[789] There was no significant opposition to changing or interfering with nature.[790] There was no aversion to experiments.[791] There was no failure to mathematize the study of nature.[792] They actually did understand natural processes mechanically rather than organically.[793] And there is no evidence of any significant 'rise' in irrationality under the Romans (at least before the 3rd century A.D.).[794] Everything else Sambursky proposes confuses the *effects* of the Scientific Revolution with its *causes*, and thus fails to explain anything even when true.[795]

So when Sambursky claims a fictional stagnation resulted from a "lack of systematic experimentation and the consequent stagnation of technology, and the failure to develop algebraic notation and to introduce mathematical symbols and procedures in the description and explanation of physical phenomena," we already know every single one of these claims is false.[796] The Romans were seeing progress in all. And even Sambursky knew he had to qualify his remarks, admitting the Romans held a "greater regard for observational evidence and an increasing demand for a more accurate description" and were conducting systematic experiments that "led to conclusions which conflicted with Aristotelian conceptions about the nature of light" and other subjects. Hence, he concludes, it was really only after the era of Galen and Ptolemy that "the combined effects of the irrational tendencies within neo-Platonism and of the anti-scientific attitude of the early Church," and the general decline of educational institutions everywhere, finally put an end to scientific research.[797] On all

789. Contrary to Sambursky 1962: 257–60. See discussion below and in chapter 4.6.

790. Contrary to Sambursky 1962: 259–62. See discussion below.

791. Contrary to Sambursky 1962: 261–70. See examples throughout sections 3.1 through 3.6.

792. Contrary to Sambursky 1962: 270–73. See examples in sections 3.3 and 3.4 (and even in biology: see discussion and sources in von Staden 1996: 88–90).

793. Contrary to Sambursky 1962: 273–74. See discussion below.

794. Contrary to Sambursky 1962: 260–61, 274. See remarks above.

795. Sambursky 1962: 254–56. On the relevance of this point see section 1.1.

796. Sambursky 1963: 62.

797. Sambursky 1963: 63–66.

that, at least, he was correct.

Similarly, Ludwig Edelstein once claimed "ancient science remained relatively useless" and "changes which in principle were within reach were actually not made" because empirical scientists were too skeptical to theorize, theorists were too disinterested in empirical research, and everyone was uninterested in controlling the natural world through technology.[798] But not one of these assertions is true, as any perusal of Galen, Ptolemy, Hero, or Vitruvius would easily reveal. More credible but still dubious is Peter Green's assertion that "quantitative methods, essential to true scientific progress, were conspicuous by their absence" among the Romans.[799] But he still never explains what he means by "true" scientific progress, or even "quantitative methods." Was all the scientific progress I just documented 'fake'? Was measuring doses of medications, angles of refraction, mechanical advantages, or velocities of planets not 'quantitative'? There were certainly many failings in the way ancient science was conducted, but an absence of quantitative methods was not among them. At most one can say such methods were not more widely exploited than they could have been, but there was no evident *decline* in this respect.

Peter Green has voiced many other absurd allegations. For example, he claims "the enormous weight of [Aristotle's] authority" did "more to hold up the progress of astronomy than any other single factor," yet progress in astronomy was not held up, and as we have seen, Aristotle's authority was not particularly great in antiquity (in fact it was *greater* at the dawn of the Scientific Revolution).[800] Green claims the Hellenistic trend toward moral philosophy "culminated in the abandonment of true research" and a "reversion" to excessive theorizing, but he never identifies any point in time when the research he has in mind was "abandoned" or when theorizing was *not* excessive. To the contrary, *Presocratic* science was heavy on theorizing and light on research, while most science after Aristotle leaned quite the other way, with moral philosophy and scientific advances increasing in tandem.[801] Green also claims scientific progress in antiquity was hindered by

---

798. Edelstein 1963: 24–27.

799. P. Green 1990: 470.

800. P. Green 1990: 459.

801. P. Green 1990: 482. It should also be noted that Green has been deceived by medieval selectivity in preserving texts, creating the illusion of a rising interest in

a "prejudice" against written texts, but there is no evidence of this, any more than lectures and internships indicate any such thing now.[802] Likewise, "the subordination of experimental science to philosophical system-building" was true all throughout antiquity (in fact reversing this was a defining feature of the Scientific Revolution), yet progress continued.[803] Likewise, the fact that, as Green says, logico-deductive conclusions are more reliable than empirical ones is a fact made much of even by Descartes and recognized still today. Though entirely true, this fact has had no effect on science now, nor did it then.[804]

In a similar fashion, Joseph Ben-David repeats one of Sambursky's indefensible claims, that ancient "scientists built their individual systems without reference to those of others and established rival schools which, like so many religious sects, did not communicate with each other."[805] Again, he is wrong on all counts. The works of Ptolemy, Hero and Galen are full of references to, adoptions from, and improvements upon the work of numerous predecessors in their respective fields, while Galen's writings are filled with evidence of a lively public interaction among contemporary scientists.[806] There is no evidence that any ancient scientist behaved differently. And while there were many "competing schools of thought" on crucial questions of method and epistemology, these were not isolated nor even dogmatic enclaves, but loosely-affiliated groups of researchers regularly engaged in improvement, intercommunication, and debate. The most successful scientists, in fact, refused to align themselves with any one school, but instead learned and borrowed from them all, a phenomenon of 'eclecticism' that typified the entire intellectual atmosphere of the Roman

---

moral philosophy at the expense of physics and logic that actually never happened in antiquity: see Carrier 2016: 102–04.

802. P. Green 1990: 457. See the more reasonable analysis of Alexander 1990 and discussion in chapter seven of Carrier 2016.

803. P. Green 1990: 481.

804. P. Green 1990: 457.

805. Ben-David 1991: 301.

806. Ptolemy's *Almagest*, *Geography*, and *Harmonics* are good examples of his discussion of predecessors and his reliance and improvement on them, as are Galen's many treatises on anatomy and pharmacology, and likewise Hero's *Pneumatics*.

period.[807] This is quite evident in Ptolemy, who merged the epistemologies of all the major schools into a practical proto-scientific system, and in Hero, who loved trumping sectarian dogmas with physical demonstrations, and in Galen, who railed against the very idea of distinct schools of medical thought and instead embraced elements of many different schools, criticized the rest, and synthesized a nearly modern combination of deductive and empirical methods of his own.[808] Galen also sought to unify formal logic by developing a comprehensive system from of the doctrines of several schools.[809]

---

807. See discussion in Carrier 2016 (index, "eclecticism") and section 3.2 above. For further discussion of the eclecticism of Galen and Ptolemy see: Gottschalk 1987: 1164–71. For Ptolemy: *DSB* 11.201–02 (in s.v. "Ptolemy"). For Galen: Hankinson 1992. For Hero: Tybjerg 2005: 214–15. Galen specifically describes and advocates eclecticism in *On the Affections and Errors of the Soul* 1.8 and 2.6–2.7 (= Kühn 5.42–43 and 5.96–103) and Seneca effectively does the same in *Moral Epistles* 33. See also the 'eclectic' credo advocated in Celsus, *On Medicine* pr.45–47.

808. For Ptolemy's scientific epistemology: Huby & Neal 1989; Long 1988: 176–207; A.M. Smith 1996: 17–18; Barker 2000. For Galen's scientific epistemology: Frede 1981; Walzer & Frede 1985: xxxi-xxxiv; Iskandar 1988; J. Barnes 1993; Hankinson 1988: 148–50, 1991a: xxii-xxxiii and 109–10, 1991b, and 1992; M.T. May 1968: 45–64. For an early summary of both: Edelstein 1952: 602–04. Ptolemy's *On the Criterion* and Galen's *On Medical Experience* are prominent examples, as also Galen's *On the Sects for Beginners* and *An Outline of Empiricism* (for all three see translations and discussion in Walzer & Frede 1985, esp. xxxi-xxxiv), as well as his synthesis of epistemologies in *On the Doctrines of Hippocrates and Plato* 9, but much more important was Galen's treatise *On Demonstration*, which was specifically devoted to scientific method, and yet medieval scribes had no interest in preserving it (Nutton 1999: 166, §P.82,3–5 lists sources containing extant fragments of it, and Hankinson 1991b attempts to reconstruct Galen's scientific method from his extant works). On Galen's related interest in mathematics, and mathematical sciences and methods, see discussion in chapter seven of Carrier 2016 and example in section 3.6.VI. For examples of his commitment to an almost modern empiricism see Galen, *On the Method of Healing* 1.4, 2.7, 3.1, and 4.3 (= Kühn 10.31, 10.127, 10.159, 10.246) and *On the Affections and Errors of the Soul* 2.3 (= Kühn 5.66–69 and 5.80–90). That Galen's epistemology was influential in the development of modern scientific method is argued in Crombie 1953: 27–28, 40–41, 74–84, and Walzer & Frede 1985: xxxiv-xxxvi. I think one could argue the same of Ptolemy's as well (e.g. consider his anticipations Occham's Razor in *Planetary Hypotheses* 2.6 and *Almagest* 13.2).

809. See Kieffer 1964.

Moreover, Hero, Ptolemy, and Galen all insisted upon the use and methodology of mathematics in the sciences.[810] And all employed systematic experiments in their work. In his *Pneumatics*, for example, Hero begins with a physical theory, describes experiments that establish its basic principles, affirms that such experiments conclusively refute all armchair philosophical arguments against the conclusions thus demonstrated, and then moves on to describe an extensive series of technological applications of the theoretical principles just demonstrated.[811]

We can see the same trends in the scientific writings of Ptolemaïs by the 1st century A.D. Though her books were not preserved, surviving quotations show her attacking those who divided her science into sectarian dogmas. She argues instead that to get to the truth one must unify the best elements of competing sectarian approaches and discard the rest. She criticizes those who rely on reason and theory and ignore or discount observations, and also those who only observe and ignore theory. She defends instead the need for a unified theoretical and observational approach to harmonics, integrating empiricism with mathematics. This is essentially what we also hear from Hero, Ptolemy, and Galen, and the generalizing nature of her remarks suggests she would have agreed with their extension of the same principles across the sciences.[812] Hence the Roman trend in ancient science was not as Ben-David claimed, but in exactly the opposite direction: toward communication, unification, and integration of the best elements of science and philosophy into an increasingly superior methodology.

So all these arguments for decline don't hold up.

Besides those, however, there are four other arguments that appear repeatedly in the literature, which purport to prove that the ancients had no conception of scientific (and technological) progress or were even hostile

---

810. See chapter 2.7, discussion in chapter 1.2.III, and relevant discussion on Galen in chapter seven of Carrier 2016.

811. Hero, *Pneumatics* 1.pr. (see discussion in Argyrakis 2011). Hero also implies here that he had demonstrated other relevant principles in his treatise on waterclocks, which is unfortunately lost. Similar patterns are visible in various works by Galen and Ptolemy (see the end of sections 3.2, 3.3, and 3.4 for examples).

812. Ptolemaïs, *On the Difference Between the Aristoxenians and the Pythagoreans*, frg. 3, quoted in Porphyry, *Commentary on Ptolemy's Harmonics* 25.3–26.5. See also supporting quotation of Ptolemaïs in chapter 2.7 and the sources for Ptolemaïs in Carrier 2016 (index).

to the idea. It is often claimed the ancient slave system discouraged interest in progress, or that progress was blocked due to the Romans being dead set against the idea of changing or interfering with the natural order, or that they never had the idea of explaining nature and natural processes mechanically (rather than, say, organically or supernaturally), or that they were so obsessed with a cyclical model of time that they were incapable of even *imagining* progress or thinking it possible or worthwhile. All false.

I. The Slavery Thesis

Was progress in ancient science and technology impeded by a dependence on slavery? It is popular to claim so.[813] But since Western society remained dependent on slavery until well into the 19th century, a fact that hindered neither the Scientific nor Industrial Revolutions, any claim that slavery had the opposite effect in antiquity is a hard sell. Hence scholars have long been challenging this argument, finding it both false and illogical.[814] Even Peter Green is rightly skeptical of the idea, but cannot bring himself to give it up, implausibly maintaining that the Romans were too dependent on slavery to think of saving money through capital investment (a notion already refuted by the Barbegal flour factory), or somehow thought labor-saving devices would leave slaves dangerously idle, or believed slaves were so cheap no one needed machines—even though the Romans must have been appallingly stupid to accept any one of these non sequiturs.[815] Fortunately for them, there is no good evidence they did. After all, by the same logic the Romans should never have built a system of aqueducts to meet their water needs, but instead arranged their slaves in systems of bucket brigades. This certainly would have solved the problem of idle slaves, and by Green's logic, it would either have been cheaper or the Romans would not have noticed it was not.

In truth, not a single ancient text expresses any concern over the possibility of idle slaves, most likely because machines would not make

---

813. Countless examples exist in the literature, the following merely typical: Zilsel 1945: 328–29; Farrington 1946: 22–23; Vernant 1983: 283–84; and sources cited in P. Green 1990: 831 n. 81 and Pot 1985: 1.51–57.

814. Challenges to the slavery thesis can be found in Edelstein 1952: 579–81, 586–87, Brunt 1987, and Greene 1994: 26; even Sambursky 1962: 256–57 (though he still defends related ideas); and most recently Temin 2004 and Rihll 2008.

815. P. Green 1990: 458 vs. 469–70.

slaves idle, but more productive. That is exactly what cranes were for: to multiply the productive output of the slaves who operated them. Likewise pole mills, bread kneaders, and many other technologies we know were used. If the Romans had been afraid of such machines, or actually believed they cost more than doing without, or were not worth the investment even if they saved money, then they would have done away with cranes and pole mills and bread kneaders and everything else, and just had slaves pulling ropes, manning pestles, and kneading dough by hand. Obviously Romans preferred to buy the machines. In fact, far from thinking slaves were *sufficient*, the Romans were busy complaining that slaves were not even *efficient*, which entails at least some interest in increasing that efficiency.[816] We can already see from everything surveyed in section 3.6 that the Romans were not uninterested in technologies that increased the efficiency of their labor force. Even something as seemingly innocuous as a padlock would have reduced the amount of labor expended on guard duty, and further increased efficiency by reducing pilferage, or even the escape of rebellious slaves—hence advanced locks became a capital investment to protect a capital investment. Aristotle was not even joking when he praised the most mundane of inventions, the child's rattle, for saving money by reducing damage to furniture from active toddlers, an example we might consider trivial, but if such a benefit was evident even in trivial cases, it surely could not have been overlooked when the effects were more substantial.[817]

All this hardly has anything to do with science. But the slave thesis is sometimes twisted in that direction. Not only is it falsely claimed that slaves eliminated the need for labor-saving technology (and thus should have prevented any scientific interest in developing cranes or watermills), but the association of handiwork with slaves is also supposed to have led the elite to despise all handiwork as servile. As a result, the argument goes, ancient scientists never did anything with their hands and never deigned to communicate with craftsmen, and thus cut themselves off from an essential source of empirical discovery. All the previous sections of the

---

816. Pliny the Elder, *Natural History* 18.4.21; Columella, *On Agricultural Matters* 1.pr.3.

817. Aristotle, *Politics* 8.6.1340b ("one must regard Archytas' rattle a good invention, which people give to children in order that while occupied with this they may not break any of the furniture").

present chapter combine to refute this notion, and we will drive the last nail into it in chapter 4.6, where we will show the reverse, that scientists readily engaged in hands-on work, and were usually in fact craftsmen themselves, or regularly communicated with them. But in the words of Lucio Russo, the Antikythera computer alone should "lay to rest once and for all old clichés to the effect that the Greeks scorned technology and that the easy availability of slave labor led to an insurmountable gap between theory and experimental and applied sciences."[818] D.L. Simms likewise challenges the claim that "the *ethos* of the Classical period" was "at best indifferent, and at times actively hostile, to technology [and] those in mechanical occupations," concluding that, to the contrary, "the assumption of a general hostility or indifference to technical advance and [of] the absence of technical enterprise in Classical Antiquity is untenable."[819]

Nevertheless, this argument is still used to explain why certain scientific fields did not develop in antiquity. For example, as Peter Green says, though "Greek mathematicians, geographers, physicists, and astronomers" (a list to which we must add engineers, biologists, and physicians) "made theoretical discoveries that would not be matched, let alone surpassed, till long after the Renaissance," other fields nevertheless stagnated because of an "intellectual elitism and acute snobbery inherent in Greek society, which (for example) stultified the advance of scientific chemistry," since it "was associated in the Greek mind with such banausic pursuits as dyeing, mining, and herbal medicine," which were "all practiced in an *ad hoc* fashion by common artisans."[820] But dissection was associated with the butcher, artillery, crane and instrument building with the blacksmith and carpenter, and architecture with the roofer and stoneworker, yet these fields received extensive theoretical interest and were actually engaged in by scientists well into the Roman period. And "herbal medicine" was hardly treated with disdain—to the contrary, discovering and describing medicinal herbs was among the most common and respected occupations of ancient medical scientists. We also hear no snobbery from Pliny the Elder when he discusses dyeing and mining and countless other 'banausic pursuits.'

---

818. Russo 2003: 130.

819. Simms 1995: 82–93 (quotes from 82 and 89).

820. P. Green 1990: 456 (similarly repeated: 473). This echoes Farrington 1946: 52–53, whose entire argument was already refuted by Edelstein 1952: 585–96.

In actual fact, any pursuit, no matter with whose occupation it could be associated, would be made respectable the moment it was articulated as a formal science, an *ars* or *technê*, which required a systematic explanation and exploration of all related phenomena in terms of fundamental and universal axioms, principles, or categories.[821] This was Columella's desired tactic for bringing the study of agriculture into the sciences,[822] and this had long been the means of bringing cartography, engineering, medicine, and other fields into a similar high status. In principle there is no reason why this could not also have been done for chemistry, metallurgy, or anything else. In other words, if the same process of theoretically grounding the banal could so easily overcome "elitism and snobbery" for every other science, there is no reason it could not also have done so for any other. After all, no one points to slavery or snobbery to explain why neither aerodynamics nor thermodynamics, nor even psychology, sociology, meteorology, or genetics, became proper sciences in the 17th century, or even the 18th, despite these subjects being well within the technical means of the time to empirically explore. One must look elsewhere for the neglect of certain fields in any given age.[823] Simply put, slavery cannot have had any more effect in antiquity than it did in modernity.

II. Changing Nature

Was progress impeded because the Romans were against the idea of changing or interfering with the natural order? Many have said so. In the words of Joseph Ban-David, "the possibility of changing nature did not enter the Greek mind," and so "they did not aim either to change or influence physical nature but were content to understand it."[824] Some have even said the Romans thought such meddling with nature would be morally or religiously taboo.[825] It is then suggested this ignorance of, disinterest

---

821. As supported by the various studies in Lévy et al. 2003.

822. See chapter seven of Carrier 2016.

823. Such as the hypotheses proposed in section 3.5.

824. Ben-David 1991: 301. Essentially the same claim appears in Edelstein 1963: 26; Finley, 1981: 180; Vernant 1983: 283; Pot 1985: 1.36–48; P. Green 1990: 458, 472; and more skeptically in Lloyd 1991: 162–63.

825. For example, "human intervention into the natural order of things was

in, or aversion to meddling with nature impeded science by turning the ancient mind away from the experimental method. This has already been soundly refuted.[826] Experiments were actually common in the medical and engineering sciences, and were in fact a prominent component of scientific arguments, just not yet a universally central one.

In respect to technology, the underlying idea here is so contrary to the facts it is hard to understand why it was ever maintained. The widespread existence of aqueduct technology alone is refutation enough: here was the radical transformation of the natural world to suit human desire, not only by filling it with man-made rivers, but, by using their scientific understanding of the siphon and the law of equilibrium, the Romans even made water run uphill! Can they have acted any more contrary to nature than that? Indeed, the entire repertoire of Roman pump technologies were specifically designed to move water contrary to its natural direction—draining mines, for instance, that nature kept trying to fill. But even apart from pumps, and the widespread construction of artificial rivers that defy gravity, the Romans also built artificial lakes—not just dams and reservoirs, but fish farms, pleasure ponds, and makeshift pools in Roman arenas for gladiatorial sea battles. They constructed artificial harbors, islands, and peninsulas, reshaping entire coastlines—often using a cement that, in even further defiance of nature, dried underwater. They filled the world with artificial roads, frequently cutting through or even leveling hills and mountains, or actually *building* hills when they needed to cross valleys.[827] They increased the area of cultivated land through systematic drainage, irrigation, terracing, and flood control—thus growing crops and vines and

---

improper" according to Reynolds 1983: 32. He presents no evidence of this. The same assertion is made by P. Green 1990: 472, offering as his only evidence the claim that "the Greeks excelled in areas (e.g. hydrostatics) where movement was not in question," ignoring their equal mastery of mechanics and pneumatics, where movement was exactly in question.

826. For example: von Staden 1975. We have already seen evidence (in sections 3.2 through 3.4) that controlled experimentation existed in antiquity even beyond what von Staden documents.

827. As observed in Strabo, *Geography* 5.3.8, they did this not only for military use but to benefit commerce, as level and straight roads eased the transport of cargoes.

trees where nature never intended.[828]

The Romans did all this, and more, both prolifically and enthusiastically.[829] Clearly no one of any significance thought intervention against the natural order was improper. To the contrary, these examples demonstrate a pervasive belief that the world *should* be changed, exactly as would suit the interests of humanity. Moreover, many of these developments were partly dependent on scientists applying scientific principles. Hence the Romans were *not* content to merely understand nature, but actually sought to use that knowledge to benefit civilization, through improvements in engineering and architecture, just as they did in medicine, geography, cartography, agriculture, and other fields. Even the systematic use of machines like cranes and pole mills to defy gravity and magnify human power is an act of defiance against nature, by endowing men (and animals) with far more strength than nature had provided them.

Hence (as we saw in chapter 2.7) the Romans actually defined the science of mechanics as the study of how things naturally move *and* how to move things contrary to their nature, and they considered the latter of great use and worthy of considerable praise. Nor had thinking on this point ever been different. Aristotle had already said that technology does what nature cannot, explicitly recognizing, for example, that even building houses and ships is acting contrary to nature (since neither houses nor ships naturally grow), and yet Aristotle expresses no worry or concern over this.[830]

Subsequent Aristotelians were even more explicit:

> Our wonder is first excited by phenomena that occur in accordance with nature when we do not know the cause, and then by phenomena

---

828. For all these examples, besides evidence in section 3.6 and below, see Russo 2003: 203, 252–54. Also see Oleson 1984: 406–08, who despite advocating the contrary, actually refutes himself by listing numerous widespread examples of defying nature for the public good in repeatedly innovative and ingenious ways throughout antiquity.

829. Pavlovskis 1973 (with Humphrey et al. 1998: 413–16) provides considerable evidence that many Romans were proud of altering nature for their convenience. For extended examples see Pliny the Elder, *Natural History* 36.14.64–36.24.125 (praising several nature-altering projects), and Cuomo 2007: 75 (linking similar pride to Roman self-image).

830. Aristotle, *Physics* 2.8.199a-b.

> that are produced by art in defiance of nature for the benefit of mankind. For nature often operates contrary to human interests, since she always follows the same course without deviation, whereas human needs are always changing. So when we have to do something contrary to nature, the difficulty of the task perplexes us, and art has to be called to our aid. The kind of art that helps us in such perplexing situations we call 'mechanics'. Hence the words of the poet Antiphon are quite true: "What by Nature defeats us, we overcome by Art."[831]

The Romans were on board with this. Cicero concludes that mankind is divine, in part, because "we fertilize the soil by irrigation, we confine the rivers and straighten or divert their courses," in addition to everything else we do to dominate and bend nature to our will (he mentions agriculture, architecture, the textiles industry, and metallurgy), "so that by means of our hands, we endeavor to produce something like a second world within the natural world."[832] Hence elsewhere Cicero agrees with Panaetius, and after listing medicine, navigation, agriculture, quarrying, mining, metallurgy, trade, transportation, and construction, he asks:

> Think of the aqueducts, canals, irrigation works, breakwaters, artificial harbours. How should we have these without the work of man? From these and many other examples it is obvious that we could not in any way, without the work of man's hands, have received the profits and the benefits accruing from inanimate things.[833]

And for this reason, he argues, man lives *above* nature and her beasts.

As Elspeth Whitney observes, Cicero "ties human dignity and power closely to man's ability to change his environment through technology and to create a 'second nature' for himself," defending the popular Stoic view "that technological arts are a product of human reason well and properly used." Similarly, after assessing his *Natural History*, and his abundant praise of useful technologies just like Cicero, Mary Beagon concludes that Pliny the Elder virtually "celebrates a contemporary situation in which all Nature,

831. Pseudo-Aristotle, *Mechanics* 1.847a.

832. Cicero, *On the Nature of the Gods* 2.60.150–52.

833. Cicero, *On Duties* 2.3.12–2.4.16.

including the sea, is subservient to man."[834] In contrast, for example, at the dawn of the 5th century A.D., the Christian writer Augustine would consider these technological achievements to be the "superfluous, perilous and pernicious" achievements of a human genius bestowed by God and misused by men to serve their own needs and comforts rather than devoting themselves to God and His gospel.[835]

Varro illustrates the earlier Roman mindset. In the middle of the 1st century B.C. he criticized the Roman senator Marcus Licinius Lucullus as careless because his fish farm "did not have suitable tidal-basins, so the water became stagnant," in contrast to his brother Lucius, who "had cut through a mountain near Naples and let a stream of sea water into his ponds, so they would ebb and flow and he would have no need to yield to Neptune himself when it came to harvesting fish." In fact, Lucius was so enthusiastic for such enterprising building projects he told his engineer to spare no expense in constructing a similar arrangement near Baiae, where he "ran a tunnel from his ponds into the sea and threw up a mole," arranging for the tides to cool and refresh his fish farms there. Here there is every defiance of nature in pursuit of luxury and profit. Even the God Neptune is defied, mountains are cut open, artificial peninsula's constructed, and miniature man-made seas manufactured to ease the harvesting of salt-water fish. Though Varro regards these projects as extravagant, he holds them up as an ideal example of superior behavior in comparison with the carelessness of Marcus who made none of these nature-defying arrangements.[836] Hence neither the emperor Trajan nor Pliny the Younger express any moral or religious concerns about a government interest in subsidizing the alteration of nature with a substantial canal project in order to improve the local economics of stone and timber transport—as well as, Pliny adds, the luxury fruit trade, even though the movement of cash crops served no direct government interest.[837]

The only sense in which Romans opposed acts in defiance of nature is when they were directed to useless or immoral ends. Hence the

834. Beagon 1992: 185.

835. Whitney 1990: 52–54 (cf. Augustine, *City of God* 22.24).

836. Varro, *On Agricultural Matters* 3.17.9. See also Plutarch, *Life of Lucullus* 39.3. On Roman tunneling technology see M.J. Lewis 2001b: 197–216, 345–46.

837. Pliny the Younger, *Letters* 10.41, 10.42, 10.61, 10.62. See Peacock 2012 for a similar Trajanic canal project connecting the Suez with the Nile.

contrast Tacitus draws between Nero's technological feats "in defiance of nature," which were vain and useless, and those that were actually useful and praiseworthy.[838] It did not defy nature for man to use his hands and intelligence to serve the common good. And yet despite such moralizing as this, even those 'vain and useless' defiances of nature continued in abundance, thus revealing how most Romans *really* felt about their mastery over nature. We can see the same in the writings of Seneca and Pliny the Elder (as we'll see in section 3.8.I), who likewise moralize against unnatural luxuries, yet in the process reveal with countless examples how common these nature-defying luxuries really were—and like Tacitus, Pliny even qualifies his condemnation with abundant praise for *useful* technological feats. A similar split in attitude exists even today, with luxury often condemned and utility praised, except in the medical field, where even some 'useful' alterations of nature are condemned—for example, installing artificial hearts or limbs is praised, despite defying nature, because they restore some imagined idea of 'man's natural state', but *electively* enhancing human abilities through cybernetics or drugs is often regarded as abhorrent or even criminal. Put simply, there is no difference between then and now in attitudes toward the value of defying nature.[839]

III. Mechanizing Nature

Did the Romans impede scientific progress because they did not think of explaining nature mechanically, rather than organically or supernaturally? This is another claim often heard. "Even Aristotle," Peter Green insists, shared "the animistic concept that the heavenly bodies were living, sentient creatures" and thus he never imagined he should explain planetary motion dynamically.[840] As Rodney Stark puts it, for the whole of antiquity, "prompted by their religious conceptions, they transformed inanimate objects into living

---

838. Tacitus, *Annals* 15.42–43. The very same contrast between praise and condemnation of technologies in Pliny the Elder is analyzed in Wallace-Hadrill 1990.

839. Indeed there was no anti-technology movement in antiquity comparable to the Luddites of the *modern* era (Binfield 2004), whose sentiment survives even to this day in such disparate movements as the Amish, Neo-Luddites, and eco-anarchists (Brende 2004, Zerzan 2005, S. Jones 2006).

840. P. Green 1990: 462.

creatures capable of aims, emotions, and desires—thus short-circuiting the search for physical theories."[841] Or as Peter Green preferred to put it, the ancients had "no sense, as we do, of mechanical causation" and thus "regarded the fixed and repetitive movements of the heavens as evidence of divinity" rather than seeing "analogies with the world of machines."[842]

As before, this is all nonsense.[843] In actual fact, as we saw quite abundantly in chapter two, all ancient natural philosophers believed that a search for physical-causal theories was both valuable and possible, and in fact their primary task. So there is no indication that any such search was "short-circuited." Nor was the idea of mechanical causation foreign to ancient scientists. To the contrary, it was a staple feature of their theories—certainly in mechanics, hydrostatics, optics, harmonics, and every other aspect of physics, but even in medicine, and astronomy as well.[844] For example, 'the planetary system as machine' was a widely recognized analogy in the Roman period. Archimedes, Posidonius, and Ptolemy all built mechanical models of the planetary system, clearly having no trouble imagining the heavens that way. Cicero thus compares the planetary system with those mechanical reproductions of it, arguing the actual cosmos is merely a superior machine.[845] The very same view was articulated by the engineer Vitruvius.[846] And Plutarch had no trouble imagining that planetary motion could be explained by dynamic principles, proposing, for instance, that the moon does not fall because its motion keeps it in orbit, using the specific

841. Stark 2003: 152.

842. P. Green 1990: 461.

843. As others have pointed out, e.g. Greene 1994: 26–27.

844. See examples in sections 3.2 through 3.4 above. Though Berryman 2002 and 2003 employ a highly restrictive definition of 'mechanical' she still finds such explanations evident in antiquity, and even the examples she rejects still qualify on any broader definition. And Berryman 2009 thoroughly demonstrates mechanization of nature was common across all disciplines.

845. Cicero, *On the Nature of the Gods* 2.34.88–2.35.88.

846. Vitruvius, *On Architecture* 9.1.2, 10.1.4. On the pervasive use of such mechanical analogies in Roman cosmology see Aujac 1993: 157–78 and Delattre 1998 (who adds the observation that even harmonic theories in astronomy equate the solar system with physical instruments).

analogy of a sling (an explanation that is effectively correct).[847]

In fact, far from being unusual, mechanical metaphors were a *typical* tool of explanation in ancient science and natural philosophy.[848] For example, in the 1st century A.D. the anonymous *Aetna* argues extensively *against* animistic and *for* mechanical explanations of natural phenomena (in this case, in volcanology), drawing numerous analogies from machines and mechanical processes.[849] Galen likewise described physiological processes with analogies to both mechanical and chemical processes observed in nature, and describes the entire human body as a machine, a mechanical tool that we use to carry out our tasks and live our lives, and argues that its physiology can be better understood by analyzing its functions with the laws of motion and mechanics.[850] Though Galen rejected a *reductively* mechanical explanation of organ function (as still promoted by his atomist opponents), he nevertheless reduced every process to the causal interaction of natural objects and forces, regarding the body as a mechanism composed thereof.[851] In fact, in this sense even Galen's arguments for the intelligent design of the human body were routinely mechanical, repeatedly conceiving the design of its organs and parts as an engineering problem that the Creator had solved superlatively.[852] At the very same time, the atomistic Erasistrateans saw machinery as the best analogy by which to explain bodily operations

847. Plutarch, *On the Face that Appears in the Orb of the Moon* 6 (= *Moralia* 923c-d).

848. See Webster 2014, Russo 2003: 146–51, and von Staden 1997b: 201–03; and even van der Waerden 1963: 51–52 and Farrington 1946: x, 6–14 (with, e.g., *DSB* 6.428, in s.v. "Hippocrates").

849. See, for example, the methodological discourse in *Aetna* 29–93. On the *Aetna*'s attack on popular animism, and advocacy of mechanical explanations in place of divine, see Paisley & Oldroyd 1979: 2–6 and Goodyear 1984: 346–47.

850. For example: Galen, *On the Natural Faculties* 3.15 (= Kühn 2.210–11); Galen, *On the Uses of the Parts* 1.2–4, 1.19, 7.14 (= M.T. May 1968: 69–71, 103, 364–66). The body as instrument or machine was a widely understood metaphor, e.g. Marcus Aurelius, *Meditations* 10.38.

851. See, for example, Galen, *On the Uses of the Parts* 14.5 (= M.T. May 1968: 627). This mechanization of physiology had already begun with Herophilus: see von Staden 1996.

852. See discussion in Hankinson 1988.

in the *absence* of a Creator.[853]

Hence what Roman scientists saw in the natural world was, at most, evidence of divine *design*, not of perambulating gods; or they saw a godless machine that was no less explicable. Incredibly, some scholars have claimed the ancients did not believe in divine design, and that *this* somehow hindered their search for natural laws and rational order in nature.[854] This is entirely contrary to the facts on both counts. Many ancient scientists explicitly argued for intelligent design (e.g. Galen, Ptolemy, Aristotle), while those who didn't (e.g. Strato, Erasistratus, Asclepiades) were for that very reason even *more* committed to finding mechanical principles in natural phenomena, not less. And as we saw in chapter two, *all* natural philosophers believed nature was orderly, intelligible, and causally explicable, regardless of their religious convictions. The most prominent example were the Epicureans, who entirely rejected divine design, and yet were entirely devoted to explaining the universe with a system of rational, mechanical principles, precisely *because* they could not resort to animistic or supernatural explanations instead.[855]

In fact, when Stark muses that "if mineral objects are animate, one heads in the wrong direction in attempting to explain natural phenomena—the causes of the motion of objects, for example, will be ascribed to motives, not to natural forces," he entirely overlooks the fact that no ancient scientist ever resorted to any such explanation of anything.[856] After the generation of Plato, all ancient natural philosophers of any stripe sought explanations in the principles of natural qualities, motions, and masses. Obviously, of course, the fact that animals, plants, and people actually *are* 'animate' did not prevent the development of a considerable scientific understanding of them, nor were their functions explained in any other terms but physical and mechanical principles. Even in humoral theory, the four humors obeyed fixed natural principles, not psychological 'motives'. And when planets and

853. See Vegetti 1995, Nutton 2013: 136–38, and discussion in section 3.2 above (although Erasistratus himself appears to have been a pantheist, cf. von Staden 1996: 95–96). As noted in section 3.2, the more atheistic Asclepiads shared a similar interest in mechanical explanation.

854. Ben-David, 1991: 301; Stark 2003: 152.

855. See the whole of Lucretius, *On the Nature of Things*, but, e.g., in 5.96 the whole of existence is explicitly called a machine (*machina mundi*).

856. Stark 2003: 154.

stars were sometimes imagined as ensouled or animate, those scientists who suggested this still regarded their movement as fixed by natural principles or even physical laws, which they then sought to describe and explain.[857] Even when organic models were employed to explain geological and meteorological phenomena as analogous to physiological phenomena, both nature *and* living organisms were thus described as causal systems of mechanisms and processes, not motives.[858]

The same myth of ancient 'animism' has also been used to argue for a non-existent technological stagnation. Lynn White, for example, advanced the view that "Christianity, by its opposition to animism, opened the door to a rational use of the forces of nature."[859] This thesis is challenged not only by the fact that Christianity did not have this effect in the East that White alleges it eventually had on the West, but also by the fact that most Western medieval innovations actually originated among heathens (not Christians) in Asia (or Gaul, Germany, or Africa), regardless of whether they were subsequently borrowed or reinvented by Christians.[860] But more importantly, this thesis is already refuted by evidence in section 3.6 and in the rest of the present section (above and below).

In fact, few among the elite believed even religious idols, much less ordinary objects in nature, were actually inhabited or controlled by spirits, and ancient scientists typically repudiated the idea. Even beyond the evidence already surveyed in this chapter and in chapter two, Plutarch's

---

857. Despite Stark's contrary assertion (Stark 2003: 154), Aristotle argues *against* the idea that celestial bodies move according to their desires, and instead insists they move because of fixed and innate tendencies (cf. Aristotle, *On the Heavens* 2.1.283b-284a; similarly proposed in Ptolemy, *Planetary Hypotheses* 2.6: see Murschel 1995). And as we saw in sections 3.3 and 3.5, Aristotle did not command the field: several different explanations of celestial motions continued to be advanced and debated well into the Roman period.

858. For example, see analysis of Seneca's use of such models in Taub 2003: 141–61.

859. L. White 1963: 282–83 (quoting R.J. Forbes). A similar 'argument from animism' is also advanced (rather ridiculously, considering the context) in Oleson 1984: 403.

860. In L. White 1963: 272–73, 275, 278, 286–91; White attempts to explain at least the first problem as resulting from differences in Christian theology between East and West.

theological discussion of the varied interpretations of Roman Isis cult (in his treatise *On Isis and Osiris*), for example, demonstrates that educated men understood natural phenomena as *not* subject to the caprice of spirits or demons, but as governed by natural principles and forces, a perspective further articulated in Plutarch's treatise *On Superstition*. In contrast, surely the average peasant in the 16th century had as much faith in the ability of angels and saints and the Holy Spirit (not to mention demons, ghosts, magic, and even God Himself) to control the forces of nature, as any of the ancients had in their gods and spirits, reducing to no significant difference in social effect.

IV. The Cyclical Time Thesis

Were Romans so obsessed with a cyclical model of time that they were incapable of imagining progress or even thinking it possible or worthwhile? This is often claimed.[861] But once again, it has already been refuted.[862] As Rodney Stark describes the idea, the ancients thought the universe was "locked into endless cycles of progress and decay" and therefore they "rejected the idea of progress in favor of a never-ending cycle of being."[863] Such a thesis is inherently illogical. Anyone who believed in "endless cycles of progress and decay" would as a result believe, by definition, that progress is not only possible but *inevitable*. Moreover, the very idea that the universe is governed by "cycles of progress and decay" can actually inspire the scientific study of progress and decay. Aristotle's treatise *On Generation and Decay* is a prime example. In fact, all ancient study of causation was a study of the nature of change, and thus of growth and degeneration. For both reasons, ancient theories of time obviously presented no barrier to scientific advancement.

Nevertheless, some claim this belief in cyclical time led some scientists

---

861. For example von Wright 1997: 2 (his only example is Parmenides, who was already obsolete even by Aristotle's day).

862. The entire idea is refuted on both facts and logic in Edelstein 1967: xx-xxiii, 29–30, 63, 121–27. That a cyclical model of time is not even incompatible with actual, anticipated, or even desired progress is also soundly argued in Burkert 1997: 34–38.

863. Stark 2003: 152, 153 (his only examples are also Parmenides, certain unnamed "Ionians," some alleged opinions of Plato, and misquotes of Aristotle).

to conclude that humanity had already progressed as far as it could. But there is no evidence of this. For instance, contrary to G.E.R. Lloyd, Aristotle never "states his belief that nearly all possible discoveries and knowledge have been achieved already," in neither of the passages Lloyd cites in support of that assertion.[864] In the passage Lloyd cites from the *Metaphysics*, Aristotle says that after the prescientific arts have been fully developed, which meet the baser needs and immediate pleasures of man, only then are scientific arts invented.[865] He does *not* say those scientific arts had been completed, or were near to being so, or even that there were no new sciences yet to be discovered. Meanwhile, in the passage Lloyd cites from the *Politics*, Aristotle is not even talking about science or technology.[866] He is *only* speaking there of the social measures by which Plato proposes to get people to behave as he wants them to (as described in Plato's *Laws* and *Republic*). Aristotle says if those social measures actually worked, then someone would have noticed by now, "for nearly all" the ways of getting people to behave "have been discovered already, although some of these have not been collected together, and others, though brought to knowledge, are not put into practice." There is nothing said here about science or technology, only about methods of social or political control, in which case Aristotle's belief that every possibility had been thought of, even if untrue, was nevertheless reasonable.

Similarly, Aristotle's belief that certain forms of poetry (such as theatrical tragedy) had reached a state of perfection has no relevance to what he may have thought about other arts and sciences.[867] In fact, he suggests the perfection of tragedy only as a *possibility*, within a chapter specifically devoted to explaining the accepted reality of progress in the art of poetry, which actually entails a belief in progress.[868] We shall see in section 3.9 that, exactly contrary to Lloyd and Stark, Aristotle had the same confidence in the future progress of the sciences. Stark can only muster the opposite conclusion by ignoring everything else Aristotle said, and

864. Lloyd 1990: 71.

865. Aristotle, *Metaphysics* 1.1.981b.

866. Aristotle, *Politics* 2.5.1263b-1264a.

867. Aristotle, *On Poetry* 4.1449a (a belief that again may have been analytically reasonable, given Aristotle's definition of tragedy and his own aesthetic ideals).

868. Aristotle, *On Poetry* 4.1448b4–1449a32.

then twice misquoting him.[869] The first of these misquotes is a remark in Aristotle's *On the Heavens* that "the same ideas recur in men not once or twice but over and over again," which Stark takes as meaning that Aristotle rejected the idea of scientific progress on the grounds that nothing new is ever really discovered. The second is a remark in Aristotle's *Politics*, that everything "had been invented several times over in the course of ages, or rather times without number," which Stark takes as meaning the same thing about technology. Neither quote in context means what Stark says.

In the first case, Aristotle argues that the heavens have never changed and therefore must be composed of a changeless element, then he discusses the empirical basis for that conclusion:

> What we learn through the senses is enough to convince us of this, at least with human certainty. For in the whole range of time past, down as far as our inherited memory reaches, no change appears to have taken place either in the whole scheme of heaven or in any of its parts. Even the name handed down from our earliest ancestors on up to the current time, seems based on the very idea we are talking about. For we must suppose the same idea comes to us not once or twice but countless times. So for this reason...they called the highest place by the name *aether*...from the fact that it "always runs" for an eternity of time.[870]

It is obvious here that Aristotle means only that previous thinkers had probably deduced the same conclusions from the same observations and therefore had coined the word *aether* accordingly, which Aristotle now uses himself. There is nothing here against the idea of progress. To the contrary, this is little more than a true statement of the fact that people often notice the same things in every era. (And of course, soon after Aristotle, Hipparchus would refute the underlying assumption by observing the formation of a new star.)

In the second case, Aristotle is not even discussing science or technology, but only the most basic aspects of political organization:

> It seems it is not a new or recent discovery among political philosophers that the state ought to be divided by class and...have public meals. ... So we must suppose these and other things were discovered many times, over a

869. Stark 2003: 153.

870. Aristotle, *On the Heavens* 1.3.270b.

> long period, or rather countless times. For it seems the necessities of life are enough to teach men what is useful, while it is reasonable to expect an increasing refinement and improvement of those things established at the start. ... Therefore, one must rely on what has already been adequately discovered, but also attempt to seek out what remains to be discovered.[871]

Aristotle had already traced the development of a class system and public meals to long past civilizations in Crete, Italy, and Egypt, and their simplicity and necessity is so great he rightly assumes many other cultures throughout history must have discovered them as well. There is nothing here against even political progress, much less scientific or technological progress. To the contrary, all he is saying is that necessity is the mother of invention, and therefore wherever a certain necessity arises, we can expect to find men inventing the same things necessary to deal with it. And far from saying these developments mark the end of political progress, he adds that there are still things left to be discovered, and that we should look for them. Hence Aristotle says exactly the opposite of what Stark would have us believe.

These remarks by Aristotle, though in each case referring to very specific items of knowledge, do draw upon a general view that relates to his theory of eternal cycles. In his *Metaphysics*, when discussing the survival of kernels of truth within generally false myths about the heavens, Aristotle speculates that this might be oral lore that has survived, in distorted form, from long lost civilizations in humanity's past. Hence "it is reasonable to suppose that each art and philosophy has been developed as much as possible and then lost again, many times over," and that relics of these past discoveries might survive in extant lore.[872] But he does not say every art and philosophy has been developed to *perfection* in the past, only that each has been discovered "as far as possible" (*heurêmenês eis to dunaton*), meaning as far as that past civilization could get before its destruction or decline. In a sense this is actually true. Tracy Rihll observes that "unwritten knowledge" in antiquity "was even more likely" than written knowledge "to get lost and be repeatedly

871. Aristotle, *Politics* 7.10.1329b.

872. Aristotle, *Metaphysics* 12.8.1074b: *kata to eikos pollakis heurêmenês eis to dunaton hekastês kai technês kai philosophias, kai palin phtheiromenôn, kai tautas tas doxas ekeinôn hoion leipsana perisesôsthai mechri tou nun*. See also Plato, *Laws* 3.677a-678c.

rediscovered," and yet the written transmission of scientific knowledge was a relatively recent idea.[873] Aristotle was well aware of these facts. We can now point to the Middle Ages as another such phase in history, when a great deal of ordinary, technical, and scientific knowledge was lost and had to be rediscovered during the Renaissance, often independently of recovered ancient texts.

This theory of an eternal rise and fall of civilizations actually entails a belief in progress. Starting from the premise (which Aristotle argues for independently) that the universe is infinitely old, it follows that human arts, sciences, and technologies should be supremely advanced by now. But Aristotle conceded this was not the case. Therefore, Aristotle must have reasoned, there must be periodic destructions. Hence it follows that the arts and sciences must have been discovered and advanced again and again, across infinite past time. If Aristotle had *not* believed in progress, then he would not have needed his cyclical theory of time. For if no progress occurs, then any civilization could be infinitely old without ever having been destroyed. Only if Aristotle expected a civilization to make continual advances would he need periodic destructions to explain why his own civilization was not infinitely advanced. Aristotle must, therefore, have believed his civilization to be in the middle of a period of progress and advancement that had not yet reached its completion, which entails he believed further progress was not only possible but, barring catastrophe, inevitable.

By the Roman period this Aristotelian theory of historical development had been taken over by the Stoics, who juiced it up with Persian theological fantasies about periodic destructions and rebirths of the entire universe, not unlike modern cyclical cosmologies and multiverse theories. Aristotle's model, also still around in the Roman era, ranged closer to modern cinematic fantasies of the apocalyptic death and rebirth of human civilization, often facilitated in contemporary imagination by nuclear, biological, or environmental disaster. But just as such theories today do not impede or discourage scientific or technological progress, neither did they then. Indeed, the cosmic cycle in the Stoic imagination was many thousands of years long, leaving ample time for any current human society to progress in its knowledge of the natural world. Accordingly, Stoic scientists avidly continued to study and learn about nature, as did many Aristotelians and

873. Rihll 2002: 14 n. 36.

atomists, who also had similar cosmic expectations. In contrast, there is not a single example on record of any ancient pagan arguing for the cessation of scientific research on the grounds that it was a waste of time because 'the end is nigh', or even because of the pessimistic understanding that it *will* eventually all end, even if only in the distant future. Instead, as we shall see in section 3.9, we hear exactly the opposite.

There are some near exceptions. The most prominent comes from the depressing field diary of the emperor Marcus Aurelius. In one of his daily thoughts to himself, he wrote that:

> The rational soul traverses the whole universe and the surrounding vacuum and surveys its form, and it extends itself into the infinity of time, and embraces and comprehends the periodical renovation of all things, and it comprehends that those who come after us will see nothing new, nor have those before us seen anything more, but. ... he who has any understanding at all, has seen, by virtue of the uniformity that prevails, all things which have been and all that will be.[874]

However, this is not something he deduces from his cyclical theory of time, but from something closer to a theory of eternal forms, in which the same universal principles apply in all eras of history, producing the same sorts of people and experiences.[875] Hence, though Aurelius could have expanded this into an argument for the futility of scientific research, he seems only to have in mind human affairs, not scientific discovery. For example:

> Consider the great changes of political supremacy in the past. You can then foresee those that will come. For they will certainly be of like form, and it is not possible that events should deviate from the order of the things which take place now. Hence, to have contemplated human life for forty years is the same as to have contemplated it for ten thousand years. For what more will you see?[876]

---

874. Marcus Aurelius, *Meditations* 11.1.3 (cf. also 8.6).

875. That it is the *form* of things that never changes: Marcus Aurelius, *Meditations* 2.14.2 (meaning the natural laws governing the universe: 9.35 and 12.21). Though he does adhere to the cyclic view of history: Marcus Aurelius, *Meditations* 7.19 (with 7.18, 7.25, 9.19, 9.28).

876. Marcus Aurelius, *Meditations* 7.49 (cf. also 10.27).

In other words, there is nothing *fundamentally* different among past and future conquerors: once you have seen one, you have seen them all. Similarly, though we might not understand how the cosmos runs, we have nevertheless seen the same things in the sky that everyone else ever has or ever will. Such depressing (and untrue) notions are not much found elsewhere in ancient literature, and can hardly be regarded as typical. And since Marcus Aurelius also wished he had more time to study natural philosophy, he clearly did not really think he knew everything already, or that there was nothing left to learn.[877] Nor was he any kind of scientist.

These isolated remarks of Marcus Aurelius resemble the same sentiments we hear from the Biblical author of Ecclesiastes, who declared that "what has been will be again, what has been done will be done again, there is nothing new under the sun," thus asking, "Is there anything of which one can say, 'Look! This is something new!'? It was here already, long ago. It was here before our time."[878] And yet, despite this having always been a canonical element of Christian scripture, it certainly has not reflected the popular Christian view of human knowledge, at least not for the past several hundred years. Similarly, while Christians have always believed, even to this day, that the universe is going to be destroyed within their own lifetime (and thus much sooner than any Stoic or Aristotelian or Aurelius himself ever imagined)—an attitude one might surely think would doom any interest in scientific progress—at least for the past several hundred years this superstitious apocalypticism has not had any impeding effect on popular interest in scientific or technological advancement. Though such apocalyptic beliefs *can* result in the abandonment of scientific research (as I will argue it once did in early Christianity), such a consequence is apparently not inevitable.[879] And as we saw already in the first half of this chapter, ancient progress remained continuous. Cyclical theories of time had no effect.

---

877. Marcus Aurelius, *Meditations* 7.67.1, 8.11, 8.26, 10.9. See discussion in chapter 4.2.

878. Ecclesiastes 1:9–14.

879. See chapter 5.6.II for evidence of this consequence in early Christianity.

3.8 Ancient Tales of Decline

That completes our survey of untrue claims about ancient notions. We now must survey a variety of true examples, actual statements by ancient authors, which have sometimes been taken to indicate science was in decline during the Roman period. These fall into three general categories of evidence: actual assertions of decline, stories that imply hostility to technological innovation, and evidence of real socio-political barriers to scientific research. We will treat each in turn. Then we will discuss when *real* scientific decline began.

I. Roman Claims of Decline

One must always follow Harry Caplan's advice, and keep in mind the universal "human tendency to find fault with one's own era," recognizing instead that "an age which produces great satirists and other writers who are alive to the faults of their civilization is on that very account itself praiseworthy." Since, as Caplan says, "we can doubtless find in every period of history, whether it be of a high or a low state of culture, some reputable observer who looks upon his day as one of decline," the fact that we can find the same among the Romans really only teaches us the loftiness of their expectations, and very little about any actual decline.[880] For in communicating their ideals through cultural critique, Roman writers not only compared the present to an imagined past that was never really so great as they thought, they also exaggerated the degeneracy of their own time.

For instance, Seneca once paused for a diatribe on how science and philosophy are supposedly not being studied anymore because of the decadence of his generation.[881] But Seneca's infamously hyperbolic rhetoric is easily seen through here. He can only muster as examples the (exaggerated) neglect of a few obscure philosophical sects, cleverly failing to mention any of the specific arts or sciences (which we know were advancing) or any of the major schools of philosophy (which we know were flourishing). At best we can see him expressing a valid wish here that more people take scientific

880. Caplan 1944: 308 (whose analysis of ancient claims of decline in Roman oratory is in many respects outdated but still useful for the general point). Tacitus was well aware of this human tendency to find fault with one's own generation (cf. Tacitus, *A Dialogue on Oratory* 11).

881. Seneca, *Natural Questions* 7.31–32.

and philosophical study seriously than the relatively small number who actually did. In fact, Seneca explicitly deploys this rhetoric to promote *more* progress, by attempting to shame his readers into pursuing or supporting it (a common rhetorical device of the time, as we'll see). But his suggestion that no one attends any lectures when a good game is on is hardly believable. Similarly, his claim that all the lecture halls are being deserted because everyone is rushing off into the kitchens and cafes is obviously ridiculous.[882] The present chapter is alone sufficient to expose Seneca's remarks as little more than the fabrications of an armchair preacher—the more so when we examine the actual evidence of the popularity of schools and public lectures in this period.[883] Similar is his claim (actually not unpopular at the time) that no advances in medical science would have been needed if not for all our vice and luxury, which had supposedly created every illness and disorder, which, though just as ridiculous, actually confirms his recognition that medical progress had been made.[884]

More ubiquitous are Seneca's many fulminations against technologies he perceives as excessively luxurious and therefore immoral. In that cause Seneca was more or less picking up and embellishing the diatribes of his stodgier predecessors, like Papirius Fabianus, who similarly pontificated against the excesses of modern conveniences, complaints that actually prove to us how evident and popular technological progress was at the time.[885] Seneca's condemnations of luxury also partly reflect a covert attack on the decadence of Nero and his cronies, which (as we saw earlier) Tacitus could later voice more explicitly.[886] For example, shortly after Nero's death (and thus not very long after Seneca's) someone wrote the *Octavia*, a play

---

882. Seneca, *Moral Epistles* 95.23.

883. Discussed throughout Carrier 2016.

884. Seneca, *Moral Epistles* 95.15–29. Others shared his belief that new diseases were caused by immoral living, cf. Nutton 2013: 36, 170 and Cuomo 2007: 2.

885. Such a tirade from Papirius Fabianus is quoted by Seneca's father (Seneca the Elder, *Controversies* 2.1.10–13; with a similar but unattributed tirade at 5.5.1–2). Such pining for simpler times was not uncommon in antiquity (cf. Pot 1985: 1.29–35) and is just as common today (see previous note on Neo-Ludditism). This could be the same Papirius Fabianus the noted natural philosopher (see section 3.5).

886. On this aspect of 1st century attacks on cultural decadence: Beagon 1992: 18.

written in Seneca's name (and style) that poetically repeats a typical Senecan diatribe against 'modern culture' and specifically links it to the tyrannical immorality of Nero.[887] But there is clearly something more to Seneca's red-faced railing against advances in technology. Such progress clearly annoyed him, in much the same way it annoys Neo-Luddites today.

Though such sentiments can be found throughout Seneca's writings, his ninetieth epistle is devoted almost entirely to the subject of technological innovation and its position in the order of Stoic values.[888] Here Seneca positions himself against an evidently strong opposition, which he sees as most ably represented by the famous Stoic scientist Posidonius, whose attitude toward technological progress was far more positive. Seneca is willing to credit technological discoveries and advances to cunning and industrious men, but only men of low mind and contemptible living, whereas Posidonius was willing to elevate inventors to the status of genuinely wise men, and clearly held the work of craftsmen in greater esteem than a snob like Seneca could bear. Ironically for such an elitist, Seneca's posture was that all civilization is contrary to nature, which far exceeded any typical view, and entailed hypocritical absurdities in Seneca's own life. It brings him to sing the praises of cavemen, for example, while cursing the work of architects, yet Seneca certainly did not live in a cave—nor would he ever have deigned to.[889] Seneca also misrepresents his opponent, as he often does, employing the Posidonian praise of technology as a springboard to attack technological luxuries arising in Seneca's own time, even though, as I.G. Kidd plausibly concludes, in his positive appraisal of inventions "Posidonius clearly did not have in mind technological luxury, but the rise of a cultural

887. Pseudo-Seneca, *Octavia* 377–436.

888. Seneca, *Moral Epistles* 90. For a partial commentary see Kidd 1988: 960–71 and D. Russell 1974: 90–95. Nearly every passage cited from Seneca in section 3.6 includes his condemnation of one technology or other.

889. Seneca, *Moral Epistles* 90.8–10. Further exposing his hyperbole, in the very next letter he praises the city of Lyons for its magnificent buildings: Seneca, *Moral Epistles* 91.2. Similarly, while Seneca says it was far better for Diogenes to abandon the cup than for Daedalus to invent the saw (Seneca, *Moral Epistles* 90.14), we can hardly believe Seneca never drank from a cup (or could ever have tolerated living in a society that never used a saw). On Seneca's characteristic hypocrisy in matters of wealth and luxury see Griffin 1976: 286–314, who sympathetically argues it was only a rhetorical device.

civilisation through the arts," having "countered the popular mythology of a Prometheus or divine dispensation as in Protagoras, with a purely human progression sprung from rationality."

As one reads this letter, or indeed any of Seneca's condemnations of modern life, we get a definite impression, again and again, of how much it annoyed him that his opinion on the matter was being ignored by his peers.[890] Indeed, it is hard to find any other Roman author echoing Seneca's rather extreme (and, we must admit, insincere) ludditism. Every complaint Seneca registers only further proves that his views were being ignored, technologies were being widely pursued and embraced, and society was moving on without him.[891] Thus the Posidonian view seems in practice to have prevailed. Stodgy old men like Seneca could only complain—complain, that is, that they had lost the argument. And yet though Seneca was relatively unhappy about Roman technological progress, in this letter as elsewhere, he consistently decouples science from technology, and places the former among the things *true* wise men bring to society. Thus his contempt for modern technology did not equal any contempt for *science*, which was for him a quest to know and understand the natural world, as I'll show in the next chapter.[892] Hence, for example, Seneca's assault on the use of mirrors for vanity and pornography actually includes an exception for using mirrors to gain scientific knowledge, such as aiding astronomical observations, which he considers an appropriate use of the same technology.[893] All other technological advances were for him mere harbingers of decadence.

Such alarmist claims of decay were so ridiculous, however, that Seneca's contemporary Petronius was happy to mock them in the *Satyricon*.[894] Ironically, modern scholars occasionally cite the relevant passage here as if it actually reflected the serious opinion of Petronius or even his generation,

---

890. Hence, for example, his own recognition of this fact in Seneca, *Moral Letters* 90.19.

891. As is essentially argued in J.-M. André 2003.

892. In section 3.9.II (see also chapter 4.4). Though Seneca's 88th epistle contains a parallel attack on the liberal arts, it also elevates natural philosophy as a worthy enterprise, and recognizes the liberal arts as its scientific tools, as we shall see in chapter 4.6.I (though we have already seen an example of this in chapter 2.7).

893. Seneca, *Natural Questions* 1.17.2–3.

894. Petronius, *Satyricon* 88.

missing the joke entirely. A closer analysis improves our understanding. One of the characters in this satirical adventure finds an occasion to pontificate on the state of the arts, declaring that the arts had once flourished because everyone was competing to make new discoveries and advance them further, but now greed and moral decay had put an end to all that. Though the occasion is an examination of some paintings, and hence his focus is on the supposed decline of painting and sculpture, he rants on, extending his theory to all the arts and sciences, and proceeds to give examples of past greatness that are not only false, but deliberately ridiculous.

The central portion of this 'history lesson' is worth quoting in full:

> I began to ask the more knowing fellow [Eumolpus] about the ages of the pictures and the topics of certain ones that were obscure to me, and at the same time searching out the cause of the present inactivity, since the most beautiful arts had passed away, including painting, which had left not even the smallest trace of itself. Then he said:
>
> "Love of money caused this turn. For in earlier times, when bare talent was enough, the noble arts thrived and there was the greatest rivalry among men that anything that will be useful for future generations may not remain hidden long. Thus, by Hercules, Democritus extracted the juices of all the plants, and so the potency of stones and shrubs would not remain unknown he consumed his life in experiments. Eudoxus grew old on the summit of the tallest mountain so he could figure out the motions of the stars and heavens. And Chrysippus, to meet the needs of his research, thrice purged his mind with hellebore. Indeed, turning back to statuary, Lysippus stuck his eyes on the features of a single statue until he died of poverty, and Myron, who almost captured the souls of men and beasts in bronze, did not find an heir. But we, buried in wine and hookers, won't dare to learn even the staid arts, but while we stand as accusers of antiquity we teach and learn only vices. Where is dialectic? Where astronomy? Where the most reflective path of wisdom?"[895]

That not even a trace of painting existed anymore was a patent absurdity. Columella, writing more seriously around the same time, actually says the reverse, that technical artistic skill was back in vogue and inspired by a Roman admiration for the works of ancient masters.[896] Equally absurd

895. Petronius, *Satyricon* 88.1–7.

896. Columella, *On Agricultural Matters* 1.pr.31. Sober histories of art from the early empire confirm the abundant survival of actual paintings and sculptures along

is the claim that no one was pursuing the study of logic or astronomy or philosophy anymore—or even rhetoric! The ridiculously contrived series of examples in pharmacology, astronomy, and philosophy would also have won a laugh from any educated reader of the time.

Modern commentaries have caught the joke. J.P. Sullivan makes the case that Petronius is specifically poking fun at Seneca's ridiculous diatribes against modern culture, or possibly a whole generation of moralizing pedants like him, and concludes that this passage is funny precisely because it is not true.[897] That was in fact the author's intended joke, which he clearly expected readers of his day to get. P.G. Walsh, focusing on the speaker, demonstrates that Eumolpus is intentionally drawn satirically, noting that his "sententious lament on the decline of the arts is the utterance of a shallow and hypocritical poseur," so "when nostalgia for the distant days of moral rectitude is here put into the mouth of a self-confessed lecher we must assume that the purpose of the author is ironical." Walsh also shows how every 'historical fact' Eumolpus enumerates to prove his point is not only false, but patently ridiculous on many levels, concluding that the character's "judgments are the exact opposite" of the truth, producing a "comically inaccurate survey" of the "geniuses of the past" followed by "an inflated condemnation of the morality and ignorance of the Neronian age."[898]

Niall Slater concurs with this interpretation of the character Eumolpus, finding that "his anecdotes are almost diametrically wrong," hence his "art history is nearly as fraudulent as Trimalchio's mythology," referring to a scene that had just preceded, in which the character Trimalchio had said absurdly false things about history and poetry, under the guise of pretending to be cultured.[899] Thus, Slater concludes, this "prefabricated myth of decline" was

---

with evidence of continuing work by contemporary painters and sculptors of exceptional quality (a fact archaeology has confirmed): cf. Callistratus, *Descriptions*; Philostratus, both the Elder and Younger, who each wrote an extant *Images*; Pliny the Elder, *Natural History* 35; and numerous passages throughout Pausanias, *Description of Greece*. See also: *OCD* 1062–64 (s.v. "painting (techniques)," "painting, Greek," and "painting, Roman") and *OCD* 1332–35 (s.v. "sculpture, Greek" and "sculpture, Roman").

897. Sullivan 1968: 202–12.

898. Walsh 1970: 94–97 (quotes: 96–97), cf. also Walsh 1996: 183–86.

899. Slater 1990: 94–95.

an obvious joke "for a reader who knows the real history," hence Petronius did not intend his critique to be taken seriously.[900] Edward Courtney agrees, "it is of course clear that after this string of slanted and rhetorical commonplaces we cannot take Eumolpus seriously as an art critic."[901]

Peter Habermehl has analyzed this passage in greater detail than anyone before him, and he comes to the same conclusion.[902] Habermehl even adds the further observation that Eumolpus really only attacks the immoral or decadent *subjects* of art, not its popularity or technical quality, even though almost everything he says is actually historically false or even absurd, since everyone of the time knew art was undergoing *eine veritable Renaissance*, "a veritable Renaissance," in the early Roman empire, hence Eumolpus's myth of decline would have been readily recognized as a satirical irony.[903] What even these scholars may have overlooked is the fact that the kind of diatribe Petronius is making fun of inherently entailed a strong appreciation and value for scientific progress, since these moralists who were condemning the sloth of their age were in fact complaining (rightly or wrongly) that art and science were not advancing as much as they would have liked. In other words, all these complaints and satires are actually arguing for scientific progress.

A good example of the moralist diatribe that Petronius was roasting survives in the *Natural History* of Pliny the Elder, who begins his history of art with a short sermon on the decline of painting, declaring "indolence has destroyed the arts" (*artes desidia perdidit*).[904] And yet as his subsequent rant makes clear, he does not mean that skill or quality has declined, but that popular tastes have gone more for stone and metalwork than had favored paintings in earlier times, and that, again, tastes have changed with respect to the subjects painted. Such changes in fashion the conservative Pliny already has to exaggerate to rail against, but they only relate to subjective and moral assessments anyway, and thus have nothing to do with any real decline. A similar aesthetic conservatism has been inferred from Vitruvius,

900. Slater 1990: 208, 223–26.

901. Courtney 2001: 133–43 (quote: 140).

902. Habermehl 2006: 126–48.

903. Habermehl 2006: 135–39 (quote: 138).

904. Pliny the Elder, *Natural History* 35.1.1–35.2.8.

a conservatism archaeology has confirmed was wholly uninfluential, and from Pliny's own examples we can see his own artistic opinions were similarly unpopular.[905] In contrast, when we look at Pliny's many discussions of the technical *skills* of artists, we find only stories that reflect continual progress and advancement right up to his own day, with Pliny himself often marveling at the technical achievements of contemporary artists.[906]

However, we do find in Pliny the Petronian exaggeration that some actual sciences have declined due to this alleged moral corruption of his age—which of course entails he believed a moral society would be devoted to scientific research. Nevertheless, contrary to how some scholars portray the matter, Pliny never says all science is in decline or that all research has ceased. For example, in his discussion of meteorological science he says the winds "obey some law of nature, even if we do not know what it is yet," which launches him into a complaint that not enough progress has been made in scientific research on this subject.[907] He says "more than twenty Greeks" had written studies on it in the Hellenistic era, even when scientific progress was least expected, as Pliny argues, due to constant war and chaos, "but now in these glad times of peace under an emperor who so delights in the advancement of literature and science," meaning Vespasian and his son Titus (notably, the regime that replaced Nero), "no addition whatever is being made to knowledge by means of original research, and in fact even the discoveries of our predecessors are not being thoroughly studied." Pliny claims that most of these past writers had made progress in their study of the winds "for no other reward at all except the consciousness of benefitting posterity," but now men seek profit instead of knowledge, forgetting "that knowledge is a more reliable means even of making a profit." Pliny thus sounds a great deal like the 16th-17th century author Francis Bacon, arguing that scientific research would benefit even industry, and that much more is needed than is being done.

However, in all this Pliny is still speaking only of scientific research on

905. On the uninfluential conservatism of Vitruvius: Rowland & Howe 1999: 10–13, 17–18. This was only in aesthetics. In practical matters Vitruvius was a progressivist (see section 3.9.II).

906. This fact is thoroughly surveyed by Jex-Blake et al. 1968.

907. This and following: Pliny the Elder, *Natural History* 2.45.116–118. Indeed, in all of this I concur with the analysis of Meissner 1999: 209–16.

the wind, not all the sciences—in fact, more specifically, he is referring to his desire that someone discover the "laws" governing wind, which we can assume would allow the prediction of their movement, just as, Pliny says, astronomers had discovered the "laws" of heavenly bodies and thus could predict *their* movement. Hence in the course of his complaint he notes how people are willing to make dangerous sea voyages for greed, but not for knowledge, and yet knowledge could assist them even in their greed, effectively implying that navigation might benefit from advances in wind science. Pliny could not know that progress in that field had probably been abandoned for the obvious reason that a predictive weather science of the sort he imagines was quite beyond anyone's means, at least until modern times (and it is barely within our means now)—much like the ability to predict earthquakes, which Pliny also thought scientific research could someday produce.[908] In this respect his hopes resemble those of William Gilbert, whose extensive scientific study of magnetism was inspired by his desire to discover a means of determining longitude by observing the declination of compass needles, a plan now known to be futile. But the futility of such designs actually emphasizes how strongly Pliny and Gilbert believed in the power of science to make useful discoveries. Hence Pliny's entire digression on wind science actually demonstrates a firm belief in the value and possibility of scientific progress—he recognizes that advances in scientific knowledge can benefit trade and industry, and are also worthwhile even besides their material benefits, and that such progress should be expected in times of peace and imperial favor, and that it is *so* possible and desirable that a *lack* of scientific progress is something a society should be ashamed of.

On two other occasions Pliny is sometimes cited as saying no further progress in agriculture was possible, when in fact what he actually says entails quite the opposite. On one of these occasions Pliny only says that, to the best of his knowledge, every possible grafting combination has been tried, so there will probably be no new varieties of fruit.[909] He did not extend this remark to any other department of agriculture, nor even to the technology of grafting itself, since he later describes, with much approval, Columella's

---

908. Pliny the Elder, *Natural History* 2.81.191–192.

909. Pliny the Elder, *Natural History* 15.17.57.

very recent invention of a new and improved grafting auger.[910] In contrast, Pliny's assumption that new fruits were not likely to turn up was more than reasonable at the time, since the New World was a long way from being discovered.[911] On another occasion, however, Pliny expands his moralizing to viticulture, arguing that greed has supplanted knowledge in that field, so that "now it is necessary to research not only discoveries" made since Hesiod (a prescientific agricultural poet who wrote near 700 B.C.), "but also those that had been made by men of old" *before* Hesiod, all because no one had preserved any pre-Hesiodic agricultural knowledge.[912] This rather silly complaint is sometimes cited as evidence of scientific decline, though it can hardly indicate such a thing, for what Pliny is lamenting the neglect of has little to do with actual science, nor is it even true, since agricultural writings far superior to Hesiod's *Works and Days* were not only available in Pliny's day, but still being written and gaining in sophistication.

But even here, as before, Pliny's unjustified disgust at the sloth of his peers demonstrates his passionate desire that progress in knowledge be made. Hence his annoyance at the loss of even antiquated agricultural lore leads him to announce proudly that he, at least, "will carry our researches even into matters that have passed out of notice, and will not be daunted by the lowliness of certain objects, any more than we were when we were discussing animals." In fact this endeavor he says is now easier, and despite his complaining, progress is continuing:

> For who would not admit that now that intercommunication has been established throughout the world by the majesty of the Roman Empire, life has been advanced by the interchange of commodities and by partnership in the blessings of peace, and that even things that had previously lain concealed have all now been established in general use?

910. Pliny the Elder, *Natural History* 17.25.116 (its invention and advantages are described in Columella, *On Agricultural Matters* 4.29.15–16).

911. Similarly, it is sometimes claimed Pliny disparaged all attempts to calculate the distance of the planets as "madness," but in actual fact he said this only of attempts to calculate the size of the entire universe (Pliny the Elder, *Natural History* 2.1.3–4 and 2.21.85–88), which was a reasonable opinion at the time—even modern science has failed to accomplish this.

912. Pliny the Elder, *Natural History* 14.1.3–7.

Only after conceding this does he add that "so much more productive was the research of the men of old" in agricultural science, "or else so much more successful was their industry" than today. In other words, Pliny is not complaining that progress has ceased, but that it *is* being made, just not as quickly as he would prefer, and clearly thinks possible. Thus even in the matter of agricultural science, Pliny's exaggerated claims of neglect reflect in fact a great value and desire for scientific progress through sustained research.

This is confirmed when Pliny echoes a similar complaint about his favorite department of medical science, praising those who publish their discoveries in pharmacology, and condemning those who greedily keep them a secret.[913] Pliny makes it very clear that he desired and expected to see progress in pharmacological research, and he knew this had to be laboriously empirical.[914] Although his claim that no one was doing this anymore is not true (as we saw in section 3.2), it is notable that he expresses disgust at armchair doctors who do not base their theories on experience nor seek new knowledge through empirical research.[915] Pliny's grasp of medical science, factually and historically, is generally poor, but this did not deter him from imagining and valuing future improvement in the field, and joining ongoing debates over the proper methods and models to employ.[916]

Pliny's exaggerated complaints about the neglect of pharmacology, wind science, and viticultural research, all share the same rhetorical function: to scold his peers into undertaking or supporting the very research Pliny wanted to see more of. Hence it is quite significant that his *Natural History* was written

913. Pliny the Elder, *Natural History* 25.1.1–3. Condemning the selfishness of not publishing one's discoveries was a trope among ancient science writers (e.g. Vitruvius, *On Architecture* 7.pr.1; Galen, *On Conducting Anatomical Investigations* 2.1 and 14.1).

914. Pliny the Elder, *Natural History* 25.5.1, with 25.1.1–25.2.4, 25.6.16, 29.1.1, and especially 26.6.11 and 27.1.1.

915. Pliny also says no one had written about medicine in Latin (Pliny the Elder, *Natural History* 39.1.1), which was also not true (Varro, Largus, and Celsus had all done so, and no doubt others as well). My analysis here of Pliny's criticism of contemporary medicine is confirmed by Hahn 1991 and French 1994: 223–25.

916. In general the Roman elite had shown some hostility to Greek scientific medicine in the 2nd century B.C. but had fully embraced it a century later: Scarborough 1993; Marasco 1995; Nutton 1993a and 2013: 160–74.

specifically for the attention of the emperor. By the time of publication this meant Titus (whom Pliny specifically identifies as the reigning emperor in his preface), though during much of the book's composition it would have meant Titus's father Vespasian (whom Pliny still praises affectionately in his preface).[917] Pliny also claims to have written his work to be read by "common folk" (*humili vulgo*) and "farmers and craftsmen" (*agricolarum [et] opificum*).[918] Though this is surely a rhetorical exaggeration, he does appear to have intended his work to influence those of his own class who shared similar values and interests—such as agriculturalists and engineers. But by singling out the emperor as his primary audience, he was also targeting the governing elite. So we can perhaps see in his moralizing hyperbole a more specific rhetorical aim: the hope of recruiting imperial and equestrian aid in supporting more research on subjects that would be useful for commerce and empire (a possibility we will consider further in chapter 4.7).

A similar sentiment appears around the same time in the anonymous *Aetna*, which complains (somewhat contrary to Pliny) that contemporary research in the earth sciences is all devoted to advancing the agricultural industry, rather than expanding that interest more broadly into researching all geological phenomena for the sake of knowledge. It, too, blames this diversion of attention on a moral failure (pointing its finger, again, at laxity and greed), and thus attempts to shame readers into advancing the pace of pure research.[919] Tacitus articulated a similar point about the high standards of rhetorical education, fabricating a myth of decline in oratory as a means to shame his peers into embracing or maintaining his loftier ideals.[920] But Tacitus was not speaking of progress in the field, only

917. The entire preface of Pliny's *Natural History* addresses Titus intimately (they were close friends and had served in combat together), though under his cognomen Vespasian. Since Pliny died unexpectedly less than two months after Titus became emperor (in June of 79 A.D.; Pliny was killed by Vesuvius in August), clearly most of the *Natural History* had been written under (and hence to) the elder Vespasian, although it must have been completed only just before Pliny's death, in time for him to craft an elegant preface to Titus. Cf. *OCD* 1162, 1487–88, 1545 (s.v. "Pliny (1) the Elder," "Titus (Titus Flavius Vespasianus)," and "Vesuvius").

918. Pliny the Elder, *Natural History* pr.6.

919. *Aetna* 252–77.

920. Tacitus, *A Dialogue on Oratory* 29–32. See discussion in chapters five and six of Carrier 2016.

its aesthetic and educational standards. In more direct contrast to Pliny and the *Aetna*, who instead blamed 'greed' for an imagined lack of new research, the conservative Diodorus blames the same 'greed' for exactly the opposite effect: the abundance and popularity of innovations in science and philosophy.[921] Ironically, this only demonstrates that innovation was pervasive and popular, and that contrary views like his were not. And yet even Diodorus was not against culturally useful advances.[922]

There is also a passage in the *Stratagems* of Julius Frontinus that is sometimes cited as evidence of stagnation in the science of artillery design, with the author in this case allegedly embracing the futility of any further research in the matter, rather than complaining for more. But this cannot have been his meaning, since the *Strategems* is not a book about artillery construction or design, but a mere collection of *past* examples of battlefield strategies excerpted from history books, organized by topic (as he explains in the preface to book one). When Frontinus gets to his chapter on past stratagems employed in siege warfare, he says:

> Having set aside siegeworks and siege engines, because their discovery was completed long ago (so I attend no further to any material from those arts), I will put together the following kinds of stratagems regarding sieges: [listing eleven categories].... then in contrast, regarding the protection of the besieged: [listing seven more categories].[923]

Dubious translations of this passage are occasionally cited in defense of the notion that Frontinus means no further developments in siege equipment or artillery are possible, therefore he won't discuss them. For example, the 1925 Loeb translation by Charles Bennett has Frontinus saying that siegeworks were left out because "the invention of which has long since

921. Diodorus Siculus, *Historical Library* 2.29.6.

922. See example in section 3.9.II below. Likewise, 'complaints' of rising scientific specialization indicate a rising complexity of scientific knowledge, which is actually a consequence of progress, not a sign of decline (von Staden 2002; e.g. Cicero, *On the Orator* 3.23.132; Galen, *To Thrasybulus* 24 = Kühn 5.846–51 and *On the Parts of Medicine* 2; Philostratus, *Gymnastics* 15; etc.).

923. Frontinus, *Stratagems* 3.pr.: *Depositis autem operibus et machinamentis, quorum expleta iam pridem inventione (nullam video ultra artium materiam), has circa expugnationem species stratêgêmatôn fecimus: ... Ex contrario circa tutelam obsessorum.*

reached its limit, and for the improvement of which I see no further hope in the applied arts," but there is no way to get this meaning from the Latin (which contains neither the word 'improvement' nor 'hope'), or from the context. And the latter is decisive.

The *Strategems* is not a treatise on machines, but tactics (decisions made by commanders in the field that affect the outcome of battles and campaigns, as Frontinus explains in his preface to book one), and it was intended as a supplement to his more systematic treatise on *Military Science*, which we do not have (so we really do not know his thoughts on war technology). More importantly, since Frontinus is explicitly *not* innovating anywhere in the *Strategems*, but only collecting past examples of actual stratagems (many dating back centuries), he cannot be referring here to innovations in siege weapons. He can only mean their subject has already been thoroughly covered elsewhere and therefore he is no longer going to attend to it. We would otherwise expect him to include among his categories of stratagems a section that lists historical examples of tactics involving siege machinery. There was certainly plenty of such material available to Frontinus, so his apologizing for not including it cannot have been because there was none, or that he knew of none, or that a field commander would have no use for it. The only reason he could have for leaving it out is that the subject was already adequately covered by other authors.[924] Therefore, from this passage we cannot infer that Frontinus believed there would be no future developments in military science.

Our last example is the author Galen, who often claimed that medical science was in some sort of crisis or decline that only his high ideals could cure. In fact, he frequently repeats the same cultural critique echoed in Pliny and mocked in Petronius, in some respects so closely one may wonder

---

924. So in *nullam video ultra artium materiam*, *video* must mean 'attend to' (and not 'see' or 'know'), *ultra* must refer only to the rest of his present chapter, and *nullam...materiam* must mean passages he could have collected on the subject, not future technologies. Both I and Bennett correctly take *artium* in the same sense as the only other plural of *ars* in the same treatise (*artibus bellandi*, cf. *Stratagems* 1.8), here referring back more specifically to the arts involved in the development of war machinery, and thus to treatises on that subject from which he could have drawn passages for his present collection. Hence most likely *expleta iam pridem inventione* probably does not refer to the discovery (i.e. invention) of machines, but the rhetorical process of *inventio*, i.e. collecting and filling out *topoi* as Frontinus does throughout the *Stratagems*.

if he had read them. For example, while Pliny used as an example those who risk dangerous sea voyages for greed but not for exploration or the advancement of meteorology, Galen uses a similar example, but adapts it into a metaphor for an imagined neglect of the exploration of "the knowledge and understanding of the nature of things," especially, of course, in medicine.[925] Galen also repeats the myth of decline in painting and sculpture, as a parallel for his own myth of decline in medicine, even using some of the same examples as Petronius, and proposing the same cause: that greed has displaced a genuine desire for truth and excellence.[926] But like Pliny, and naively playing right into the Petronian joke, Galen greatly exaggerates the reality, for instance claiming that "no one studies medicine anymore except slaves and scoundrels, and even emperors and all the wealthy elite look on scientific medicine with disdain," which is so far from being true it is already refuted by ample evidence in Galen's own writings.[927] Likewise, though Galen correctly argues that science will decline unless enough people prefer knowledge to fame, he goes too far in pretending how dire the current situation supposedly was, claiming to find "not even five people who actually want to be wise, rather than merely appear to be so," presumably meaning in the city of Rome, but even with that qualification the hyperbole is palpable.[928]

Underlying all these examples (and one could perhaps adduce more) is Galen's own idea of a decline in medicine, which he deploys to represent himself as leading a 'revival' of medical science in Rome. Vivian Nutton has

---

925. Galen, *On Conducting Anatomical Investigations* 14.7 (compare with Pliny the Elder, *Natural History* 2.45.118; another similar seafaring analogy appears in Seneca, *Natural Questions* 5.18.11–14). Other parallels (though unrelated here) include the fact that both Pliny and Galen marvel at the anatomy of a flea and discuss it at some length (Pliny the Elder, *Natural History* 11.1.1–4 and Galen, *On the Uses of the Parts* 17.1 = Kühn 2.448–49 = M.T. May 1968: 731–32; there is also a reference to the study of flea anatomy in Tertullian, *Treatise on the Soul* 10, possibly inspired by either).

926. Galen, *That the Best Doctor Is Also a Philosopher* 2 (= Kühn 1.57).

927. Galen, *On Examinations by Which the Best Physicians Are Recognized* 1 (esp. 1.4; cf. also 9.2 and 13.2–3). See relevant discussion in chapter 4.3 for evidence in Galen exposing the hyperbole. Of course, many trained doctors were slaves, but by no means all, nor was the quality of their education necessarily inferior.

928. Galen, *On the Therapeutic Method* 2.5.16 (cf. also 1.1).

already noted the mythical nature of the picture Galen draws.[929] But it is clear enough from the fact that he imagines as his 'abandoned past' an ideal of anatomical education that never really existed, which Galen sets up as a model of what all modern doctors should be doing: beginning extensive empirical study of anatomy from an early age under the hands-on guidance of accomplished experts.[930] His ambition was not unrealistic (Galen himself exemplified it, and no doubt many others did, too), while the abundance of practicing doctors who ignored it certainly explains his rhetoric.[931] But what Galen was actually proposing was the novel idea that exceptional cases be made the norm, by inventing a glorious past as an 'example' to follow. Hence Galen's myth of 'decline' reflects his own passionate desire for progress in medical science, and in science education as a whole. Like Pliny, his rhetorical aim was to shame and alarm his respectable peers into working harder to support sound empirical science against a perceived onslaught of charlatans and hacks.

Of course, like many ancient authors, Galen also weaves his myths to attack the vices of luxury, laziness, and greed, and then praise in their stead the virtues of austerity, discipline and industry, all in the hopes of molding professional medical standards. Hence what Galen *really* means to say is that scientists like him are superior and accomplished because they are morally virtuous, while hacks and quacks remain ignorant because they are depraved. The same idea that a scientific mind will always be a moral mind was articulated by Cicero, who made the point that moral depravity and the quest for scientific knowledge involve incompatible desires, since the pursuit of science requires a discipline and sacrifice and love of truth that contradict the interests of the greedy, lazy, or sycophantic.[932] Galen clearly agreed.

---

929. Nutton 1972 and 2013: 205–06, 211–13, 220–21.

930. Galen, *On Conducting Anatomical Investigations* 2.1 (= Kühn 2.280–83) and *On My Own Books* Kühn 19.9.

931. See discussion of Roman medical education in chapter seven of Carrier 2016.

932. Cicero, *On the Boundaries of Good and Evil* 5.20.57. See discussion in chapter 4.2.

II. Stories Implying Resistance to Innovation

So much for allegations of decline made by ancient authors. They either in fact demonstrate progress was desired and ongoing after all, or fabricate a myth of decline in defense of an oddball conservatism even they admit is unpopular. Next is the accusation that, according to Thomas Africa, "the Roman state intervened twice to suppress inventions which seemed detrimental to the public good."[933] There are in fact *only* two such stories, in over three hundred years of Roman history. That is hardly sufficient to constitute a 'trend'. Modern scholars nevertheless repeat the same two examples over and over again as if they were representative of ancient culture. In fact, in both cases these stories were told because they were *not* representative, but in fact remarkable.

The first of these two stories is related only by Suetonius, who is usually cited as claiming that Vespasian rejected the use of a labor-saving machine because he feared it would lead to unemployment (an anachronistically modern concern). Unfortunately, Suetonius is frustratingly vague, but it is doubtful he said what is claimed. Completing a list of generous rewards Vespasian gave to other teachers and artists, Suetonius concludes:

> And though an engineer made an offer to bring huge columns up to the Capitol at a minimal cost, Vespasian gave him an exceptional reward for his scheme but turned down his employment, prefacing his dismissal with the remark, 'let me feed the poor'.[934]

In a textbook example of catastrophically misleading translation, in the still-popular Penguin edition Robert Graves 'inserted' after 'minimal cost' the words "by a simple mechanical contrivance" (nowhere in the Latin) and 'expanded' Vespasian's brief remark into the elaborate "I must always ensure that the working classes earn enough money to buy themselves food," which far exceeds Vespasian's mere four words.[935]

---

933. Africa 1967: 72.

934. Suetonius, *Vespasian* 18: *mechanico quoque grandis columnas exigua impensa perducturum in Capitolium pollicenti praemium pro commento non mediocre optulit, operam remisit praefatus sineret se plebiculam pascere.*

935. Graves & Grant 1979: 288. This is an example of a pervasive problem throughout Graves' translation of Suetonius, which Michael Grant laments in the introduction to the revised edition (ibid.: 10), and claims to have corrected with

Lionel Casson is the first to have drawn attention to this fact, in an article that launched a debate between him and P.A. Brunt over what Suetonius and Vespasian actually *did* mean in this story.[936] Casson argues that no machine was even proposed, but only a plan for hiring the free poor as laborers, which Vespasian turned down because he wanted to maintain the grain dole instead, rather than putting welfare recipients to work. Brunt, on the other hand, defends Graves' 'interpretation' against Casson, and rightly corrects him on various points of Roman labor history, but neither of them seems to stake out a plausible position on the meaning of this passage. Brunt insists that a "'commentum' in this context is far more likely to represent a device that an engineer might suggest, i.e. a mechanical invention."[937] But that is hard to maintain. It is absurd to think any contractor for hauling stone would not *already* be employing all available machinery, so the choice Vespasian faced cannot have been between machinery or strongbacks, especially since "giant stone blocks" would be hauled by animals, not men, and Vespasian did not say "let me feed my oxen." The choice is also unlikely to have been between better machinery or worse. For what could possibly have been built 'more cheaply' that would outperform machinery already in use? Proposing, for example, the building of some sort of water-powered conveyor belt would require a capital outlay far exceeding any other contractor's bid for the same work. Even a water-powered crane would cost more to arrange than it was worth, and at any rate the bid was not for *lifting* stone, but hauling it to the Capitol.

The evidence more strongly supports Casson's interpretation: the *commentum* was not a machine, but a 'plan', in other words a contract offer. It was not necessarily the exact plan Casson proposes, but Blunt's case is hardly more secure, since neither labor nor wages are ever mentioned in the anecdote. Perhaps the engineer's proposal involved wage or labor cuts that Vespasian thought were excessive (and yet merely suggesting them was worth rewarding?), but there is no way to know. In defense of Casson's interpretation of *commentum*, the surrounding vocabulary all matches that of a contractor making a low bid on a public works project, which as Casson

---

his revision, evidently missing this one.

936. Casson 1978 and Brunt 1980. Kevin Greene sides with Casson (Greene 2000: 49–50).

937. Brunt 1980: 81.

shows was routine, hence Suetonius is remarking on the generosity of giving a contractor a ton of cash even after rejecting his lowball bid.[938] What exactly the bid involved cannot be known. But there is no comparable support for machinery being meant instead of a contract offer.

Even if we follow Graves and Brunt and assume some new machine was involved, Vespasian's action does not demonstrate a general Roman opposition to technological innovation. The story is clearly portrayed as exceptional, both in place (Rome) and time (the remarkably unique action of a single emperor). More importantly, Vespasian did not discourage the inventiveness of the engineer, but actually rewarded him for it, which would indicate *support* for innovation, not resistance. Suetonius tells the story to praise the character of Vespasian, hence Suetonius is also praising the values Vespasian's actions embody. Thus, if anything, this event represents a Roman belief that innovative engineers deserve to be handsomely compensated. Furthermore, it is impossible that any machine this offer may have involved would have been abandoned simply because the emperor did not want to use it. Other engineers, contractors, and benefactors across the empire would have seen the value of any cost-cutting measure in their own building projects, and few would have shared Vespasian's concern to "feed the mob" instead of cutting their own costs. So even if the story is true, and is about a new invention, this engineer's machine would certainly have seen use somewhere, unless the reasons for rejecting it were more practical than Suetonius is aware—that is, it might have been dangerous or unpredictably expensive to implement, and therefore rejected because it was a bad idea.[939] In fact, Vespasian's remark might simply have been a joke—the engineer's bid being so low as to ensure his poverty, hence Vespasian chose to 'feed the poor', i.e. the engineer, by paying him without requiring him to undertake a job he could not have afforded to complete.

Either way, this tale of Vespasian's generosity did not involve the

938. For example: *pollicens* means "bid, promise, offer"; *exigua impensa* means "low cost, small expense"; *opera* means "work, job, employment"; and *mechanicus* would here mean "engineer," who was often the contractor arranging an entire works project, not just building the machinery.

939. For example: Vitruvius, *On Architecture* 10.2.13–14 relates a similar story about the engineer Paconius, who developed a new method of hauling blocks that was supposed to be cheaper but turned out to cost a great deal more than established methods.

suppression of technology as Africa claimed. The only other evidence Africa (or anyone) can offer is a story told about Tiberius. A story invented by a comedian. In his satirical novel the *Satyricon*, once again Petronius describes a scene in which a wealthy pretender named Trimalchio is boasting of his tableware, in the process relating stories that are embarrassingly false (mentioning, for example, that he has a bowl showing Daedalus shutting Niobe inside the Trojan horse), thus revealing his humorous ignorance of history and literature.[940] In the midst of all this Trimalchio explains that:

> If glassware were unbreakable, I would prefer it to gold, though now glassware is very cheap. In fact there was once a craftsman who made a glass bowl that was unbreakable. He was given an audience with the emperor, bringing along his gift. He had the emperor hand it back and threw it to the floor. The emperor was as frightened as he could be, but the man picked the bowl up from the ground, and it was dented just like a vessel made of bronze. He took a little hammer from his shirt and fixed it perfectly without any problem. By doing so he thought he had made his fortune, especially after the emperor said to him, 'No one else knows how to temper glassware like this, do they?' Just see what happened! After he said 'No', the emperor had his head chopped off, because if this invention were to become known, we would treat gold like dirt.[941]

From the context we can be certain of two things: the story is either untrue or wildly incorrect, and Trimalchio is being made to look like an idiot for telling it. Thus his belief that unbreakable glass would make gold worthless (or indeed that *anyone* of sense would think such a thing) is a fiction designed to communicate to the reader Trimalchio's shocking stupidity. We can thus conclude that no emperor ever did what he reports, or certainly not for any such reason. We can also assume no such thing was ever invented. Though flexible glass is a staple of modern fiber optics, and modern transparent plastics would have been described as 'glass' in

940. The bowl remark is in Petronius, *Satyricon* 52. If it needs to be said, Daedalus, Niobe, and the Trojan Horse all existed at different times in mythic history, at different places, and in completely unrelated stories (which makes this comparable to saying Abraham Lincoln had sealed Robin Hood in Al Capone's vault).

941. Petronius, *Satyricon* 51. Notably, the name of the emperor is not given. He is only identified as Tiberius by later authors (as we shall see). Petronius may have intended it to be Nero.

antiquity, it is very unlikely anything comparable to these was ever made in ancient Rome.[942]

After a decade or two this story became an urban legend, soon reported in Pliny's *Natural History*. After relating the most recently invented technologies in the glass industry, Pliny adds:

> There is a story that in the reign of Tiberius there was invented a method of blending glass so as to render it flexible. The artist's workshop was completely destroyed for fear that the value of metals such as copper, silver and gold would otherwise be lowered. Such is the story, which, however, has been circulating a long time now more through frequent repetition than being true. But this is of little consequence, seeing that in Nero's principate there was discovered a technique of glass-making that resulted in two quite small cups...[which] fetched a sum of 6000 sesterces...[and] for making drinking vessels the use of glass has indeed ousted metals such as gold and silver.[943]

Pliny's version of the story is different in several details, but he admits he has it from oral lore, and is aware of the fact that it is not true. Moreover, Pliny sandwiches this fable between true accounts of recent new glass technologies that were far from being suppressed, but were actually enormously successful. In fact, from what Pliny says, Trimalchio's fictional worry had come true: metalware was no longer as popular as glassware. By the time the joke had become a legend, it had altered considerably, being twisted to the point of almost making sense. Now, for example, the emperor's motive was not that gold would become worthless, but only worth *less*, since demand for goldware (and silverware and copperware) would fall while demand for glass rose. Though in reality changing fashions in tableware would hardly have affected the market value of precious metals, this was at least slightly more believable to the economically naive. As a result, the story had come to resemble comparable urban legends today about oil

---

942. Though blown glass (which had only recently been invented) is indeed flexible and repairable in its melted state, even the nitwit Trimalchio distinguishes his imagined 'unbreakable' glass as something notably different from ordinary blown glassware (though the behavior of melted glass could have inspired someone to imagine the possibility of it retaining this property when cooled). Attempts to argue that aluminum was meant (or other substances) are refuted in Eggert 1995.

943. Pliny the Elder, *Natural History* 36.66.195 and 36.67.199.

corporations buying out or assassinating the inventors of cars powered by tap water or hemp. And like those tales, this story represents values exactly opposite to those Africa infers: Tiberius is being portrayed as a villain, and his action condemned, not elevated as sound government policy. And besides being entirely untrue, it is also entirely unique, never once being represented as typical, but to the contrary, as wholly *atypical*.

There may be a genuine story behind this, of a very different character entirely, a real event whose details Petronius had Trimalchio get laughably wrong. Centuries later, the historian Cassius Dio relates his own version of what happened, which shows no awareness of the details in Pliny or Petronius and thus may derive from a more authentic source. Dio explains how Tiberius started out making praiseworthy decisions, but then his behavior swerved into appalling injustice and cruelty, which Dio demonstrates with a list of examples, all of which he clearly assumed his readers would agree were crimes, in principle if not in fact.[944] It is among these villainous acts that the following story appears. A portico in Rome had begun to lean, and was righted "by an architect whose name no one knows, because Tiberius, jealous of his wonderful achievement, would not permit it to be entered in the records," something Dio (and his sources) clearly considered reprehensible.[945] Nevertheless, "Tiberius both admired and envied him, so for the former reason he honored him with a present of money, and for the latter he expelled him from the city," curiously mixing the same generosity alleged of Vespasian (of rewarding a scientist for a clever idea or achievement), with the characteristic villainy of a bad emperor (booting an innocent out of Rome). It is only *then*, that:

> The architect approached Tiberius to crave pardon, and while doing so purposely let fall a crystal goblet. And though it was bruised in some way or shattered, yet by passing his hands over it he promptly exhibited it whole once more. For this he hoped to obtain pardon, but instead the emperor put him to death.[946]

944. Cassius Dio, *Roman History* 57.19–21.

945. In fact Hero, *Mechanics* 3.10–12 (cf. Drachmann 1963: 108–09) discusses several techniques for righting walls, columns, and porticos that have begun to pitch or lean, so this was evidently a common and expected skill among ancient engineers, and not remarkable as Dio claims.

946. Cassius Dio, *Roman History* 57.21.5–7. Whether this story bears any

Dio does not say why Tiberius killed him, though the implication is that Tiberius was simply capriciously executing a man he did not like. Moreover, as Dio describes the scene, it sounds like a rather mundane magic trick of the sort one might see in a Vegas show. Dio knows nothing about a workshop or a new technology or any imagined economic threat, and as far as he knows the engineer's reason for approaching Tiberius had nothing to do with presenting him with any invention.

It is safe to conclude from all three accounts that no emperor suppressed any invention, much less a mythical flexible glass. Several scholars have come to the same conclusion.[947] And since the very different story told of Vespasian also does not support such an idea, we can dismiss Africa's claim that the Roman government ever suppressed scientific technologies.

III. Impediments to Research

That leaves evidence of broader social impediments to scientific research, by which I mean laws or customs that directly got in the way of research that we know ancient scientists wanted to conduct. There are really only two examples of this, and only one of them was genuine. The other, an alleged law against mapmaking (which would have directly hindered advances in geography and cartography), is a modern fiction. But there *was* something getting in the way of the dissection of human cadavers, though not absolutely, and of course the scientific dissection (and vivisection) of animals as proxies for humans went unimpeded by any law or custom.[948]

---

connection to other ancient stories about glassbreaking is a question for another time (e.g. Seneca, *On Anger* 3.40; Pliny the Elder, *Natural History* 37.10.29; Cassius Dio, *Roman History* 54.23.2–4; etc.).

947. For example: Trowbridge 1930: 110–12; Simms 1995: 108 n. 7.97; E.M. Stern 1999: 441–42; Greene 2000: 46–47.

948. I will not discuss incidental impediments that have no cultural or institutional explanation, such as a need for a telescope or printing press, which were happenstance inventions that could have occurred at any time (see chapter 1.1 and introduction to the present section). But the claim that "scientific advance in the ancient world was hampered generally by the lack of a reliable means of calculating time" (M.R. Wright 1995: 127) is untrue. Astronomers routinely used a standard equinoctial hour and knew how to accurately convert seasonal hours to standard hours, and had several fairly sophisticated technologies for telling

First is the alleged law against mapmaking. Mettius Pompusianus is sometimes said to have been executed for owning a map, and therefore, it is argued, the private possession of maps must have been illegal.[949] But neither claim is true. Suetonius reports that Mettius 'allegedly' passed around a map of the world, and that this was used along with several other supposedly incriminating facts to prove he was conspiring against the emperor Domitian, who thus exiled and eventually executed him. But there is no mention here of any crime against keeping or publishing maps. In fact, the context suggests there was nothing justified in Domitian's action. The story appears in a section listing Domitian's intemperate cruelties, among which are several murders for entirely trivial reasons, including having Mettius killed "because the people were saying" certain signs portended "an imperial birth, and because he supposedly passed around a globe painted on parchment, and a collection of speeches of kings and generals extracted from Livy, and because he named his slaves Mago and Hannibal."[950] Suetonius even puts the verb *circumferret* in the subjunctive, and thus is not even conceding that the accusation of passing around maps and extracts from Livy was true. But even if it was, since naming your slave Hannibal was no actual crime, nor was publishing excerpts of Livy, neither would carrying a map have been. It is not even certain a map is meant. The clause "*depictum orbem in membrana*" can mean a map of the world, or an astrological chart, or simply a drawing of a sphere, representing the earth or the cosmos, a common symbol of power.

Cassius Dio records the same event and, like Suetonius, includes it in a series of the trivial and unjustified crimes of Domitian. Dio reports somewhat differently that Mettius "was accused of having the inhabited world painted on the walls of his bedroom, and having recorded the speeches of kings and other prominent men extracted from Livy and reading them."[951] Pascal Arnaud spends a great deal of ink speculating about this map and what threat it posed, ignoring the more obvious point that if Domitian can simply execute a boy for looking too much like a famous actor

---

time, including sundials, waterclocks, and diopters (see Hannah 2009; and general sources in section 3.3).

949. See Arnaud 1983, with *OCD* 951 (s.v. "Mettius Pompusianus").

950. Suetonius, *Domitian* 10.3.

951. Cassius Dio, *Roman History* 67.12.2–5.

(which Suetonius includes in the same list of misdeeds), and then condemn Mettius for naming a slave Mago, clearly his alleged map need not have posed any real threat either—it was just another spurious rationalization for killing someone Domitian did not like.[952] This tells us something about Domitian, but nothing at all about Roman law or policy. In fact, Suetonius and Dio both expected their readers to regard this execution to be shockingly absurd. Hence the fate of Mettius lends no support to Arnaud's contention that Roman emperors generally regarded 'monumental cartography' to be an imperial monopoly.

Which leaves us with the vexing question of human dissection.[953] As already noted in section 3.2, we know human cadavers were dissected by Herophilus and Erasistratus a century or two before the Roman period, in circumstances believed to be exceptional, and there is no evidence of a comparable practice continuing afterward. This appears to be the first and last time in history, until the Renaissance, that human bodies would be systematically dissected. In contrast, for example, all of Galen's anatomical works clearly depend a great deal on the autopsy of apes and other animals, which entails that, at best, it was unusual for him to get his hands on a human cadaver. There is no reason to believe his circumstances differed from any other scientist. Nevertheless, Galen was aware of the fact that ape anatomy often disagrees with human anatomy, requiring caution when drawing analogies between them, and despite his awareness he did make some mistakes in this regard. But there is *also* evidence he had anatomical experience with humans, and that he and his contemporaries had in fact dissected them.[954]

So why wasn't human dissection carried out more frequently? Heinrich von Staden has tried to make some sense of this, but since all our sources are

---

952. Domitian's capricious execution of an acting student is reported in Suetonius, *Domitian* 10.1.

953. This question has been widely discussed: Edelstein 1935; M.T. May 1968: 22–24, 40–41; Kudlien 1970: 21; P. Fraser 1972: 1.348–52 and 2.504–14; Lloyd 1973: 75–78; Ferngren 1982; Fischer 1984; Iskandar 1988: 165–67; von Staden 1989: 139–53; Debru 1994: 1725–26 n. 40; Byl 1997.

954. On his caution: Galen, *On the Uses of the Parts* 1.22 (= M.T. May 1968: 107–08). Evidence of human anatomical knowledge: C.J. Singer 1956: xxi-xxiii. Evidence Galen and his contemporaries had dissected humans: Hankinson 1994a: 1786; and evidence to follow.

vague, he admits he can reach no definite conclusions.[955] He locates some cultural and moral taboos against touching or cutting open human cadavers, but none of these were insurmountable, since we have ample evidence they were violated in the name of science. The only text that comes anywhere close to offering a plausible reason is Pliny the Elder's off-hand remark that cannibalism is surely depraved, since even "inspecting human organs is considered a crime," but this is unlikely to mean dissection, because surgery was not illegal, yet that certainly involved handling and observing human organs.[956] Pliny is more likely referring to necromancy, or ritual human sacrifice, and the nefarious magical or divination practices that could be associated with them. Although a scientist could be accused of such crimes by the malicious or superstitious, even for dissecting a fish, doing this "for science" was evidently considered to be a successful legal defense.[957]

Meanwhile, in medical authors, the only passages that directly suggest an inability to dissect human cadavers give no indication of why it had become uncommon. In the 1st century A.D., Rufus remarks in passing that "in old times" anatomy was "more suitably taught" by dissecting human cadavers, while in his day it was done less successfully using apes and monkeys, and the 'surface inspection' of slaves. But he never says why this had changed or when, or what was stopping anyone.[958] Similarly, Galen said a century later that Herophilus had greatly advanced anatomical and physiological knowledge "not by dissecting irrational animals like most do, but by dissecting actual human beings," but again giving no hint why anything had changed.[959] In fact, neither Galen nor Celsus, who specifically

---

955. See von Staden 1992.

956. Pliny the Elder, *Natural History* 28.2.5: *aspici humana exta nefas habetur*. But contrast this with Cicero, *Prior Academics* (= *Lucullus*) 2.39.122.

957. Apuleius, *Apology* 38. And see the Senecan legal case discussed below.

958. Rufus, *On Naming the Parts of the Body* 1.

959. Galen, *On the Anatomy of the Uterus* 5 (= Kühn 2.895). Conspicuously absent here is any mention of Herophilus having *vivisected* humans (in fact, in all his extant writings, Galen never mentions anyone ever having done this), which I think casts doubt on the tale (see section 3.2). Human vivisection was certainly not going on in the Roman period, despite irrational fears of it, as reflected in a mock trial of a Roman doctor accused of murder by vivisection in Quintilian, *Major Declamations* 8 (cf. Ferngren 1982 and 1985).

and extensively discuss the dissection of human cadavers more than any other author, ever mentions it being illegal or immoral or religiously prohibited, or gives any reason why it was not done.

Celsus does not even mention that it had ceased.[960] When he presents in detail the arguments for and against *vivisecting* human beings, he explains, quite reasonably, why it was regarded as morally repugnant, and thus, we can assume, why it was no longer allowed or approved (if it ever really had been). Yet oddly he never presents any comparable arguments against dissecting cadavers, other than methodological disputes that had no bearing on what anyone was permitted to do, and which instead show that a significant segment of the scientific community was entirely supportive of dissecting the dead, including Celsus himself. The only hint he offers of any reason this might be disallowed or shunned is neither moral, nor legal, nor religious, but merely aesthetic: it was *foedus*, "filthy, disgusting, hideous," a term that merely describes the fact of guts and mutilation, which was no less true of dissecting animals, and hardly anything that would deter a scientist.[961]

Though the word *foedus* could carry the connotation of disfigurement and defilement, which might underlie any moral or legal barriers that may have existed, it is strange that Celsus would never call it immoral or illegal or irreligious, or even oppose it at all—to the contrary, he gives it his unqualified approval:

> Butchering the bodies of the living is both cruel and unnecessary, but butchering the bodies of the dead is necessary for students, for they need to know the position and organization [of the organs], which a cadaver shows better than a living, wounded man, while the rest, which can only be discovered in living bodies, actual practice will demonstrate in the course of treating the wounded, in a little slower but much gentler way.[962]

Remarks like this imply human bodies were being dissected without opposition.

---

960. Celsus, *On Medicine* pr.23–26, pr.40–44, and pr.74–75 survey various facts and opinions on the matter of human dissection.

961. Celsus, *On Medicine* pr.44. Aristotle said essentially the same of all dissection in his *Parts of Animals* 1.5.645a27–37.

962. Celsus, *On Medicine* pr.74–75.

There is some uncertain evidence Roman scientists did have (at least occasional) access to human cadavers.[963] But we have at least one good source confirming it. Galen explicitly mentions how anatomists under Marcus Aurelius had botched the scientific dissection of cadavers taken from German war casualties, only because of their inexperience, while others "have frequently dissected many bodies of exposed children" or "have often rapidly observed whatever they wished in bodies of men condemned to death and thrown to wild beasts, or in brigands lying unburied on a hillside."[964] He presents no apology for any of these occasions, and in fact seems unaware of any specific law or taboo standing in the way.

It is admittedly curious that Galen only explicitly describes dissecting humans for his study of osteology, presenting the only descriptions he ever gives of *his own* inspections of human bodies in contexts that did not require actually touching the corpse:

> Make it your serious endeavor not only to acquire accurate book-knowledge of each bone but also to examine assiduously with your own eyes the human bones themselves. This is quite easy at Alexandria because the physicians there employ visual demonstration in teaching osteology to students. For this reason, if no other, try to visit Alexandria. But if you cannot, it is still possible to see something of human bones. I, at least, have done so often on the breaking open of a grave or tomb. Thus, once, a river inundated a recent hastily made grave and broke it up, washing away the body. The flesh had putrefied, though the bones still held together in their proper relations. It was carried down a stadium's length and, reaching marshy ground, drifted ashore. This skeleton was as though deliberately prepared for such elementary teaching. And on another occasion we saw

963. A late medical source (Vindicianus, *Gynecology* pr.) lists among those who dissected human corpses several Roman anatomists, including Rufus, Lycus, and Pelops, though he might have assumed this more than known it (von Staden 1989: 52, 60, 189). Augustine, *City of God* 22.24, and others in late antiquity, refer to continuing dissections of human cadavers, though in contexts that could be rhetorical (cf. von Staden 1989: 188–89, 235). Even more inconclusive is a visual depiction of what appears to be a human dissection in an early 4th century catacomb painting (see note in chapter 1.3).

964. Galen, *On Conducting Anatomical Investigations* 3.5 (= Kühn 2.384–86). Celsus had said convicts were also the source for Herophilus and Erasistratus. Aborted or miscarried fetuses might also have been dissected in the Roman period, as they had before (cf. Nutton 2013: 220).

> the skeleton of a brigand, lying on rising ground a little off the road. He had been killed by some traveler repelling his attack. The inhabitants would not bury him, glad enough to see his body consumed by the birds which, in a couple of days, ate his flesh, leaving the skeleton as if for demonstration. If you have not the luck to see anything of this sort, dissect an ape.[965]

These are Galen's only references to his own 'dissections' of humans, and yet in both cases he seems to imply that he did not need to touch the bodies, just inspect them where they lay. But he does say he found such opportunities "often," and that at Alexandria human skeletons were kept for school use, though again these might not have been touched (except perhaps by the slave, undertaker, or embalmer who prepared them).[966] Still, it is strange that Galen does not warn his readers or students about any laws or taboos such touching might violate. Nor does he explain why skeletal displays were only available in Alexandria.

As von Staden suggests, popular belief in the religious pollution that results from contact with a corpse may have presented a problem for medical practice, since patients might then refuse a doctor's services if they knew he had touched the dead. But why does Galen never mention this? Surely he would warn his readers, especially in a textbook urging them to inspect human bodies at every opportunity, if there were concerns and cautions to observe in contacting or handling corpses, lest they anger gods, offend locals, lose clients, or (literally) get the axe. Moreover, Galen still recommends or refers to hands-on dissections of humans performed by others, still without offering apologies, warnings, or reservations. In addition to the examples noted above, he also discusses veins that can only be seen on dissection, and though he recommends apes for this, he adds

965. Galen, *On Conducting Anatomical Investigations* 1.2 (= Kühn 2.220–22; inspecting human skeletons is repeatedly referred to as something that happens only by chance: 1.2 = Kühn 2.223–24; 14.1 = Kühn 2.229).

966. Galen, *On Bones for Beginners* 4.750, 6.754, and 11.762, which mentions examining both human and ape bones, and speaks generically of the fact that for research use all bones are typically boiled and dried (cf. C.J. Singer 1952). But according to Galen, *On Conducting Anatomical Investigations* 14.1 (= Kühn 229), they are typically not cut out but rotted out, as for example by burying apes "for four months or more in earth that is not dry" and then exhuming them. So it is worth noting that burying and exhuming human bodies, even for science, would have been illegal (see later note).

that "if you have the luck to dissect a human body, you will be able readily to bare each of the parts" because of your prior practice on apes.[967] He also adds that "this is not everybody's luck," but is usually an opportunity that arrives only on short notice, and thus he warns that mistakes can be made from the hasty or inexperienced dissection of human cadavers unless one has long practiced on apes. There does not seem to be any concern here about contacting or cutting into human corpses.

The only consistent impression we get is that access to cadavers was limited to chance opportunities (in which the deceased had no family or burial rights, like enemy war dead, executed criminals, and abandoned babies), and occasionally to brief periods of time (such as the corpses of executed criminals, which had to be inspected "quickly," suggesting they were not turned over to doctors for medical use, but formally buried in relatively short order). Only once does Galen mention needing "permission" or "authority" (*exousia*), but only to dissect enemy casualties, which we would expect, given the variability of the diplomatic situation (e.g. a surrendering foe might expect to receive the corpses of their dead undefiled, so the treatment of bodies would have to remain at the discretion of the commanding officer).[968] Though digging up bodies was sacrilege and certainly illegal, most of the examples in Galen and Celsus involve bodies being dissected before burial, or after being exhumed accidentally.[969] And in the instances Galen mentions of 'accidental' exhumation, he implies avoiding contact, and he might not have said so more explicitly because it was commonly understood.

There was at least one Roman legal writer who regarded scientific dissection as a legitimate defense for cutting open the dead.[970] Seneca the Elder, writing shortly before or after the turn of the era, describes a fictional trial set in Athens in the 4th century B.C., in which the famous painter

---

967. Galen, *On Conducting Anatomical Investigations* 3.5 (= Kühn 2.385).

968. Galen, *On the Composition of Drugs by Class* 3.2 (= Kühn 13.604).

969. That exhuming the buried was both sacrilegious and illegal under the Romans (as it probably had always been): Gaius, *Institutes* 2.2–10 and *Digest of Justinian* 47.12.1 (cf. 47.12.3.2–7, 47.12.8, 47.12.11).

970. Seneca the Elder, *Controversies* 10.5.17 ("Parrhasius and Prometheus"). For scientific analysis: Fischer 1984. For historical analysis: Rouveret 2003. For literary analysis (and technical background on Parrhasius): Morales 1996.

Parrhasius is on trial for crucifying a slave as a model for his painting of the crucifixion of Prometheus. Seneca discusses the handling of the scenario by previous law professors in their own casebooks, observing that one of them proposed as a defense "how much license the arts have always had," giving the specific example that "doctors have laid bare the vital organs so they will know the hidden power of a disease, and even today (*hodie*) the limbs of cadavers are opened up so the position of sinews and joints can be ascertained." Unfortunately the name of the professor he is quoting is missing, but scholars conjecture it to have been Seneca's contemporary, Marcus Porcius Latro.[971] The casebooks Seneca quotes go all the way back to the Greeks, with this example simply taken up by Roman law schools and treated in their own way, yet Seneca says none of the Greek professors dared even propose a defense.[972] So the argument from human dissection certainly came from a Roman, and whether Latro or not, all the Romans quoted by Seneca on this case date from the 1st century B.C.

Klaus-Dietrich Fischer argues that Seneca would surely have criticized this defense had it been out of touch with reality, which means educated Romans of the first century must have believed human cadavers were being dissected by scientists in the 1st century B.C., and thought this was not only appropriate, but admirable enough to cite in defense of a painter murdering a slave for his art. Though Latro could be speaking within the historical context of the case, Fischer argues against this, noting that it is set in 348 B.C., when it could not plausibly be said that doctors in the *past* had dissected human bodies for discovering pathology, nor even that they were "now" doing so for anatomical knowledge, and neither ever happened in Athens. Moreover, Latro would more likely have assumed a present condition obtained in the past, than have known (even erroneously) such obscure details of medical history. But either way, no one would propose such a defense unless Latro and his peers embraced the idea that dissecting cadavers for science could be an allowable exception to something that might generally be condemned. There is certainly no way Latro would propose such a thing if scientific dissection were *illegal*.

It would seem, then, that scientific research was only somewhat impeded by a limited access to human cadavers, and that this limitation

---

971. *OCD* 1190 (s.v."Porcius Latro, Marcus").

972. Seneca the Elder, *Controversies* 10.5.19.

did not come from any actual law against dissection, or any elite disdain or disapproval of it, but only from the practical realities of ancient burial law (which required exploiting loopholes in the care of bodies to gain scientific access to them), and perhaps, at least to a lesser extent, religious taboos that were evidently of no real concern to the educated elite, but probably of enough concern to their lower class patients that doctors would not want to flaunt their contact with corpses (as for example by dissecting cadavers in public theatres, or developing any regular arrangement for access to bodies that would draw the undue attention of a superstitious public). Similar problems vexed Renaissance and early modern attempts to secure human cadavers for study, and these barriers were cleverly (though gradually) overcome.[973] Roman medical scientists seem to have been more content with their situation; although if Roman society had continued to flourish, Galen's influential recommendations to study human cadavers might have inspired more interest in changing the status quo. After all, the Alexandrians had already managed to arrange a steady supply of human skeletons for the same purpose.

However, skeletons and cadavers were not the only point of access to the study of human anatomy. Galen reports that during a particular epidemic of a severe skin disease, "many people presented parts of their body stripped of skin and even of flesh" and "all of us, who saw Satyrus demonstrating on exposed parts, recognized them explicitly and completely," because of their prior study of apes and cadavers. As a result:

> We were telling the patients to make this movement or that, such as we knew was effected by this or that muscle, sometimes contracting or displacing the muscles a little to observe a large artery, nerve, or vein lying beside them. We then saw some students, as though blind, unable to recognize the parts, uselessly raising or displacing the exposed muscles (which needlessly distressed the patients), or even making no attempt to observe. Yet others, who had had more practice, knew how to direct the patient to move the part appropriately.[974]

Galen also says anatomy can be learned by inspecting the wounds of the living, and by observing healthy bodies closely. He even recommends

---

973. Sawday 1995.

974. Galen, *On Conducting Anatomical Investigations* 1.2 (= Kühn 2.224–25).

how best to choose and prepare subjects for these observations.[975] Galen discusses frequent occasions like this, of more harmless experimentation on live humans.[976]

Finally, although Gary Ferngren suggests a lay fear of medical experimentation also hindered progress in medicine, his evidence only argues for an opposition to novel treatment among many (by no means all) patients of the time, not opposition to experimentation or new treatments achieved without risk to human life.[977] Nor can it be said that hindering unbridled experimentation was necessarily bad for the advancement of medicine.

IV. The Real Decline

Though the rate of progress varied, and was always slow, there are no signs of a decline in scientific progress until after the 3rd century A.D. But then we do see "markers" of "a pattern of overall decline," such as "an increasing tendency toward preservation of an existent body of knowledge instead of its ongoing expansion" or even improvement, coupled with a pervading "scepticism concerning the possibility of discovering the true causes of phenomena" in the first place.[978] In other words, the markers of decline are an actual loss of what we would call the 'scientific spirit': a belief that the natural world *can* and *should* be increasingly understood by studying it. The abandonment of this ideal led to the corresponding cessation of original research—and, in its place, the ossification of natural philosophy as a "received tradition," which is the opposite of treating it as a body of knowledge in constant need of correction, expansion, and improvement.[979]

Before this decline, A.C. Crombie concludes, "the Greeks introduced

975. Galen, *On Conducting Anatomical Investigations* 3.5 (= Kühn 2.383–84 and 2.386).

976. Surveyed in Debru 1994: 1725–30.

977. Ferngren 1985.

978. Cohen 1994: 253 (cf. Lloyd 1973: 154–78).

979. Accordingly, Collins 1998: 501–04 summarizes the markers of decline as (a) a significant loss or neglect of the knowledge previously achieved, (b) an ossification of the knowledge preserved, and (c) an obsession with the logical refinement of ideas rather than a search for new evidence. The middle ages qualifies on all three counts.

an exclusive form of rationality based on two fundamental ideas: universal, self-consistent and discoverable natural causality and, matching this, formal proof" which together made science, and thus scientific progress, possible, so long as enough people maintained "confidence in the capacity of their scientific methods and the desirability of their results."[980] As we have already seen, and shall see more clearly in the next section, and the next chapter, there were numerous intellectuals who embraced all these ideals in the early Roman empire. Not so much afterward.

To see what *real* decline looks like, one need merely compare the relatively brilliant and detailed technological and mechanical treatises of Hero, which ground technology in scientific principles and practical experience, with the absurdly naive *On Matters of War*, an anonymous text from the late 4th century A.D. that proposes, among other things, a series of impractical or absurd 'inventions' with which to save a drowning empire.[981] The author of this bizarre book never claims to have invented or built anything he describes, and none of his descriptions show any technical knowledge of how to actually make them, and indeed some of them border on the ridiculous. His treatise includes no instructions for making the strange machines, discusses no scientific principles, and exhibits no real experience with their construction or use. Welcome to the dark ages.

To illustrate how ignorant this unnamed (though probably pagan) author was, consider his proposal (the first on record) of an ox-powered paddle boat. Had he ever spoken to a real engineer, he would never have believed that a mere six oxen, driving one paddlewheel each, could ever propel "a ship so large it cannot be operated by men" with "such furious strength that it easily crushes and destroys" every ship it collides with.[982] Contrast this six

980. Crombie 1997: 49, 52.

981. Better known as the *De Rebus Bellicis*, cf. E.A. Thompson 1952. Even the earlier hack, Athenaeus the Mechanic, whose treatise *On War Machines* is also relatively weak, nevertheless produces superior work (see note on him in section 3.4).

982. *On Matters of War* 17 (cf. E.A. Thompson 1952: 51–54, 77–78, 119–20). Though the text does not specify how many pairs of wheels, the drawing in the manuscript depicts three, and these drawings originate with the author—and even if poorly copied the number of wheels is unlikely to have changed. See also Landels 2000: 15–16, who likewise finds the description impractical, and Russo 2003: 139, who rightly suggests that the idea may have had more realistic precedents

oxen with a standard trireme complement of 170 oarsmen and you might start to get the picture.[983] Hence when comparable paddleboats were built in the 19th century, they had no military application—one of these, powered by four horses, was much smaller and slower than a trireme, and was only good for ferrying passengers.[984] The only practical paddle-wheel warships ever fielded prior to steam power were the Chinese dragon boats, most of which were neither larger nor faster than a trireme or quadrireme, and all of which had more appropriate propulsion and design.[985]

---

now lost. Otherwise, an engineer like Hero would have heaped ridicule on this presentation, just as Galen had on a ridiculous invention proposed by a similarly inexpert fantasist (cf. Galen, *On Conducting Anatomical Investigations* 7.16 = Kühn 2.643–44), though Galen pokes fun not so much at the impracticality of the tool (an Erasistratean resector) as the fact that it was never made or used.

983. If one man produces an effective power output of 75 Watts, then 170 men = 12,750 Watts. If an ox produces an effective power output of 450 Watts, then 6 oxen = 2,700 Watts, barely a fifth of a trireme's power train (for the relative power output of man and ox: Hicks 1997). A trireme had about 170 oarsmen, but a quadrireme about 220 and a quinquireme 300 (hence 16,500 and 22,500 Watts). Though men and animals can triple their power output in short bursts (thus tripling all Wattage figures, e.g. Landels 2000: 166 estimates a trireme's ramming power at over 30,000 Watts), this makes no relative difference. I use cruising power for ease of comparison, though notably even at *maximum* power six oxen cannot equal even a trireme's *cruising* power. For the oarage of triremes and polyremes see Meijer 1986: 36–37, 119–20 and Foley & Soedel 1981: 154–57. Even larger warships were deployed in the Greco-Roman period, though with less frequency, cf. Meijer 1986: 115–46 and Foley & Soedel 1981.

984. E.A. Thompson 1952: 54.

985. Needham 1971: 4.2.413–35 (cf. also 4.3.476 and 4.3.688–89, with 4.3.681 note b). The most common dragon boat had four wheels powered by 42 men with an estimated ramming speed of 4 knots (half that of a trireme). Smaller two-wheel models were powered by 28 men. Larger 11-wheel versions were known, but the largest (and rarest) had 23 wheels powered by more than 200 men. Though at a length of 360 feet this monster was so large it could never have outrun or outmaneuvered a trireme, it would certainly destroy any it managed to hit. Needham also says dragon boats were not seaworthy and thus were deployed on inland waters (effective small-river deployment may have been one advantage over a trireme) and, though used to ram, were mainly deployed as platforms for artillery. Needham's estimates of dragon boat tonnage are implausible so I won't consider them, but regardless of their displacement their power output in Watts would be 2,100 (small), 3,150 (medium), 7,500 (large), and 15,000 (giant), so a 6-oxen

Another example of the way things were going can be seen by comparing the work of Galen and Dionysius the Great, two 3rd century 'natural philosophers' who both sought to defend biological creationism.[986] The Christian Dionysius flourished only a generation or two after Galen, yet his (partly) extant *On Nature* defends the theory of intelligent design entirely by armchair reasoning, making no contribution to the relevant sciences, nor exhibiting any profound grasp of what had already been achieved in them. When Galen, however, sought to defend the very same thesis in his book *On the Uses of the Parts*, he engaged in extensive and meticulously thorough empirical research, which contributed to advances in scientific knowledge, and produced the most brilliant and thorough textbook on human anatomy ever attempted until the Renaissance. In other words, Galen took the label "Creation Science" seriously, and actually did real science.[987]

Both Galen and Dionysius were trying to refute the same people: philosophers and scientists who argued that no God had designed the world or the things in it (like the human body). And both were using the same tactic: presenting evidence of intelligent design in nature and arguing that there is no other plausible explanation for it. But there the similarities end. Unlike Dionysius, Galen did not sit in the armchair and just prattle on about how things appeared to be designed. Galen sought to directly refute the hypotheses of more atheistic scientists, often very specifically, through detailed anatomical investigation, demonstrating the immense complexity of human organs and parts, and their functions, often presenting carefully constructed arguments from irreducible complexity. The result was an extraordinary work in anatomical science and physiology which was, until Darwin produced the first effective challenge, the most decisive defense of intelligent design ever written. Galen thus exhibited a commitment to the most essential of scientific values: curiosity, empiricism, and the

---

paddleboat would have barely outperformed a small dragonboat (see previous note).Another serious defect of ox power in combat is that it would be impossible to reverse without a complex variable transmission, a problem clearly unimagined by the medieval ox-boat fantasist.

986. For background see *ODCC* 484 (s.v."Dionysius (4) the Great, St.").

987. On 'intelligent design' driving scientific research in Galen's work see Hankinson 1988. Galen was agnostic about whether there had been a singular moment of creation in the past, but he fervently believed the Mind of God had arranged the universe and its contents intelligently.

advancement of knowledge. Dionysius, not so much.

The difference between Galen and Dionysius parallels the difference between ancient and medieval intellectual society as a whole. But it was not a difference between pious religion and godless science, but between a religious attitude that was indifferent or even hostile to scientific values, and a religious attitude that fully embraced them. A telling example is the fate of Archimedes' treatise *On the Method of Mechanical Theorems*. This was specifically written on how to discover new theorems in mathematics using the principles learned from mechanical apparatuses, embodying all at once the ancient passion for curiosity, progress, and empiricism. In the middle ages no one read or copied this much anymore. Instead, a Christian scribe scraped the ink off one of the last surviving copies so he could recycle the papyrus for a hymnal, which was written over it, embodying all at once the medieval disinterest in those same three scientific values, their replacement with an evidently greater passion for singing praises to God, and the most paradigmatic symbol of a degenerate society: desperately cannibalizing and thus destroying the achievements of the past rather than preserving them.[988] Apparently, new material to write a hymnal on had become so scarce or inaccessible it was necessary to erase an existing book instead, in much the same way that ancient buildings were cannibalized to build peasant walls, and ancient bronze statues, machinery, and equipment were melted down for mundane use, rather than repaired or replaced.

Other ancient books on how to make new scientific discoveries similarly vanished, such as Galen's books *On Demonstration*, *On Dissection*, *On Vivisection*, even *On the Therapeutic Method*, largely devoted to the question of how to make new discoveries in medical science, which was barely preserved only in Arabic.[989] There were no doubt many works like these in other sciences, which are not even known to us now, so thoroughly were they erased from history, as were so many works in ancient science, far more having been discarded than were saved, and even those that survived often

988. On such palimpsests see note in section 3.4.

989. Galen, *On the Therapeutic Method* 1.4.1–3 explains its relevance to making scientific progress. This is similar to Galen's *On Conducting Anatomical Investigations*, which is entirely devoted to the question of how to advance the field of anatomy: only half of this survived in Greek, the other half in Arabic (*On Dissection* also appears to have survived in Arabic, cf. Ormos 1993). And no one appears to have followed its advice.

remained rare and obscure, and little read.[990] Hence medieval intellectuals and institutions were barely interested in preserving scientific knowledge, much less advancing it.

A similar decline can be traced in art, reflecting the same shift in values. And once again, this was a *real* decline, as archaeology abundantly confirms. Society retreated from the disciplined skill and curiosity entailed by the realistic and observational art of the Classical, Hellenistic, and Roman eras, into increasingly simplistic forms of art, which replaced accuracy and skill with symbolism and minimal representation. This development *also* arose, probably not coincidentally, on the other side of the 3rd century.[991] The decline in science and art thus occurred at exactly the same time, and appear to track the same change of mind: a loss of interest in observing and accurately describing nature.[992] It appears that society retreated from reality into mysticism in both science *and* art.

3.9 Ancient Recognition of Scientific Progress

The differences in attitude between pagan antiquity and early Christianity I have touched on so far are also reflected in the frequency with which ancient scientists, and even many laypeople, recognized the reality and value of scientific progress. There are no such statements among Christians of the same period. After examining modern discussions of this subject, I will survey what ancient authors actually said.

990. Sections 3.1 through 3.5 document countless examples of known losses, which entail many more unknown losses, since it is usually only by accident that we know even of lost titles. In respect to technology the same is argued in Oleson 2004 and Greene 1992.

991. A transformation documented in Elsner 1995. Paralleling science and the visual arts, P. Harrison 1998 establishes a similar shift in the literary arts (including biblical interpretation), moving away from realism (e.g. literalism and naturalism) in the 4th century, and back again in the Renaissance (thus contributing to the restoration of scientific progress).

992. See, for example, Russo 2003: 59–60, 224–27. Other indicators of decline in the medieval period are surveyed in J. Russell 1987 and Mazzini 2012 (and on technological and economic decline, see discussion and references in section 3.6.IV above).

I. THE MODERN DEBATE

Though some say there was no optimistic idea of scientific progress in antiquity, such a notion has already been soundly refuted.[993] One scholar even claimed "it was far from the thought of classical scientists to speak of their publications as 'contributions' to science" and then for proof cited only "exceptions" to his alleged rule, which turns induction from the particular to the general quite on its head.[994] Getting it the right way around, Ludwig Edelstein found enough evidence of *The Idea of Progress in Classical Antiquity* to fill a whole book. He did not even finish his planned survey of evidence for the Roman period, yet he extensively documents a widespread belief in various kinds of progress all the way from Classical Athens to the Roman empire.[995] Some years later a short study by E.R. Dodds challenged some of Edelstein's conclusions regarding belief in *social* progress, but even on that limited question Dodds did not come anywhere near examining the full scope and depth of Edelstein's evidence, and in any comparison between them, Dodds fares the worse.[996]

Edelstein found that previous scholars could only deny the ancients believed in progress by "citing the Roman testimony out of context or with neglect of contrary assertions by the same authors" and by having "overlooked certain statements that unambiguously testify to the Greek belief in future advance," whereas "if one collects the material in a more systematic way" and "does not rely on passages selected at random and discussed again and again," then "it becomes apparent that there is abundant and unimpeachable evidence for ancient progressivism."[997] He finds examples among actual

993. Claiming none: Zilsel 1945; Stark 2003: 153. Refuting with examples: Edelstein 1952: 575–76 and 1963: 19–22, and esp. Edelstein 1967; Cuomo 2000: 96; Russo 2003: 211–12.

994. Zilsel 1945: 327.

995. Edelstein 1967. According to the foreword by Jack Goellner (ibid.: vii), Edelstein only completed four of eight planned chapters before his death, bringing his analysis up to 30 B.C. (quite short of his goal of 500 A.D.), though in fact citing evidence up to 80 A.D.

996. Dodds 1973. Dodds did not claim to 'refute' Edelstein as Oleson 1984: 401 implies, but in fact often confirms Edelstein's findings, especially for the idea of scientific progress.

997. Edelstein 1967: xiv-xv.

scientists, in the philosophies of Plato and Aristotle, and from Roman philosophers of all the leading schools.[998] In fact, in science and technology, Edelstein concluded that "progressivism was a living force that could not easily be resisted by anyone who saw the world as it was."[999] At most, Dodds brings greater attention to a parallel thread of pessimism about mankind's future betterment, but the very same sentiment can be found today, still at war with a more enthusiastic futurism. In fact, Dodds concedes the two are probably inseparable in all eras, as progress entails change, and change always elicits anxiety and resistance from some, even sometimes the very same people who admire its prospects.

There are, nevertheless, many different kinds of progress, and one's awareness of, or attitude toward, any particular one can differ from any other.[1000] Antoinette Novara surveys Latin literature of the late Republic and early Empire for ideas of moral, artistic, and political progress, and finds some pessimists, but concludes that on balance there was a belief in both progress and its positive value in all three domains.[1001] Unfortunately she barely addresses scientific and technological progress.

Edgar Zilsel, on the other hand, tried to restrict his examination to what he called "the ideal of scientific progress," which he defined as the following combination of ideas:

> (1) [T]he insight that scientific knowledge is brought about step by step through contributions of generations of explorers building upon and gradually amending the findings of their predecessors; (2) the belief that this process is never completed; [and] (3) the conviction that contribution to this development, either for its own sake or for the public benefit, constitutes the very aim of the true scientist.[1002]

---

998. Edelstein 1967: 142–48 (scientists), 101–18 (Plato), 19–29 (Aristotle), 158–80 (Romans).

999. Edelstein 1967: 176. That any keen observer of the time would have seen abundant progress in these areas is demonstrated in the first half of the present chapter.

1000. For discussion of conceptual issues related to identifying different ideas of progress in various periods of history see Podlecki 2005: 16–27; G. Campbell 2006: 39–60, 164–66; and Mazlish et al. 2006 (with Carrier 2007).

1001. Novara 1982.

1002. Zilsel 1945: 326. Like most work of such an early date, Zilsel's understanding

As we have seen, or soon will see, many Roman intellectuals embraced all three, despite Zilsel's assumption to the contrary. But this relates only to scientific progress. Georg Henrik von Wright argues the very different thesis that "the Great Idea of Progress never dawned upon the Ancients" and hence "is no part of our Greco-Roman legacy," by which he means a much broader ideal of social progress, "according to [which] the road to the future is a progressive, unending improvement of the human condition, in spite of occasional and temporary set-backs."[1003] I agree this is probably too ambitious an ideal to be found in antiquity, and is more likely a consequence of the Scientific and Industrial Revolutions.[1004]

According to von Wright, this "Great Idea of Progress" is the idea that "progress in science and technology" has "an instrumental role in promoting" two other kinds of progress, "improvement of the material well-being of individuals and societies" and the "moral perfection" of humanity. Though scientific, moral, and material betterment all had their advocates in antiquity, he suggests the linking of them was an idea born in the Renaissance, then "crystallizing" in the Enlightenment, and that may be, though such a development was gradual and complex, and its origins bound up with ongoing revolutions in science and industry, making questions of causes or chronology difficult to evaluate.[1005] I will set that question aside as beyond the scope of the present work. My concern is only with the first of von Wright's triplex: scientific progress (and, to a lesser extent, technological progress).

As Walter Burkert observes, after Aristotle "there is unquestionable progress" in science, technology, and even "in the organization of mass society," a fact that certainly came to the attention of some.[1006] Though

---

of ancient science and technology is obsolete, and effectively refuted by the rest of the present chapter.

1003. Quoting von Wright 1997: 2–3.

1004. As suggested in section 3.6.VI.

1005. See von Wright 1997. The same argument is concisely made in P. Harrison 2007, while Pot 1985 provides an extensive survey of theories of how and why this link was eventually made, as well as positive and negative responses to technological progress in modern times (and a wide variety of modern ideas of progress through technology).

1006. Burkert 1997: 38 (cf. 38–43).

Burkert thought "awareness of these manifestations of progress is limited or even lacking" in the sources, or at least "was restricted to a few specialists," and though A.C. Crombie thought Roman "writers on science and technology" did "expect advances" but "belief in general progress was not characteristic of the ancient world," Edelstein found to the contrary "that the progressivists were not a negligible group of isolated thinkers out of touch with their own world but were the representatives of a movement that lasted almost from the beginning to the end of antiquity."[1007] In fact, there are so many enormous gaps in the source record that arguments from silence to the contrary are hardly weighty even when they can be proposed, and even with that limitation Edelstein's evidence is far more pervasive than either Berkert's or Crombie's.

Though there is now a general agreement that some progress was sufficiently evident in antiquity to be noticed and even remarked upon, long ago J.B. Bury claimed "there had been no impressive series of new discoveries suggesting either an indefinite increase of knowledge or a growing mastery of the forces of nature." Yet by simply removing that subjective word "impressive," Bury's statement becomes obviously false.[1008] Perhaps a suitably rapid burst of new discoveries was necessary to spark a scientific revolution, or perhaps all it does is cause a slow process to evolve more quickly, but such an amazing phenomenon is hardly necessary for observant intellectuals to notice past scientific and technological progress, or to believe there can and should be more, or then produce it. As Edelstein concludes, "in antiquity, science advanced far enough and new discoveries were numerous enough to permit belief in future progress."[1009]

II. The Ancient Evidence

Just as Edelstein observed, there is a clear and consistent thread of belief and support for scientific progress in ancient literature throughout our period of interest. The fact that this sentiment continues to be found in many authors, century after century, without apology or complaint, implies it was a popular view among the elite in the early Roman empire. To demonstrate this I will

1007. Burkert 1997: 41; Crombie 1997: 53; Edelstein 1967: xxxiii.

1008. Quoted in Edelstein 1967: xvii.

1009. Edelstein 1967: xviii (against the idea that fatalism or pessimism ruled the day: xxiii-xxvii). See also Zhmud 2006: 16–22, 77–81, 210–13.

survey many relevant passages in chronological order.

All Romans who wrote on technology were aware of new inventions in their own lifetimes that had become commonly employed, and were aware that development had occurred over time in several technological fields. Hence all were aware of ongoing technological progress.[1010] That this was more widely noticed is evidenced in Tertullian, who remarks:

> Surely it is obvious enough, if one looks at the whole world, that it is becoming daily better cultivated and more fully peopled than anciently. All places are now accessible, all are well known, all open to commerce. The pleasantest farms have obliterated all traces of what were once dreary and dangerous wastes. Cultivated fields have subdued forests. Flocks and herds have expelled wild beasts. Sandy deserts are sown. Rocklands are planted. Marshes are drained. And where once were hardly solitary cottages, there are now large cities. No longer are islands dreaded, nor their rocky shores feared. Everywhere are houses, and inhabitants, and settled government, and civilized life.[1011]

Though Tertullian does not entirely approve of this, as he proceeds to argue that all it has done is overpopulate the world, which he says God will surely cure with "pestilence, famine, wars, and earthquakes," the fact that he regarded signs of material progress as *obvious*, and still ongoing, through advances in agriculture, industry, commerce, urbanization, navigation, and hydrological engineering, indicates that there was a widespread recognition of this fact. And this is but an example. In general, as John Peter Oleson says:

> Most historians of ancient technology now recognize that both the craftsmen and the elite of Greek and Roman society were aware of the benefits of technological innovation, and that what we would call "progress" took place in many technologies even during the Roman empire.[1012]

1010. The relevant works of Vitruvius, Hero, Seneca, and Pliny the Elder, for instance, are filled with examples (such as the many references to them already in section 3.6). The prospect of new technologies had been voiced at least as early as Theophrastus (cf. Theophrastus, *On Stones* 60, with Caley & Richards 1956: 198–204) and had been identified as a valuable result of experimentation and theoretical knowledge as early as Philo (cf. Philo of Byzantium, *Siegecraft* 4.1–5, 4.19 = Marsden 1971: 107–09, 121–23 = DeVoto 1996: 5–9, 28–29).

1011. Tertullian, *On the Soul* 30.

1012. Oleson 2004: 65.

Nevertheless, as I've noted before, science and technology were different affairs. Hence I will now set aside the matter of technological progress as peripheral to my real interest: science.

Ideals of scientific progress began in Classical Greece, and are most clearly evidenced in Aristotle, our usual starting point in the history of science, whose ideas continued to influence Roman authors. But even when Thucydides remarked that "in politics as in any skill it is always necessary to keep up with new developments," he was clearly aware of ongoing progress in the arts and sciences, and assumed others were as well.[1013] Plato outright said his contemporaries knew significant progress was being made in the arts and sciences, adding only that it was not too openly praised in order to avoid causing envy among the living and anger among the dead.[1014] Thus it is no surprise to find Aristotle on board with the idea. In fact, Leonid Zhmud argues that Aristotle and his students recorded the history of sciences and philosophical doctrines specifically to demonstrate the reality and possibility of progress in knowledge, especially through the application of Aristotelian methods.[1015] Aristotle himself said he had witnessed in his own lifetime enormous and rapid progress in science and mathematics.[1016] But above all, the entire Aristotelian ethic elevated the scientific values of curiosity, progress, and empiricism. His *Metaphysics* begins with the declaration that "all men naturally desire knowledge," and proceeds to argue that it is a human's ability to advance their knowledge of the arts and sciences through reason and empirical observation that sets them above the animals—and other humans.[1017]

Aristotle was certainly a fan of scientific progress. He says he approves of what has "happened in regard to rhetoric and to practically all the other arts: those who invented them at first made progress in them only a little, but the renowned contributors today are, in a sense, the heirs of a long succession of men who each advanced the arts little by little," and therefore,

---

1013. Thucydides, *Peloponnesian War* 1.71.2. Similarly in Polybius, *Histories* 10.47.12 (in the context of a discussion of the value of progress in the development of telegraphy: 10.43–47).

1014. Plato, *Hippias Major* 281d-282b.

1015. Zhmud 2003 and 2006.

1016. In a lost treatise quoted by Iamblichus and Proclus: cf. Burkert 1997: 32–33.

1017. Aristotle, *Metaphysics* 1.1.980a-982a.

he says, in a much newer art like dialectic his peers should expect it to be only a little advanced by now and in need of much future improvement.[1018] Aristotle adds elsewhere that recent progress in medicine and other arts and sciences is so obvious that it justifies pursuing more, for "in general all men really seek what is good, not what was customary with their forefathers."[1019] Hence on another occasion he says:

> It appears to be in the reach of anyone to move further on and to improve on what is well outlined, and in such efforts time is the discoverer or at least a good helper. In such a way progress has been made in the arts, and it is within the reach of anyone to add what is still missing.[1020]

Aristotle also says one must endeavor to correct the errors of past experts, improve on their work, and study and pass on what they got right.[1021] And though "no one is able to attain the truth completely, we do not collectively fail," for "everyone says something true about the nature of things, and while individually we contribute little or nothing to the truth, by the union of all a considerable amount is amassed."[1022] Thus, for example, when discussing the question of why the heavens rotate in the particular direction they do, Aristotle remarks that "when anyone shall succeed in finding proofs of greater precision, gratitude will be due to them for the discovery, but at present we must be content with a probable solution," hence expressing an expectation of the value and possibility of scientific progress that he echoes several times in his works.[1023]

Aristotle's awareness and appreciation for progress continued into the Roman period. Several authors in the 1st century B.C. hint at the fact. We see this even in the Epicurean Lucretius, who cited it as a respectable reason

---

1018. Aristotle, *Sophistical Refutations* 3.34.183b-184b.

1019. Aristotle, *Politics* 2.8.1268b-1269a.

1020. Aristotle, *Nicomachean Ethics* 1.7.1098a.

1021. Aristotle, *Metaphysics* 13.1.1076a.

1022. Aristotle, *Metaphysics* 2.1.993a-b.

1023. Aristotle, *On the Heavens* 2.5.287b-288a. For another example (in apiology) see section 3.1. For these and other examples of Aristotle's enthusiasm for progress see Edelstein 1967: 19–29. On passages suggesting the contrary only when taken out of context see section 3.7.IV.

to adopt new philosophies that are superior to those that came before, noting as obvious that:

> Even now some arts are being improved, even now some are developing. Today many improvements are being made to ships, and only recently organists have devised musical tunes. Even [the Epicurean] order and theory of the world was discovered in recent times, and I am the first to describe it [in Latin].[1024]

Around the same time, or a little later, though not specifically referencing science, even the conservative Diodorus is aware of the value of progress in knowledge, and confident it will continue. He argues eloquently and at length that historians do a service to humanity by extending the collective memory far beyond what the elders of any community can remember, which greatly increases our knowledge of past successes and failures, which betters society by improving political decision-making.[1025] In other words, the general or statesman of today has more information at his command than his predecessors, and so will make better decisions—if he pays attention to historians who have done their job.

It is clear this would entail an unending process of improvement, which can only go wrong if generals and statesmen ignore history, or if historians shirk their responsibilities. Diodorus also argues that history benefits society by encouraging innovation and accomplishment. Because history is the only real source of immortality, he argues, good men strive to outdo each other in achievements, so they can win the attention of historians and thus be immortalized (while, conversely, villains will be immortally punished by historians). Here Diodorus specifically recognizes scientific advances as among the achievements that dutiful historians encourage: "In hopes of having the memory of their good deeds recorded by historians," he says, "some have sought fame by building cities, some have produced laws

---

1024. Lucretius, *On the Nature of Things* 5.332–37 (cf. 5.1448–57). Various aspects of the Lucretian attitude toward progress are discussed in Novara 1982: 313–84. (Lucretius was not actually the first to describe Epicurean philosophy in Latin: see note in chapter 1.2.1.)

1025. Diodorus Siculus, *Historical Library* 1.1–1.2. For his more conservative side see example in section 3.8.1, but his generally positive view of cultural progress is argued in Sacks 1990.

defending the common good, and many strive to discover arts and sciences for the good of the human race."[1026]

Hence, for example, Diodorus says the practical technologies scientifically invented by Archimedes are worthy of universal admiration and praise, for "one rightly marvels at the inventiveness of this craftsman, not only" for his waterscrew, which Diodorus calls "exceptionally brilliant," specifically because it saves a tremendous amount of labor, "but also for many other even greater" inventions than the waterscrew, "which are celebrated throughout the inhabited world."[1027] As Polybius had already said of the same man a century before, "So true it is that the genius of one man can become an immense, almost miraculous asset, if it is properly applied to certain problems."[1028] That this was practically an aphorism in antiquity is suggested by Philo of Byzantium, who declares, not of Archimedes but as a universal principle, that "to have an original idea and put it into practice is the work of a superior genius," but for later men to improve upon what has already been invented is even easier.[1029] Respect for invention entails respect for progress.

Just as Diodorus believed progress could be made through history, Strabo argued a generation later that progress could also be made in and through geography as a scientific field. In his introduction to the subject, Strabo recognizes and describes scientific progress in geography, and offers the need to make further progress as his reason for writing, in order to make

---

1026. Diodorus Siculus, *Historical Library* 1.2.1, using *epistêmas kai technas*, "[systematic] knowledge and [formal] skills."

1027. Diodorus Siculus, *Historical Library* 5.37.4, using *technitês*, "craftsman," and *philotechnon kath' hyperbolên*, "exceptionally brilliant." Dalley & Oleson 2003: 19–22 suggests the possibility that 'craftsman' refers to an unknown inventor of the waterscrew, which Archimedes only 'discovered' already being used in Egypt (and he was only the first to analyze it mathematically in terms of basic physics), but Oleson is rightly skeptical of this reading (as am I) since the context is certainly the mechanical genius of Archimedes and not his luck or his books.

1028. Polybius, *Histories* 8.7 (referring to Archimedes' mechanical defenses of Syracuse). See further examples of the valorization of Archimedes in chapter 4.2, 4.5, and 4.6.1.

1029. Philo of Byzantium, *Siegecraft* 4.19 (= Marsden 1971: 121–23 = DeVoto 1996: 28–29). Philo's attitude toward progress is discussed in Cuomo 2007: 51–52.

additions and improvements, as well as correct past errors.[1030] And he says more still remains to be done. Hence he was fully conscious of the reality and value of scientific progress, and how it is accomplished: through the accumulated efforts of experts who build on and correct each other over time. But it is Cicero, Strabo's predecessor and Diodorus' contemporary, who first finds occasion to discuss scientific progress in a broader sense, and since Cicero was held in considerable esteem by Roman intellectuals of subsequent centuries, his views on the matter had a definite prospect of being influential, even regarded as patriotically Roman.

Cicero recognized that the same concepts could be applied across the whole range of arts and sciences, since "in every field, continued observation over a long time brings incredible knowledge," and though early natural philosophers got a lot wrong and found discoveries difficult to make, even "if those old thinkers found themselves floundering like babies just born in a new world, do we imagine that all subsequent generations and their consummate intellects and elaborate investigations have not succeeded in making anything clearer?"[1031] Just as astronomers can predict the movements of the planets, moon, and sun, "the same thing may be said of men who, for a long period of time, have studied and noted the course of facts and the connection of events," and yet, Cicero argues, such progress has never and will never arise from revealed knowledge or divine inspiration, but only from laborious and extensive observation.[1032] Cicero even criticizes Aristotle for thinking progress would ever end:

> Aristotle upbraids the philosophers of old for thinking, according to him, that thanks to their genius philosophy had reached perfection, and says they were guilty of extreme folly or boastfulness. And yet even he adds that he saw that, as a consequence of the great advances made, in a short time philosophy would be absolutely complete.[1033]

Though we have no evidence Aristotle actually said what Cicero

1030. Strabo, *Geography* 1.2.1.

1031. Cicero, *On Divination* 1.49.109 and *Prior Academics* (= *Lucullus*) 2.5.14.

1032. Cicero, *On Divination* 1.49.111 (quote; cf. also 1.56.128 and 1.57.131) and 2.3.9–2.5.12 (that observation, not revelation, will lead to progress in knowledge).

1033. Cicero, *Tusculan Disputations* 3.28.69: *brevi tempore philosophiam plane absolutam fore.*

alleges, the notable point is that Cicero thought it silly for anyone to think philosophical progress would soon end, just as we now criticize optimists of the late 19th century who thought physics would soon be completed.

Cicero was also a fan of technological progress, at least when it served the public good.[1034] For him, just as "all are great men" who study the stars and solve the mysteries of planetary motion, so are all who invent new skills and technologies useful to humanity.[1035] But it is his younger contemporary, the engineer Vitruvius, who more directly combines an awareness of scientific and technological progress, understanding that it is through scientific principles that technology can be most directly improved. In fact, in both science and technology, Rowland and Howe find that "in general" Vitruvius "tends to favor innovation," sees "the value of innovative progress," and "is aware of the importance of experimental method and direct observation in the cumulative growth of science."[1036]

Hence in his introduction to machinery Vitruvius argues that observant men had "made some things more convenient with machines, and others with instruments," and "what they found useful in practice they took care to improve, step by step, with the help of research, craftsmanship, and established principles," which he then demonstrates with several examples of how machinery has improved human life.[1037] He also articulates a broad vision of scientific and technical progress with an imaginative analysis of the function of technological (and even scientific) progress in the advancement of human life throughout history.[1038] Later he adds an appreciative account of recently accumulated scientific and technical knowledge, in which we can hear an echo of the same ideals voiced by Diodorus:

1034. Argued in Novara 1982: 257–70.

1035. Cicero, *Tusculan Disputations* 1.25.61–1.26.65.

1036. Rowland & Howe 1999: 16–17.

1037. Vitruvius, *On Architecture* 10.1.4: *ut essent expeditiora, alia machinis...nonnulla organis, et ita quae animadverterunt ad usum utilia esse studiis, artibus, institutis, gradatim augenda doctrinis curaverunt.* He follows with examples in 10.1.5–6.

1038. Vitruvius, *On Architecture* 2.1.1–9. A similar recognition of the arts and sciences as a cumulative and beneficial process of discovery on which civilization was founded is argued even by the Platonist philosopher Maximus of Tyre in *Orations* 6.2, and is a notion repeated by other ancient authors (see Lovejoy & Boas 1935 and G. Campbell 2006: 39–60, 164–66).

> Our ancestors, not only wisely but also usefully, established the practice of transmitting their ideas to posterity through the reports of treatises, so that these ideas would not perish, but instead, grow with each passing age, so through publishing books they could arrive, step by step, at the highest refinement of learning. Thus it is not moderate but infinite thanks that should be given those who did not jealously let their ideas pass in silence, but rather took care to hand on to memory their thoughts of every kind, preserved in their writings.[1039]

Vitruvius goes on to list examples in natural and moral philosophy, and (like Diodorus) history. He later explains that he intends to add to this chain of progress with his own synthesis and ideas, and follows with an example of cumulative progress in the field of optics and scenography and its role in improving art and architecture.[1040]

Even Vitruvius' more inept contemporary Athenaeus announced that he had "taken personal pride in enlarging the resources of what is useful for building machines" because "one must not only know the fine inventions of others, but when someone has an agile mind, he must invent something himself."[1041] Though his own "inventions" were not entirely well conceived, it remains notable that he believed it was not only expected of an accomplished engineer to add to existing technology, but that such innovation was something of which one could be acceptably proud. That this was a general view is confirmed by Apollonius of Citium, probably of the same century, who criticized certain of his opponents for failing to do research that would lead to advances in medical treatment.[1042]

Thus in the middle of the next century, and regarding the science of pneumatics, which in antiquity was explicitly pursued to advance technologies, the engineer Hero said:

1039. Vitruvius, *On Architecture* 7.pr.1.

1040. Vitruvius, *On Architecture* 7.pr.10 (example: 7.pr.11–16).

1041. Athenaeus the Mechanic, *On War Machines* 31.12–32.2, following with an entire section devoted to innovation (see discussion in Whitehead & Blyth 2004: 36–39). This attitude is evident across the Roman engineering profession: Kolb 2015.

1042. Apollonius of Citium, *On Joints according to Hippocrates* 23–24 (cf. Potter 1993: 119). For more on this Apollonius see relevant note in section 3.2.

> Because research in the science of pneumatics was deemed worthy by both the philosophers and engineers of old (some having demonstrated its power theoretically, others through the action of its observable effects), we concluded it was necessary to arrange in order what had been handed down by our predecessors, and to add what we have discovered ourselves. For this will benefit those who want to delve into further studies.[1043]

He likewise said he had improved the designs of his predecessors in programmable robotics.[1044] And in his textbook on *Mechanics* he says "we have, in our opinion, proved more than those that came before us."[1045] Hence, like Athenaeus, Hero regarded innovation as something expected of every contributor to the field. In fact, the similarity of these ideas and expectations and assumptions, already apparent over time (and, as we shall see, even more apparent in subsequent generations) suggests we are observing a zeitgeist, and not just the isolated notions of unique individuals.

This faith in progress is certainly evident all throughout Hero's writings. In her analysis of his treatise on the history of artillery development, Serafina Cuomo demonstrates that he specifically uses this history to argue the benefits of making continual progress in science-based war technology, emphasizing the dangers of abandoning research in the field. He thus clearly expected progress to continue in weapons technology, and knew science had to play a part in that development. Moreover, he regarded this scientific pursuit of technological progress to be more useful to the pursuit of human happiness than any of the armchair debates of philosophers.[1046] In another treatise, Hero describes the importance of progress in geometry in a similar way, explaining that this had begun with land measurement, but "since this science was useful to men, it was advanced still more," from the study of areas to the study of volumes, "and since the first theorems invented were not sufficient, further research was needed, and to this day some of them remain incomplete," even though the greatest mathematicians had worked on them, and "since the research we mentioned is needed, we think it

1043. Hero of Alexandria, *Pneumatics* 1.pr.1.

1044. Hero of Alexandria, *On Constructing Automata* 2.20.1–5 (= Murphy 1995: 3–4). See discussion in Tybjerg 2005: 211 and Rausch 2012.

1045. Hero of Alexandria, *Mechanics* 3.1 (cf. Drachmann 1963: 94).

1046. Cuomo 2002. See related discussion in chapter 2.7.

worthwhile to collect as much useful material as was written before us, plus as much as we have examined ourselves."[1047] There can be no doubt that Hero believed in scientific progress, and assumed even his readers would agree it was worthwhile.

Several authors of the same century echo the same values and expectations. Columella laments how agriculture has lagged behind the other sciences, and says it is amazing "that the matter of the highest importance to our physical welfare and the needs of life should have made, even up to our own time, the least progress," despite the fact that it should be the most respectable means of increasing a family's wealth and inheritance.[1048] He actually thinks his peers have gotten too snobby about it and attacks those who think the study beneath them, praises hands-on labor and hard work, and believes there is a great deal of progress that could be made from empirical research in the field.[1049] Though his complaints are no doubt exaggerated and largely rhetorical, they certainly entail a more widespread value for scientific progress. For he would not otherwise have thought it a sufficiently chastising rebuke to claim that his fellow Romans had failed to achieve a more rapid progress in so important a subject. Likewise, his use of the mathematical sciences as a point of comparison suggests his peers already found scientific advancement respectable.[1050]

Medical progress was also a respectable example. Written earlier that same century, the entire proem to Celsus's *On Medicine* is a discourse on the reality and value of scientific progress in the field of medicine, from searching for new treatments to making advances in methodology. He later adds a similar survey of progress in surgical knowledge and technique.[1051] Celsus says some were even arguing that "it is not cruel, as some say, to vivisect a few condemned criminals to discover cures for innocent people of future ages," which is a rather bold elevation of the moral value of scientific progress—though clearly exceptional in degree, it is not exceptional in its direction. Celsus more sensibly agreed it was pointless and cruel to vivisect human

1047. Hero of Alexandria, *Metrics* 1.pr.1.

1048. Columella, *On Agricultural Matters* 1.pr.7.

1049. Columella, *On Agricultural Matters* 1.pr.10–21 and 1.pr.29.

1050. Columella, *On Agricultural Matters* 1.pr.3 and 1.pr.5.

1051. Celsus, *On Medicine* pr. and 7.pr.

beings, but not to dissect human cadavers, from which valuable knowledge and progress could result (as discussed in section 3.8.III).[1052] Similarly, the physician Dioscorides, also writing in the 1st century A.D., explains that he wrote on drugs because of the need to advance pharmacology with a new and more accurate treatment of the facts. Since he attacks other writers who rely on hearsay and books rather than actually observing and testing things themselves, or questioning informants carefully, he clearly understood how progress was to be made, and was in effect advertising how it should be made.[1053]

Such ideals are voiced most explicitly by Seneca in the early 60's A.D. Even though he was not a scientist, and otherwise not very fond of technological progress, he was immensely fond of scientific progress, which he saw as a quest for knowledge that was valuable and rewarding in itself. Seneca clearly recognizes the reality of scientific progress when he says "everything was new for those who first attempted to understand" natural phenomena, but "later their conclusions were refined" because in every subject "nothing is completed while it is beginning," in fact "the first research is always a long way from being complete," and therefore "even when a lot has been done, every generation will have something to do."[1054] Accordingly, he believed the whole of philosophy (and thus natural philosophy as well) could and should be advanced with every generation:

> The truth will never be discovered if we rest contented with discoveries already made. Besides, he who merely follows another not only discovers nothing but is not even investigating! What then? Shall I not follow in the footsteps of my predecessors? I shall indeed use the old road, but if I find one that makes a shorter path and is smoother to travel, I shall open the new road. Men who have made these discoveries before us are not our masters, but our guides. Truth lies open for all. It has not yet been monopolized. And there is plenty of it left even for posterity to discover.[1055]

Carrying this further, and echoing Columella's call for progress in

1052. Celsus, *On Medicine* pr.26 (cf. pr.40 for those who argue the contrary; and pr.74 for Celsus' own opinion on the matter).

1053. Dioscorides, *On Medical Materials* 1.pr.

1054. Seneca, *Natural Questions* 6.5.2–3.

1055. Seneca, *Moral Epistles* 33.10–11.

agricultural science as a means to increase the inheritance we pass to our descendants, Seneca calls for progress in the whole of philosophy for the same reason, though instead of a growing estate, it is a growing body of wisdom our heirs inherit:

> The very contemplation of wisdom takes much of my time. I gaze upon her with bewilderment, just as I sometimes gaze upon the heavens themselves, which I often behold as if I saw them for the first time. Hence I worship the discoveries of wisdom and their discoverers. To enter, as it were, into the inheritance of many predecessors is a delight. It was for me that they laid up this treasure. It was for me that they toiled. But we should play the part of a careful householder. We should increase what we have inherited. This inheritance shall pass from me to my descendants larger than before. Much still remains to do, and much will always remain, and he who shall be born a thousand ages hence will not be barred from his opportunity of adding something further. But even if the old masters had discovered everything, one thing would always be new: the application, study, and classification of the discoveries they made.[1056]

Thus, when urging his friend Lucilius to make a scientific expedition to mount Aetna and write up something on volcanology, Seneca adds:

> It matters a lot whether you approach a subject that has been exhausted, or one where the ground has merely been broken. In the latter case, the topic grows day by day, and what is already discovered does not hinder new discoveries.[1057]

Seneca thus asks Lucilius to investigate whether the legend of the whirlpool Charybdis is true, and if so what is actually going on there, and to climb Aetna and gather data on whether its elevation has changed as some lately had reported, and what the natural causes are of either this change or the appearance of it, requests that clearly imagine a continuing need for scientific research.

Seneca directly applies these generalizations to the science of astronomy, concluding, for example, that the debate between heliocentrism and geocentrism will only be resolved with future research:

1056. Seneca, *Moral Epistles* 64.4–8.

1057. Seneca, *Moral Epistles* 79.6. See discussion in chapter 4.4 of this expedition as well as the poem *Aetna*, which may have been the result of it.

> It will be relevant to investigate [the nature and orbits of comets] so that we may know whether the universe travels around while the earth stands still, or whether the earth turns while the universe stands still. For there have been some who say that we are the ones whom nature causes to move, even though we are unaware of it, and that rising and setting does not happen from the motion of the sky but we ourselves rise and set. This deserves study so we may know what our status is, whether we possess the most inactive abode or a very swift one, whether god causes all things to move around us, or causes us to move around.[1058]

Seneca also recognizes that we need a lot of accumulated data over several ages to develop a good theory of cometary orbits, then notes that this research is still young yet, so we need many more generations of it, which he fully expects and considers worthwhile.[1059] As he says even more generally:

> The time will come when diligent research over very long periods will bring to light things which now lie hidden. A single lifetime, even though entirely devoted to the sky, would not be enough for the investigation of so vast a subject, and yet we do not even divide our few years equally between study and vice! And so this knowledge will be unfolded only through long successive ages. There will come a time when our descendants will be amazed that we did not know things that are so plain to them.[1060]

Therefore:

> Some day there will be a man who will show in what regions comets have their orbit, why they travel so remote from other celestial bodies, how large they are and what sort they are. Let us be satisfied with what we have found out, and let our descendants also contribute something to the truth.[1061]

Seneca even expects that new planets, now too faint to see, will someday

1058. Seneca, *Natural Questions* 7.2.3.

1059. Seneca, *Natural Questions* 7.2–3, 7.25.3–7 and 7.30.2.

1060. Seneca, *Natural Questions* 7.25.4–5.

1061. Seneca, *Natural Questions* 7.25.7.

be discovered.[1062] Meanwhile, "we can only investigate such things and grope in the dark with hypotheses, not with the assurance of discovering the truth, and yet not without hope" that we, or someone, will.[1063]

Seneca extends this value for progress to all the sciences, using them as examples of how all of the them make progress over time:

> How many animals we have learned about for the first time in this age! How many are not known even now! Many things that are unknown to us, the people of a coming age will know. Many discoveries are reserved for ages still to come, when memory of us will have faded. Our universe is pathetic if it does not have something for every generation to investigate.[1064]

Hence in every scientific matter, Seneca says, it takes a long time to get all the facts and answer difficult questions about the natural world.[1065]

In fact, Seneca imagines scientific research as a religious enterprise, equating the discovery of nature's secrets to the sacred mysteries, in which initiates advance gradually by stages, learning a little each time. Just as in the holy mysteries, so in science, Seneca says, every generation will advance a little further than the last.[1066] Though he chastises his generation for being so consumed with vice that they are neglecting to advance the sciences as quickly as they ought, this only demonstrates how highly he placed scientific research in his order of values, and how much he expected his peers to agree.[1067] Hence Seneca concludes:

> There is no interest in philosophy. Accordingly, so little is found out from those subjects the ancients left partially investigated that many things which were discovered are being forgotten. But, by Hercules, if we applied ourselves to this with all our might, if the young seriously devoted themselves to it, if the elders taught it and the next generation learned it,

1062. Seneca, *Natural Questions* 7.30.3–4.

1063. Seneca, *Natural Questions* 7.29.3.

1064. Seneca, *Natural Questions* 7.30.5.

1065. Seneca, *Natural Questions* 7.30.2, 7.30.6, 7.31.1.

1066. Seneca, *Natural Questions* 7.30.6 (cf. also 3.pr.1). This equation of scientific research with the sacred mysteries was not unique: see chapter 4.2 and further discussion in 4.4.

1067. See discussion in section 3.8.

> we would scarcely get to the bottom where truth is located, which we now seek on the surface of the earth and with slack effort.[1068]

Though his claims of neglect were not entirely true, there can be no doubt that Seneca held scientific progress to be of great moral value, and believed many of his readers would agree.

Writing a decade or two later, Pliny the Elder was also in favor of scientific progress.[1069] In his encyclopedia of the natural world he says his aim is to survey everything written before him, and to "add a great number of other facts that were either ignored by our predecessors or have been discovered by subsequent experience," and yet he has "no doubt there are many things that have still escaped" us.[1070] Pliny later observes that he and many others agree that a great deal remains to be discovered in pharmacology, and he discusses some of the presumed hindrances to progress in this field with the clear intention of their being overcome once recognized.[1071] In astronomy he approves of the fact that Hipparchus had observed the appearance of a new star and thus began a project to chart the stars, developing instruments for this purpose and leaving records so future generations could check if any other stars vanish or appear, or move, or change their brightness—another clear conception of the value and possibility of scientific progress, and of taking steps to facilitate it.[1072] Likewise, Pliny says the 'modern' understanding of the planets "differs in many points from that of our predecessors," and though "credit must be given to those who first demonstrated the methods of investigating" the heavens, at the same time, "no one should lose hope that every generation makes progress" in this research.[1073] Pliny even proposes a 'new' causal theory for the motion of the inner planets that is far fetched, but

1068. Seneca, *Natural Questions* 7.32.4.

1069. For a broader discussion of Pliny's faith in progress see Beagon 1992: 56–63, 183–90 and related discussion in section 3.8.1.

1070. Pliny the Elder, *Natural History* pr.17–18.

1071. Pliny the Elder, *Natural History* 25.5.15–25.6.17.

1072. Pliny the Elder, *Natural History* 2.24.95. Pliny held a very high opinion of Hipparchus and his project (cf. also *Natural History* 2.112.247).

1073. Pliny the Elder, *Natural History* 2.13.62–63.

still notable as evidence of a recognized need for progress in the subject.[1074]

These ideals continued their influence in the 2nd century A.D., when we hear expressions of the same set of values and expectations from Ptolemy and Galen. In astronomy, though clearly imagining the same applying to every science, Ptolemy says:

> We constantly strive to increase a love of contemplating [the eternal truths of nature]...by studying those sciences which have already been mastered by those who approached them with a genuine spirit of enquiry, and by ourselves attempting to contribute as much advancement as has been made possible by the additional time between those people and ourselves.[1075]

Like Seneca, Ptolemy often refers to the fact that observations gathered over a long time help advance the sciences, especially astronomy, where he is explicitly aware of the fact that by combining the collected observations of previous astronomers with those of our own generation, we will arrive at much more accurate results, and so will those who come after us.[1076] Ptolemy also believed in the value of *methodological* progress, arguing that:

> Those who approach this science in a true spirit of enquiry and love of truth ought to use any new methods they discover, which give more accurate results, to correct not merely the old theories, but their own, too, if they need it. They should not think it disgraceful, when the goal they profess to pursue is so great and divine, even if their theories are corrected and made more accurate by others besides themselves.[1077]

Elsewhere, echoing what we heard from Hero, Ptolemy says he found past work on astrolabe construction useful but in need of improvement, and

1074. Pliny the Elder, *Natural History* 2.13.71–2.14.76. Accordingly, omens portending an age of "geniuses and learning" were regarded as auspicious (Pliny the Elder, *Natural History* 2.23.93).

1075. Ptolemy, *Almagest* 1.1 (= Toomer 1984: 37).

1076. Ptolemy, *Almagest* 13.11 (with 3.1, 7.1, and 7.3) = Toomer 1984: 647 (137, 321, 329).

1077. Ptolemy, *Almagest* 4.9 (= Toomer 1984: 206). Similarly, Frontinus, in *Stratagems* 1.pr. (a book serving the very purpose Diodorus had imagined), says he welcomes others improving his work with the addition of more stratagems and examples from history.

wrote his own treatise accordingly.[1078]

Likewise in his geographical work, Ptolemy says "the first step in any systematic scientific inquiry" is to collect all the scientific data accumulated by your predecessors, and then combine this with your own observations (which in the case of geography including surveying and astronomy).[1079] He then elaborates on the need for more and better scientific observations to improve on his own work, repeatedly recognizing the defects in the data available to him and their prospects for improvement, while discussing how his predecessor Marinus set the standard for exactly that, having corrected and revised even his own work many times, and having treated critically the data provided by scientists before him.[1080] Hence Ptolemy concludes that:

> In all subjects that have not reached a state of complete knowledge, whether because they are too vast, or because they do not always remain the same, the passage of time always makes far more accurate research possible, and such is the case with global cartography, too.[1081]

Not only is it clear how well Ptolemy understood the need for progress in science, and the means of achieving it, his sentiments are also voiced in a way that assume his readers would agree.[1082]

Finally, we come to Galen, who was an enthusiastic fan of Hippocrates, and of the Hippocratic belief that progress in medicine was both possible and necessary. In fact, our earliest evidence of a belief in scientific progress comes from Hippocrates, whose work remained immensely influential in the Roman era.[1083] Though the exact authorship of Hippocratic books was debated even then, they had all been collected as a generally respected database of examples and values in medical science. In one of the most

1078. Ptolemy, *Analemma* 1 (= Edwards 1984: 79).

1079. Ptolemy, *Geography* 1.2.

1080. Ptolemy, *Geography* 1.7–11 (Ptolemy's situation) and 1.4–6 (example set by Marinus).

1081. Ptolemy, *Geography* 1.5.

1082. For further discussion of Ptolemy's ideas of scientific progress see Cuomo 2001: 183–85.

1083. On Hippocrates' influence in the Roman period: Nutton 2013: 207–21. For sources on Hippocrates and the origins of Hippocratism see Appendix B.

influential of these, Hippocrates is made to declare that "medicine has had for a long time a method of discovery" that works quite well, and by following that method, "many excellent discoveries have been made over a long time, and the rest will be discovered, if someone competent who knows the discoveries already made will build on those by conducting his own research."[1084] This same treatise also says one should not speculate about hidden organs and functions (though other Hippocratic texts did so), and yet this skeptical concern eventually inspired the empirical investigation of those very things, through anatomy and vivisection, in an apparent effort to end the need for mere speculation.[1085] This agreed with the general Hippocratic opinion that "to discover something that has not been discovered yet, or to improve what has already been discovered, would seem to be the ambition and task of an intelligent mind."[1086]

Galen agreed. "We are more fortunate" than those who advanced the arts and sciences before us, Galen says, because "we can learn in a short time the useful discoveries that cost them much time, effort, and concern" to discover, while "if in the time that remains in our lives we practice the arts not as a diversion but with constant attention," and the proper method, then "there is nothing to prevent us from advancing beyond the men who came before us," a sentiment Galen clearly regarded as applying to all the sciences.[1087] Elsewhere he says that if medical scientists pursue their art as true philosophers, then "there is nothing to prevent us, not only from

---

1084. Hippocrates, *On Ancient Medicine* 2.

1085. On this point, alongside more expressions of the value of scientific progress, see Hippocrates, *On Ancient Medicine* 1, 2, 8, 12, 14, etc. Erasistratus thus justified anatomical research by appealing to the Hippocratic ideal of tireless inquiry leading to inevitable progress (cf. Nutton 2013: 136). In contrast, the Empiricist sect consisted of those who stuck to the original Hippocratic injunction against excessive theorizing (cf. Nutton 2013: 145).

1086. Hippocrates, *On the Art of Medicine* 1. Similarly, Diocles (the 'Second Hippocrates') argued that one should trust that new discoveries will be made over time, according to Diocles, *On Health to Pleistarchus*, as quoted in Galen, *On the Powers of Foods* 1.1 (= Kühn 6.456 = Mark Grant 2000: 69). On Diocles as the 'Second Hippocrates' see section 3.2.

1087. Galen, *On the Doctrines of Hippocrates and Plato* 9.1.23–26 (Galen later quotes Hippocrates as saying he had discovered things unknown to his elders: 9.6.49).

reaching a similar attainment" as, for example, Hippocrates, "but even from becoming better than him, for it is open to us to learn everything which he gave us a good account of, and then to find out the rest for ourselves."[1088]

Galen even considered it the moral duty of medical scientists not to neglect or forget what has been methodically discovered before them, and also to compete with each other "in practicing, perpetually increasing, and attempting to complete the science" of medicine.[1089] As we saw earlier (in section 3.8.I), Galen often expressed this moral value by criticizing his peers. Providing the most direct example, Galen said:

> The fact that we were born later than the ancients, and have inherited from them arts which they developed to such a high degree, should have been a considerable advantage. It would be easy, for example, to learn thoroughly in a very few years what Hippocrates discovered over a very long period of time, and then to devote the rest of one's life to the discovery of what remains. But it is impossible for someone who puts wealth before virtue, and studies the art for the sake of personal gain rather than public benefit, to have the art itself as his goal. It is impossible to pursue financial gain at the same time as training oneself in so great an art. Someone who is really enthusiastic about one of these aims will inevitably despise the other.[1090]

Hence while Christians preached the aphorism that "you cannot serve both god and mammon," Galen had very much the same idea, only putting the advancement of science in the place of God.[1091] Galen goes on to explain that in his opinion too many doctors were chasing after money and not going out and testing theories and gaining useful experience, such as by treating the poor and traveling to many foreign environments. One can hear similar criticisms of doctors today.

Further emphasizing the value of empirical methods in making progress, Galen says "Hippocrates discovered a lot, but those who followed after have not discovered less, and one finds up to the present day that some

---

1088. Galen, *That the Best Doctor Is Also a Philosopher* 4 (= Kühn 1.63).

1089. Galen, *On the Therapeutic Method* 1.1.7 (quote), 2.6.5 (duty).

1090. Galen, *That the Best Doctor Is Also a Philosopher* 2 (= Kühn 1.57).

1091. For the Christian doctrine, almost identical to Galen's, see Matthew 6:24 and Luke 16:13 ("mammon" was an Aramaic word for wealth, in any form, e.g. money and valuables).

things have already been discovered, and other things it is hoped to discover later," hence we need "the empirical method" in order to gradually discover "what has not yet been discovered in the past."[1092] Likewise, echoing the ideas of both Diodorus and Seneca, Galen says "the empiricist makes use of historical data," which "we need to do because of the vastness of the science, since one man's life will not suffice to find out everything," and so "we accumulate these data and collect them from all sources, turning to the books of our predecessors," but "we cannot just simply believe what has been written down by our predecessors" but must test it all.[1093] Hence as Galen says elsewhere, any good doctor must "learn thoroughly all that has been said by the most illustrious of the ancients" and "when he has learnt this, then for a prolonged period he must test and prove it, observing what part of it is in agreement, and what in disagreement with obvious facts," and thus he will progress in knowledge and advance the field.[1094] Drawing on the methods of the Empiricist sect, Galen clearly agreed with these ideals, adding to them his own respect for the discovery of physiological theories.

Galen repeated these ideals more specifically in the field of anatomy. He noted that even in his own lifetime dissection had greatly advanced knowledge of anatomy, and yet "even those who have devoted much time to anatomy have been unable to bring it to perfection," and therefore much remained to be done.[1095] Hence Galen explained that in his own work he sought to correct past errors in anatomical knowledge, to discover and publish new useful facts, to make advances in theory thereby, and especially to criticize conclusions reached from the armchair by making factual, empirical observations instead, and by this means test, confirm, or

---

1092. Galen, *On Medical Experience* 10. This is all couched in a convoluted contrafactual rhetorical question, the point of which is that this is Galen's view, and not that of certain of his opponents.

1093. Galen, *An Outline of Empiricism* 7.

1094. Galen, *On the Natural Faculties* 3.10 (= Kühn 2.179–80). Galen even wrote a treatise *On Disagreement in Anatomy* (now lost), explaining why it is that some anatomists disagree with each other in their reported findings (cf. ibid. 3.11 = Kühn 2.182).

1095. Galen, *On Conducting Anatomical Investigations* 1.3 and 2.3 (= Kühn 2.234, 2.289).

refute what had been claimed.[1096] Galen even says the more dissections he performs the more he discovers, especially by discovering things he had missed before, and he uses this as the central point in an argument that continual anatomical research is necessary for medical science generally.[1097] Galen elsewhere says one should never hesitate to revise medical knowledge in light of new evidence, and he extends this expectation even to the mathematical sciences.[1098]

From all this evidence and more, R.J. Hankinson concludes that despite his conservative nature, "Galen certainly saw himself as an innovator in medical science," having taken from the Hippocratic writings a strong idea of the need for continuing scientific progress.[1099] As Hankinson explains, "Galen's method was 'the method of Hippocrates', but that method allowed for, indeed perhaps expressly involved, progress," hence "to be a true Hippocratic was not merely to mouth the words of the sacred text: it was to follow out the master's precepts in practice, and carry on the business of accumulating knowledge, and completing the science of medicine," which also meant for Galen that we "must not accept the doctrines of the great men of the past uncritically."[1100] Galen frequently argued that, as Hankinson says, "only by a combination of logical theorizing and the empirical method can medical science both be properly grounded and make progress," a fact that continues to be true to this day.[1101]

The only time Galen expresses doubt about the possibilities of scientific progress are when answers are inherently undiscoverable because we lack

---

1096. Galen, *On Conducting Anatomical Investigations* 1.3 (= Kühn 2.227–28 and 2.232).

1097. Galen, *On Conducting Anatomical Investigations* 7.10 (= Kühn 2.621–22).

1098. Galen, *On the Powers of Foods* 1.1 (= Kühn 6.479–80 = Mark Grant 2000: 78); in mathematical sciences: Galen, *On the Affections and Errors of the Soul* 2.5 (= Kühn 5.86–87). For these and other examples of Galen's views on scientific progress see Hankinson 1994a (and sources in Hankinson 1991a: 86) and Meissner 1999: 226–45.

1099. Hankinson 1994a: 1779 (cf. 1779–81).

1100. Hankinson 1994a: 1784 (Hankinson also identifies several areas where Galen made scientific progress and then applied those advances to practical effect: 1784–89).

1101. Hankinson 1994b: 1836.

the means to find them, such as how God creates anything or whether the universe is infinite, and other questions like that.[1102] However, Galen only says we should not pursue these questions because they cannot be accessed empirically. It is clear from the mode of his argument that as soon as we had any empirical access to them, he would agree that answering them would then be within our grasp after all. For example, when it comes to debates raging in his day about the nature of gravity and its role in astrophysics, Galen remarks:

> Let us return to those philosophers who make rash declarations regarding the issue of bodies placed in the void, either remaining in one place or moving downwards. Now, an engineer would not have declared himself on this issue before making a personal expedition to that part of the universe where there is void, putting the matter to the test empirically and making a definite observation as to whether any object placed there does remain in one place or moves elsewhere. Certainly that is the type of starting-point an engineer uses in his demonstrations—matters which can be universally agreed to be evident and indisputable.[1103]

Though traveling to outer space is a rather ambitious standard (probably the only obvious approach to the problem at the time), Galen would certainly have accepted the procedures and conclusions of Galileo and Newton. He already revered the methods and findings of astronomers of his own time, and held precise empirical-mathematical demonstration in the highest regard.[1104] Similarly, Galen argues that everything on which empirically competent doctors do not agree should be rejected as unreliable speculation—which entails that once reasonable men agree, science can advance.[1105]

Similarly, though Galen included the substance of the soul among the great unanswered questions of science, he did not regard even that subject

1102. For example: Galen, *On the Uses of the Parts* 15.1 (= M.T. May 1968: 658), *On the Affections and Errors of the Soul* 2.7 (= Kühn 5.98–103 = P.N. Singer 1997: 147–49), *On the Doctrines of Hippocrates and Plato* 9.6.20–22.

1103. Galen, *On the Affections and Errors of the Soul* 2.7 (= Kühn 5.98 = P.N. Singer 1997: 147).

1104. See note in chapter 1.2.III.

1105. Galen, *On Medical Experience* 11.

as beyond hope, nor did he consider our ignorance as grounds not to learn as much as we can about related physiological questions, such as why we need to breathe. To the contrary, Galen argues:

> If life is an action of the soul and seems to be greatly aided by respiration, how long are we likely to remain ignorant of the way in which respiration is useful? As long, I think, as we are ignorant of the substance of the soul. But we must nevertheless be daring and must search after the truth, and even if we do not succeed in finding her, we shall at least come closer than we are at present.[1106]

Like Seneca's expectations in astronomy, Galen was substantially correct about the science of respiration: discovering the substance of the 'vital soul' *was* essential to uncovering the purpose of respiration. He failed at this task, but he inspired modern scientists to complete it.

Galen had speculated that human bodies were governed by three 'souls' or what we might call 'command systems', one involving pulse and respiration that maintains a living body through the lungs, arteries and heart, which he called the 'vital soul', another based in the liver that operates the nutritive system through the veins, and finally what we now usually call a 'soul', the human mind, which, as Galen and others had demonstrated, resides in the brain, which in turn depends on the other two 'souls' for its operation and survival.[1107]

In this tripartite theory we can see Galen's empiricism and allowance for progress. For he said that of the mind's existence and location he was empirically certain, and of the vital soul's centrality in the heart he was less certain but fairly sure, but the nutritive soul's existence and location were uncertain to him and only hypothesized. But he was not very far from correct. The 'substance' of the nutritive soul could best be described now as an array of chemicals and minerals, and these nutrients do in fact pass from the digestive system to the liver through the veins—Galen only fell short of fully recognizing their connection with the arteries, though he did understand that nutrients were transported through the blood and went

1106. Galen, *On the Function of Respiration* 3 (= Kühn 4.472).

1107. Galen, *On the Therapeutic Method* 12.5 (= Kühn 10.839–40). Galen's tripartite theory (and its connection to his research on respiration) is discussed in Nutton 2013: 238–40.

into the construction and operation of the whole body.[1108] Meanwhile, the corresponding 'substance' of what he called the 'vital' soul turned out to be oxygen, a discovery that was in fact key to understanding life, exactly as Galen predicted, though the answer eluded him. And though the 'substance' of the mind-soul turned out to be the flesh of the brain (at least according to modern neurophysiology), oxygen and nutrient transport was still key to understanding how and why this organ worked. It is clear that all the efforts and methods that led to these discoveries would have been fully approved by Galen. These were exactly the kinds of advances he anticipated and hoped to encourage.

3.10 Summary & Conclusion

The evidence is fairly conclusive. Even from Seneca's remarks alone Samuel Sambursky had to agree that the Romans understood the reality and value of scientific progress, concluding:

> It was no doubt the great scientific period of the third and second centuries B.C. which brought about the first beginnings of a more permanent and more widespread cognizance of scientific progress and its significance. Seneca's words faithfully reflect the feeling of his generation, which, like us to-day, saw the understanding of the cosmos as an historical process, an endless task passed on from generation to generation.[1109]

That task was worked on rather slowly by modern standards. The reasons for its acceleration after the 15th century continue to be debated, but it cannot have been because Europeans only then appreciated the value

---

1108. Furley & Wilkie 1984: 40–69 survey Galen's attempts to improve contemporary Erasistratean theories of circulation and respiration, the errors he made in the process, and how Harvey even more soundly integrated Galenic and Erasistratean theories to develop the first correct theory of circulation (contributing to the eventual discovery of chemical respiration). See also Siegel 1968: 27–134 (circulation) and 135–82 (respiration). It seems arguable that had Galenists and Erasistrateans continued their debate (by refining each other's experiments and research), the discoveries of the 17th century would have been achieved by the 5th.

1109. Sambursky 1962: 252.

and possibility of scientific progress. We have shown that many Roman intellectuals were already in agreement with that goal. Though these attitudes were not universal (see chapter 4.8), the opinions of the more indifferent or antagonistic elements of Roman society had no effective power to determine the amount or course of scientific inquiry, so their opinions could have no effect on it.

Hence not only did many Romans know science had advanced and could continue advancing, expanding and perfecting their knowledge of the natural world, but as far as we can tell, all who recognized the fact of such progress perceived it as valuable and good, and thus worth moral or material support. Nevertheless, we must still separate this from other ideas of progress. For science can be pursued and promoted for many different motives, not only those that operate today. The idea that all other forms of progress, from the social to the economic, are linked in some way to progress in science, is a relatively modern development, and not essential to the pursuit of science as such. Modern ideas of progress may even be partly or largely explained by a prior commitment to scientific advancement for entirely different reasons, an advancement that eventually produced such results as to awaken minds to new possibilities.

G.E.R. Lloyd's evaluation of ancient ideals is therefore correct, and aptly qualified. Though the Romans saw, valued, and pursued progress in both science and technology, they did not see the pursuit of this progress in and of itself as a means to power and wealth. Many ancient scientists, in all fields, "recognized the practical importance of some aspects of their theoretical inquiries," and knew that scientific and technological progress benefitted humanity, commerce, industry, and the state, but "their efforts were uncoordinated," at least in the sense that "no *systematic* attempt to explore the practical applications of science was made" (emphasis mine), hence the idea that science "could be of practical use, while not totally absent, took second place to the idea that the study of nature contributed to knowledge and understanding" as ends in themselves. Consequently, there was no "*sustained* attempt to justify scientific inquiry in terms of the increased material prosperity to which it might lead" (emphasis again mine).[1110] Such an idea only occurred to the Western mind after tremendous upheavals in society, caused by scientific and technological advances that were in turn the

---

1110. Lloyd 1973: 174–75 (less coherently echoed in Vernant 1983: 284–85).

products of mutually unrelated happenstance (such as the development of the cannon, the compass, the printing press, the telescope, and the discovery of the New World) all within a few centuries, combined with explosive force to punch humankind in the face with undeniable evidence of the value of progress itself as a means to the acquisition of power and wealth.[1111]

Though this rather modern idea of progress is sometimes credited to the rise of Christianity, such a hypothesis struggles against its own implausibility, since the acquisition of power and wealth were not Christian values, but pursuits quite the reverse of any distinctly Christian message. Instead, these were goals just as avidly sought in pagan antiquity as in any other age. Had the Romans thought of it, surely they would have pursued intensive scientific and technological research to these same ends as avidly as anyone. Much the same could be said of von Wright's "moral" and "material" betterment. All Roman philosophers called for the former, believing it could be acquired through knowledge and reason, and most Romans pursued the latter, with the elite even promoting it through urban development and acts of philanthropy.[1112] In contrast, the central doctrine of Christianity had long been that moral improvement could only be achieved through God and his Gospel, certainly not through science and technology (much less unaided reason), and that material betterment suggested an immoral attachment to this world, when one ought to embrace instead an austere journey to the next.[1113] Pursuing any of these goals through science or technology thus required a revolution or accommodation in Christian thinking. The Romans were already nearer to the required mindset. They just hadn't fully completed the thought. Neither would the Christians for well over a thousand years.

Nevertheless, from all the evidence we have seen (and will see in the next chapter), it is clear that ancient interest in technological progress,

---

1111. See discussion in chapter 1.1 (and remarks at the end of 3.9.1 and 4.7).

1112. That ancient philosophy promoted moral improvement through reason and knowledge (including the scientific), see Trapp 2007, Bryant 1996, and Nussbaum 1994. For Roman examples see the *Moral Epistles* of Seneca and Galen's *On the Affections and Errors of the Soul.*

1113. See chapter five, although both propositions are ubiquitously displayed throughout the New Testament, and repeatedly exemplified by later ascetic and monastic movements (cf. e.g. Fox 1987: 293–335).

and to some extent also scientific progress, was driven by a desire and appreciation for what makes life easier or better. Even those who scorned such values replaced them with what they regarded as even loftier motives: either the pleasure of knowledge for its own sake, or the moral benefits that accrue to one who truly knows the world and how it works. Many embraced these values in combination, as we see from Galen's discussion of the many laudatory uses of anatomical research, or Hero's discussion of the value of studying robotics or artillery.[1114] But there is another, unexpected example illustrating everything we have seen so far.

In the 2nd century A.D. a professional dream interpreter named Artemidorus of Daldis sought to make a science of his art. As documented in his lengthy treatise *Interpretation of Dreams*, he researched the science of 'dream interpretation' by consulting not only every book on the subject he could find, but the "much-despised" street diviners as well, thus showing no aversion to interacting with craftsmen in his pursuit of knowledge (see chapter 4.6), nor any isolation from philosophers or competing schools of thought.[1115] He continually added to this 'science' with his own research, conducting countless interviews of live subjects in order to build a database, and then test hypothesized correspondences between the content of dreams and a dreamer's subsequent fortunes. And from this he sought to develop an empirically-based system of divination, which he believed should be increasingly freed of superstitious nonsense, and could be improved over time with ever more research.

Artemidorus imagined himself both building on and improving the work of his predecessors, and all of this he regarded as valuable because such an understanding of dreams was useful to present and future generations. His approach was almost modern and surprisingly empirical, and I think reflects the scientific zeitgeist of the time. The fact that he was chasing a phantom is not relevant to the point. Even modern scientists have done that, and still do on occasion. It is far more important to observe that even a *diviner* thought cautious, extensive, and organized empirical research was necessary to his field and would lead to worthwhile improvement in its accuracy and usefulness over time. Artemidorus obviously held this

1114. For Galen see examples in chapter 2.7. For Hero see chapter 4.6.II.

1115. For this and the following see discussion in R. White 1975: 1–11 and *EANS* 164, along with Artemidorus, *Interpretation of Dreams* 1.pr and 5.pr.

attitude because it was increasingly respectable, and even expected. He thus reflects everything we have argued in this chapter: ancient scientists, and many others among the educated elite, believed there had been and would continue to be progress in scientific knowledge, and that this was a valuable, useful, and desirable thing. And they believed the way to accomplish this was through more, and more accurate, empirical research, and the testing of theoretical models against observed evidence.

4. In Praise of the Scientist

In *Science Education in the Early Roman Empire* I explored to what extent science and natural philosophy held a relatively marginalized position in ancient education. I found they were still well-enough represented there that the most educated among the elite were familiar with what had been achieved, and with the basic outlines of the sciences themselves, and were impressed by what they knew. And in the previous chapter here, you have seen that the value of advancing the sciences was also widely enough understood that scientific progress continued, and continued to be lauded, up to the 3rd century A.D. We have already surveyed occasions in ancient literature when the specific issue of scientific progress is mentioned. Now we will broaden our attention to see how widely science and natural philosophy were praised in general.

In every section that follows, we will survey only a small but representative sample of what could be said, enough to establish the general point to be made in each case. There is already enough evidence here to confirm that many among the elite held science in great esteem, and regarded its pursuit as noble and morally beneficial, or practical and useful, or both. There are almost no comparable sentiments from their Christian contemporaries (as we shall see in the following chapter).

4.1 Philosophers for Science

Galen would often invite to scientific discussions a wide range of guests, and their interaction could produce heated argument. He describes a typical affair:

> Sometimes I request three or four Platonists, three or four Epicureans, the same number each of Stoics and Aristotelians, and three or four Academics or Skeptics to be present at the discussion, so that we have about twenty men of philosophy. I also ask a similar number of persons who have studied academic subjects and thus have an ability to reason, but have no familiarity with philosophical arguments...and sometimes doctors might also be present, and other educated professionals, or men with a good education who are neither practitioners of any craft (as they have private means) nor slaves to any philosophical sect.[1116]

Galen claims these two groups typically did not get along, and uses this fact to argue that slavishly following a sect, and ignoring sound empirical standards of argument, were shameful faults that everyone ought to correct in themselves. Yet it is clear that he arranged such gatherings with some frequency—and everyone continued to come. Such was the ease of arranging interaction among diverse philosophers and professional craftsmen, where, Galen reports, they would discuss astrophysics, gravity, matter, cosmology, and other subjects. This is only a typical example of how much interest was shown in the sciences by philosophers of every major school. Conspicuously absent from Galen's guest list are any Cynics, Rabbis, or Christians.

Galen goes on to poke fun at the philosophers who do not construct a more eclectic philosophy but defend instead the teachings of only one school, dogmatically and without regard for the actual findings of the sciences or even everyday observation. As Galen attests, there were definitely natural philosophers ignorant of even some of the established science of their day, as was also the case in Galileo's time. But interest in the subject of explaining and understanding the phenomena of nature was strong and widespread, and continual debate arose among the sects and professions as to the truth in such matters, and the methods of finding it. Though one could fill an entire book with an analysis of the complex attitudes each sect held toward science and natural philosophy, and the various values and methodological commitments each brought to the debate, that will not be my concern here. My aim is only to show that the most popular schools of thought advocated several motives for studying and advancing the sciences, which, if sincerely

1116. Galen, *On the Affections and Errors of the Soul* 2.5 (= Kühn 5.92–93). One of the subjects discussed (gravity) was quoted at the end of chapter 3.9.II; another (matter) will be quoted in section 4.6.III below.

embraced, would have continually driven them toward increasingly superior methods of getting at the truth, as they had in Galen's case.

Elsewhere I have already discussed how, and to what extent, all the major schools supported an education in science and natural philosophy.[1117] Though the various sects, or even individual philosophers, may have advocated ideas unconducive to a fully sound or successful science, eclecticism was the offered solution to this, as scientists sought to navigate a sound philosophical position from among the many debated options.[1118] But in one way or another, all the major philosophies valued the rational and empirical pursuit of the causes of natural phenomena.[1119] Even the skeptics valued the empirical pursuit of cause-effect relationships in nature, even when they reject theorizing about the hidden processes responsible.[1120] As Jonathan Barnes explains, for example, the Pyrrhonists were "ready to welcome certain arts and sciences," in fact any, "insofar as it is *useful*," which was to say, "an art must ameliorate the conditions of everyday life."[1121] Though Pyrrhonists were overly skeptical of the more theoretical methods and elements of the sciences, they valued (and gave assent to) the predictive and practical abilities that the sciences bring to society—astronomy and medicine were the most prominent examples, with their ability to predict eclipses or heal the sick, though mechanics would certainly have been granted the same status, as an art definitely of benefit to society and life, and based on clearly evident patterns of cause and effect.[1122] They even embraced weather prediction as a valid science.[1123]

The less skeptical Academics went even further in favor, valuing science not only as a probable and often useful source of knowledge, but

1117. In chapter seven of Carrier 2016. For background: *OCD* 657–58 (s.v. "Hellenistic philosophy").

1118. On the rise of eclecticism see discussion and notes in the introduction to chapter 3.7 and Carrier 2016 (index, "eclecticism").

1119. As implied already here throughout chapters two and three.

1120. See Hankinson 1998 and further discussion in section 4.8.

1121. J. Barnes 1988: 63–67.

1122. In astronomy, for example, they had no quarrel with the work of Hipparchus (Sextus Empiricus, *Against the Professors* 5.1–2, cf. also 1.50–51 and 5.103–05), and in medicine they supported the Empiricist sect.

1123. Taub 2003: 39–40.

also as a valuable source of pleasure that elevates the mind toward God. As Cicero explains this view, here in the role of an Academic responding to astronomical efforts to calculate the size of the sun:

> I put no confidence in this measurement...but such physical investigations should not be stopped. For the study and observation of nature is a kind of natural pasturage for the spirit and intellect. We are uplifted, we seem to become more exalted, we look down on what is human, and while reflecting upon things above and in the heavens we despise this world of our own as small and even tiny. There is delight in the mere investigation of matters at once of supreme magnitude and also of extreme obscurity, while if a notion comes to us that looks like the truth, the mind is filled with the most humanizing pleasure. These researches therefore will be pursued both by your wise man and ours, but by yours with the intention of assenting, believing and affirming, by ours with the resolve to fear forming such rash opinions, and yet still concluding it to be a good thing if in matters of this kind he has discovered what looks like the truth.[1124]

Thus, even with skepticism, science is useful, valuable, pleasurable, and worthy, as long as its findings are accepted as only probable, and not certain or infallible. Cicero himself was most fond of this perspective, as being the most moderate and defensible.[1125]

On the other side of the spectrum, the Aristotelians were the most obvious in their support for the sciences, and in fact they emphasized the value of theoretical understanding above all else, and to this end supported the need for both empirical research and the application of mathematics to natural problems.[1126] In fact they embraced knowledge for its own sake as a supreme value, thus not limiting inquiry to what was thought to be useful. Aristotle argued that those who know the causes and reasons of everything were wiser and thus superior to those who do not know the reason or cause of anything, and that the greatest wisdom, the wisdom nearest to God, lay in the study of the most fundamental principles of the universe, which he argued could only be accomplished by studying the entire chain of causes from everyday phenomena back to their ultimate beginning, which entailed

1124. Cicero, *Prior Academics* (= *Lucullus*) 2.41.126–28.

1125. See quotes and examples in chapter 2.5 and 2.6.

1126. See discussion and scholarship cited in chapters 2.2 and 3.1.

the pursuit of many sciences.[1127] Aristotle also advanced as a fundamental tenet of his ethics that philosophy (hence natural philosophy) is the greatest and most respectable source of pleasure, and therefore philosophical (and thus also scientific) research is a supreme moral good.[1128]

Accordingly, as we have seen, Aristotelians, and those sympathetic to their philosophy, showed the widest and most avid scientific interests throughout antiquity.[1129] Cicero described with obvious admiration the Aristotelian achievements in natural philosophy resulting from this:

> When it comes to thoroughness of material on this subject, we shall find the Stoics thin in comparison with the Aristotelians, who are copious in the extreme. Look how much they observed and recorded regarding the classification, reproduction, morphology, and life-history of animals of every kind! Look how much regarding the products of the earth! Look how numerous and wide-ranging their explanations of the causes why and demonstrations how! All these facts supply them with copious and conclusive arguments to explain the nature of each particular thing.[1130]

As Aristotle put it, "the causes, principles, and elements of substances are the object of our research," which for him entailed a study of "the things of nature, such as fire, earth, water, air, and all the fundamental elements, and secondly, plants and their parts, and animals and the parts of animals, and finally the physical universe and its parts."[1131] In other words, everything.

Platonists had more restricted interests, but they accepted the utility of predictive science, and believed social peace and personal happiness would arise from knowledge and the pursuit of philosophy and learning, and even held that anyone who contemplates the higher truths of natural

---

1127. As argued in Aristotle, *Metaphysics* 1.1.980a-1.3.983b (with 2.1.993a and 8.1.1042a).

1128. This is the argument of Aristotle, *Nicomachean Ethics* 10.1.1172a-10.8.1179a. Aristotle also elaborated on all these reasons for pursuing philosophy in his *Exhortation to Philosophy*. Though lost, relevant quotations from this survive in Iamblichus, *Protrepticus* 6–12.

1129. As shown here, for example, in chapter 3.1 and 3.5 (Galen, Ptolemy, and Hero, for example, though eclectics, all held Aristotelian sympathies).

1130. Cicero, *On the Boundaries of Good and Evil* 4.5.12–13.

1131. Aristotle, *Metaphysics* 8.1.1042a.

philosophy (especially at its most mathematical, such as in astronomy and harmonics) will approach the divine and be morally improved by it.[1132] They also saw natural philosophy as a path away from superstitious fear and toward a proper reverence for God.[1133] And in addition to all that, scientific discovery was supremely pleasurable.[1134] Nevertheless, Platonists were less empirical in their commitments than the Aristotelians, and did not produce much scientific research, except insofar as their ideals inspired more eclectic thinkers to pursue it.

Similarly, Epicureans appear to have produced much less scientific research than the Stoics, though both Stoic and Epicurean ideas inspired many eclectically-minded scientists, many of whom had strong atomist commitments without becoming dogmatically Epicurean, though almost none of these scientific writings survived the middle ages.[1135] Nevertheless, even the staunchest dogmatists valued knowledge of the natural world. For both Epicureans and Stoics science was as essential to ethics (the truth of which required knowing the nature of humanity and the world) as to epistemology (since the only way to know when perception and reason can be trusted, is to know how they are produced and deceived and what relation they bear to the actual facts of the world).

Cicero provides the most convenient summary of the Epicurean and Stoic motives to pursue the study of nature. For the Epicureans, the pursuit of natural philosophy was less scientific an enterprise by modern standards and yet, ironically, often more prescient in its conclusions, while for the Stoics the same pursuit entailed sounder scientific interests and activities, which, just as ironically, often lead them to promote what we now know to be bogus sciences (such as astrology and divination). But both believed in the importance of getting the facts right, and both believed the most important

---

1132. For Roman Platonists on these points see Maximus of Tyre, *Orations* 6, 10, 13, and 27 and Alcinous, *Epitome of Platonic Doctrine* (= *Didaskalikos*) 3–4 and 7–8 (quoted in chapter 2.5). On Platonist (and Pythagorean) interest in mathematics: Horky 2013.

1133. At least as implied in the judgment of one Roman Platonist: Plutarch, *Pericles* 6.1.

1134. Plutarch, *That Following Epicurus is Unpleasant* 11 (= *Moralia* 1093e-1094d).

1135. Several prominent scientists inspired by Stoic or Epicurean principles are mentioned here in chapters 3.2, 3.3 and 3.4.

facts to get right were those pertaining to the causes and principles of the natural universe and its contents.

This is how Cicero summarizes the motives of the Epicureans:

> Epicurus bet everything on natural philosophy. For with such knowledge we can discern the power of words, the nature of speech, and the logic of consequents and contradictions. Moreover, by understanding the nature of all things we are relieved of superstition, freed from our fear of death, and not confused by an ignorance of the facts, which is often in itself a cause of horrible terrors. We also become morally better when we learn what nature's real requirements are. Moreover, only if we embrace a well-established knowledge of the facts of the world...can we remain unshaken by the eloquence of any man, just as without a full understanding of the nature of the world it is impossible to maintain the truth of our sense-perceptions. ...
>
> Thus natural philosophy supplies us with courage against the fear of death and confidence against the terror of religion, along with peace of mind (by removing our ignorance of every obscure thing), moderation (by teaching us the nature of our desires and their different kinds), and... regulation and judgment in our understanding (by teaching us how to distinguish the true from the false).[1136]

So Cicero found in Epicurean philosophy four general reasons we should pursue the study of natural philosophy: (1) it teaches us how to think better; (2) it frees us from superstition, fear, and ignorance; (3) it improves our moral character; and (4) it defends us against manipulation, deception, and error.

Though the Stoics distinguished logic from physics, the reasons they gave for studying both were similar. Cicero summarizes their motives thus:

> Among the virtues the Stoics also add both Logic and Natural Philosophy. Both of these they call virtues, in the first case because it conveys a method that guards us from giving assent to any falsehood or ever being deceived by specious probability, and enables us to retain and defend the truths that we have learned about good and evil—for without the art of logic they hold that any man may be seduced from truth into error. If therefore rashness and ignorance are in all matters fraught with mischief, the art that removes them is correctly entitled a virtue.

1136. Cicero, *On the Boundaries of Good and Evil* 1.19.63–64. Giovacchini 2012 provides the most recent discussion of the quasi-scientific empiricism of Epicureans.

> The same honor is also bestowed with good reason upon natural philosophy, because he who is to live in accordance with nature must base his principles upon the system and government of the entire world. Nor again can anyone judge truly of things good and evil except by a knowledge of the whole plan of nature, and also of the life of the gods, and of the answer to the question whether the nature of man is or is not in harmony with that of the universe. Moreover, without natural philosophy, no one can discern the value (and their value is great) of the ancient maxims and precepts of wise men, such as 'obey occasion', 'follow God', 'know thyself', or 'moderation in all things'. Also, this science alone can impart a conception of the power of nature in fostering justice and maintaining friendship and the rest of the affections. Nor again without unfolding nature's secrets can we understand the feeling of piety towards the gods or the degree of gratitude that we owe to them.[1137]

So, in addition to three of the reasons for studying logic that the Epicureans also held for studying physics (a reliable use of the senses, security against the manipulation of demagogues, and immunity to the evils of ignorance), the Stoics embraced five moral reasons to pursue natural philosophy: (1) to live according to nature requires knowing how nature really works; (2) to discern what is actually good or evil requires knowing the nature of each thing and its place in the natural order; (3) natural philosophy is needed to understand what moral propositions really mean; (4) one ought to know nature's role in making our greatest goods possible (like friendship and justice); and (5) we will never truly appreciate God's providence and wisdom unless we truly understand the world he made.

The ultimate Stoic dictum was "to understand nature and follow her," hence their entire moral philosophy depended on a sound natural philosophy, which ultimately entailed a commitment to scientific education and research.[1138] Whereas the Epicureans were devoted to defending a coherent system that they already believed must be true, the Stoics were more devoted to following wherever the evidence leads. The facts of nature were for them more primary, ethics followed. Though as subjects they could be taught in reverse order, natural philosophy held first place in logical

---

1137. Cicero, *On the Boundaries of Good and Evil* 3.21.72–3.22.73.

1138. M. Clarke 1971: 87–88. For some of the underlying ideas see Diogenes Laertius, *Lives and Opinions of Eminent Philosophers* 7.86–89.

importance.[1139] And unlike the Epicureans, who saw natural philosophy as having more directly practical ends, the Stoics, like the Aristotelians, considered research in natural philosophy to be an act of deep piety toward God. Thus Cicero concludes:

> Natural philosophy is pursued by both the Aristotelians and by the Stoics not merely for the two reasons recognized by Epicurus, of banishing superstition and the fear of death. For besides those benefits, the study of the heavens bestows a power of self-control that arises from the perception of the consummate restraint and order that obtain even among the gods, and a loftiness of mind is inspired by contemplating the creations and actions of the gods, and justice is inspired by realizing the will, design and purpose of the supreme Lord and Ruler, to whose nature, as these philosophers tell us, the true reason and supreme law are conformed. The study of natural philosophy also affords the inexhaustible pleasure of acquiring knowledge. It is the only pursuit that can provide an honorable and elevated occupation for us [in our spare time].[1140]

We also know the Stoics were interested in the sciences for other theological and epistemological reasons, such as their interest in finding predictable regularities in nature to justify a rational system of astrology and divination, or their rather demanding logic of demonstration, which placed a healthy emphasis on the need for evidence.[1141]

4.2 Literary Praise

Having established the kinds of motivations for studying natural philosophy promoted by the leading schools of philosophy, we now have a context for understanding the abundant praise and admiration for it in Roman literature. We have already seen many examples of this throughout chapters two and three.[1142] Here we shall add only a sample of what remains, more or

1139. A distinction unfairly mocked as an alleged contradiction in Plutarch, *On the Contradictions of the Stoics* 9 (= *Moralia* 1035d).

1140. Cicero, *On the Boundaries of Good and Evil* 4.5.11–12.

1141. See related quotes from Cicero in chapter 2.7.

1142. With further examples throughout Carrier 2016.

less in chronological order (with a special section later for Seneca and the *Aetna*).

As I've shown elsewhere, like other Romans (from Quintilian to Tacitus), Cicero insisted upon the study of natural philosophy as a fundamental aspect of any good education in rhetoric and oratory, which was the expected course of study for any actual or aspiring member of the governing elite.[1143] Even the critics Antonius and Scaevola in Cicero's dialogue *On the Orator* "cannot deny" that a man who studies the sciences is "a remarkable kind of man and worthy of admiration," as long as he is *also* an accomplished speaker.[1144] But Cicero's passion went even further than that. He believed everyone who is free of immoral desires, both now and in the afterlife, will devote themselves to the investigation of nature, examining things in heaven and on earth.[1145] Though he said he expected to learn more scientific truths in the afterlife (when he would have more time and freedom and clarity of vision), Cicero proposed this as a model to follow in the present life, and not as a reason to set aside our scientific curiosity until we are dead.[1146]

As I noted earlier (near the end of chapter 3.8.I), Cicero argued that any real scientist will have an honest, frugal, and disciplined character, because science requires such a love for the truth, and so much effort for so little material gain, that only good men would ever pursue it. So, rather than seeking a life of vain pleasure:

> The devotees of learning are so far from making pleasure their aim, that they actually endure worries and cares and loss of sleep, and in exercising the noblest part of man's nature, the divine element within us (for so we must consider the keen edge of reason and the human intellect), they ask for no pleasure and avoid no toil, but are ceaselessly occupied in marveling at the discoveries of those of old or in pursuing new researches of their own. Insatiable in their appetite for study, they forget all else besides, and harbor not one base or mean thought. ... [Instead] they spend their whole lives in investigating and unfolding the processes of nature.[1147]

1143. See chapters five and six of Carrier 2016.

1144. Cicero, *On the Orator* 1.17.75–76 (Scaevola), 1.18.80–81 (Antonius).

1145. Cicero, *Tusculan Disputations* 1.19.43–21.49.

1146. Unlike the Christian Origen, who argued it may be better to wait until the afterlife to learn the facts of science (see chapter 5.9).

1147. Cicero, *On the Boundaries of Good and Evil* 5.20.57.

He then ranks the activities of man's mind in order of worth, putting natural philosophy first, politics second, and ethics third.

Though Cicero says the scientist forsakes pleasure, he does not mean absolutely, for inquiry itself is, as Aristotle argued, the highest pleasure, an end for which we were clearly designed. Hence, Cicero says "we are designed by nature" for "the contemplation and study of heavenly bodies and of those secrets and mysteries of nature which reason has the capacity to penetrate."[1148] In fact, Cicero argues, acquiring knowledge of every kind brings us so much pleasure that God must have designed us with the intention of our pursuing learning and research.[1149] "So great is our innate love of learning and of knowledge," Cicero says, "that no one can doubt that human nature is strongly attracted to these things even without the lure of any profit," and for that reason "it must be deemed the mark of a superior mind to be led on by the contemplation of high matters all the way to a passionate love of knowledge."[1150]

Cicero also believed the evidence of design in nature communicates to us the truths of God, which we should thus learn from studying his creation. Accordingly, Cicero is awed by the beauty of the universe and praises the many wonders in it, on earth and above it, as deserving of our closest attention and investigation, even chastening those who take the natural world for granted and fail to study its diverse contents. Cicero concludes that all these amazing things we see in the world "ought to arouse us to inquire into their causes."[1151] Hence, just as Aristotle had said the measure of wisdom is how much one knows the causes of things, Cicero calls upon everyone to investigate the causes of things:

> It is necessary that whatever arises, of whatever kind, has its cause from nature, so that even if it happens contrary to what is usual, it still cannot happen contrary to nature. Therefore, investigate the cause for every new

---

1148. Cicero, *On the Boundaries of Good and Evil* 5.21.58 (in context: 5.18.48–21.60).

1149. Cicero, *On the Boundaries of Good and Evil* 5.19.50–54.

1150. Cicero, *On the Boundaries of Good and Evil* 5.18.48–49 (in context: 5.19.50–54).

1151. Cicero, *On the Nature of the Gods* 2.38.96 (in context: 2.34.87–61.153).

and remarkable event, if you can, and if you discover none, you must still consider it certain that nothing can happen without a cause.[1152]

Though he was speaking of the unusual, and in the context of claims of omens and divination, it is clear from all we have seen that Cicero applied this sentiment to all natural phenomena that interest, frighten, or affect us.

Cicero's admiration and respect for scientific pursuits is thus quite evident. Others of his century shared the same view. For example, like Marcus Aurelius centuries later, Virgil pines for the idea of studying natural philosophy, arguing that the next best thing is only to love the simple life of the farm and country, away from the evils of civilization:

> Indeed first and before all things may the sweet Muses, whose priest I am and whose great love hath smitten me, accept me and show me the pathways of the sky, the stars, and the diverse eclipses of the sun and the moon's travails, and whence is the earthquake, by what force the seas swell high over their burst barriers and sink back into themselves again, why winter suns so hasten to dip below the ocean, or what hindrance keeps back the lingering nights. But if I cannot approach this aspect of nature... may the country and the streams that water the valleys content me, and, lost to fame, let me love stream and woodland. ... Otherwise, happy is he who is able to comprehend the causes of things, and has laid under his feet all fears and inexorable fate and the roar of ravenous Acheron.[1153]

Utility was also a recognized motive for respecting and pursuing the sciences. Vitruvius, for instance, argues throughout his book *On Architecture* that science and technology had significant moral value and social utility.[1154] In the following century Columella made the same point in a different context, insisting that "one who would profess to be a master of agricultural science must have a shrewd insight into the works of nature," and since nothing, he argued, was more useful or important than agriculture, natural philosophy was just as important.[1155]

1152. Cicero, *On Divination* 2.28. That aetiology (the study of causes) was widely perceived to be the defining preoccupation of natural philosophy was shown in chapter 2.7.

1153. Virgil, *Georgics* 2.4.90. For the similar pining of Marcus Aurelius see below.

1154. See the analysis of J.-M. André 1987 and examples in chapter 3.5.

1155. Columella, *On Agricultural Matters* 1.pr.22.

Likewise, Pliny the Elder says "those who discover the laws of such great divinities" as the sun and moon "are great men, beyond the nature of mortals, freeing the miserable mind of men from fear," especially of dreaded celestial events, and we "should all be awed by the genius of those who interpret the heavens and grasp the nature of things, who discover the evidence that defeats gods and men."[1156] Accordingly, Pliny included a chapter on eminent men in the sciences, declaring that "countless men have shined in the knowledge of the various arts, some of whom must at least be mentioned by anyone who would collect the flower of humankind."[1157]

In the first half of his survey Pliny lists a brief selection of famous scientists, while in the second half he devotes roughly as much space to famous artists. Fame, in the form of state or public recognition, was Pliny's apparent standard for inclusion, rather than scientific accomplishment (or even his own preference, since Hipparchus does not appear, though we know Pliny praised him above all). But the fact that scientists could become famous for their work also tells us something about Greco-Roman society. His short list includes an astronomer, a philologist, four doctors, and five engineers. Why an astronomer would be honored for his predictions and doctors for their cures hardly needs explanation. But more interesting are the reasons Pliny includes engineers, in fact more of them than any other class of scientist:

> Archimedes received a remarkable testimony to his knowledge of geometry and mechanics when Marcus Marcellus captured Syracuse and Archimedes was exempted from all harm, except for the ignorance of a soldier who failed to heed the order. Others praised include Chersiphron of Gnossus for constructing the amazing temple to Diana at Ephesus, Philo [of Eleusis] for building the dockyard at Athens housing 400 warships, Ctesibius who discovered the theory of pneumatics and invented hydraulic machines, and Dinochares for building the city of Alexandria in Egypt at Alexander's command.[1158]

1156. Pliny the Elder, *Natural History* 2.9.54.

1157. Pliny the Elder, *Natural History* 7.37.123–38.127.

1158. Pliny the Elder, *Natural History* 7.37.125. Chersiphron's Temple of Artemis had become one of the seven wonders of the world. Archimedes is discussed further in sections 4.5 and 4.6.1.

We know there were other lists of famous scientists in circulation, though none survive intact. A fragment of such a list on papyrus even includes scientists otherwise unknown to us, such as Abdaraxos "who built the machines in Alexandria," a feat that was evidently quite famous at the time.[1159] Pliny's list likewise emphasizes technological sciences and accomplishments, not merely amazing military or architectural feats, but even the invention of new sciences (pneumatics), which is all a nod to the study of natural philosophy for useful ends.[1160]

A bit later Plutarch would reveal something else of the status of natural philosophy, when he complains that "practically all beginners in philosophy are more inclined to pursue those forms of discourse which make for repute," providing an example for each branch of philosophy, and drawing an analogy between puppies and birds.[1161] Some ambitious beginners are "like puppies" and go for scraps and quibbles—most jump into logic and argument and thence to sophistry, while some simply collect sayings, and as a result both miss the deeper meaning of ethical philosophy. But *other* beginners are "like birds" and "are led by their flightiness and ambition to alight on the resplendent heights of natural philosophy." The comparison is notable. Not only is the study of nature a "resplendent height" (and it is perhaps more flattering to be equated with birds than dog-babies), but the very idea of jumping into the study of nature is presented here as an object of ambition and desire for renown. That would only make sense as a rebuke if it were in some sense true, which entails that a natural philosopher could attain a certain degree of prestige in Roman society.

Thus, as we saw in chapter two, a Roman father could imagine his son's pursuit of natural philosophy as at least respectable, but his pursuit of

---

1159. See note on Abdaraxos (and this fragmentary list) in chapter 3.4.

1160. Pliny's fondness for engineering (and its benefits to mankind) is also reflected in his survey of the architectural wonders of the known world, where he includes the pyramids of Egypt and every comparable marvel, yet he concludes it is the sewers of Rome that were *opus omnium dictu maximum*, "the greatest achievement of all," and the aqueducts of Rome that were *vera aestimatione invicta miracula*, "a marvel unsurpassed in value" by anything else ever built: Pliny the Elder, *Natural History* 36.24.104–08 (sewers), 36.24.121–24 (aqueducts).

1161. Plutarch, *How a Man May Become Aware of His Progress in Virtue* 7 (= *Moralia* 78e).

Cynic philosophy was so shameful he would rather disown him.[1162] This is a notable reversal of the status science seems to have held in Classical Athens, when Socrates could be condemned in court (in part) for teaching natural philosophy, which was then regarded by many as dangerously impious, hence part of his defense was that he had never taught natural philosophy but only moral philosophy, taking a position close to what would become the basis of Cynic philosophy years later.[1163] Yet in Rome, Socrates' defense would have become grounds for his prosecution, and the original charges against him his best defense!

This turnaround is clearest of all in Plutarch's treatise on curiosity. How should we escape the sins of meddling, snooping, and gossiping? "By a process of shifting and diverting our curiosity," Plutarch argues, "turning our soul to better and more pleasant subjects. Direct your curiosity to heavenly things and things on earth, in the air, in the sea." He says the physics and astronomy of the sun and moon, for example, are the true spectacles that deserve our awed attention, and thus are more proper objects of our prying and questioning. Plutarch's entire argument, in fact, is that we should divert our curiosity from immoral concerns (like the miseries and private affairs of our neighbors) to discover, instead, "the secrets of Nature, because Nature is not vexed with those who find them out." Even if astronomy does not interest us, Plutarch says, we should look below and study something like botanical science. In fact, only if we delight in witnessing depravity should we *then* turn our curiosity to history as a safe outlet. Otherwise, scientific inquiry is the most respectable occupation for the curious and inquisitive.[1164] While this was clearly a common attitude among the Roman elite, as we shall see in chapter five the Christians preferred returning to a more Socratic distrust of curiosity about nature, flipping social values back around.[1165]

1162. See the end of chapter 2.5.

1163. On Classical Athenian attitudes see note in chapter 1.3. On Socrates' proto-Cynical rejection of natural philosophy (a fact cited by Christians in defense of their own rejection of it) see relevant notes and examples in chapters 5.2 and 5.4 (and in chapter seven of Carrier 2016). For the charges against Socrates and his defense see (and compare) the *Apology* of Plato and the *Apology* of Xenophon.

1164. Plutarch, *On Curiosity* 5 (= *Moralia* 517c-e).

1165. This socially positive view of scientific curiosity, as found in Pliny and Seneca, is surveyed in Beagon 1992: 60–63. For the Christian reversal see chapter

Even the Emperor Marcus Aurelius declared that "just because you have lost your hope of becoming a dialectician and a natural philosopher, do not for this reason renounce the hope of being both free and modest and social and obedient to God," thus implying that the study of logic and physics was something to be admired and sought after, and that (as any true Stoic would agree) these studies would benefit our pursuit of modesty, friendliness, and piety—as well as our personal liberty, for by being acquainted with the causes of things we can exercise more control over our fate, or accept what fate we cannot control, and even tell the difference between the two, a sentiment found throughout Aurelius's diary.[1166] However, his attitude overall was more depressing than most, trapped as he was in public life. He says he is thankful he had not turned away from his responsibilities "to emulate sophists or write about theoretical matters or deliver cute little exhortations" or otherwise show off, and that he had learned to "keep away from rhetoric and poetry and fine literature," and to write only in a simple style, while he tries to persuade himself that he should stop reading books, as he had no more time for them.[1167] But all this he said with regret, lamenting that he had put off too many opportunities to study and understand the universe more.[1168]

As Aurelius explains, "when I set my heart on philosophy I did not fall into the hands of any sophist, nor did I stop to figure out books or syllogisms or dwell on meteorology, for all these things require the help of gods and fortune," in other words, he was not lucky enough to have the time or talent for such studies.[1169] So he did not imagine himself as a scientist. But he admired those who were, listing several as among the many heroes of the past, "Eudoxus, Hipparchus, Archimedes, and other men of acute

---

five and evidence in P. Harrison 2001 (and in chapter nine of Carrier 2016).

1166. Marcus Aurelius, *Meditations* 7.67.1 (already mentioned in chapter 2.5). His belief that a knowledge of nature increases moral character and control is repeated throughout his *Meditations* (e.g. 2.1, 2.9, 3.11, 4.29, 8.52, 10.9, 11.5, etc.).

1167. Marcus Aurelius, *Meditations* 1.7.1–2 (he struggles with books in 2.2 and 2.3).

1168. Marcus Aurelius, *Meditations* 2.4.

1169. Marcus Aurelius, *Meditations* 1.17.9. Though he might also have had some ideas at odds with a proper study of natural philosophy (as discussed in chapter 3.7.IV).

talents, great minds, lovers of labor, ingenious, confident, mockers of the very perishable and ephemeral life of man."[1170] And he reveals a hint of the scientific spirit within him when he says that as long we have enough intelligence and drive to pursue "knowledge of the human and the divine" (a Stoic periphrasis for 'natural philosophy') then we should also take delight in observing the natural world in meticulous detail, seeing beauty in every little facet of even bread and plants and animals.[1171] Such phenomena of the natural world, he argues, will "inspire the soul" of any who really notice them, especially "if one should have a passion and deeper understanding for the things produced in the universe."[1172] We can see in this attitude the same spirit that inspired not only scientific inquiry among his fellow Romans, but also the art of his era, which likewise attended to the realistic details of the natural and human world.[1173]

While pleasure and wisdom and personal and public utility were recognized as valid motives for the pursuit of science, so was fame and glory, which, like curiosity, were only respectable when attached to respectable goals. Though Galen attacked the pursuit of glory as incompatible with the scientific spirit, there was room for disagreement.[1174] Praising his native city of Carthage, Apuleius gives the following example of respectable scientific glory:

> The best reward is what they say was suggested by Thales. Thales of Miletus was easily the most remarkable of the famous seven sages. For he was the first of the Greeks to discover the science of geometry, was a most accurate investigator of the nature of things, and a most skillful observer of the stars. With the help of a few small lines he discovered the most momentous facts: the revolution of the years, the blasts of the winds, the wanderings of the stars, the echoing miracle of thunder, the slanting path of the zodiac, the annual turnings of the sun, the waxing of the moon when young, her waning when she has waxed old, and the shadow of her eclipse.

1170. Marcus Aurelius, *Meditations* 6.47.

1171. Marcus Aurelius, *Meditations* 3.2 (quote comes from the leading point in 3.1).

1172. Marcus Aurelius, *Meditations* 3.2.3.

1173. On this point about ancient art see relevant remarks in chapters 1.1, 3.8.I, and 3.8.IV.

1174. For Galen's attitude see chapter 3.8.I.

> Even when he was far advanced into the vale of years, he evolved a divinely inspired theory concerning the period of the sun's revolution through the circle in which the sun moved in all his majesty. This theory, I may say, I have not only learned from books, but have also proved its truth by experiment. This theory Thales is said to have taught soon after its discovery to Mendratus of Priene. The latter, fascinated by the strangeness and novelty of his newly acquired knowledge, bade Thales choose whatever recompense he might desire in return for such precious instruction. "It is enough recompense," replied the wise Thales, "if you will refrain from claiming as your own the theory I have taught you, whenever you begin to impart it to others, and will proclaim me and no other as its discoverer."
>
> In truth that was a noble recompense, worthy of so great a man and beyond the reach of time. For that recompense has been paid to Thales down to this very day, and shall be paid through all eternity by all of us who have realized the truth of his discoveries concerning the heavens. Such is the recompense I pay you, citizens of Carthage, through all the world, in return for the instruction that Carthage gave me as a boy.[1175]

In other words, in alignment with what we heard the historian Diodorus declare earlier, eternal fame is the greatest of rewards, and scientific discoveries are worthy of eternal fame.

It does not matter whether any of his story about Thales is true. What matters for us is the high opinion Apuleius clearly held for what was then understood as the scientific enterprise, and the fact that he assumed his audience would not only agree, but even be flattered by such a comparison between their beloved city and the achievements of a legendary scientist. For Apuleius, as for Romans generally (as noted in chapter 2.4), Thales was the progenitor of natural philosophy, and for that reason the 'most remarkable' of the seven sages, a man 'so great' his discoveries will never be forgotten. Why? Because, according to Apuleius, he developed geometry, turned astronomy into a theoretical science, and investigated nature 'most accurately'. Above all, by uniting all three, Thales developed a theory of solar motion that was subsequently confirmed by observation, even by Apuleius himself—which seems to have impressed him the most, a sterling example

1175. Apuleius, *Florida* 18.19–36 (see section 4.5 for his similarly heroic portrayal of the medical theorist Asclepiades). For commentary on this passage: S.J. Harrison 2000: 69–72, 122–25. On this as a popular motivation for discovery and invention in antiquity: Meissner 1999: 89–91 (in the context of a broad survey of many other expressed motives to the same end in ancient literature: 37–122).

of the very thing Cicero found so attractive in natural science: the joy of discovery.

Just as Thales could be held in such high esteem for having introduced natural philosophy to the world (as even the Christian Eusebius had to admit, he "was the first natural philosopher among the Greeks" and as a result "became the most distinguished man among them"),[1176] so could others achieve fame for making remarkable contributions to natural philosophy. We saw such respect extended to Posidonius for that same reason, and to Strato before him.[1177] Of the latter Diogenes Laertius says, "Strato was a man deserving of extensive approbation, because he excelled in every branch of learning," but "most of all in the one called 'physics', which is a branch of philosophy more ancient and more worthy of serious attention" than any other.[1178]

Galen did not (ostensibly) agree with the idea of seeking fame through scientific accomplishment, but he did argue, like Cicero, that such pursuits will be the passion of any moral man devoted to the truth—and the more devoted, the more respectable he would be. Accordingly, Galen describes the admirable devotion a true scientist embodies:

> It is essential that anyone, who wants to know anything better than the ordinary run of humanity, must far outshine them, both in natural endowment and in the quality of their early training. As a lad he must develop an almost erotic passion for the truth, so that day and night, like someone possessed, he will not let up in his desire and effort to learn what was propounded by the most illustrious of the ancients. And when he has learnt these things, he must spend a great deal of time testing and justifying them, seeing what accords with the observable facts and what does not, and on the basis of this he will accept some doctrines and reject others.[1179]

Clearly Galen considered such hard work not only worthwhile, but the mark of a superior man.

Similarly, Ptolemy declared that scientific inquiry leads us to what is both

1176. Quoting Eusebius, *Preparation for the Gospel* 10.14.10.

1177. For Strato's prestige see chapter 2.5; for Posidonius see chapter 3.3.

1178. Diogenes Laertius, *Lives and Opinions of Eminent Philosophers* 5.64.

1179. Galen, *On the Natural Faculties* 3.10 (= Kühn 2.178–80).

beautiful and useful.[1180] For example, "to exhibit to human understanding through mathematics the heavens themselves in their physical nature," and "the earth as well," with the sciences of geography and cartography, is a labor that "belongs to the loftiest and loveliest of intellectual pursuits."[1181] Ptolemy agreed that pursuing the sciences had moral value and could even improve one's moral character.[1182] For example, much as Cicero had argued centuries before, Ptolemy declares:

> Of all studies, astronomy will especially prepare men to recognize nobility of action and character: when the constancy, good order, proportion and freedom from arrogance of divine things are contemplated, such study makes those who follow it lovers of this divine beauty, and instills, and as it were makes natural, the same condition in our soul.[1183]

We find something similar in a proverb from a lost play (possibly from Euripides, which would mean the same sentiment spanned almost the entire history of antiquity), quoted less comprehendingly by the Christian theologian Clement (who assumed, rather implausibly, that it spoke of contemplating the Platonic Forms rather than an empirical study of nature):

> Happy is he who has an education in inquiry. He is not moved to commit crimes or harm the people, but explores the eternal order of immortal nature, how it comes together and in what way or form. With such interests as these he never clings to any concern for immoral deeds.[1184]

Thus science had profound moral and religious status in the pagan mind, at least among the educated elite. As Ptolemy declares in his famous epigram: "I know I am mortal and fleeting, but when I search the densely

---

1180. See Ptolemy, *Harmonics* 3.94.

1181. Ptolemy, *Geography* 1.1.

1182. Argued in Taub 1993: 146–53.

1183. Ptolemy, *Almagest* 1.1.7.

1184. Quoted without attribution in Clement of Alexandria, *Stromata* 4.25:155.1. It is conjectured to derive from the *Antiope* of Euripides (frg. 910) and appears to contain allusions comparing scientific research to the sacred mysteries (cf. Kambitsis 1972: 130–34), a common notion among the pagan elite (see chapter 4.2 and 4.4 and example in chapter 3.5).

revolving spirals of the stars, I no longer touch the earth with my feet, but dine with Zeus himself, and take my fill of ambrosia."[1185]

This was a common sentiment among astronomers. But Vivian Nutton has demonstrated that the same religious reverence attended the pursuit of medical science (as we shall see from Galen below).[1186] We have already seen evidence of how the religious beliefs of the pagan elite could actually support, justify, or motivate scientific inquiry. Another example of this is an evolution in the theological role of the Nine Muses. Plutarch says the 'ancients' only recognized three Muses, which were later tripled into nine, and he gives the following account of this (in the voice of his brother at a dinner party):

> Once they observed that all branches of knowledge and crafts that attain their end by the use of reason belong to one of three kinds—the philosophical, the rhetorical, and the mathematical—they considered them to be the gracious gifts of three goddesses, whom they named Muses. Later, in Hesiod's days in fact, by which time these faculties were being more clearly seen, they began to distinguish different parts and forms. They then observed that each faculty in its turn contained three different things. The mathematical category includes music, arithmetic, and geometry. The philosophical comprises logic, ethics, and physics. And for the rhetorical, it is said that the original laudatory kind was first joined by the deliberative, and finally the forensic. Thinking it wrong that any of these subjects should be without its god or Muse or deprived of such higher control and guidance, they naturally discovered (for manufacture they surely did not) the existence of as many Muses as there are branches.[1187]

Consequently, there had to be a Goddess for each of the ensuing nine branches of learning. Despite Plutarch's historical 'theory', Hesiod shows no awareness of such a sophisticated development. He only knows all nine Muses as patronesses of music and storytelling. And most sources, even in the Roman period, assign them to various forms of song, dance, music, and poetry, including two for comedy and tragedy and one for history; though

1185. From the *Palatine Anthology* 9.577. Compare Eusebius's attitude towards this epigram (quoted and discussed in Carrier 2016: 147–48).

1186. Nutton 2013: 280, 288.

1187. Plutarch, *Questions at a Party* 9.14.3 (= *Moralia* 744c-e).

one (Urania) was at some point named the patron goddess of astronomy.[1188]

Plutarch is the first we know to suggest the Muses had been paired as well (or instead) with the formal arts and sciences (because, his brother says, "correctness of discourse about valid truth is a unity and therefore the common property of the Muses"). So by the Roman period natural philosophy had been assigned an honored Goddess as its holy patron. Plutarch thus shuffles all forms of poetry under music (a department of mathematics) and assigns only a single Muse to this, while geometry he then associates with astronomy (and thus assigns to Urania, as others had already done).[1189] Plutarch's brother then explains the absence of a Muse for medicine with the excuse that this science already has its divine patrons in Apollo and Asclepius (and likewise the absence of a Muse for agriculture is explained by farmers already having Dionysus and Demeter), and though he makes no mention of other sciences (like mechanics, which could also have been subsumed under geometry), we know the mechanical arts had their own divine patrons in Hephaestus, Athena, Minerva, and Prometheus.[1190] But natural philosophy spanned and grounded all the sciences, and for this Plutarch declares (now in his own voice) that "everyone would assign to Euterpe the study of the facts of nature, and would reserve no purer or

---

1188. *OCD* 974 (s.v."Muses").They are first named in Hesiod, *Theogony* 76–79 (as the daughters of Zeus and Mnemosyne, the Goddess of Memory: Hesiod, *Theogony* 915–17), but the specific arts each governed are known only from other sources (including Hellenistic and Roman art).

1189. Plutarch, *Questions at a Party* 9.14.3 (= *Moralia* 744e).

1190. Plutarch, *Questions at a Party* 9.14.4 (= *Moralia* 745a). Hephaestus was the Greek god of fire, blacksmiths, and artisans: *OCD* 660–61 (s.v."Hephaestus"); Vulcan, the closest Roman parallel, had no direct association with craftsmen, only life-threatening fires: *OCD* 1563 (s.v."Volcanus (Volkanus,Vulcanus)"); Minerva, the Roman goddess of craftsmen, was a third of the Capitoline Triad at Rome, and thus one of the three principal Gods of the City, after Jupiter Optimus Maximus and his wife Juno: *OCD* 957 (s.v. "Minerva"); Minerva's annual festival was a day for honoring all craftsmen (including doctors), who celebrated by marching in a parade: Ovid, *Fasti* 3.809–21, *OCD* 1251 (s.v."Quinquatrus"), and Graf 2001;Athena, the closest Greek parallel, was also a goddess of craftsmen and technology in her aspect as Athena Erganê ("Athena the Maker"), while in her more general aspect, of course, she was the chief and eponymous goddess of Athens (and by Hephaestus the mother of the first Athenian): *OCD* 194 (s.v."Athena"), cf. Cuomo 2007: 38–39 and Deacy 2008. For Asclepius and Prometheus see section 4.5.

finer enjoyments and delights to any other kind of activity," thus reflecting the elevated status of natural philosophy, and the fact that its pursuit was a respectable pleasure that deserved its own goddess.[1191]

We have also seen (from Cicero above and from Seneca in chapter 3.9.II) that scientific inquiry was equated with the sacred mysteries, and thus comparably pious and holy. Galen says the same of discovering the secrets of human and animal bodies through dissection. Mentioning his groundbreaking work on the physiology of speech, for example, he says "I was the very first to discover this mystery which I now practice" and so "fix your mind now on holier things, make yourself a listener worthy of what is to be said, and follow closely my discourse as it explains the wonderful mysteries of Nature."[1192] As he also says in his final book of the same treatise, in a chapter he entitles his "Ode" to God:

> Such work [on anatomy and physiology] is useful not only for the physician, but much more so for the philosopher who is eager to gain an understanding of the whole of Nature. And I think all men of whatever nation or status who honor the gods should be initiated into this work, which is by no means like the mysteries of Eleusis or Samothrace. For feeble are the proofs these give of what they strive to teach, but the proofs of Nature are plain to be seen in all animals.[1193]

And as we shall see, he saw this not only in animals, but in the order of

1191. Plutarch, *Questions at a Party* 9.14.7 (= *Moralia* 746e). Euterpe (a name that means "the well-pleasing") was more traditionally the goddess of the flute. Also identified as patron of advanced knowledge was Hermes (Mercury among the Romans): *OCD* 668–69 and 935 (s.v. "Hermes" and "Mercurius"). Ancient astrologers placed all craftsmen, including doctors, engineers, and astronomers (as well as rhetors, priests, and prophets), under the sign of Mercury: Nutton 2013: 259.This is loosely echoed in Galen, *Exhortation to Study the Arts* 3–5 (= Kühn 1.5–8; compare Philostratus, *Life of Apollonius of Tyana* 8.7.3), who calls Hermes "the lord of reason and every art."

1192. Galen, *On the Uses of the Parts* 7.14 (= M.T. May 1968: 367).

1193. Galen, *On the Uses of the Parts* 17.1 (= M.T. May 1968: 731). On this chapter as a Hymn to God see below. Equations of scientific research with sacred mysteries might connect with an Orphic belief that those who make useful discoveries will go to heaven (e.g. Cicero, *On the Republic* 6.13 and Virgil, *Aeneid* 6.663 and 847–53), on which see Habinek 1989.

the heavens as well, and we can reasonably assume he would have included the whole world order, from botany to elemental physics, as affording proofs in Nature of these mysteries of God into which all men should be initiated. And yet, as Galen had said earlier, one must not be admitted to those mysteries until he has learned and embraced a sound logical, empirical, and "almost mathematical" method of inquiry.[1194] As he says in physiology "we ought never to leave alone what we have sought out before testing it thoroughly in dissection," and we know he believed this need for empirical testing held for all other sciences as well.[1195]

Galen's religious passion for scientific inquiry is starkly contrary to the religious passions of his Christian peers (as we shall see in chapter five). For Galen, the scientist is "seeking to discover great and noble beauty in the works of Nature," and therefore is not undertaking any vain or improper pursuit.[1196] To the contrary, Galen says it is "impious toward the Creator to leave unexplained a great work of his providence for animals," and thus he endeavored to discuss every detail he could.[1197] As he says most explicitly:

> This sacred discourse I am composing as a true hymn of praise to our Creator. And I consider that I am really showing him reverence not when I offer him unnumbered hecatombs of bulls and burn incense of cassia worth ten thousand talents, but when I myself first learn to know his wisdom, power, and goodness, and then make them known to others.[1198]

Thus, Galen's creationism actually inspired a love of empirical inquiry and the pursuit of scientific understanding. At one point he equates anatomy and physiology with astronomy, as praiseworthy sciences that each discover the wisdom of god by investigating the way things work and why. He then develops the analogy that while common people only see beauty in the material (e.g. the gold of a statue), the true artist sees beauty in the art (e.g.

1194. Galen, *On the Uses of the Parts* 12.6 (= M.T. May 1968: 558–59).

1195. Galen, *On the Uses of the Parts* 15.1 (= M.T. May 1968: 657–58).

1196. Galen, *On the Uses of the Parts* 7.15 (= M.T. May 1968: 369).

1197. Galen, *On the Uses of the Parts* 10.12 (= M.T. May 1968: 491).

1198. Galen, *On the Uses of the Parts* 3.10 (= M.T. May 1968: 189–91). That this entire book was a worshipful Hymn to God (apart from being obvious here, and throughout its last chapter, and scattered passages elsewhere) is also explained in ibid. 17.3 (= M.T. May 1968: 733).

the craftsmanship and skill of the sculptor as revealed in his works), and therefore the scientist does not obsess over material things, but rather the principles underlying how and why they work. And so, Galen exhorts his readers, "Come, let us make you skillful in Nature's art so that we may call you no longer a person without culture, but a natural philosopher instead," obviously something much better to be.[1199] Hence, like the lists in Aurelius and Pliny of history's greatest geniuses, worthy of admiration and fame, Galen produces his own list, though his choices appear more methodical, naming "Plato, Aristotle, Hipparchus, and Archimedes," representing (or so we can infer) cosmology, biology, astronomy, and mechanics—hence for Galen, the greatest of minds, launching all the greatest sciences of antiquity, are the greatest of role models.[1200]

Thus, as the Christian Tertullian once complained, pagans "defend the authority of natural philosophers like valuable property."[1201] And we have relevant examples in Roman law. When Diocletian and Maximian wished to stamp out astrology in 296 A.D., more decisively than any emperor had before, they passed a decree announcing that "to teach and practice geometry serves the public good, but damnable mathematics is prohibited," thus taking care to distinguish between astrology and geometric astronomy as two predictive arts, one "serving the public good," and the other damnable.[1202] A different contrast had already been made between 'scientific' and 'supernatural' knowledge when deciding who counts as a 'doctor' for protections and privileges in Roman law:

> Some will perhaps regard as doctors those also who offer a cure for a particular part of the body or a particular ill, as, for instance, an ear doctor, a throat doctor, or a dentist. But one must not include people who make incantations or imprecations or, to use the common expression of imposters, exorcisms. For these are not branches of medicine, even though people exist who forcibly assert that such people have helped them.[1203]

1199. Galen, *On the Uses of the Parts* 3.10 (= M.T. May 1968: 189–91), cf. also 17.1 (= M.T. May 1968: 730).

1200. Galen, *On the Uses of the Parts* 17.1 (= M.T. May 1968: 730).

1201. Tertullian, *To the Nations* 2.2.

1202. *Codex of Justinian* 9.18.2 (cf. Cuomo 2000: 38–39).

1203. From Ulpian, *All Seats of Judgment* 8 (early 3rd century A.D.), via the *Digest of Justinian* 50.13.1.3.

Medicine was only legitimate, then, if it was based on empirical methodology or some system of natural philosophy, which thus held more respect among the governing elite than popular religion. We will see this order of values reversed among the Christians in chapter five.

I must set aside the complex question of whether anyone imagined a connection between natural philosophy and existing standards of rationality, as there appears to be very little agreement (then or now) as to what those standards were.[1204] But from two treatises specifically written on 'superstition' in the 1st century A.D. (one by a Platonist, another by a Stoic), combined with several powerful asides on the subject by Lucretius (an Epicurean), we can infer that many elite Romans believed a logically and (more or less) empirically grounded natural philosophy was the only source of a proper and reverent religion, while all religion cut free of such a foundation was to them no more than an irrational, and thus reprehensible, 'superstition.'[1205] Notably, descriptions of the evils of superstition focused on the disturbing 'fear' (of both God and Nature) that typified it, which was believed to drive men to adopt irrational behaviors and convictions—a consequence actually thought worse than atheism. We have already seen how natural philosophy was seen as freeing men from all such fears—and yet (as we shall see in chapter five) Christians embraced such religious fear as the basis of their faith, thus reversing pagan elite values again.[1206]

1204. See, for example, Lloyd 1979, Barton 1994a, Frede & Striker 1996, Schmid 1998, Buxton 1999, and Nutton 2013: 272–78 (who specifically discusses the issue of medical rationalism vs. supernaturalism and superstition in antiquity).

1205. Plutarch, *On Superstition*, which is extant (= *Moralia* 164e-171f); Seneca, *On Superstition*, which is lost (but a sizable quotation, enough to grasp the gist of its argument, appears in Augustine, *City of God* 6.10–11, confirmed by a brief description of the same book in Tertullian, *Apology* 12); Lucretius, *On the Nature of Things* 1.62–135, 5.1161–1240, 6.35–95. A fragment of a lost work on the same subject confirms the same sentiments as these (cf. *P.Oxy.* 2.215).

1206. On the ancient concept of 'superstition' as 'bad religion' built on irrational fear see discussion and sources in *OCD* 1413–14 (s.v. "superstitio") and Janssen 1979, M. Smith 1981, Salzman 1987: 172–75, and Aubrion 1996.

4.3 Evidence of Elite Interest

Besides direct articulations of their values, attitudes are further indicated by behaviors. The following are prominent examples of this. We have already seen examples of elite interest throughout chapter two, and I have demonstrated this elsewhere—from the popularity of public anatomical demonstrations and lectures in science and philosophy, to the interest shown in questions of natural philosophy at dinner parties, sparking conversations on sciences as diverse as harmonics, medicine, and astronomy.[1207] The fact that the most educated and accomplished speakers were expected to have a basic background in science and natural philosophy also indicates the same.[1208] We also noted evidence (in chapter 3.5) of a rising interest in biology, botany, geology and other sciences in the early Roman empire. Here we will add more examples to secure the point, but by no means comprehensively.

The most general evidence comes from Dio Chrysostom, who described the intellectual atmosphere of his age when he addressed an audience at Tarsus in the 2nd century A.D.:

> It seems to me you have listened frequently to marvelous men, who claim to know all things, and, regarding all things, to be able to tell how they have been appointed and what their nature is, with their repertoire including not only human beings and demigods, but gods, and even the earth, the sky, the sea, the sun and moon and other stars—in fact the entire universe—and also the processes of corruption and generation, and thousands of other things.

For which, he says, audiences "are full of admiration," to the point that these lectures...

> ...become a spectacle like the exhibitions of so-called 'doctors' who seat themselves conspicuously before us and give a detailed account of the union of joints, the combination and juxtaposition of bones, and other topics of that sort, such as pores and respirations and excretions. And the

1207. Some of these points are demonstrated in Carrier 2016 (index, "lectures"), others here in chapter two.

1208. Demonstrated in chapter five of Carrier 2016.

crowd is all agape with admiration and more enchanted than a swarm of children.[1209]

Dio's remarks, which are intended to be mocking, nevertheless entail these kinds of lectures were common, and public reaction to them positive.

Judging from extant literature, astronomy and cosmology were the most popular of subjects. This is not a secure standard, however, given that what survives reflects medieval more than ancient values, but even if extant astronomical and cosmological literature is disproportionate, this only entails *a fortiori* an even stronger and broader interest in scientific subjects. But astronomy might have been genuinely more popular. We have already seen, for example, that we have more ancient textbooks on astronomy than for any other science (though that could again be a result of the selective bias of medieval antiquarians).[1210] And we have seen that the most popular science-related school text was the astronomical *Phenomena* of Aratus, which enjoyed widespread use and several translations into Latin, although poetry on other subjects in natural philosophy were not unknown.[1211] To these could be added the Latin poet Marcus Manilius, whose epic poem on astrological astronomy includes a didactic section on Stoic natural philosophy, in part a response to the popularity of the Epicurean poem of Lucretius *On the Nature of Things*.[1212]

Cicero had already said the greatest orators of old "both lectured and wrote" a great deal on natural philosophy, thus linking such interests to popular elite role models, while Cicero himself contributed to the subject, most notably in his philosophical treatises on theology and divination, and his unfinished 'translation' of Plato's *Timaeus*.[1213] This latter is most telling. Plato's *Timaeus*, his only treatise on natural philosophy, appears to have

1209. Dio Chrysostom, *Discourses* 33.4–6.

1210. See relevant note on Geminus in chapter 3.3.

1211. See discussion and notes in chapter four of Carrier 2016. Poets in this genre included Dionysius, Oppian, and (to a lesser extent) Mesomedes (see note on latter two in chapter 3.5). See also section 4.10.

1212. Taub 2003: 138–41. On Lucretius and Roman-era Epicurean poetry see note in chapter 1.2.1. On Manilius see *OCD* 892–93 (s.v. "Manilius, Marcus").

1213. Quote from Cicero, *On the Orator* 3.22.128. On Cicero's unfinished (or partially lost) *Timaeus* see Lévy 2003.

been among the most popular Platonic treatises in the Roman period, at least judging from the number of Roman commentaries on it.[1214] Similarly, among the most popular Aristotelian writings in the Roman period was the pseudonymous *On the Universe*, which was in a sense a response to the *Timaeus*.[1215] And like Manilius, though biology and other subjects were not excluded, these works did emphasize astronomy and cosmology. Further interest in astronomical subjects is indicated by more practical or creative endeavors. Sulpicius Gallus wrote a textbook on eclipses (see section 4.5), Julius Caesar wrote an astronomical weather almanac (which the scientist Ptolemy thought a worthy source, according to the concluding paragraph of his *Phases*), and the emperor Titus wrote a poem on a comet he observed from Rome.[1216]

The most impressive example of this Roman interest in astronomy is Plutarch's dialogue *On the Face that Appears in the Orb of the Moon*, half of which records (or fictionally reconstructs) a series of conversations on lunar astrophysics.[1217] Those engaging in the debate seem all to be friends and acquaintances of Plutarch who have dined and spoken together on other occasions.[1218] Those speaking represent a typically diverse gathering of the Greco-Roman elite, resembling the same kind of mixed discussions we saw Galen said he was organizing a century later—so the frequent assumption that several of the characters in the dialogue are fictional is unwarranted, though their presence all on the same occasion could still be a fiction of convenience.[1219] Those present according to Plutarch were: Sextius Sulla (a

---

1214. Besides Cicero's (highly interpretive) translation, at least ten commentaries are known to have existed (cf. Runia 1986: 55–57), including one by Galen (Larrain 1992). Further evidence of the popularity of the *Timaeus* is discussed throughout Reydams-Schils 2003.

1215. Gottschalk 1987: 1132.

1216. On this lost poem: Pliny the Elder, *Natural History* 2.22.89.

1217. Plutarch, *On the Face that Appears in the Orb of the Moon* 1–23 (= *Moralia* 920b-937c) constitutes the science portion (the remainder discusses mythology).

1218. See discussion in Cherniss 1957: 2–8.

1219. For Galen's gatherings see section 4.1. Plutarch's inclusion of a famous contemporary astronomer (Menelaus) may be comparable to Athenaeus's inclusion of a famous contemporary doctor (Galen) among the participants of his (probably) fictional discussion of all things gastronomical in *The Dinnersages* (cf.

wealthy Roman from Carthage), Plutarch's brother Lamprias (a Delphian politician who dabbled in Aristotelian and Academic philosophy), an Aristotelian aptly named Aristotle, the Stoic philosopher Pharnaces (possibly of Persian descent), a philosophy student named Lucius, Theon the grammarian and bibliophile, and two astronomers, Apollonides and Menelaus (*the* Menelaus, discussed in chapter 3.3). This group goes on to discuss advanced questions in optics and astronomy, and theories of matter, inertia, and gravity. However real or fictional this dialogue may be, like Plato's dialogues it would have reflected a socio-intellectual reality familiar to the author and his readers, enough to warrant the conclusion that these kinds of parties and conversations really did happen and really were popular among many of the Greco-Roman elite, as evidence from Galen attests.

As Tracey Rihll rightly says, "What this discussion really shows, and should be emphasized, is that what then passed for astrophysics is a suitable subject of conversation at a dinner party, and that everyone present knows a fair amount about it, and other topics in natural history."[1220] As Harold Cherniss explains:

> It is not a technical scientific treatise and is not to be judged as if it were meant to be such; but it is all the more significant that in a literary work intended for an educated but non-technical audience towards the end of the first century A.D. Hipparchus and Aristarchus of Samos are familiarly cited and a technical work of the latter is quoted verbatim, the laws of reflection are debated, the doctrine of natural motion to the universal centre is rejected, and stress is laid upon the cosmological importance of the velocity of heavenly bodies.[1221]

Indeed, the extant portion of the dialogue opens with the declaration that many "opinions concerning the face of the moon" are "current and on the lips of everyone."[1222] What follows was thus intended to reflect popular debates on the subject, which entails that the scene described, the knowledge and interests represented, were *not* atypical among the educated elite.

Related to all this is the strangely exceptional appearance of a coin

Nutton 2013: 235).

1220. Rihll 1999: 77.

1221. Cherniss 1957: 18–19.

1222. Plutarch, *On the Face that Appears in the Orb of the Moon* 1 (= *Moralia* 920b).

honoring the astronomer Hipparchus. I have not examined the extensive evidence of doctors and medical subjects on ancient coins for the reasons explained in chapter 1.3, and the same could be said of astronomical subjects, since they also shared religious, political, and symbolic meanings rarely linked, at least explicitly, to natural philosophy or scientific research. But it is still hard to explain why, suddenly, near the middle of the 2nd century A.D. and continuing over a century, the astronomer Hipparchus came to be honored on bronze coins of his native Nicea, over two hundred years after he was dead. These "depict a seated man contemplating a globe, with the legend HIPPARCHOS" ('of the Niceans') but offer nothing to explain the use of this motif at that time.[1223] Overlapping this, by spanning the first two centuries A.D., was an issue of coins from Cos honoring their own native hero, Hippocrates, widely regarded as the father of scientific medicine, though again with no indication of why suddenly then.[1224] The reasons in both cases might have been the same—to celebrate the prestige of each city as the home of a great scientist, which would indicate that a scientist could make a city famous and that a city would be proud to say so. It also entailed such scientists had to be renowned enough for handlers of these coins to know who they were, or easily find out.

We have only one historian who purports to tell us why the Nicean coins started honoring Hipparchus, but he is not among the most trustworthy writers, and his story cannot be entirely true—and might not be at all true. At any rate, according to Aurelius Victor in the late 4th century A.D., "Marcus Aurelius had punished the Niceans with a fine because they had been unaware that Hipparchus, a man of outstanding genius, had been a native" of their city.[1225] This 'fine' consisted of an "extraordinary requisition of grain and oil" that was such a heavy annual burden for Nicea that it was abolished by Constantine in the early 4th century, or so Victor claims. No modern

1223. For scholarship see *DSB* 15.207–08 and 15.222 (in s.v. "Hipparchus") and Schefold 1997: 418–19 and 543 (Abb. 302), and Evans 1999: 297–99, legend reading HIPPARCHOS NIKAIEÔN, always with the reigning emperor (bust, name, and titles) on the obverse (with known examples dating from the reign of Antoninus Pius to Gallienus).

1224. Nutton 2013: 212 (with other doctors on Roman coins discussed in Nutton 2013: 261, and see note on the school of Zeuxis in chapter 3.2).

1225. Sextus Aurelius Victor, *On the Caesars* 41.19–21.

scholar seems inclined to believe this story, and as the coin series begins under Pius, not Marcus, at least that much of the account cannot be true. A voluntary issue made by the Niceans, boasting of an admired connection to their scientific past, is more likely. But if Victor's account is even the gist of what actually happened, then it represents a marked appreciation for the heroes of ancient science from the highest levels of Roman society.

Such is the example of astronomy. Galen provides us with similar evidence of elite interest in medical science. As we noted before, no comparably chatty author survives to tell us about ongoing public affairs in the astronomical or engineering sciences, much less ancillary fields like botany or zoology or any of the earth sciences, but from what we have seen for astronomy, and what Galen tells us about medicine, combined with Dio's more general observations, and evidence we surveyed in other chapters and will present in the next section, it seems more than reasonable that the picture Galen draws held as much for other sciences as for his own. Galen names several elite Romans who eagerly attended Galen's lectures and anatomical displays and were keenly interested in his books and scientific findings, and their number and rank is considerable. His audiences "included not only fellow physicians, medical students, and masses of ordinary citizens, but also the political and intellectual elite," including Alexander of Damascus, imperial chair of Aristotelian philosophy at Athens (this may possibly be Alexander of Aphrodisias), "famous sophists" like Aelius Demetrius and Hadrian of Tyre (who held the imperial chair of rhetoric at Athens), Roman prefects (like Lucius Sergius Paulus), consuls (like Flavius Boëthius, Gnaeus Claudius Severus, Marcus Vettulenus Barbarus, and "perhaps" Lucius Calpurnius Piso), even "members of the imperial family, and so on."[1226] It is clear that many of even the highest ranking Romans took an active interest in the affairs of scientific discovery and debate.

Galen reports, for example, that the Roman consul Flavius Boëthius was "as keen an anatomist as ever lived," who conducted his own dissections and avidly read Galen's advanced works on the subject.[1227] Boëthius also kept close company with several leading Aristotelian philosophers. Hence we hear that Galen made "many anatomical demonstrations for Boëthius" who

---

1226. See von Staden 1995: 58 (quoted) and 1997: 36 and 47–49, and Nutton 2013: 230–32, 245.

1227. Galen, *On Conducting Anatomical Investigations* 1.1 (= Kühn 2.215–16).

was "constantly accompanied by Eudemus the Aristotelian, by Alexander of Damascus, official professor of Aristotelian doctrines in Athens, and often," Galen adds, "by other important officials, such as Sergius Paulus the consul, present governor of Rome, a man as distinguished in philosophy as in affairs."[1228] Galen also relates an occasion when an Erasistratean scientist "was always promising to exhibit the great artery empty of blood," relating to a key empirical dispute at the time, "but never did," and "when some ardent young men brought animals to him and challenged him to the test, he declared he would not make it without a fee." Not willing to be outdone by such an excuse, "they laid down at once a thousand drachmas for him to pocket should he succeed," yet he still made excuses. Eventually, under pressure, he attempted to perform the demonstration and failed. So the young men made fun of him, and then conducted the required vivisection experiment themselves, empirically refuting his claim before an interested crowd.[1229]

As we also showed near the end of chapter 3.2, there were many more scientists engaged in such activity than Galen and his colleagues. Galen reports, for example, that "many physicians crowded around" to observe him and his pupils dissecting an elephant killed in the arena, indicating a broader interest in the empirical study of animals.[1230] Galen even complains that dissection for use in natural philosophy—like seeking to answer questions about the purpose and functions of various organs and parts—was becoming increasingly popular at the expense of dissection for surgical and medical utility, which is probably an exaggeration (Galen was occasionally fond of hyperbole), but it nevertheless indicates a rising popularity of the empirical pursuit of knowledge for its own sake, and confirms a widespread interest in scientific dissection.[1231]

For all the reasons noted earlier, there is no reason to think similar interest

---

1228. Galen, *On Conducting Anatomical Investigations* 1.1 (= Kühn 2.218).

1229. Galen, *On Conducting Anatomical Investigations* 7.16 (= Kühn 2.642–43). There are many other examples of such occasions in Galen's writings (e.g. Galen, *On Conducting Anatomical Investigations* 7.16 = Kühn 2.644–46, and see related note in chapter seven of Carrier 2016).

1230. Galen, *On Conducting Anatomical Investigations* 7.10 (= Kühn 2.619–20). See similar examples relating to human dissection in chapter 3.8.III.

1231. Galen, *On Conducting Anatomical Investigations* 4.1 (= Kühn 2.417, 2.420).

was not also shown in other sciences besides medicine, not only by actual scientists but also among the Roman elite. Plutarch, for example, attests that Romans were discussing at dinner parties the technological applications of harmonic theory, and this was considered such a delight that he offers as an argument against the Epicurean life the fact that Epicureans didn't talk about such things. He had already made the same point about mathematics and astronomy.[1232] Likewise, during a trial Apuleius assumed a sitting judge would understand the scientific reasons for dissecting fish, and that it would be obvious to any educated person that even Epicurean physicists would own and experiment with mirrors, and then he pokes fun at a prosecutor for not having read the *Catoptrics* of Archimedes, an advanced scientific treatise (now lost) on the science of reflection.[1233] As S.J. Harrison points out, this indicates "the importance of recondite learning" for "the culture of the period in general," and such expected knowledge included science and natural philosophy to a significant degree.[1234] Indeed, that Apuleius would think to defend himself by calling attention to the scientific nature of his tools and experiments, and that he could thus embarrass a prosecutor who was attempting to accuse him of magic, entails not only that his scientific activities were respectable, but that ignorance of them was shameful.

That interest in natural philosophy ranged much wider in subject than medicine and astronomy is also indicated by a rising tide of reference books in science and natural philosophy. Though original and more advanced science was still written in Greek, even by Romans, Latin was the language of popularization in the West, used to that end by Cicero and Lucretius and Manilius, for example, but most notably by a series of famous encyclopedists.[1235] I've discussed this trend already as an indication

1232. Plutarch, *That Following Epicurus is Unpleasant* 13 (= *Moralia* 1095c-1096c); mathematics and astronomy: ibid. 11 = 1093d-1094d.

1233. Apuleius, *Apology* 16.7 (see related note in chapter 3.5).

1234. S.J. Harrison 2000: 56. Examples of similar expectations of significant medical knowledge among laymen are presented in Nutton 2013: 258 (further evidence, for example, in Ballér 1992, Durling 1995, and Renehan 2000). And see discussion and evidence in chapters five through seven of Carrier 2016.

1235. On the continuing use of Greek as the language of the sciences by a bilingual Roman elite see discussion in chapter three of Carrier 2016.

of a rising interest in science and natural philosophy.[1236] I will end with it again here. The earliest example is a lost encyclopedia of the sciences by Varro in the 1st century B.C., the *Disciplines*, to which he also added an extant treatise on agriculture, and extensive books on many other subjects, from history and literature to law and language, almost all of which have been lost, including a monumental encyclopedia of culture and religion, the *Antiquities of Things Human and Divine*.[1237] All these works reflect the same trend to organize and popularize knowledge in all fields. The *Disciplines* in particular included volumes on grammar, logic, rhetoric, arithmetic, geometry (both the theory and its associated sciences), harmonics, astronomy, medicine, and engineering. A similar interest is reflected in the handbook *On Architecture* by Vitruvius near the end of the same century, which covered mainly the science of building, with an additional book devoted to machinery and another on sundials, clocks, and associated astronomy, but Vitruvius begins the whole by surveying the encyclopedic knowledge required of a competent engineer, practically reproducing a summary list of the encyclopedic subjects addressed by Varro.[1238]

More impressive was the Latin encyclopedia of Aulus Cornelius Celsus, written in the early 1st century A.D. His *Arts* covered multiple sciences in better detail, with many volumes devoted to each. Unfortunately only his volumes on the history and practice of medicine survive. What other subjects were surveyed in the lost volumes, and how many, is not entirely known, although it is certain rhetoric, agriculture and military science were among them, and that his work was highly prized for its excellence.[1239] Agricultural and military science were particularly suited to recently fashionable Roman interests and national character, while medicine and rhetoric were already

1236. Here in chapter 4.5, and previously in chapters three and ten of Carrier 2016.

1237. On Varro's *Disciplines* see *EANS* 774–78 and discussion in chapter five of Carrier 2016.

1238. On Vitruvius and his encyclopedic standards see discussion and notes in chapter seven of Carrier 2016 (with Vitruvius, *On Architecture* 1.1). There was an earlier Latin writer on the subject of architecture, Publius Septimius (Vitruvius, *On Architecture* 7.pr.14), but little is known of his works.

1239. Quintilian, *Education in Oratory* 12.11.24. His *On Agriculture* filled five volumes and was among the best surveys of the subject according to Columella, *On Agricultural Matters* 1.1.14.

topics treated by Varro, so it is conceivable that Celsus treated the same nine arts as Varro and added two or three others.[1240] His extant treatment of medicine is so excellent and incorporates such a quantity of first-hand reports that scholars still debate whether Celsus was himself a doctor.[1241] Though most conclude in the negative, all agree he was superbly educated in the field for a layman, and judging from the opinion of his peers he seems to have been as well versed in all the other sciences he wrote on.

Celsus began his section on medicine with a transition from the prior (lost) volumes on agricultural science, "just as agriculture promises nourishment to healthy bodies, so medicine promises health to the sick," which suggests he justified the inclusion of various sciences in his collection according to the benefit they provided humanity—as we can expect he opened with similar statements for rhetoric and military science, and whatever other arts he surveyed.[1242] This accords with our earlier observation (in chapter 3.5) that the sciences experiencing the most advancement in antiquity were those perceived to be the most useful in practical affairs. But as we saw in section 4.1, such 'utility' could include the moral value of scientific knowledge and inquiry, which appears to have driven another encyclopedia produced later in the first century, this time on meteorology in its broadest sense (hence including the earth sciences), the *Natural Questions* of Seneca. Though all of this does not survive, we can tell it was a popularizing compendium of natural science and philosophy, with an underlying Stoic theme of demonstrating the moral value of studying them. We will discuss this and Seneca in the next section. But we should note here that it included volumes on rivers; on the Nile; on clouds, rain, snow, and hail; on winds; on earthquakes; on comets; on meteors, rainbows and other lights in the sky; on thunder and lightning; and possibly also volumes on the

---

1240. The logic of his transition from agriculture to medicine (as arts that nourish and heal the body) in his preface to the latter might suggest a twelfth and final topic was planned or completed, on philosophy (as the art that nourishes and heals the soul), although other subjects are possible (such as gymnastics or the graphic and plastic arts).

1241. *DSB* 3.174–75, *EANS* 217–19, *OCD* 377, with a handy survey of the debate in Scarborough 1970: 298–302.

1242. Celsus, *On Medicine* pr.1.

sea and volcanoes.[1243]

Finally, a bit later in the same century Pliny the Elder composed what he believed to be the first encyclopedia covering the entire range of natural phenomena, the *Natural History*.[1244] I have discussed this work here before (most notably in chapters 3.8.I and 3.9.II). Not only does it indicate a broadened Roman interest in documenting and understanding the whole of the natural world, it is also clear from its contents that Pliny greatly valued the pursuit of natural studies and considered his encyclopedic work to be useful to humankind. Though Pliny was not himself a scientist but only an avid layman, his enthusiasm for science and natural philosophy cannot have been unique, as we have already seen it was not. Joyce Reynolds argues that Pliny's *Natural History* "was an effort that reflected what seems to have been an enormous contemporary curiosity, an appetite for knowledge" especially of nature and technology, the two most common themes of the *Natural History*.[1245] John Healy even finds that "Pliny and Lucretius share a number of themes," in fact "both have two main aims, namely (1) to explain the Universe and its phenomena in rational terms and (2) to free the minds of men from superstitious fear through a greater understanding of the world," yet Pliny's approach to this same end is more empirical and practical.[1246] The same contrast can be drawn between the more factual approach of Pliny and Celsus, and the more theoretical treatment of Seneca and Lucretius, revealing a continuum of diversity in the way natural philosophy was explored in the early Roman empire.

The contents of the *Natural History* are primarily connected to the earth (like botany, zoology, and geology), and the information included in it was of mixed quality even by the standards of its own day, but the values and interests it reflects are what matter most.[1247] For all its faults, its empirical,

---

1243. On the order and contents of the lost and extant books see scholarship in Taub 2003: 141–42, with 221–22 (notes 60–61 and 66), and Codoñer 1989 and Lausberg 1989.

1244. That he believed he was the first to do this: Pliny the Elder, *Natural History* pr.14.

1245. French & Greenaway 1986: 7.

1246. Healy 1999: 75.

1247. On the merits of his work see related note in chapter 3.5. Greek encyclopedias existed, though more specialized and sometimes of inferior quality,

practical, and exhaustive character is hard to deny, and Pliny's enthusiasm for the subject is everywhere evident. His first volume contained a preface, a table of contents, and a vast list of authorities consulted. The remaining thirty-six volumes were divided as follows: one volume on the universe as a whole (including astronomy and elemental and cosmological theory), four volumes on geography ("in which is contained an account of the situation of the different countries, the inhabitants, seas, towns, harbours, mountains, rivers, dimensions, and populations") plus a more general volume on human beings (and their abilities, limits, inventions, and achievements), one volume on land animals, one on aquatic animals, one on birds, and one on insects (making four volumes on zoology altogether), three volumes on trees, three on vines and fruits, three on gardens and orchards (and the arts of cultivation), and one on flowers and other decorative plants (making ten volumes on botany altogether), six volumes on botanical medicines and five volumes on zoological medicines (making eleven volumes on pharmacology altogether), two volumes on metals and metallurgy, one volume on the art of painting and its materials and techniques, one on sculpture and stone, and one on gems (making five volumes of geology and mineralogy altogether).

This was by no means a comprehensive survey of the subjects in natural philosophy he could have covered, even from the writings available to him, and the reasons for his selection of subjects are not known.[1248] But within the subjects he did cover his aim was to document only what he regarded as established or credible (while occasionally criticizing what he thought was not). Hence he was not aiming to produce an original scientific work, even though he does occasionally report some new fact or speculation. For example, in his zoological survey he eventually arrives at some unanswered questions about insect physiology, where he concludes that his treatise will not concern itself with unsolved questions (like whether insects breathe or have blood in them) since "our purpose is to indicate the manifest nature

e.g. Aelian's *On the Characteristics of Animals* (3rd century A.D., cf. *EANS* 32–34 and *OCD* 18) and the lost works of Apollonius (and/or Alexander) of Myndus (1st century A.D. or B.C., cf. notes in chapter 3.3 and 3.5).

1248. For example, given Pliny's occupation as a naval commander, books devoted specifically to oceanography, navigation, and seafaring (e.g. ship and naval technology) are conspicuously absent. Similar omissions are suggested by the topics of Seneca's *Natural Questions* and other subjects we know many other authors discussed (cf. chapter 3.5).

of things, not investigate their questionable causes."[1249] This essentially describes the *Natural History*, which was not intended to contain ground-breaking scientific research but a catalogue of 'established' facts and observations—just like encyclopedias today. Pliny's information mostly came from original scientific observations and theories drawn from a vast array of books, almost all lost to us now, which is why we cannot treat Pliny's *Natural History* as representative of the best scientific knowledge of his age, but only the average awareness of it among the most educated laypeople of the time. But even with that qualification, the interests and knowledge it represents is considerable.

4.4 Seneca and the Aetna

All the above is enough to establish the existence of a widespread interest in science and natural philosophy among the pagan elite, at least to a degree notably different from the Christians, as we shall see in chapter five. But since the best extant examples of praise and interest come from, or are linked to, the Stoic philosopher Seneca, we must devote particular attention to this. Seneca exemplifies everything we have shown so far, from elite interest in science and natural philosophy, to Roman literary praise of it, and the motivational support it received from the most popular elite philosophies. We have already seen considerable evidence of Seneca's attitudes in chapter 3.8.I and 3.9.II. Here we will add evidence from Seneca that confirms the findings of all three sections before this.

I. Seneca

Seneca was born into an equestrian family near the dawn of the 1st century A.D. (his father being a renowned and successful legal scholar of the previous generation), but then achieved Senatorial rank later in life. After being exiled for eight years by the megalomaniacal Caligula, he was later restored to become the tutor and advisor to Nero, until Nero's tyrannical character became incompatible with Seneca's loftier expectations, resulting in his voluntary retirement, gradual loss of wealth, and eventual forced

1249. Pliny the Elder, *Natural History* 11.2.8.

suicide under Nero's orders in 65 A.D.[1250] We already discussed (in section 4.3) his most ambitious (and only surviving) work on natural philosophy, the *Natural Questions*. This he completed in his retirement, working on it in the years just before and after 62 A.D.[1251] Although this was hardly scientific, more a layman's summary of the field from a largely Stoic perspective, and of mixed quality at that, it is notable for documenting Seneca's rare contribution to astronomical science: his own personal observations of comets, which he used to confirm to his own satisfaction, from among a variety of competing theories, what we now know to be the correct account of them.[1252] He says he had long been fascinated by science and natural philosophy since as early as 41 A.D., having written a study of earthquakes in his youth (now lost), and he regretted the fact that he waited until his retirement to research the subject further.[1253]

Throughout his writings Seneca describes several moral reasons for pursuing scientific study and research, including the pleasure of it. His earliest praise comes from an epistle to his mother during his exile in the early 40's A.D. Among the things he said should console her in his absence is that his exile allows him ample time to study astronomy, an occupation that makes him happy no matter where he may be. Thus he concludes:

> I am as happy and cheerful as when circumstances were best. Indeed, they are now best, since my mind, free from all other engrossment, has leisure for its own tasks, and now finds joy in lighter studies, and being eager for the truth, mounts to the consideration of its own nature and the nature of the universe. It seeks knowledge, first, of the lands and where they be, then of the laws that govern the encompassing sea with its alternations of ebb and flow. Then it takes ken of all the expanse, charged with terrors, that lies between heaven and earth, this nearer space, disturbed by thunder, lightning, blasts of winds, and the downfall of rain and snow and hail.

1250. *DSB* 12.309–10, *EANS* 84–85, *OCD* 92–95 (s.v. "Annaeus Seneca (2)"), and Griffin 1976.

1251. Cf. Seneca, *Natural Questions* 6.1.1 and 7.28.3.

1252. See discussion in chapter 3.3.

1253. Cf. e.g. Seneca, *Moral Epistles* 20.2 and *Natural Questions* 3.pr.3 and 6.1.1. Regarding his lost *On Earthquakes* see Lausberg 1989: 1926–27. He may also have written a lost work *On the Structure of the Universe* (Lausberg 1989: 1928–29), though other works in natural philosophy attributed to Seneca are less likely to be his.

> Finally, having traversed the lower spaces, it bursts through to the heights above, and there enjoys the noblest spectacle of things divine, and, mindful of its own immortality, it proceeds to all that has been and will ever be throughout the ages of all time.[1254]

This almost sounds like a program for the *Natural Questions* written twenty years later. It is clear from this and many other remarks among his writings that Seneca regarded scientific study as respectably pleasurable, and to be as deeply religious as any Christian would consider the study of scripture.[1255]

Nearer the end of his life Seneca argued that:

> It is helpful to study nature, first, to pull us away from sordid matters, and next, to free the mind from the concerns of the body, as we must in order to attain what is great and loftiest of all, and finally, by exercising our intellect on obscurities we will be no less capable of understanding the more obvious.[1256]

In other words, natural science trains and improves the mind. We also gain moral improvement, he goes on to say, because studying nature leads us away from superstition, teaches us the true mind and intentions of the Creator, and eventually helps us to recognize and correct our flaws and vices.[1257] Not only does it thus teach us which behaviors lead to misery and which to joy, but the study itself "arouses our minds to greatness" and as a result "we strive for grand accomplishments," rather than base ones.[1258] Hence "to have seen the universe in your mind and to have subdued

---

1254. Seneca, *To Helvia on Consolation* 20.1–2 (further praise of astronomy occupies chapter 8).

1255. See, for example, Taub 2003: 143, 148–51, 159, 161. Extensive analysis of Seneca's justifications for valuing science and natural philosophy (and linking them to moral philosophy) is provided by Chaumartin 2003 and Scott 1999, who confirm the conclusions presented here.

1256. Seneca, *Natural Questions* 3.pr.18.

1257. These are amply demonstrated as themes throughout the *Natural Questions* in Scott 1999.

1258. Seneca, *Natural Questions* 3.pr.3.

your vices—no victory is greater than this."[1259] In fact, he says, "the mind possesses the full and complete benefit of its human existence only when it spurns all evil, seeks the lofty and the deep, and enters the innermost secrets of nature."[1260]

Likewise to be gained, again, was freedom from fear:

> Since the cause of fear is ignorance, is it not worth a great deal to know in order not to fear? It is much better to investigate the causes and, in fact, to be intent on this study with the entire mind. For nothing can be found worthier than a subject to which the mind not only lends itself but spends itself.[1261]

Thus Seneca concluded his *Natural Questions* with the assurance that "when we enter into the secrets of nature," especially when we are devoted natural philosophers "who deal with this study exclusively," our minds are uplifted and strengthened and can rise above and face any dangers or ills.[1262] This was especially the result of scientific *knowledge*. Borrowing a common cliché, he argues that eclipses were held to be terrifying prodigies only because they seemed contrary to the natural order by those who were ignorant of the way things really worked.[1263] Hence teaching astronomy to kings and emperors would be an admirable gift to them worth more than any gift they could ever bestow in return, because it rescues them from their enslavement to superstitious fears.[1264] Likewise, understanding the natural causes of earthquakes and lightning, and the random inevitability of all such threats, brings the courage to face and accept them.[1265]

1259. Seneca, *Natural Questions* 3.pr.10.

1260. Seneca, *Natural Questions* 1.pr.7.

1261. Seneca, *Natural Questions* 6.3.4.

1262. Seneca, *Natural Questions* 2.59.1–3, that this was the meaning, and originally the end of the treatise, see sources and discussion in Taub 2003: 141–42 and 221–22 (notes 60, 61, and 66).

1263. Seneca, *Natural Questions* 7.1 (cf. also 6.3.2–3).

1264. Seneca, *On Benefits* 5.7. Similarly, Plato saved Dio from fear by teaching him the astronomy of eclipses, according to Plutarch, *Nicias* 23.4. As we'll see in a following section, Cicero and Frontinus told similar stories.

1265. See Scott 1999: 60–62.

But even beyond its utility in purging ignorance, superstition, and fear, scientific research and the philosophical examination of nature are *intrinsically* rewarding:

> What, you ask, will make it all worthwhile? To know nature—no reward is greater than this. Although the subject has many features that will be useful, the study of this material has nothing more beautiful in itself than that it involves us in its own magnificence and is cultivated not for profit but for its marvelousness. Let us examine, then, why natural phenomena occur.[1266]

In fact, Seneca continues, such "investigation is so appealing to me" that though he studied earthquakes himself in his youth, he is eager to study them again in his retirement, to see if anything new has been learned. The value of astronomy was also not limited to understanding scary things, like eclipses of the sun or moon, for when he says "no one could study anything more magnificent or learn anything more useful than the nature of the stars and planets," he includes learning or contemplating what they are made of and why they shine.[1267] He further praises astronomy and meteorology as the loftiest of studies because they "rescue us from fog and darkness," and not only bring us closer to god, but bring us pleasure as well.[1268] In fact, apart from moral philosophy, no intellectual occupation was worthier than natural philosophy. It was, he says, far better than the more popular pursuits of his day, rhetoric and history.

As Seneca recommends:

> Live with Chrysippus, or Posidonius, they will make you acquainted with things earthly and things heavenly, they will encourage you to work hard on something more than neat turns of language and phrases crafted for the entertainment of audiences.[1269]

Elsewhere he says the same of history, much as Plutarch had in his essay on curiosity.[1270] Like Plutarch the Platonist, Seneca the Stoic also defends

1266. Seneca, *Natural Questions* 6.4.2.

1267. Seneca, *Natural Questions* 7.1.6.

1268. Seneca, *Natural Questions* 1.pr.1–2 and 1.pr.12.

1269. Seneca, *Moral Epistles* 104.22.

1270. Seneca, *Natural Questions* 3.pr.5–6.

curiosity about nature as an admirable virtue, necessary to the good life. He clearly believed it was a noble enterprise to "survey the universe, uncover its causes and secrets, and pass them on to the knowledge of others."[1271] He says that "as a curious spectator the mind separates details and investigates them—and why not do this?"[1272] In fact, he says, "If I had not been admitted to these studies it would not have been worthwhile to have been born," for "man is a contemptible thing unless he rises above his human concerns" by studying and understanding the natural world.[1273]

We already noted the connection often made between the study of nature and the sacred mysteries (in section 4.2). Like Galen and Cicero, Seneca says "I, for one, am very grateful to nature, not just when I view it in that aspect which is obvious to everybody, but even more when I have penetrated its mysteries." This, he says again, teaches him about god's powers and designs, and improves his character by teaching him that the universe is vaster than petty human concerns, and that real beauty lies in nature, not in silver and gold.[1274] He also says past scientists deserve our admiration, because "it was the achievement of a great mind to move aside the veil from hidden places and, not content with the exterior appearance of nature, to look within and to descend into the secrets of the gods," and in fact "the man who had the hope that the truth could be found made the greatest contribution to discovery," and therefore this is a model we, too, should follow.[1275] Notably, this religious rationale for pursuing empirical science, is never to be seen in any extant Christian writing for a thousand years. Christianity as a religion, simply did not come packaged with this idea. Renaissance Christians had to borrow it from *pagan* religion, and try to repackage it as Christian.[1276]

Seneca ties all these ideas together in an essay he wrote on how one should spend his free time or retirement. Among the activities we ought

---

1271. Seneca, *Natural Questions* 3.pr.1.

1272. Seneca, *Natural Questions* 1.pr.12.

1273. Seneca, *Natural Questions* 1.pr.4–5.

1274. Seneca, *Natural Questions* 1.pr.3 (with the whole of 1.pr).

1275. Seneca, *Natural Questions* 6.5.2–3. Seneca draws a lengthier analogy between science and the sacred mysteries in *Moral Epistles* 90.26–29.

1276. On which point see Carrier 2010: 399; and chapter five here.

to occupy ourselves with, he says, is natural philosophy and the study and examination of the natural world, because this is what God intended. Seneca lists many examples of questions to explore, from the nature and behavior of matter to that of the oceans and skies and heavens, all the way to fundamental questions of creation and cosmology. In the midst of all this he says:

> And what service does he who ponders these things render unto god? He keeps the mighty works of God from being without a witness! We are fond of saying that the highest good is to live according to Nature. But Nature has begotten us for both purposes—for contemplation *and* for action. Let me now prove the first. But what more need I do? Will this not be proved if each one of us shall simply think to himself, and ponder how great is his desire to gain knowledge of the unknown, and how this desire is stirred by tales of every sort? Some sail the sea and endure the hardships of journeying to distant lands for the sole reward of discovering something hidden and remote. It is this that collects people everywhere to see sights, it is this that forces them to pry into things that are closed, to search out the more hidden things, to unroll the past, and to listen to the tales of the customs of barbarous tribes.
>
> Nature has bestowed upon us an inquisitive disposition, and being well aware of her own skill and beauty, has begotten us to be spectators of her mighty array, since she would lose the fruit of her labour if her works, so vast, so glorious, so artfully contrived, so bright and so beautiful in more ways than one, were displayed to a lonely solitude. That you may understand how she wished us, not merely to behold her, but to gaze upon her, see the position in which she has placed us. She has set us in the centre of her creation, and has granted us a view that sweeps the universe. And she has not only created man erect, but in order to fit him for contemplation of herself, she has given him a head on top of his body, and set it upon a pliant neck, in order that he might follow the stars as they glide from their rising to their setting, and turn his face about with the whole revolving heaven. And besides, guiding on their course six constellations by day, and six by night, she left no part of herself unrevealed, hoping that by these wonders which she had presented to man's eyes she might also arouse his curiosity in the rest. For we have not beheld all her wonders, nor the full compass of them, but our vision opens up a path for its investigation, and lays the foundations of the truth so that our research may pass from revealed to hidden things and discover something more ancient than the world itself. ...
>
> Man was born for inquiring into such matters as these. ... Consequently I live according to Nature if I surrender myself entirely to her, if I become

her admirer and worshipper. For Nature intended me to do both—to take action and to have leisure for contemplation.[1277]

Seneca thus finds religious reasons to fully integrate science and natural philosophy into the Stoic scheme of values, bringing it almost as high in its importance as the contemplation of ethics and virtues. The Stoic addage "Live according to Nature" here becomes a reason to study nature, not only so that we may live according to her, but the very activity of studying her *is* living according to her. Thus curiosity and the value of advancing our knowledge about nature are elevated as religious virtues. Seneca did believe this should be done empirically as much as possible, and though we know he was unaware of the best method for it, we can be sure his Stoic values and expectations would have led him to accept increasingly valid empirical demonstrations on any point of natural science, if they were logically sound and based on confirmable observations.[1278]

We have one example of these values that may be linked to Seneca and his books. In one of his letters Seneca asks his equestrian friend Lucilius to investigate first-hand various natural phenomena in and near Sicily (as I mentioned in chapter 3.9.II).[1279] This is similar to when Pliny the Younger requested help from 'the most learned' Sura in explaining a strange spring in Italy, and later some poltergeist phenomena.[1280] But in this case we have

1277. Seneca, *On Leisure* 4.1–5.8 (= *Dialogues* 8.4.1–8.5.8).

1278. As argued in Scott 1999: 59 and evident throughout Seneca's treatment of subjects in the *Natural Questions*, in which priority is given to empirical observations when they are available (which for him always trump authorities, armchair logic, or dogmatic requirements).

1279. Seneca, *Moral Letters* 79. Cf. *OCD* 862 (s.v. "Lucilius (2) (Iunior), Gaius"). This Lucilius is also the addressee of Seneca's *Natural Questions* and *On Providence* (= *Dialogues* 1). Though Seneca's letters are literary creations (generally composed after-the-fact), the events and communications they refer to are probably genuine (cf. D. Russell 1974, with Griffin 1976: 3–6, 349–50, 416–19). In this case, for example, Seneca and Lucilius could already have exchanged correspondence on this matter years before, which Seneca then conveyed and summarized in this literary creation (quite possibly after the publication of the *Aetna*, a poem discussed below).

1280. Pliny the Younger, *Letters* 4.30 (this spring was briefly mentioned by his uncle: Pliny the Elder, *Natural History* 2.106.232) and (for ghosts and poltergeists) 7.27. On Sura see *OCD* 835 (s.v. "Licinius Sura, Lucius").

more vivid detail on Roman interests and methods in natural philosophy. As Seneca writes:

> I am expecting letters from you, in which you were to inform me what new things were revealed to you during your trip all round Sicily, especially more reliable information about Charybdis itself. I know very well that Scylla is a rock—and indeed a rock not much dreaded by mariners. But with regard to Charybdis I would love to have a full description, to see whether it agrees with the fabulous stories about it. If you have by chance observed it (and it is certainly worthy of your taking a look!), please give us a better report. Is it driven into a whirlpool entirely by a single wind, or does every storm twirl the sea around there as much as any other? And is it true that anything seized by the whirlpool in that strait is pulled under and dragged for many miles, and then emerges on a beach near Tauromenium?
>
> If you will write me a full account of these matters, I will then have the boldness to ask you to climb Aetna, too, at my special request. Some people are inferring that it is being consumed and gradually sinking, because at some time or other it used to be visible to sailors from a greater distance. It is possible this is happening not because the height of the mountain is decreasing, but because the flames have dimmed and its eruptions have become less violent and robust, and for the same reason is producing less smoke during the day. Neither explanation is incredible—that a mountain that is daily being devoured would shrink, or that it would remain the same size because it is not eating itself up but, instead, new material is seething up from some infernal chasm below, and the mountain's eruptions are continuously fed by this—the mountain in that case not being its own food, but a mere passageway. In Lycia, for example, there is a well-known place, which the natives call Hephaestion, where the ground is broken through in many places, where a harmless fire burns all around without causing any harm to what grows there. Hence the area is fertile and lush, since the flames do not scorch but merely shine with a force that is mild and weak.
>
> But let us postpone this discussion, and look into it when you have given me a description of just how far the snowline falls from the mouth of the volcano—I mean the snow which does not melt even in summer, so safe it is from the adjacent fire. There is no need for you to charge this project to my account. For you were going to satisfy your own mad passion without a commission from anyone. What do I have to offer you, not merely to describe Aetna in your poem, nor just to touch briefly on this solemn tradition among all the poets? Ovid was no less prevented from treating this subject because Virgil had already fully covered it, nor could either of them discourage Cornelius Severus from doing the same. To the

> contrary, this topic has happily surrendered to them all, and those who have gone before seem to me not to have forestalled all that could be said, but merely to have opened the way.
>
> It makes a big difference whether you approach a subject that has already been exhausted, or only just tilled, when material grows day by day and what is already discovered does not hinder discoveries anew. Besides, the last to write is in the best situation. He finds words already prepared, which when arranged a different way have a whole new look. He is not stealing words, as if they belonged to someone else, for they are public property. At any rate, if Aetna does not make your mouth water, I do not know you![1281]

Here we have two elite Romans communicating with each other about natural phenomena in oceanography and volcanology, exhibiting skepticism and seeking empirical data to judge by, working together to gather and examine that data, and exhorting each other to write up their findings to win fame and satisfy their personal passion for knowledge. Though Seneca concludes here by discussing a poem about Aetna that Lucilius was working on, not an original scientific treatise, we know there were such treatises, and even for this poem, first-hand observation is considered a requirement, and Seneca asks his friend not to write some fluff piece or mere epigram, but to really cover the subject in substantial detail. These two men cannot have been unique in their interests, communications, activities, and expectations. Hence this letter presents evidence of the kind of interest that was taken and research conducted, even by laypeople, in questions of natural philosophy in the early Roman empire.

II. The Aetna

As it happens, we probably have the result of Seneca's request. The *Aetna* is an extensive Latin poem that survives among the manuscripts of the writings

1281. Seneca, *Moral Letters* 79.1–7. For those curious, Charybdis (now dubbed Gerofalo) is technically real; it manifests sometimes as a small (and thus quite harmless) maelstrom in the Strait of Messina (between Sicily and Italy), caused by the *real* danger to navigation there: unnaturally strong tidal currents, which can rapidly pull ships into unexpected courses (Bignami & Salusti 1990). And Hephaestion is a real place: natural gas burns from the ground at Yanartash (modern "Burningstone," in the province of Lycia, now Turkey), which in antiquity was named after a nearby temple to Hephaestus (the gods' blacksmith).

of Virgil. Scholars agree it cannot be by Virgil, and that it ended up bundled with his writings sometime in the middle ages. The text has been transmitted terribly, by a long series of scribes of little competence, but modern textual analysis of surviving manuscripts has remedied this somewhat, and much of it has been restored. From internal evidence it has rightly been concluded that the *Aetna* was written after the *Natural Questions* (which must have been completed between 62 and 65 A.D.), as it alludes to it in several ways, but also before the eruption of Vesuvius in 79 A.D., which it never mentions yet certainly would.[1282] Its apparent date of composition, its connection to the *Natural Questions* not only in style but also factual and philosophical content, and the evidence from Seneca that Lucilius, the dedicatee of the *Natural Questions*, was planning an epic poem on this very subject, *and* had made a personal expedition to view and explore the mountain described, all combine to argue rather strongly for Lucilius as its author. Especially since the poem shows many examples of vivid first-hand observation of the mountain and its surroundings, even of experiments performed there, such as striking lava rocks with blades to inspect the effects.

The *Aetna* treats the volcanology of the Sicilian mountain in over six hundred lines, with vivid naturalistic descriptions as well as theoretical explanations of the phenomena thus described, and with content that is very Senecan and Posidonian in character. It builds on the basic scientific principle, defended by Seneca and so many others, that the secrets of nature can and should be learned by observing the effects of hidden mechanisms, and then developing explanatory theories about those hidden causes. "Only let your mind guide you to an understanding of cunning research," the poet sings, "and from things clearly seen, derive faith in things unseen."[1283] Most relevant to our present concern is a lengthy passage devoted to praising

1282. *OCD* 30 (s.v. "*Aetna*" and "Aetna (1)") and *EANS* 39. For an accessible introduction, Latin edition, and English translation of the *Aetna* see Duff & Duff 1935: 351–419. For the most recent scholarship on the *Aetna*'s authorship and date see W. Richter 1963: 1–8, Paisley & Oldroyd 1979, and Goodyear 1984; and on its content and purpose, Taub 2008: 30–55. For critical editions of the Latin text (with commentary) see W. Richter 1963 and Goodyear 1965. Regarding its date, some note the *Aetna* also does not mention an earthquake that struck Pompeii in 63 A.D., but there is no reason to expect that a poem on volcanoes would mention an earthquake that, at the time, was not linked to volcanic activity.

1283. *Aetna* 144–45.

the value of natural science, over against the myths and superstitions of the foolish rabble, as well as those among the elite who would prefer only practical scientific applications over a broader theoretical understanding of the natural world.

When asking what the causes are of the volcano's behavior, the poet declares:

> I shall follow up the inquiry. Infinite is the toil, yet fruitful, too. Just rewards match the worker's task. Not to gaze on the world's marvels merely with the eye like cattle, not to lie outstretched upon the ground feeding a weight of flesh, but to grasp the proof of things and search into doubtful causes, to hallow genius, to raise the head to the sky, to know the number and character of natal elements in the mighty universe.[1284]

Immediately he follows with a long string of examples, a widely diverse selection of questions in natural philosophy, which he says are of value to explore and answer, ranging from the fundamentals of elemental theory to various issues in astronomy, climatology, and planetary theory, including knowing the distance of the moon and sun and the nature of comets.[1285] "In fine," the poet continues, "to refuse to let all the outspread marvels of this mighty universe remain unordered or buried in a mass of things, but to arrange them each clearly marked in the appointed place—all this is the mind's divine and grateful pleasure."[1286]

Though he had just emphasized astronomy and meteorology and questions of more fundamental physics, the poet troubles to emphasize that the earth sciences are just as important:

> Yet this is man's more primary task—to know the earth and mark all the many wonders nature has yielded there. This is for us a task more akin than the stars of heaven. For what kind of hope is it for mortal man, what madness could be greater—that he should wish to wander and explore in God's domain and yet pass by the mighty fabric before his feet and lose it in his negligence?[1287]

1284. *Aetna* 222–29.

1285. *Aetna* 229–47.

1286. *Aetna* 248–51.

1287. *Aetna* 252–57.

The poet then attacks those who pursue research in mining and agriculture only out of greed, but not deeper subjects in the same domain, which are even more valuable to know.[1288] He concludes:

> Everyone should imbue himself with noble accomplishments. They are the mind's harvest, the greatest reward in the world—to know what nature encloses in earth's hidden depth, to give no false report of her work, not to gaze speechless on the mystic growls and frenzied rages of the Aetnaean mount, not to blench at the sudden din, not to believe that the wrath of the gods has passed underground to a new home, or that hell is breaking its bounds. To learn what hinders the exhalations, what nurtures them, whence their sudden calm and the silent covenant of their truce, why their furies increase...[1289]

And, he goes on to add, countless other things regarding how and why the mountain both behaves as it does and produces its local effects on the terrain and sky.[1290] Though the author does not have the scientific tools and methods we now know are needed to answer most of his questions—hence his activity is still squarely in the domain of ancient natural philosophy and not what we now call science—we can still see skepticism, questioning, curiosity, and empirical observations are all important components of his reasoning, and it is clear that he believes progress could be made on these questions, and that he would readily accept as valuable and credible any more scientific approaches to discovering the causes and underlying mechanisms of the phenomena that fascinate him, were anyone to propose or carry them out. The passion is there. The values are there. All he needed was the right scientific tools and methods.

4.5 The Scientist as Hero in the Roman Era

Another category of evidence for elite values are portrayals of natural philosophers as heroes or role models, as men or women who use science

1288. *Aetna* 258–73.

1289. *Aetna* 274–82.

1290. *Aetna* 282–95 (blending into the rest of the poem, which explores many of these questions).

to save the day, or whose scientific pursuits are portrayed as lofty and noble. There are many examples, but I will only dwell on a few, and summarize some others. I will also not ask whether any of the stories here told are true, since their truth is irrelevant to their social meaning, and it is the latter that concerns us here. Even though Prometheus, for example, almost certainly never existed, and certainly never did anything ancient stories claim of him, the fact that such tales were widely told still tells us a lot about ancient values.

It is as important to note that there are no comparable stories of scientific villainy in the same period, apart from vague fictions about vivisection (mentioned in chapter 3.8.III), and ample criticism of the greed and incompetence of doctors who did not practice their science as they ought. Though the more hostile lower classes may have had a different perspective (as we have seen hints of in chapter 2.6 and will see again in coming pages), we will not explore that question further (except briefly in section 4.8), as our interest is on the values of the dominant elite. Elite Romans were apparently not much moved to disparage science or scientists with tales of their evils or faults—in stark contrast to trends in early *modern* literature.[1291] But they were often keen to valorize them. Even when doctors came in for criticism, it was usually for failing to be *better scientists* (or failing to have genuinely scientific motives), not for being scientists as such.[1292] Likewise, even when astrologers were attacked, it was for *not being scientists*, hence their first defense was to argue the scientific merits of their art.[1293]

There were of course gods and heroes of mixed reputations that could be associated with various sciences, but with one exception they were not regarded as *scientists*. We already discussed the Muses and other patron deities of the arts and sciences in section 4.2. But two other gods in particular draw the most attention. Asclepius, son of the god Apollo and mythical progenitor of the art of medicine, was struck down by Zeus for resurrecting the dead, but then rose from his own death to become a god himself, the leading patron of healing and medicine throughout antiquity well into the Roman period, whose temples became in turn the leading civic centers of healing and healthcare for nearly eight hundred years.[1294]

1291. For example, see Crouch 1975 and Lougee 1972.

1292. As noted in chapter 3.8.I.

1293. See references on ancient astrology in Carrier 2016: 109, n. 285.

1294. See *OCD* 180–81 (s.v. "Asclepius"); Edelstein & Edelstein 1945; Hart 2000;

Likewise, the trickster god Prometheus—creator of humankind, mythical progenitor of the arts and sciences, and patron of crafts and technologies—was repeatedly killed and resurrected by Zeus for conning the gods into a cult contract more favorable to mortals, until he was saved by Hercules, and then stole fire from the gods and gave *that* to mortals, along with all the arts and tools of civilization (and by some accounts it is *this* that angered Zeus into unleashing his ghastly punishment). As for Asclepius, the portrayal of Prometheus is routinely favorable, as having heroically benefitted humanity at his own cost.[1295] And just as Asclepius came to be revered or worshipped by practitioners of medicine, Prometheus was worshipped by potters and other craftsmen, and his theft of fire and other gifts to humanity were celebrated with an annual torch race at Athens.[1296]

Though these gods represent values favorable to the arts and sciences and their associated technologies, they were not portrayed as scientists or natural philosophers. Likewise, the mythic hero Daedalus, "legendary artist, craftsman, and inventor," who built many marvelous things, and whose successful use of his own handcrafted wings to escape from Sicily to Greece became a popular object of Roman art and poetry, was never explicitly associated with the sciences, not even by engineers, nor was he deified.[1297] Although, the Roman engineer Marcus Aurelius Ammianus, who in the 3rd

and Nutton 2013: 104–12, 162–64, 282–90.

1295. Notably, certain Jewish sectarians who later influenced the Christians reversed this sentiment, replacing Prometheus with the Watchers, fallen angels in league with Satan, portraying their bringing to humanity knowledge and technology as an evil deed and a betrayal of God that ruined the world: Portier-Young 2014.

1296. Hesiod, *Theogony* 506–616; Pausanias, *Description of Greece* 1.30.2, 2.19.5–8, 5.11.6, 9.25.6, 10.4.4; [Pseudo-?]Aeschylus, *Prometheus Bound* (esp. lines 442–525; this was part of a trilogy, with *Prometheus Unbound* and *Prometheus Firebringer* extant only in fragments); etc. Modern scholarship: *OCD* 1217 (s.v. "Prometheus") with Lovejoy & Boas 1935: 200–03, Edelstein 1967: 6–7 and 43, Vernant 1983: 237–48, Kreitzer 1994: 11–49, Podlecki 2005: 15–16, Dougherty 2006, and Calame 2010.

1297. See *OCD* 409–10 (s.v. "Daedalus") and Frontisi-Ducroux 2000. It is sometimes forgotten today that it was only his son, Icarus, who flew too close to the sun and thus melted his manmade wings and fell to his doom. Daedalus himself successfully completed the flight. Moderns similarly tend to forget that the Titanic's sister ship, the Olympic, was breached many times but never sank, and her other sister ship, the Britannic, was only sunk by a mine in WWI after being pressed into military service. Our selective memory tells us more about us than the ancients.

century built a water-powered sawmill at Hierapolis, did proudly compare his own inventiveness with that of Daedalus—not just famously a mythic inventor, but also famously the mythic inventor of the saw—which comes close.[1298]

But there is one (probably) mythic 'hero' who was unquestionably a scientist: the lady Hagnodike.[1299] As the story goes, Athenian women (like, also, slaves) were prohibited by law from learning or practicing medicine (though as far as we can tell, this was never true), and as a result, excessively modest women were dying because they were refusing medical treatment by men. So Hagnodike disguised herself as a man and studied medicine under the famous medical scientist Herophilus (the same fellow discussed in Chapter 3.2), then set up practice at Athens. She was then so successful that she was accused by jealous competitors of seducing her female patients, so at trial she disrobed, offering in her defense physical proof that, being a women, she could do no such thing. This threw her onto the horns of a dilemma: either she was guilty of one crime (seduction) or another (practicing medicine), and so she was condemned to death. Then, like something out of a Frank Capra movie, all the leading women of Athens stormed the courthouse in protest, and the men were shamed into not only acquitting Hagnodike, but amending the law to allow women to study and practice medicine. And that is why we now have educated female physicians. Or so this story would have us believe. It is not likely to be true. But this is definitely an example of a mythically heroic scientist, with a positive message about her science: that it saves lives, and that it was respectable for women to learn and practice it.

Such ancient fancy for myth and legend also encouraged admiring stories about the Presocratic philosophers, who were especially revered as the progenitors of natural philosophy (as we saw in chapter 2.4). One common example is the 'olive monopoly' myth. When Pliny says we could prevent or compensate for weather disasters, even improve the yield of our crops, "if mankind did not prefer slandering Nature to benefitting themselves" by studying her (see chapter 3.8.I for an analysis of this rhetorical device), he

1298. Ritti et al. 2007.

1299. The only extant version of her story appears in the Roman author Hyginus, *Fabulous Stories* 274.10–13. See King 1986, von Staden 1989: 38–41, and *EANS* 354 and *OCD* 39–40 (s.v. "Agnodice"), with Irby-Massie 1993: 364–67.

offers to prove his point by telling a heroic tale about a natural philosopher who did just that:

> Democritus was the first person to realize and point out the alliance that unites the heavens with the earth. The story goes that, when the wealthiest of his fellow citizens despised his devotion to these studies, Democritus foresaw, on the principle we have stated and shall now explain more fully, that the rising of the Pleiads would be followed by an increase in the price of oil, which at the time was very cheap because of the crop of olives expected. So he bought up all the oil in the whole of the country, to the surprise of those who knew that the things he most valued were poverty and learned repose. But when his motive had been made manifest and they had seen vast wealth accrue to him, he gave the money back to the anxious and covetous and now repentant landlords. For he was content to have proved that riches would be easily within his reach whenever he chose. A similar demonstration was later given by Sextius, a Roman student of philosophy at Athens. Such is the opportunity afforded by learning, which it is my intention to introduce, in treating the operations of agriculture, as clearly and convincingly as I am able.[1300]

Pliny thus conveys a double lesson: that the study of nature is valuable because it can increase our wealth, and yet a true natural philosopher is not greedy but content to study a science that is rewarding enough in itself.[1301] A very similar story was earlier told of Thales, who was more commonly regarded as 'the first person' to study nature and thus to link the phenomena of the heavens with those of the earth.[1302] But Pliny hints that the same story was being told of the Roman philosopher Sextius, too, thus linking his own Roman heritage with a more established legendary past. This Sextius was famous for having founded a uniquely Roman philosophical sect in the reign of Augustus, having as his most famous pupils Seneca the younger and his own son, Sextius Niger, a renowned natural philosopher and botanist,

---

1300. Pliny the Elder, *Natural History* 18.68.272–274.

1301. A similar sacrifice of wealth to study astronomy is told of Anaxagoras: Diogenes Laertius, *Lives and Opinions of Eminent Philosophers* 2.7.

1302. Diogenes Laertius, *Lives and Opinions of Eminent Philosophers* 1.26; Cicero, *On Divination* 1.49.111–50.112; and at the earliest: Aristotle, *Politics* 1.11.1259a (who is already aware of its legendary nature, correctly noting that it is too ridiculous to be true). A number of other heroic legends accumulated around Thales (cf. Dicks 1959).

thus bringing us right back to the study of nature and agriculture.[1303] None of these stories are credible. But they reflected elite values in the mere telling of them.

Another example is the 'heroic prediction'. Many Presocratic philosophers were heroically portrayed as saving towns by predicting earthquakes, and though few really believed such stories, it is notable that they would be contrived and repeated, along with tales of predicting eclipses, meteorites, plagues, bad harvests, and other scary things, all as rather fantastic exaggerations of the benefits of studying natural philosophy.[1304] For in all these stories it is an understanding of natural causes and correlations, acquired from extended study and reasoning, that makes the predictions possible, never supernatural powers, divinatory rituals, or revelations from God (though the knowledge and wisdom thus accumulated and applied is equated with the knowledge and wisdom of God).

As a typical instance, Pliny says:

> If we are to credit the report, a most admirable and immortal spirit, as it were of a divine nature, should be ascribed to Anaximander the Milesian, who, they say, warned the Spartans to beware of their city and their houses. For he predicted that an earthquake was at hand, when both the whole of their city was destroyed, and a large portion of Mount Taygetus, which projected in the form of a ship, was broken off, and added further ruin to the previous destruction. Another prediction is ascribed to Pherecydes, the teacher of Pythagoras, and this was indeed divine: by a draught of water from a well, he foresaw and predicted that there would be an earthquake in that place. And if these things be true, how nearly do these individuals approach God, even during their lifetime! But I leave every one to judge these stories as he pleases.[1305]

1303. Cf. *OCD* 1358 (s.v. "Sextius, Quintus"), Capitani 1991, and Griffin 1976: 37–42. Celsus the encyclopedist is also thought to have been a pupil.

1304. For example: Cicero, *On Divination* 1.50.112 and 2.13.30–32; Pliny the Elder, *Natural History* 2.59.149 and 7.37.123; Plutarch, *Pericles* 6; Maximus of Tyre, *Orations* 13.5; Diogenes Laertius, *Lives and Opinions of Eminent Philosophers* 1.116; Philostratus, *Life of Apollonius of Tyana* 1.2; Ammianus Marcellinus, *Deeds of the Divine Caesars* 22.16.22; Iamblichus, *Life of Pythagoras* 136; etc.

1305. Pliny the Elder, *Natural History* 2.81.191–92. See my related discussion of Pliny's high hopes for natural philosophy in chapter 3.8.I.

With such fantastic tales natural philosophers were depicted as godly, useful, and morally admirable.[1306]

After the Presocratics, about whom almost any legend could be woven, there continued to arise heroic tales told of scientists more recent and renowned. Every science had its hero. The three most prominent examples are Asclepiades (medicine), Gallus (astronomy), and Archimedes (engineering).[1307] I will say something of each in turn.

Pliny attests to various quasi-heroic tales told of Asclepiades, claiming that he was in Pliny's time the most famous of all scientists:

> The greatest fame goes to Asclepiades of Prusa, who founded a new school, spurned the envoys and promises of King Mithridates, discovered a method of using wine to cure the sick, and brought a man back from the dead and saved his life, but most of all, he made a bet with posterity that he would not be regarded as a doctor if he was ever in any way sick himself, and he won: he was killed at an extreme old age by falling down some stairs.[1308]

So his claims to glory and admiration were not just allegedly new and improved medical treatments, or miraculous applications of his art, but also a feat of moral courage: refusing the wealth and status offered him by an enemy of Rome, and bringing his gifts to the Roman people instead. Pliny paints a more antagonistic picture of Asclepiades elsewhere, criticizing his methodology, but there is none of that antagonism here, where Pliny evidently realizes that general opinion held him in great esteem.[1309]

And thus we find Apuleius paints an even more heroic picture of the man. Though just as fantastical as the legends of the Presocratics, his reasons build on the same assumption that it is through recognizing and

1306. Cicero reports that the Pherecydes-earthquake myth was commonly taught in Roman-era schools (despite it being, Cicero is quick to note, unbelievable): Carrier 2016: 77.

1307. Asclepiades was discussed in chapters 2.3 and 3.2; Archimedes in chapter 3.4.

1308. Pliny the Elder, *Natural History* 7.37.124 (cf. also 26.8.15).

1309. Pliny the Elder, *Natural History* 26.7.12–9.20. Galen likewise criticized Asclepiades and his pupils, even more harshly, but again on grounds of methodology and scientific merit.

understanding natural causes that real miracles are accomplished:

> The famous Asclepiades, who ranks among the greatest of doctors—indeed, if you except Hippocrates, as the very greatest—was the first to discover the use of wine as a remedy. It requires, however, to be administered at the proper moment, and it was in the discovery of the right moment that he showed special skill, noting most carefully the slightest symptom of disorder or undue rapidity of the pulse. It chanced that once, when he was returning to town from his country house, he observed an enormous funeral procession in the suburbs of the city. A huge multitude of men who had come out to perform the last honors stood around the bier, all of them plunged in deep sorrow and wearing worn and ragged apparel. He asked whom they were burying, but no one replied. So he went nearer to satisfy his curiosity and see who it might be that was dead, or, it may be, in the hope to make some discovery in the interests of his profession. Either way, he definitely snatched the man from the jaws of death as he lay there on the verge of burial.
>
> The poor fellow's limbs were already covered with spices, his mouth filled with sweet-smelling unguent. He had been anointed and was all ready for the pyre. But Asclepiades looked upon him, took careful note of certain signs, handled his body again and again, and perceived that the life was still in him, though scarcely to be detected. Right away he cried out, "He lives! Throw down your torches, take away your fire, demolish the pyre, take back the funeral feast and spread it on his board at home!" While he spoke, a murmur arose: some said they must take the doctor's word, others mocked at the physician's skill. At last, in spite of the opposition offered even by his relations (perhaps because they had already entered into possession of the dead man's property, perhaps because they did not yet believe the doctor's words), Asclepiades persuaded them to put off the burial for a brief time. Having thus rescued him from the hands of the undertaker, he carried the man home, as it were from the very mouth of hell, and soon revived the spirit within him, and by means of certain drugs called forth the life that still lay hidden in the secret recesses of his body.[1310]

This legend has all the appearance of a holy man resurrecting the dead, yet is attributed to an application of 'scientific' medicine—at least in the ancient sense, of an art based on natural philosophy, empirical observations, and entirely human actions.

More realistic, though possibly just as fictional, is the heroic tale of the

1310. Apuleius, *Florida* 19. For commentary see S.J. Harrison 2000: 125–26.

Roman field commander (and eventual consul) Gaius Sulpicius Gallus, who was renowned for studying astronomy, and for writing a book on the science of eclipses (which may be the origin of the ensuing legend).[1311] Cicero lavishes praise on Gallus for his tireless research in astronomy, including his production of mathematical diagrams and predictions and measurements.[1312] According to legend his opportunity for scientific heroism occurred at the battle of Pydna in 168 B.C., resulting in a decisive Roman triumph over the Macedonians, which effectively secured Roman dominance in the East for centuries to come. In a fictional dialogue, Cicero has Scipio Aemilianus recollect his acquaintance with Gallus:

> I myself loved the man, and I was aware that he was also greatly esteemed and beloved by my father [Aemilius] Paulus. For in my early youth, when my father, then consul, was in Macedonia, and I was in camp with him, I recollect that our army was on one occasion disturbed by superstitious fears because, on a cloudless night, a bright full moon was suddenly darkened. Gallus was at that time our lieutenant (it being then about a year before his election to the consulship), and on the next day he unhesitatingly made a public statement in the camp that this was no miracle, but that it had happened at that time, and would always happen at fixed times in the future, when the sun was in such a position that its light could not reach the moon.[1313]

At this point Tubero, another character in Cicero's dialogue, asks in astonishment whether Gallus could actually have educated an army of illiterate hicks on such a complex point, and Scipio answers that indeed he did, and with great success, "for his speech showed no conceited desire to display his knowledge, nor was it unsuitable to the character of a man of the greatest nobility," and as a result "he accomplished a very important result

1311. See Pliny the Elder, *Natural History* 2.9.53–55, with *OCD* 1412 (s.v. "Sulpicius Gallus, Gaius"), Rawson 1985: 162, and Bowen 2002 (though Bowen provides a useful survey of the evidence, his conclusions are undermined by a flawed treatment of both the chronological problems and the sources, e.g. he ignores the Antikythera computer as evidence of eclipse prediction as a going concern of the 2nd century B.C., and assumes modern computer calculations of solstice dates correspond to ancient identifications of solstitial days).

1312. Cicero, *On Old Age* 14.49.

1313. Cicero, *On the Republic* 1.15.23–24.

in relieving the troubled minds of the soldiers from foolish superstitious fear."[1314]

Thus, as Valerius Maximus would later write, "because Sulpicius Gallus was very enthusiastic about studying every type of learned work, he thereby did a great service to the Republic," for "it was Gallus' knowledge of the liberal arts that paved the way to the famous victory of Paulus, because if Gallus had not vanquished the panic of our soldiers, our general could not have vanquished our enemies."[1315] Here we have scientific astronomy not only defeating superstition but heroically saving the Roman Empire. Notably different accounts are provided by Livy, who has Gallus announce the coming eclipse *in advance* (thus exaggerating the achievement), and Zonaras (summarizing Cassius Dio), who does not mention Gallus but says Aemilius Paulus learned of the coming eclipse in advance (from Gallus?) and informed the troops himself (Dio's original text may have been more accurate in the details).[1316] But Plutarch's version of events, which may have been based on a lost account by the contemporary (and usually very reliable) Polybius, not only fails to mention Gallus, but says though Paulus knew the astronomical facts, he did not inform his troops but played along with their superstitions.[1317] Since this might be the truth, it is worth noting that the legendary Thales, the recognized 'founder' of natural philosophy, was said to be the first to predict an eclipse, on an occasion *also* coinciding with a major battle, and similar stories were told of Agathocles and Pericles.[1318]

1314. On the symbolic use of science in dispelling a fear of eclipses see section 4.4. For an attempt by the emperor Claudius to use such a tactic on a grander scale see Carrier 2016: 133–35.

1315. Valerius Maximus, *Memorable Deeds and Sayings* 8.11.1, which recounts essentially the same story in Cicero, as also in Pliny the Elder, *Natural History* 2.9.53–55; Frontinus, *Stratagems* 1.12.8; and Quintilian, *Education in Oratory* 1.10.47.

1316. Livy, *From the Founding of the City* 44.37 and Zonaras, *Epitome of History* 2.316–17 (paraphrasing a lost section of Cassius Dio, *Roman History* 20).

1317. Plutarch, *Aemilius Paulus* 17.7–12. On the competing strategies of exploiting the superstitions of one's troops vs. teaching them science instead, see Carrier 2016: 83.

1318. Thales: e.g. Herodotus, *History* 1.74. Agathocles and Pericles: e.g. Cicero, *On the Republic* 1.16.25; Plutarch, *Pericles* 35.2 and *Nicias* 23; Frontinus, *Stratagems* 1.12.9–10; Quintilian, *Education in Oratory* 1.10.46–48; Valerius Maximus, *Memorable Deeds and Sayings* 8.9.ext.2.

Thus, we may have here yet another attempt to update Greek heroes into Roman, just as Pliny had hinted that Sextius had also done the same thing as Democritus and Thales.

The legendary heroism of Gallus the astronomer was widely told. But many more tales of amazing feats and admirable genius were attributed to Archimedes, too many to survey them all here. Yet what is most remarkable is that he was always praised and admired by the Roman elite, despite the fact that he ended his life on the wrong side of a war with Rome and was actually responsible for inflicting horrifying casualties on the Roman army and navy.[1319] Archimedes was for most of his life an ally of Rome, and also a Sicilian, and thus, though a Greek, he may still have been regarded as practically one of their own. But his genius as a scientist seems to have been the principal factor inspiring this unquenchable Roman admiration. Cicero, for example, argues that it is better to be a scientist like Archimedes, and delight in theoretical inquiry and discovery, than to be an all-powerful tyrant—such was the order of values among the noble-minded.[1320]

Among the heroic tales told of Archimedes, the most famous was his miraculous feat of military engineering against the Romans at the battle of Syracuse in 212 B.C. Regarding his accomplishments there, in one broad stroke, in the late 1st century A.D. the Latin poet Silius Italicus captured the general Roman sentiment of the time:

> It was the ingenuity of a Greek, and cunning more powerful than force, that kept Marcellus and all his threats at bay by sea and land, and the mighty armament stood helpless before the city walls. For there was living then in Syracuse a man who sheds immortal glory on his city, a man whose genius far surpassed that of other sons of earth. He was poor in this world's goods, but to him the secrets of heaven and earth were revealed. He knew how the rising sun portended rain when its rays were dull and gloomy. He knew whether the earth is fixed where it hangs in space or shifts its position. He knew the unalterable law by which Ocean surrounds the world with the girdle of its waters. He understood the contest between the moon and

1319. For examples of this praise see chapter 3.9.II and sections 4.2 above and 4.6.I below. For scholarship and a summary of his life and achievements see chapter 3.4 (with related discussion of his orreries and armillary spheres in chapter 3.3).

1320. In Cicero's account of his own 'heroic' discovery of the lost tomb of Archimedes: Cicero, *Tusculan Disputations* 5.23.64–66 (discussed in Carrier 2016: 135, n. 373; and: Simms 1990, Cuomo 2001: 197–98, and Jaeger 2002).

> the tides, and the ordinance that governs the flow of Father Ocean. Not without reason men believed that Archimedes had counted the sands of this great globe. They say, too, that he had elevated ships and carried high great buildings of stone, though drawn by women only. This man wore out by his devices the Roman general and his men.[1321]

Thus we have a military engineer whose application of scientific knowledge was so heroic even his *enemies* revered him. Though Marcellus knew Archimedes was responsible for everything suffered by the Romans in their attempt to capture Syracuse, "he was nevertheless delighted with the exceptional intelligence of this man and decreed that his life should be spared, believing that he would win as much glory by saving Archimedes as he had won by destroying Syracuse."[1322] But, as the story goes, a greedy or hasty (or perhaps vengeful) soldier killed Archimedes in the ensuing sack of the city, a fact that enraged Marcellus, who allegedly regarded this soldier as a murderer, then honored the family of Archimedes and personally ensured his proper burial.[1323] True or false, such legendary reverence for a great scientist was not unique—it is also attested for the Greek scientist Posidonius, who by all accounts was treated with remarkable deference by

1321. Silius Italicus, *Punic Wars* 14.338–53 (following his account of the ingenious weaponry deployed against the Romans: 292–337). Note that Silius's reference to Archimedes counting the sands of the earth is a deliberate allusion to *The Sandreckoner*, in which Archimedes more or less does exactly that. Possibly Silius's other statements also allude to books of Archimedes known at the time but now lost. There are many extant accounts of Archimedes' mechanical defenses of Syracuse: e.g. Polybius, *Histories* 8.3–7, Livy, *From the Founding of the City* 24.33–34, and Plutarch, *Marcellus* 14–17. Polyaenus, *Stratagems* 8.11.1 merely mentions the fact, while the accounts in Cassius Dio, *Roman History* 15 and Diodorus Siculus, *Historical Library* 26.18 are lost, though more or less paraphrased in John Tzetzes, *Book of Ages* 2.103–49 and Zonaras, *Epitome of History* 9.4–5 (= 2.262–65).

1322. Valerius Maximus, *Memorable Deeds and Sayings* 8.7.ext.7.

1323. Various accounts of Archimedes' death are given in some of the accounts of the siege of Syracuse (see earlier note) and also in: Cicero, *On the Boundaries of Good and Evil* 5.19.50, *Against Verres* 2.4.58.131, and *Tusculan Disputations* 5.23.64–66; Livy, *From the Founding of the City* 25.31; Pliny the Elder, *Natural History* 7.37.125; Plutarch, *Marcellus* 19. On a mosaic depicting it (though a suspected forgery) see chapter 1.3.

Cicero and Pompey and others among the Roman elite.[1324]

A very different story of Archimedes' scientific 'heroism' is related by Vitruvius, who writes that "Archimedes in his infinite wisdom discovered many wonderful things, but one in particular seems to convey his boundless ingenuity," which he then describes. King Hiero of Syracuse suspected he had been cheated by a craftsman contracted to manufacture an elaborate gold grown for him, having heard a report that this craftsman had debased the metal with silver in order to abscond with some of the gold the king had given him for the project. So Hiero asked Archimedes to figure out a way to test the metallic content of the crown—without, of course, destroying it.

In what might be the first application of science to criminal forensics, Archimedes pondered and eventually solved the problem. According to Vitruvius:

> Now Archimedes, once he had charge of this matter, chanced to go to the baths, and there, as he stepped into the tub, he noticed that however much he immersed his body in it, that much water spilled over the sides of the tub. When the reason for this occurrence came clear to him, he did not hesitate, but in a transport of joy he leapt out of the tub, and as he rushed home naked, he let one and all know that he had truly found what he had been looking for—because as he ran he shouted over and over in Greek, *Eureka! Eureka!* ("I found it! I found it!").[1325]

Whether there is any truth to this story is unknown. Vitruvius is not our only source for it (Plutarch mentions it), and it is clearly drawn from some other source now lost to us, and it's not incredible for a scientist to have behaved that way. But it can just as easily be a legend. Either way, the outcome would have been Archimedes' treatise *On Floating Bodies*—and a conviction of the craftsman for fraud. But whether true or only a fiction—indeed, even *more* so if a fiction—this story communicates respect and admiration for scientific discovery through reason and empirical observation, elevating these virtues above even modesty and decorum. For the Roman Vitruvius

1324. For example: Cicero, *Tusculan Disputations* 2.61; Pliny the Elder, *Natural History* 7.30.112; Plutarch, *Pompey* 42; cf. Rawson 1985: 106, Kidd 1988: 22–30 and 1999: 38–41, and (more cynically) P. Green 1990: 642–43. On Posidonius see chapter 3.3.

1325. Vitruvius, *On Architecture* 9.pr.9–12, abbreviated in Plutarch, *That Following Epicurus is Unpleasant* 11 (= *Moralia* 1094c).

this event ranked among the discoveries for which Archimedes should have received the greatest public rewards—even elevation to the status of a god.[1326]

Tales of scientific heroism were also attached to Romans of the first two centuries A.D. In one case the story comes from the 'hero' himself, the Roman engineer Lucius Nonius Datus, who set up an inscription in 157 A.D. telling the tale of how he 'saved the day' by employing his mathematical and leadership skills to fix a mistake his predecessors had made on a difficult tunneling project in Numidia (now Algeria).[1327] The inscription begins with a fragmentary letter from an official at Saldae begging the emperor to send Datus to help set right what had gone wrong. Then Datus narrates what happened:

> I set out on the journey, and was attacked by brigands. Naked and wounded, my men and I managed to escape. I arrived at Saldae. I met Clemens the provincial governor. He took me to the mountain, where they were uncertain and weeping about the tunnel, on the point of giving up the whole thing, because the tunnelers had covered a distance greater than that from side to side of the mountain. It turned out that the cavities diverged from the straight line, to the point that the upper end of the tunnel was leaning to the right southwards, and analogously the lower end of the tunnel was leaning to its right northwards: so the two parts were diverging, deviating from the straight line. But the straight line had been marked off with stakes on the top of the mountain, from east to west. [...] When I assigned the work, to make them understand how to do the tunneling, I set a competition between the team from the navy and the

1326. Vitruvius, *On Architecture* 9.pr.1–3. Other quasi-heroic feats attributed to Archimedes are discussed in Russo 1996: 25–26 and Simms 1991, 1995, and 2005. Plutarch, in *Marcellus* 16.3 and 17.1–4 relates how Marcellus and his soldiers also regarded his achievements in military engineering to be godlike and heroic in scale. That those who do anything of benefit to mankind should be deified as a reward is also suggested in Pliny the Elder, *Natural History* 2.5.18 (since "for mortal to aid mortal, that is God, and the path to eternal glory"). Accordingly, some engineers considered their skills divine: R. Taylor 2003: 10 and Tybjerg 2003: 457–62.

1327. *CIL* 8.2728 (= *ILS* 5795). Translation and discussion in Cuomo 2001: 158–59 and Cuomo 2011. For the broader context of similar inscriptions by builders and engineers see Kolb 2015. Though we can't be sure of his ethnicity, it is notable that Datus is an African name; so his story may belong to black history as well as the history of science (Cuomo 2011: 158–59).

> team from the javelin division, and in this way they met in the middle of the mountain. [...] Having completed the work, and released the water, the provincial governor Varius Clemens inaugurated it. [...].

Datus clearly spoke with pride about the trials he endured in his quest to solve the distressing problem of the Saldaeans, and the ingenuity of his solution, all to secure the public benefit of a new water supply for the locals. His account almost reads like the plot to an adventure movie, and it is clear he imagined himself as something of a hero. Though nowhere near as momentous as any of the tales we have heard so far, his story reflects the same ideals, and is an example of the sort of event from which more exaggerated legends could begin.

Pliny the Elder was also valorized—having died, like Archimedes, in the grip of his devotion to natural philosophy. Though a layman in the sciences, he was always keen to observe what he could of natural phenomena, being a great fan of science and natural philosophy as we have seen; and his nephew (and adopted son) clearly saw him as a heroic naturalist.[1328] For inclusion in a history Tacitus was writing (the relevant section of which is now lost), Tacitus asked his good friend Pliny the Younger to tell him about the circumstances of his uncle's death, which he certainly knew from being there.

Like the inscription from Nonius Datus, the letter Pliny wrote back is so exciting in detail, and its content so revealing, it must be quoted at length.

> Gaius Pliny, to his friend Tacitus, greetings.
>
> Thank you for asking me to send you a description of my uncle's death so you can leave an accurate account of it for posterity. I know that immortal fame awaits him if his death is recorded by you. It is true that he perished in a catastrophe which destroyed the loveliest regions of the earth, a fate shared by whole cities and their people, and one so memorable that it is likely to make his name live forever. Of course, he also wrote a number of books of lasting value. But you write for all time and can still do much to perpetuate his memory. The fortunate man, in my opinion, is he whom the gods have granted the power either to do something which is worth recording or to write what is worth recording, and most fortunate

1328. All evidenced in chapter 3.8.1 and 3.9.11 and in sections 4.2 and 4.3 above. Pliny the Younger was also (eventually) his uncle's son by adoption: *OCD* 1162–63 (s.v. "Pliny (2) the Younger").

of all is the man who can do both. Such a man was my uncle, as his own books and yours will prove.

My uncle was stationed at Misenum, in active command of the fleet. On 24 August, in the early afternoon, my mother drew his attention to a cloud of unusual size and appearance. He had been out in the sun, taken a cold bath, and reclined for lunch, and was then working at his books. He called for his shoes and climbed up to a place that would give him the best view of the phenomenon. It was not clear at that distance from which mountain the cloud was rising (it was afterwards known to be Vesuvius). Its general appearance can best be expressed as being like a pine rather than any other tree, for it rose to a great height on a sort of trunk and then split off into branches, I imagine because it was thrust upwards by the first blast and then left unsupported as the pressure subsided, or else it was borne down by its own weight so that it spread out and gradually dispersed. Sometimes it looked white, sometimes blotched and dirty, according to the amount of soil and ashes it carried with it. My uncle's scholarly acumen saw at once that it was important enough for a closer inspection. So he ordered a fast boat to be made ready, telling me I could come with him if I wished. I replied that I preferred to go on with my studies, and as it happened he had himself given me some writing to do.

As he was leaving the house he was handed a message from Rectina, wife of Tascius, whose house was at the foot of the mountain, so that escape was impossible except by boat. She was terrified by the danger threatening her and implored him to rescue her from her fate. He changed his plans, and what had begun in a spirit of inquiry he completed as a hero. He gave orders for the heavy warships to be launched and went on board himself with the intention of bringing help to many more people besides Rectina. For this lovely stretch of coast was thickly populated. He hurried to the place everyone else was hastily leaving, steering his course straight for the danger zone. He was entirely fearless, describing each new movement and phase of the dire event to be noted down exactly as he observed them. Ashes were already falling, hotter and thicker as the ships drew near, followed by bits of pumice and blackened stones, charred and cracked by the flames: then suddenly they were in shallow water, and the shore was blocked by debris from the mountain. For a moment my uncle wondered whether to turn back, but when the helmsman advised this he refused, telling him "Fortune favors the brave—head for Pomponianus," who lived at Stabiae across the bay.[1329]

1329. Pliny the Younger, *Letters* 6.16.1–12 (written c. 106 A.D.). The word for 'hero' in the text (at 6.16.9) is *maximo*, "in greatness," i.e. heroically. Contrast this heroic narrative with how the first generation of Christians (didn't) write about Jesus: Carrier 2014a: 510–28.

The account continues in detail from there, but in short the elder Pliny reaches and rescues his friend Pomponianus (what the other ships in the fleet accomplished the younger Pliny does not say) but finds himself trapped by fire, dense ash, and a sea roughed by earthquakes and debris, with the eruption worsening into the night. He and his friend and companions try to ride out the worst of it where they had landed (about four miles south of Pompeii, which had been completely buried), but during the following day, "engulfed in a darkness blacker and denser than any night that ever was," Pliny was gradually overcome by ashes and fumes, and "when daylight returned on the 26th—two days after the last day he had seen—his body was found."[1330]

Though his death was tragic, and a direct consequence of his daring, there is no hint of folly in his nephew's account. Pliny and Tacitus clearly endorsed the ideal espoused by Diodorus, that history should reward heroes in the arts and sciences by immortalizing them with praise.[1331] And so Pliny's uncle became a hero. Notably, Pliny's letter reveals the scientific curiosity and naturalistic attention to detail not only of the elder Pliny, but of the younger Pliny as well. But above all, his narrative portrays his uncle as a fearless and devoted inquirer, and compassionate savior, who died doing the right thing, both as a scientist and humanitarian.

Dying for science could have become grounds for reproach rather than praise, and among the more superstitious masses that may have been the fate of many potential tales of heroism. The writings of the Roman natural philosopher Demostratus are lost, but some quotations and paraphrases survive.[1332] One of these involves an interesting incident in which a would-be scientist became the alleged victim of divine wrath. After describing "tritons" as real creatures with human torsos and tails like fish, Aelian relates the following:

1330. Pliny the Younger, *Letters* 6.16.17 and 6.16.20. Pliny says he included in this letter every detail he witnessed himself or heard immediately afterward "when reports were most likely to be accurate" (6.16.22), then gives an account of what happened to himself in *Letters* 6.20.

1331. See chapter 3.9.II. That Tacitus agreed with Diodorus on the function of historical writing is explicit in Tacitus, *Agricola* 1.1 and *Annals* 3.65.

1332. On Demostratus see note in chapter 3.5.

> Demostratus says in his treatise *On Fishing* that in Tanagra he saw a fully-preserved triton. Its appearance was, he reports, pretty much like in statues and pictures, but its head had been so marred by time and was so indistinct that it was not easy to make it out or recognize it: "And when I touched it there fell from it rough scales, quite hard and resistant. Someone from the Greek assembly honored with the government of Greece and entrusted with the presidency for a year, intending to test and examine the nature of what he saw, removed a small piece of the skin and put it over a fire, whereupon a heavy smell from the burning object assailed the noses of those who were present. But," he says, "we were unable to conclude whether the nature of the animal was terrestrial or marine." The experiment was also not useful in another way, judging from his reward. For shortly afterwards he lost his life while crossing a small, narrow strait in a shallow, six-oared ferry boat and, as Demostratus says, the Tanagrians maintained that he suffered this because he profaned the triton, claiming as proof that when he was taken lifeless from the sea he belched a fluid that smelled like the hide of the triton when he had ignited and burned it.[1333]

Though Aelian seems inclined to believe the marvelous explanation of the councilman's death, Demostratus instead seems to have distanced himself from the superstitious claims of the locals. Their account of what happened was clearly more hostile to the values of curiosity and empiricism, and certainly did not come from anyone who believed in seeking explanations in terms of natural causes. In fact, the councilman's death has a different kind of smell to it: it looks ominously like a pious assassination. Lucian had faced exactly the same threat—a ferry crew paid off by local religious nuts to drown him over the side for daring to profane a local holy man.[1334] This contrast between elite values and the values of the predominately

1333. Aelian, *On the Characteristics of Animals* 13.21. It is hard to discern what is from Demostratus and what from Aelian here, but I take the peculiar clause *hôs ekeinos legei* ("as Demostratus says") as indicating the rest of that sentence is from Aelian, and no longer a direct quote of Demostratus (as I have indicated with punctuation accordingly). The same 'Triton of Tanagra' is also discussed in Pausanias, *Description of Greece* 9.20.4–9.21.1, though with no mention of these events.

1334. Lucian, *Alexander the Quack Prophet* 55–58. Similarly, devoted followers of the physician Thessalus allegedly plotted to assassinate Galen (cf. Iskandar 1988: 154 §P.60,12), and members of the Sanhedrin to assassinate Paul (Acts 23:12–35), in each case for skeptically criticizing what could be described as a popular religious business enterprise.

uneducated masses, throws into starker relief the other depictions we have seen of scientific 'heroism' from admiring elite authors, and provides some background for understanding how Christians could become so popular by embracing instead a hostility to scientists and the very scientific values that guided them.[1335]

4.6 The Scientist as Craftsman in the Roman Era

Everything presented throughout this chapter so far will aim to show, by stark contrast, how very different Christian literature of the same period is in its treatment of natural philosophers (as we will see in chapter five). Such a rich variety of praises and admiration for science and natural philosophy can be found in the Roman period only among the pagan elite, though we've seen glimpses of a more ominous hostility among the masses. But before we turn to the Christian evidence we must first address two issues that are often cited as indicating a pervasive disrespect for science among the pagan elite, which are markedly different from the occasional (and obviously uninfluential) hostility or indifference one might otherwise chance to meet. The first of these 'indicators' is the claim that all elite respect for natural philosophy stopped short of respect for actual scientific experiments and research, due to a snobbish disdain for craftsmen and shopwork, which also created (or so it is said) an unbridgeable divide between the bookish 'scientist' and the 'men of experience' who actually knew how things worked. This claim will be addressed in the present section (though we have already brought arguments against it in chapter 3.5 and 3.7.I). The second 'indicator', which we will address in the next section, is the absence of direct institutional support for scientific research.

I. Did Snobbery Impede Science?

Benjamin Farrington states the thesis most broadly: there was a "prejudice against manual labor" that came as a result of "the decline in social status of the manual labourer which accompanied the growth of civilization."[1336]

1335. A fact also explored in chapter 2.6.

1336. Farrington 1946: 28 (cf. Cohen 1994: 248). This remains a standard view on ancient attitudes towards craftsmen: *OCD* 178 (s.v. "artisans and craftsmen").

Insofar as civilization led to an increasing division of labor and a class-based system of social organization, this is true. Although it had already occurred a thousand years or more before the time of Aristotle, and never appreciably diminished until the 20th century, long after the Scientific and Industrial revolutions had already transformed society.[1337] Hence claims of aristocratic snobbishness do not work very well to explain why those revolutions did not occur in antiquity.

Nevertheless, scholars try. Peter Green, for example, claims to see a "deep, radical split in sensibility (and consequent lack of communication) between theorist and craftsman" in the Hellenistic period, following Moses Finley, who declared that "the ancient world was characterised by a clear, almost total, divorce between science and practice," but there is no evidence for these claims, and plenty of evidence to contradict them.[1338] Green even claims there was "a corresponding lack of contact" between *medical* theorists and practitioners, yet once again all the evidence of the Roman period indicates the contrary.[1339] He likewise alleges that a "habit of keeping company with technicians and artisans" was considered "ungentlemanly" for a natural philosopher, but whatever may have been true of Hellenistic royalty (his only real example), for actual philosophers there is again ample evidence to the contrary.[1340] We have already seen some of that evidence in this and previous chapters. In this and the next section we'll see a lot more.

We should look at some examples of what inspired these ideas in the first place. There certainly was elitist snobbery in antiquity, just as in the 19th century and before, but just as in more recent centuries, in antiquity this snobbery came in many shades and depended a great deal on the individual. Hence for every Seneca there was a Posidonius and for every Plutarch a Pliny. Someone like Pliny or Columella may have seen fit to defend their work against perceived snobbery among some of their peers, but then so did an 18th century German historian, Johann Beckmann, who thought he would be ridiculed for deigning to discuss the history of soaps and detergents: "I shall here take occasion to remark," he wrote, "that there is no subject,

1337. See discussion in chapter 1.1.

1338. P. Green 1990: 481 and Finley 1981: 180. Also argued by others, e.g. Zilsel 1945: 328–29; Sambursky 1962: 257–70; etc.

1339. P. Green 1990: 482.

1340. P. Green 1990: 471.

however trifling, which may not be rendered useful, or at least agreeable, by being treated in a scientific manner; and to turn such into ridicule, instead of displaying wit, would betray a want of judgment."[1341] Indeed he seems to have been more worried about this than Pliny or Columella or any ancient authors ever were. Pliny, for example, was only concerned about readers being bored or put off by alien vocabulary, and Columella was only worried he might be ignored, and both were no doubt rhetorically exaggerating their actual situations (since both their works appear to have been well and widely received).[1342]

Surely the nobility of the 17th and 18th centuries were no less averse to tinkering in workshops or hobnobbing with mechanics than any Roman senator would have been. And even as recently as the 1950's "manual or clerical work was considered demeaning for scholars" at Oxford, so finding any similar attitudes in antiquity is hardly significant.[1343] Hence the issue is not whether there were aristocrats who had no interest in doing anything mechanical or laborious, but whether *all* aristocrats held such an attitude, and whether even those who did never communicated with craftsmen to learn or get things done. Moreover, even those questions are moot if scientists did not even come from the aristocracy (such as the senatorial elite), but from the far more numerous professional class—such as actual and aspiring decurions and equestrians or the growing population of successful freedmen. For scientists were not likely to be shackled by any aristocratic attitude if they were not even aristocrats to begin with.

But first we must understand the nature of the ancient aristocratic attitude. A typical passage repeatedly quoted on this point comes from the *Economics* of Xenophon, writing in the time of Aristotle, which purports to present the words or sentiments of Socrates:

> The so-called 'banausic arts' are much maligned, but it is entirely reasonable that they should be held in no esteem among the citizens. For they utterly

1341. Beckmann et al. 1846: 2.97 (n. 1). The first edition was published by Beckmann in 1797, Beckmann died in 1811, and his work was translated and updated several times after that by additional contributors, but this note remains his.

1342. Pliny the Elder, *Natural History* pr.12–16 (discussed below) and Columella, *On Agricultural Matters* 1.pr.10–29 (discussed in chapter 3.9.II).

1343. Arnander 2007.

> destroy the bodies of those who work at and manage them, forcing them to sit all day indoors, some even by a furnace, their bodies weakened and their souls quite enfeebled. Worst of all, the so-called 'banausic arts' leave no time to attend to friends or the interests of state.[1344]

And yet, he goes on to argue, warfare and agriculture—in fact, actual fighting and tilling—are preeminently respectable, even though these activities involve more grueling labor, and are just as dirty and time-consuming and occupationally harmful as any craft. As another author of the era put it, "Agriculture contributes greatly to manly character," for "unlike the banausic arts, which make the body weaker, agriculture accustoms the body to outdoor living and hard physical labor, and makes it all the more capable of undergoing the dangers of war," for any respectable citizen must be able and willing to take up arms in defense of his country.[1345]

Thus, what is being condemned in Xenophon is not physical labor, nor getting one's hands dirty, but being forced to get no sun or exercise and instead ruining and weakening your body, and having no free time. I doubt there has ever been or will ever be any civilization without a good number of people who would despise such a fate, or prefer one better. Xenophon's Socrates' judgment may have been misplaced or unsympathetic (not all craft work need be so ruinous), but the same sentiment is just as frequently heard today, in attacks against the pallid, beer-bellied couch-potato who never leaves time for friends, politics, or learning, or denunciations of the destructive, soul-crushing life of a sweat-shop. Ask anyone today whether they would prefer going to college or working in a coal mine and the answer would be the same now as it would have been then.

Hence ancient hostility to the banausic arts cannot be bootstrapped into some fictional aristocratic disdain for making or doing things. As Aristotle explains:

> Some of the duties one can undertake do not differ in regard to the work done but in regard to what it is done for. Hence much of what is considered

1344. Xenophon, *Economics* 4.2–3. The word *banausos* (and its cognates) derives from *baunos*, "furnace, forge" and thus denoted workmen associated with furnace work, and then by association all similarly crippling shopwork (cf. *LSG* 305, s.v. "*banausia*," "*banausikos*," "*banausos*," etc. and *LSG* 311, s.v. "*baunos*").

1345. Pseudo-Aristotle, *Economics* 1.2.1343b.

> menial work is attractive even for free men in their youth. For activities are not attractive or distasteful in and of themselves, but only in regard to their ultimate aim and the reasons they are pursued.[1346]

Hence:

> It is easy to see that the young must be taught only what is necessary among the useful arts, not all of them, obviously distinguishing between the activities of free men and those of men who are not free. Moreover, they must take part in such activities as are useful to do, but not as a 'banausic' does. We must consider work to be banausic, both the art and the knowledge of it, insofar as it leaves the mind or body of free men useless for the employment and performance of virtue. Hence such arts as worsen the body we call 'banausic', as also work done for hire, because it keeps one's thoughts preoccupied and low.[1347]

Thus, it is not a craftsman's work or manual labor or "getting your hands dirty" that Aristotle is objecting to. Rather, he objects only to pursuing a craft so much or in such a way that it actually cripples the body or distracts us from more important studies and affairs. Hence Aristotle immediately adds that even arts *fitting* for free men can be pursued too far, even to the point of having the same harmful effects of damaging the mind or body or distracting us from higher pursuits. Hence "it makes a big difference *why* you are engaging in some activity or study," for "what is done for the sake of yourself or your friends or because of a virtue is not servile, but doing the very same thing for other reasons" can easily be perceived as "menial and slavelike" (*thêtikon* and *doulikon*).

In other words, doing anything for hire (*thêtikon*) or in a submissive, subservient role (*doulikon*), is what is unbecoming a free and virtuous man, not 'getting your hands dirty' or anything like that. To the contrary, 'getting your hands dirty' and building or making things, or anything else alike, is entirely acceptable and respectable, as long as you are doing it for your own reasons, or as a gift or favor for your friends, or because it serves some virtuous end—like, say, scientific discovery. Hence Aristotle approves

1346. Aristotle, *Politics* 7.14.1333a.

1347. Aristotle, *Politics* 8.2.1337b.

of scientific dissection, no matter how messy or distasteful it may be.[1348] Aristotle likewise defends the study of music against snobs who claim it is improper for men of leisure. Though "we tend to call those who sing or play music 'banausics' and consider it unmanly to do it ourselves, except when drunk or having fun," it is only wasting your life away as a professional musician that is unbefitting a free man—otherwise cultivating music in your own free time, and for pleasure and intellectual study, is entirely appropriate.[1349] Obviously the very same arguments could be extended to cover any craft or labor of use to a scientist. As long as you pursue a virtuous goal, and are in command of your own labor, and do not become a slave to any man or wage, then almost any activity was acceptable, mechanical or otherwise.[1350]

Likewise, as I demonstrated for Isocrates in *Science Education in the Early Roman Empire*, when Aristotle characterizes as 'banausic' the sort of man who gives vulgar displays of wealth—merely to show off and without any appropriate tact, culture, or taste—it is the lack of those virtues that he is criticizing, not any underlying trade.[1351] In his social reality artisans obviously tended to be uncultured and pretentious, hence 'banausic' became an epithet meaning the latter, but this does not entail that all artisans were like this or that merely being an artisan or mastering a craft produced such failings of character. In other words, Aristotle did not imagine that working with tools or constructing instruments or setting up experiments or dissecting animals *caused* people to become vulgar or pretentious, and he certainly did not think such activities were *in themselves* vulgar or pretentious. Moreover, if you strove to become educated and cultured and tasteful, it did not matter what trade you pursued, you would then be a man of virtue regardless.[1352]

1348. Aristotle, *Parts of Animals* 1.5.645a.

1349. Aristotle, *Politics* 8.5.1339b. His defense of music occupies 8.3.1338a-8.5.1341b.

1350. As explicitly stated, for example, in Aristotle, *Politics* 3.4.1277b. See my related discussions in chapters 3.5 and 3.7.1.

1351. Aristotle, *Nicomachean Ethics* 4.2.1123a. Compare my discussion of the related attitudes of Isocrates and subsequent authors in chapter five of Carrier 2016.

1352. As implied in Plato, *Gorgias* 512b-e (which attests a diversity of opinions regarding craftsmen among the aristocracy of Classical Athens, ranging from

Accordingly, Aristotle could look down his nose at the 'mere craftsman' who does not understand the underlying technical and scientific theory of what he does, and at the same time hold in high regard the 'master craftsman' who, like a scientific engineer, is wise and respectable—not least because he is educated (and presumably cultured) and knows the underlying causes and reasons for what he does.[1353]

Hence in the 1st century B.C. the philosopher-poet Automedon could use the word *banausos* as an insult to someone who went out in public looking dirty and unkempt, which implies that if his target had cleaned himself up before going out he could have avoided the reproach.[1354] We can assume, for example, if Aristotle went out in public all covered in blood from dissecting animals in the lab, he could easily be derided as an uncouth butcher, but only because a cultured man would know better than not to clean up before going out, not because dissecting animals was inherently vulgar. Likewise, in the 2nd century A.D. the poet Strato would use the same word to mean "whore" (both metaphorically and literally)—in other words, someone who is willing to do whatever they are paid to, and who could thus be derided for selling themselves (we often use the word "whore" today in the very same way, or indeed even more broadly, e.g. 'media whore'). But this means if someone did the very same work on his own initiative, and for a scientific cause or as a favor to a friend, then the insult would no longer stick.[1355] Hence perhaps a mercenary could be derided as a whore—but a soldier taking a salary to defend his country, not so much. Likewise, getting drenched in mud and blood in combat would not be grounds for reproach—but going to a fancy dinner in such a condition, certainly.

Hence Cicero identifies 'engineer' and 'doctor' as "honorable" professions, even though they obviously involved a lot of messy labor and were certainly craftsmanlike occupations. A typical representative of his time, he says the *most* respectable occupation is agricultural land management. Next after that is any honest large-scale trading enterprise, and next after that

---

snobbery to appreciation).

1353. Aristotle, *Metaphysics* 1.1.981a-b.

1354. *Palatine Anthology* 11.326.

1355. *Palatine Anthology* 12.237. Hence note the advice on respectable vs. inappropriate ways of procuring pay and employment in Vitruvius, *On Architecture* 6.pr.4–6.

are "professions in which either a higher degree of intelligence is required or from which no small benefit to society is derived—like medicine, or engineering, or teaching respectable subjects," which he says "are suitable for those of honorable rank," hence such trades would be unsuitable to those of lesser rank, which entails doctors and engineers enjoyed a respectable status.[1356] The only occupations he considers beneath a gentleman are those that incur ill will, require deceit, serve only base pleasures, rise no higher than petty trade, involve no skill, or completely tear us away from a noble life:

> Now in regard to trades and means of income (*artificiis et quaestibus*), which ones are to be considered appropriate to a gentleman and which ones are vulgar, we have been taught, in general, as follows: First, those means of livelihood are rejected as undesirable which incur people's ill-will, like tax-gatherers and usurers. Vulgar and unbecoming a gentleman, too, are the means of livelihood of all hired workmen whom we pay for mere manual labour and not for artistic skill. For in their case the very wage they receive is a pledge of their slavery. We must also consider vulgar those who buy from wholesale merchants to retail immediately, for they would get no profits without a great deal of downright lying, and there is no action meaner than misrepresentation. And all craftsmen exhaust their time in a vulgar craft, for the workshop cannot have anything noble about it. Least respectable of all are those trades that cater to sensual pleasures: 'fishmongers, butchers, cooks, poulterers, and fishermen', as Terence says. Add to these, if you please, the perfumers, dancers, and acrobats.[1357]

None of these relate to the scientist, except possibly the expressed disdain for "the workshop," but since Cicero had previously made an exception for those paid for their artistic skill, he clearly has in mind allowances for master craftsmen just as Aristotle did. Hence his remark that there is nothing noble in a workshop more likely means something closer to what we would call a sweatshop, in which to work was harmful to the body and tantamount to slavery, a situation that could never be appropriate for a

1356. Cicero, *On Duties* 1.42.151. The key passage here reads: *artibus aut prudentia maior inest aut non mediocris utilitas quaeritur (ut medicina ut architectura ut doctrina rerum honestarum) eae sunt iis quorum ordini conueniunt honestae.* The latter phrase literally reads "these [professions] are for those to whose honorable rank [such professions] are suited."

1357. Cicero, *On Duties* 1.42.150.

free man (*ingenuum*).

In other words, we can assume that here Cicero is echoing the very same dislike we saw in Xenophon and Aristotle, of those who spend all their time (*versantur*) in a workshop and thus destroy their bodies and have no freedom, and all this only to make a living.[1358] He is thus not describing a private workshop used at leisure by a scientist to make his own instruments, for instance. Nor does he say there was anything wrong with a nobleman hiring a craftsman to build such an instrument for him. Instead, in Cicero's mind it is the rarity of a man's genius and skill that increases his value, not birth or wealth or even prestige of occupation. For elsewhere he argues that field commanders are more useful and important than trial lawyers, just as roofers are more useful and important than sculptors, yet Cicero says he would rather be a brilliant sculptor than the most talented roofer (and thus, by analogy, rather an orator than an officer), because "a man's value should be weighed not by how useful he is but how rare he is," and there are not many exceptional painters and sculptors, but porters and workmen are in endless supply.[1359]

Finally, Cicero, and most Romans of the more privileged classes, would certainly have measured a man's worth in terms of his manners and education, just as Aristotle and Isocrates did. For example, when Galen makes respectable birth and education essential for a respectable scientist, and thus attacks pretenders who meet neither condition, there is an element of snobbery in this, but nothing of a science-crippling kind. Galen certainly derides his opponents for being unlearned men from vulgar trades, but since having a decent education and a proper family upbringing were enough to attain respectability, it is not in fact *craftsmen* whom Galen is scorning, but unmannered and uneducated craftsmen—and even then they are only attacked for *pretending to be scientists.* Otherwise, Galen does not say it is unseemly to associate or work with them, and elsewhere says a noble birth is irrelevant if one masters a genuine art. In fact, Galen regards the rich and the poor as equally useless if they learn no art—or worse, do nothing useful at

1358. The key sentence reads: *opificesque omnes in sordida arte uersantur nec enim quicquam ingenuum habere potest officina.* An *opifex* is anyone who makes things professionally, and an *officina* anywhere such manufacturing is done.

1359. Cicero, *Brutus* 73.255–257.

all.[1360] Hence Galen reserves his worst contempt for athletes, declaring that "I should rather like to see them digging or harvesting or sowing or doing anything of practical value on a farm," or taking up military service, for either would raise his opinion of them, thus demonstrating some measure of elite respect for manual labor.[1361]

Beyond the general atmosphere of this value system among the elite, one can still find diverse opinions. Platonists, for example, held the physical world in greater disdain and were more fond of theoretical armchair reasoning than the average Stoic, Aristotelian, Skeptic, Epicurean, or Eclectic, who were quicker to embrace empirical real-world experience. Members of any given sect could also be split on fundamental issues—as among Stoics we saw Seneca was at odds with Posidonius (in chapter 3.8.I). And there were more casual questions of popular fancy. Pliny the Elder, for example, says history and oratory and storytelling were more entertaining and thus more popular, and that science is by contrast dull (a claim still frequently made even today). Science also requires discussing the lowest of subjects (*sordidissima*), and Pliny must apologize for sometimes resorting to the terminology of hicks and foreigners (lacking any alternative). But as with the rest of his preface, the contrast he builds is hyperbolic (as we also saw in chapter 3.8.I), and of course completely belied by the passion he shows for his subject throughout, which is anything but "sterile" (as he says the study of "life" would be, an obvious play on words).[1362] Hence Pliny outright says his project is "abundantly beautiful and magnificent" and our "particular motive for research" should be "the utility of overcoming difficulties" exactly like those he just listed, rather than pursuing the "popularity of giving pleasure" instead.[1363]

1360. For example: Galen, *On the Therapeutic Method* 1.1 vs. *Exhortation to Study the Arts* 5–9 (= Kühn 1.7–22). On this point see commentary in Hankinson 1991a: 85–86.

1361. Galen, *Exhortation to Study the Arts* 13 (= Kühn 1.32–33). An even higher opinion of agricultural labor (and laborers) is expressed in Musonius Rufus, *Sermons* 11, and though this was by no means a universal sentiment, it was not uncommon (cf. Xenophon above).

1362. Pliny the Elder, *Natural History* pr.12–14.

1363. Pliny the Elder, *Natural History* pr.15 (*abunde pulchrum atque magnificum*) and pr.16.

To an extent Pliny's concerns all have modern parallels, though now it is technical and high-brow vocabulary that must often be apologized for; movies, sports, and television that compete for interest; and the seemingly trivial (or occasionally distasteful) nature of most scientific work that relegates popular interest mainly to what is more exciting and grand than what is actually most frequently done, known, or studied by scientists themselves. But the degree of each then was greater, e.g. no one now would feel the need to apologize for discussing a 'low' subject like cleansers as Beckmann had to do in the 18th century. And yet, as I noted before, Pliny still seems less worried than Beckmann was about the consequences to his reputation of discussing even the 'lowest' of subjects, and his defense is more interesting: that even such subjects were beautiful, magnificent, and useful. Pliny, and his intended audience (the emperor and his court, and professional men of Pliny's own class), seem to be among those who might agree with the emperor Marcus Aurelius that, for example, even the cracks in baked bread were beautiful and worthy of examination.[1364] Not everyone would have agreed, though not everyone would have been impressed by Beckmann's excuses either. Yet both found accepting audiences.

But among those who would be most put off were the Platonists. Plutarch is often cited as a paradigmatic example, though it is curious that it is always him who is cited, which leaves us wondering just how typical his views could have been. But even if typical of a Platonist, they were not typical of a Roman of any other philosophical persuasion. It is thus notable that Neoplatonism is the pagan philosophy that triumphs after the third century, and precisely then science begins its decline.

Two passages are most commonly cited as representing Plutarch's Platonizing snobbery, though only one definitely captures the relevant attitude. The other is questionable, but illuminating all the same. In one of several essays on zoology, Plutarch says "I do not accept those who, to make a complete study of anthills, inspect them in a sense 'anatomically,'" and thus section and excavate them to find out and report how they are constructed.[1365] This is sometimes offered as evidence of a Roman rejection of scientific 'dirty work', but it actually proves the opposite: not only are sentiments like this very hard to find and thus quite rare (Plutarch is hardly

1364. Marcus Aurelius, *Meditations* 3.2 (discussed in section 4.2).

1365. Plutarch, *Whether Land or Sea Animals are Cleverer* 11 (= *Moralia* 968a-b).

a 'typical' Roman in this respect, nor even a scientist), but in registering his objection Plutarch is confirming there *were* actual scientists 'getting their hands dirty' by making empirical investigations in field zoology—and despite Plutarch's seeming 'disapproval' he goes on to summarize their findings. Moreover, it is not so clear that Plutarch actually disapproved of such field work. For he does not in fact say he "disapproves" of it, but that he does not accept it. He never says *why* he does not "follow" or "accept as teachers" those who do this, but what he does *not* say here (at least not explicitly) is that it was distasteful to dig up anthills.[1366]

Plutarch is much clearer when he attempts to usurp the fame of Archimedes to promote his own Platonic worldview in his biography of Marcellus (whose siege, as we saw in section 4.5, eventually ended Archimedes' life). As Plutarch tells it:

> Marcellus had not reckoned on Archimedes and his machines, whose construction Archimedes had not even considered worthy of serious attention, but mainly as a byproduct of his geometric amusement, until the ambitious king Hiero persuaded him to direct some of his skill from intellectual applications to physical ones, and in some way or other make his theories more evident to the people, by displaying to the senses how they can be applied to useful ends.
>
> It was actually those linked to Eudoxus and Archytas who began to employ this prized and celebrated use of mechanical instruments, cleverly working out geometrical problems in elegant ways, taking problems not easily solved through logical and diagramatic demonstration and supporting their solutions instead with proofs that depend on instruments and the senses. For example, to solve the problem of finding two mean proportionals, which is necessary for the construction of many other geometrical figures, both resorted to using mechanical instruments, adapting for this use certain instruments called 'mesographs', which were employed in drawing conic sections. Plato was enraged by this and attacked both men for having corrupted and destroyed the beauty of geometry, as they had fled from disembodied and abstract thought into the realm of the senses, returning instead to bodily instruments that required much base and burdensome labor. As a result, mechanics became divorced from geometry, and was for a long time ignored by

1366. Cf. *LSG* 196 (s.v. "*apodechomai*"): "accept," "accept advice from," "accept as a teacher," "follow," "admit to one's presence," "admit to the mind," "receive favorably," "approve," "accept the statement or story that," "receive or accept from," "believe or agree with," "acknowledge," "understand."

philosophy, becoming one of the military arts instead.[1367]

...

As for Archimedes, he possessed so great a mind and so deep a soul and such a wealth of theories that, although he had gained from his machines a name and fame not even human, but more like the genius of a god, he did not want to leave behind anything written about them, but considered the business of mechanics, and all art that as a whole aims to be useful, as lowborn and banausic, so he directed his ambition only to those things in which the beautiful and the extraordinary are not mixed with the necessary.[1368]

This account of what Archimedes did and thought is entirely false. It appears nowhere else, and is belied by abundant evidence to the contrary. We know Eudoxus, for example, eventually solved the problem of mean proportionals with a formal geometric proof, so clearly such solutions were not 'too intricate' for words, and Eratosthenes says neither Eudoxus nor Archytas had instruments for the purpose, in the very inscription where Eratosthenes, a good friend and colleague of Archimedes, celebrates his *own* invention of the mesolabe long after Plato was dead.[1369] Likewise, the very attitude Plutarch attributes to Plato, and thence to Archimedes, is refuted in the latter's case by the fact that Archimedes wrote a book, *On the Method of Mechanical Theorems*, arguing for exactly the reverse: that such mechanical means of discovering theorems were useful and not only appropriate, but well worth encouraging.[1370] His friend Eratosthenes even proclaimed using a mesolabe was *superior* to the more laborious method of using written calculations to achieve the same result.[1371] And appropriately, Archimedes addressed *On the Method* to Eratosthenes, establishing their common bond.

More importantly, the idea that mechanics became divorced from geometry, and then was relegated only to military use, is ridiculous. It is already refuted by the extant work of Archimedes himself, wherein he

1367. Plutarch, *Marcellus* 14.7–12.

1368. Plutarch, *Marcellus* 17.5–7.

1369. See sources on Eudoxus in Appendix B. For Eratosthenes' inscription about (and invention of) the mesolabe see sources in the relevant note in chapter 3.5.

1370. See discussion in chapter 3.8.IV.

1371. In his inscription celebrating the fact (see relevant note in chapter 3.5).

meticulously links formal geometric proofs with the mechanics of levers and the principles of floating bodies, and we know he extended these explorations to the whole of formal mechanics. Plutarch himself includes an account of how Archimedes applied geometry to the construction of wheels and pulleys in order to accomplish an astonishing public display of moving a great ship with his own body, thus refuting any notion of a divorce between geometry and mechanics, or of mechanics having ever been limited to military applications—or, in fact, of Archimedes ever thinking labor and mechanical operations were beneath him.[1372]

It is also impossible that Archimedes could have produced the described defenses of Syracuse without considerable hands-on experience and a craftsman's knowledge of their construction and operation—and, indeed, of their application in the field, since even Plutarch attests that Archimedes had anticipated several tactical realities in his arrangement of the city's defenses.[1373] Some of the mechanical planetariums Archimedes had constructed also survived into the Roman period, which further entails he had no qualms about shopwork or working closely with craftsmen.[1374] And though most accounts of his death depict him as so engrossed in a geometric problem he did not take any measure to save himself, one account—ironically also related by Plutarch—said he was carrying scientific instruments, in fact "mathematical instruments, dials, spheres, and angles, by which the size of the sun might be measured," which was mistaken for loot by a greedy soldier. If true (and it sounds the more plausible), this does not sound like a man averse to using instruments, but a man quite proud of them.[1375] And even if that account is false, it still entails a very different tradition about Archimedes' values and interests.

Likewise, Plutarch's fantasy that Archimedes did not leave any writings on his machines may be true of his war technology (and probably for the more obvious reason of state secrecy), but it is not at all true otherwise.[1376]

1372. Plutarch, *Marcellus* 14.12–13. See chapter 3.4 for the actual development of mechanics.

1373. Plutarch, *Marcellus* 15–16.

1374. See note on armillary spheres in chapter 3.3.

1375. Plutarch, *Marcellus* 19.11–12.

1376. That a lot of war technology was not published due to concerns of national security (to ensure a competitive edge against state enemies) is soundly argued in

We have evidence that he wrote not only on the construction of armillary spheres, but on the five basic machines as well as ship hull design, and possibly on the construction of waterclocks, odometers, water organs, and mirrors—which also casts doubt on the claim, made by another author, that Archimedes only wrote one book on the actual *construction* of machines, otherwise deigning only to write on theory, which is a bit more nuanced than Plutarch's hyperbole, but not much more plausible.[1377] As even Plutarch admits, Archimedes was celebrated for his inventions, many of which saw widespread use (like the helical water pump), which entails he must have written or taught on them in some sufficient detail, while Vitruvius outright says he had consulted books by Archimedes on machinery.[1378] We can also see that *extant* writings by Archimedes on the lever and the properties of floating volumes directly tie the real-world design principles of machines and ships to underlying geometric theory. These provide only a glimpse of what has otherwise been lost, yet are already enough to show that Archimedes imagined no split between mechanics and geometry, but in fact appears to have fully united them.

Hence Plutarch's conclusion that Archimedes disdained all mechanics, shopwork, or anything useful as low and vulgar, and only directed himself to geometric theory, is certainly untrue. Thus, as several scholars have now concluded, his account of Archimedes appears to be a complete fabrication, invented to promote the Platonic values it glorifies by attaching them to a

---

Russo 2003: 198–99 (cf. also 110, 114–16, 137–38, 165 and Tybjerg 2004: 44–46) and attested as actual policy at Rhodes, Cyzicus, and Marseilles (Strabo, *Geography* 14.2.5). In addition, Simms 1995: 67–68 shows that by all reliable accounts nothing Archimedes employed militarily was of novel invention (thus he might have written nothing because he designed nothing); although Rihll suggests the opposite, that Archimedes' lost treatises or teachings on the subject influenced those of Philo of Byzantium (Rihll 2007: 122). Though one should perhaps add that, the competing views of Simms and Rihll aside, Archimedes might have written on his defenses of Syracuse had he survived the siege.

1377. See section 4.5, discussion in chapter 3.4, example in chapter 3.9.II, and relevant notes in chapter 3.6. The more nuanced claim was made by Carpus of Antioch, a Roman-era astronomer, mathematician, and engineer (see chapter 3.4), according to Pappus, *Mathematical Collection* 8.3.1026, who nevertheless could not recall exactly where Carpus had said it.

1378. Vitruvius, *On Architecture* 7.pr.14 (cf. also 1.1.17).

much-revered hero.[1379] Though this at least attests to there being a more mystical, anti-banausic attitude among some Romans (as Plutarch evidently endorses it), one can just as easily find opposing values in the same period, and everything in between.

For example, another Platonist would complain:

> Most people only admit geometry, which is in fact the noblest part of philosophy, as a low-grade activity aimed at low-grade ends. For they restrict it to practical necessity, for example for measuring out land or setting up walls. In short, they approve of all its contributions to the manual crafts, but they see no further.[1380]

Cicero confirmed the same sentiment (though not indicating it as his own).[1381] But as we shall see, the dominant view among Roman scientists lay between these two extremes, valuing both mathematical theory *and* its practical applications, regarding neither as 'low-grade' or 'vulgar'. Thus Peter Green's remark that the "attitude" Plutarch attributes to Archimedes "is consonant with all we know" of "the Greek intellectual tradition" and therefore credible, is itself not credible.[1382] Even apart from all the evidence above and below, we already know many other Hellenistic kings of the time were employing scientific engineers to research and develop both military and practical technologies, just as Plutarch says Hiero was doing with Archimedes, so in fact what seems to be "consonant" with the zeigeist of the time is the *practical* engineering interests of kings and court scientists, *not* Plutarch's fantasies of elite disinterest. Just as Seneca's railing against 'modern technology' only proved how popular those technologies really were (and thus how uninfluential Seneca's distaste for it was), so also the occasional example of Platonist moralizing about the 'higher' value of theory and the 'base' nature of its practical uses only ends up indicating how unpopular

1379. Simms 1995: 71–96, Authier 1995, Cuomo 2001: 192–211, and Russo 2003: 198 all identify reasons to reject Plutarch's account as in almost every respect the exact reverse of the truth. That ancient mathematicians *in general* were very pro-instrument and practical (proving Plutarch's view unusual) is demonstrated in Cuomo 2000: 83, 91–109 and Cuomo 2001 *passim*.

1380. Maximus of Tyre, *Orations* 37.7.

1381. Cicero, *Tusculan Disputations* 1.2.5.

1382. P. Green 1990: 850–51, n. 33 (cf. also 458 and 465).

their opinion was (and how far they had to distort reality to make it seem otherwise).[1383]

Which brings us back to Seneca. Even in his ninetieth epistle (discussed in chapter 3.8.I), which most directly deploys his own peculiar anti-banausic snobbery, his overall attitude is still nearer that of Isocrates than Plutarch, as his principal complaint is that cultivating a moral intellect is what defines a good man, not merely technical studies, achievements, or skills. So we can still expect that anyone who excelled in both domains would probably annoy Seneca a great deal less. After all, though he refuses to agree with the camp of Posidonius that philosophers invented the arts and crafts (even using craftsmen as assistants), Seneca still allows that admirable men can invent tools and technologies and remain respectable, provided they also cultivate real wisdom, first and foremost.[1384] Otherwise, Seneca maintains, there is no wisdom in "any discovery that must be gained with a bent body and a mind gazing at the ground," an allusion to the crippling work of the shop and the workman's lack of loftier thoughts, the same complaints of Xenophon and Aristotle.[1385] The entire gist of Seneca's letter is that such discoveries may still be clever and intelligent, even useful, but wisdom does not consist in them, and they do not produce virtue or improve character. Again, views more nuanced than Plutarch's.

In his eighty-eighth letter, like Cicero, Seneca had already objected to arts pursued for monetary gain and for mere luxury rather than scientific knowledge or patriotic duty, yet he agrees even "these petty and commercial arts, which depend on the hands, contribute a great deal to the instruments of life, they just do not pertain to virtue."[1386] Here Seneca also attacks an

1383. On Seneca's assault on modern technology see chapter 3.8.I.

1384. Seneca, *Moral Epistles* 90.31–33.

1385. Seneca, *Moral Epistles* 90.13.

1386. Seneca, *Moral Epistles* 88.1–2 (against money-seeking professions), 88.18–19 (against luxury professions), 88.20 (quoted remark). It should be noted that all such railing against money-making pursuits is mostly for show: the elite (Seneca included) routinely pursued profits through often massive enterprises in trade and industry and even the much-maligned usury. They relied on trusted intermediaries of lower status to manage their commercial and industrial assets and banking contracts, but there is no reason to believe they were in any way aloof to how their business managers were disposing of their capital investments. This is extensively discussed in D'Arms 1981. See also sources on ancient economic attitudes and

excessive devotion to the liberal arts.[1387] But as he also strives to make clear there, he does not attack *studying* them, only studying them to the exclusion of philosophy—including natural philosophy, as he makes quite clear. In fact, he actually *defends* the liberal arts here, as an essential tool for use by natural philosophers to discover and understand the nature of the universe. Hence the liberal arts (including geometry, harmonics, and astronomy) should be used in the service of philosophy, and hence to pursue scientific knowledge, rather than being pursued only for themselves—whether for gain or boorish erudition. Seneca concludes by emphasizing his contempt for the latter, adding a new spin on the 'aristocratic' attitude: snobbery against the snob. "The desire to know more than enough is a type of intemperance," Seneca argues, "because such an overzealous pursuit of the liberal arts turns men into annoying, wordy, tactless, self-satisfied bores, who fail to learn the essentials because they have already learned the superfluous."[1388] By the 'superfluous' Seneca means what he considers useless knowledge (like debating where Homer was born). But overall he is expressing the same view as Isocrates: *any* education to the neglect of cultivating good manners and a gentlemanly character is contemptible. Which entails that a cultured gentleman, who is agreeable, plain-spoken, tactful, congenial, and interesting to be around, could practice or study whatever he pleased and it would be no detriment to him. Moreover, as even Seneca admits, there are worse things than being an uncultivated bore, "for it is better to know useless things than to know nothing."[1389]

So though Seneca disparages an excessive pursuit of the liberal arts, he

---

realities in the introduction to chapter 3.6.

1387. On Seneca, *Moral Epistles* 88, see the commentaries of Stückelberger 1965 and Kidd 1978. On the content of the liberal arts, see chapter five of Carrier 2016 (as noted there, these arts sometimes included gymnastics and drawing—as even Seneca attests in 88.18, though he opposes their inclusion).

1388. Seneca, *Moral Epistles* 88.36–37 (in context: 88.33–36): *quod ista liberalium artium consectatio molestos, verbosos, intempestivos, sibi placentes facit et ideo non discentes necessaria, quia supervacua didicerunt.* Seneca offers a similar attack on the aimless pursuit of trivia (in the context of accumulating libraries) in *On Tranquility* 9.5 (= *Dialogues* 9.9.5). Galen also attacks the boorishness of 'trivia hounds' who avoid a serious pursuit of the sciences, in Galen, *On Examinations by Which the Best Physicians Are Recognized* 13.2–3.

1389. Seneca, *Moral Epistles* 88.45.

never once attacks their scientific applications, only their trivial or avaricious ones. For example: he attacks grammar for its obsession with verbal, literary, and other trivia, not its ability to help us write and think clearly; he attacks music for teaching us how to listen and play, but he never derides the scientific study of harmonics; he attacks arithmetic for its use in counting money, but not any of its scientific applications; and he attacks geometry for its vain use by the rich to survey their estates, but not its applications in mechanics, optics, or astronomy; and when he does attack astronomy, it is only *astrology* that comes under his gun, not any other scientific aims or content.[1390] In fact, Seneca explicitly defends the scientific uses of geometry and astronomy in this very letter (see my discussion of that very fact in chapter 2.7), and he accepts the study of mathematics for scientific use elsewhere, such as in the employment of optics to explain rainbows.[1391] In other words, as long as science and mathematics are pursued in the service of philosophy, and not to the exclusion of it, Seneca has no complaint, but only praise.

Of course Seneca still put moral philosophy first. But though Alfred Stückelberger argues that Seneca wants here "to replace the Middle Stoic preference for scientific research with strictly moral thought," that is too extreme a conclusion.[1392] For it generates far too enormous a contradiction with what Seneca abundantly says elsewhere in praise of scientific research, a problem with his argument that even Stückelberger acknowledges.[1393] In fact, Seneca nowhere mentions scientific *research* in this letter, or the advancement of scientific understanding in any relevant sense. Though he does tend toward hyperbole and the occasionally ridiculous diatribe—here as elsewhere (according to his usual style)—he still leaves enough qualification and precision of point to clarify what he is really arguing, which is not the abandonment of scientific research, but its subordination to moral improvement as every man's primary and ultimate goal. As Seneca

1390. Seneca, *Moral Epistles* 88.3–8 (grammar), 88.9 (music), 88.10–13 (geometry and arithmetic), 88.14–17 (astronomy).

1391. Seneca, *Natural Questions* 1.5.13.

1392. Stückelberger 1965: 54: *die Überwindung der wissenschaftlich forschenden Haltung der Mittelstoa durch ein streng moralisches Denken.*

1393. Stückelberger 1965: 79. As demonstrated in chapter 3.9.II and section 4.4 above, there can be no doubt what Seneca really felt about scientific research.

says elsewhere, though "it is more important to be brave than learned, the one does not occur without the other, for strength of mind comes only from the liberal arts and the study of nature."[1394] So even *Seneca's* snobbery did not get in the way of science. And Seneca's negativity in this regard was already idiosyncratic among his fellow Stoics. Medieval Christians selectively preserved the writings of the likes of Seneca or Epictetus, and destroyed the writings of other Stoics like Posidonius, thus skewing our perception of what was typical for that sect, and for the Roman elite generally (as we've already seen ample evidence of). In fact Roman Stoics were *more* passionate about logic and science than ethical theory, making Seneca atypical.[1395]

II. Roman Scientists and Middle-Class Values

Needless to say, other scholars have noticed the discrepant evidence, and thus argued against any sort of split between scientists and craftsmen, or between learning and handiwork.[1396] And I have already surveyed abundant evidence that ancient scientists were themselves fully capable craftsmen without apology (e.g., Chapter 3.4 and 3.6.IV). But a more common mistake made by proponents of such a split is to assume that ancient scientists all came from the aristocratic elite. To the contrary, they predominately came from the upper middle class, a social stratum just below that of the wealthiest and most powerful citizens who at the time constituted the closest thing to a Roman aristocracy. Thus, noting the attitudes of *the aristocracy* will not necessarily relate to the attitudes *of scientists*. For though I class both groups among the 'elite' of that era, there were still important differences between them.

Of course, ancient society did not break down in quite so simple a way as to have a distinct lower, middle, and upper class as such.[1397] Slaves, for

1394. Seneca, *Natural Questions* 6.32.1–2.

1395. Carrier 2016: 102–04.

1396. From Edelstein 1952: 579–85, to K.D. White 1993, to Rihll & Tucker 2002 and Cuomo 2007.

1397. Scholarship on the Roman social system is extensive, but for a summary supporting the following assessment see Atkins & Osborne 2006: 4–11, MacMullen 1974, and Garnsey 1970, with a particular focus on the elite in Mratschek-Halfmann 1993. J. Clarke 2003: 4–9 provides a useful discussion of the application of class terminology to the Roman period, but expanding on his scheme, I position the

example, were not categorically at the bottom, as they could enjoy a status at many levels depending on their education and position, and yet were always considered beneath free men of the same level, and could be sneered at even by free men of lesser education and authority. Likewise, even among the free. Though meeting specific benchmarks in wealth and property could determine official rank, status was not entirely a measure of wealth—a freedman, for example, could be the richest man in the world and yet still not be recognized as the equal of anyone of equestrian or senatorial rank, while crucial distinctions of local or Roman citizenship could cross all other social boundaries among the freeborn, and various social distinctions could exist in local districts that did not carry as much weight at Rome or among the more cosmopolitan elite elsewhere.

Nevertheless, every social system can be divided according to a member's access to power and resources. Hence here and throughout, I mean by the 'upper class' all those with an inherited wealth so great that they have no need of employment and can easily compete on the stage of conspicuous consumption; by 'lower class', I mean all uneducated or uninfluential slaves and all free men who must work merely to live (or who even depend on charity to survive); and by 'middle class', everyone else. In this scheme the middle class consists of all those who have enough means to increase their quality of life beyond a basic or subsistence level (possibly even considerably beyond), but who still have to work in some fashion to maintain that quality of life (even if they could retire at a *lower* level of maintenance by taking a regular income from their accumulated properties instead of continuing to work). In other words, the middle class consists of persons of means who must still seek employment to improve their finances (whereas aristocrats can do so just by managing their investments).

By this definition, as I shall argue, the ancient 'middle class' corresponds to the social group most responsible for producing doctors and engineers and astronomers—and thus scientists—which is significant because this is the class whose members have enough wealth to enjoy some leisure and education, and yet still depend on a strong labor ethic to maintain or improve their elevated station, and who retain strong ties to their working class peers while developing strong ties to their aristocratic superiors. Hence this was a class straddling both worlds, whose members would have

middle class as straddling Clarke's binary demarcation between elite and non-elite.

many friends, neighbors, and relations practicing arts and trades, and yet who would also be advising, dining, and working closely with members of the upper class, and even going to school with them. The most successful would be patrons with clients of their own, and yet even they could still be the clients of aristocratic patrons in turn. If this is the social group Roman scientists came from, it means their values will likely differ from those of the aristocratic elite. And yet we know they often mixed and associated with that elite, ensuring continual communication and interaction between their respective social spheres.[1398]

To illustrate the general point we should revisit the situation of Lucian, who was destined to be a sculptor and stonemason but preferred to become an orator instead.[1399] His family, clearly of middle class status by the above definition, had definite social expectations. There were trades even they considered unsuitable to a free man. But working with his hands in a craft or trade—and for money, no less—was not among them.[1400] To the contrary, learning a manual skill and working in a shop to bring an income was considered respectable, even desirable. His uncle was a sculptor, for example, and easily persuaded his father that Lucian's training in the same art would be the best thing for him. It was only Lucian's determination to defy his family and escape the drudgery of an occupation that did not interest him that kept him from it. Everyone else whose opinion mattered to his family was happy to see him a stoneworker. Even Lucian shows no shame at becoming a sculptor, admitting he looked forward to it and saw many advantages in it. It was only when he actually tried it, and did not like getting beat to hell for making a mistake on his first effort, that he decided

---

1398. For discussion of the social status of artisans and the values unique to their class see Burford 1972: 185–218 (and throughout) and Cuomo 2007 (with also Cuomo 2011: 159). J. Clarke 2003 likewise finds evidence in extant art of the difference in values with respect to work between the Roman 'aristocracy' and the lower and middle classes, a conclusion supported by the epigraphic analysis of Joshel 1992.

1399. I discuss this in chapter ten of Carrier 2016.

1400. Lucian, *On the Dream* or *Lucian's Career* 1 (his family recommended *tina technên tôn banausôn toutôn ekmathoimi*, "that he be educated in some art of the banausic variety") and yet 2 (this art still had to be *andri eleutherô prepousa*, "appropriate for a free man," hence being a stoneworker was judged to be such).

he would prefer to become an orator instead.[1401]

We are clearly looking here at a different social world than we saw described by Cicero or Seneca. Lucian's story unveils a social context wholly alien to the leisured elite, unthinkable to someone like Cicero or Seneca. Yet his would have been the reality for far more people in antiquity than would share the fortunate situation of such as them. Pursuing an occupation in astrology or engineering or medical science would have been an admirable step up for Lucian, and hence for most people in the empire—far more in fact than could ever be counted in the social class of Cicero or Seneca. And yet, as we have seen, even *they* held scientific pursuits in high esteem. Scientists could and sometimes did come from the ranks of the wealthy elite, who had no need of making a living, just as they often came from middle class families of substantially higher station than Lucian's (Galen's family, for example, was not of Senatorial status, and possibly not even Equestrian, but was certainly much wealthier than Lucian's), and yet the latter seem to have more in common with Lucian's system of values than Cicero's or Seneca's.

Of course, Lucian's dream of a battle for his soul between the Goddess of Sculpture and the Goddess of Education reveals what also remains true today: that a clean white-collar job is often thought preferable to a dirty blue-collar one, an education more often preferred to ignorance, and good grammar respected more than bad.[1402] Hence becoming a well-educated, well-spoken, but hard-working doctor or engineer—or, as in Lucian's case, a traveling orator—would be seen by everyone of his station as a considerable step up, even while someone like Cicero or Seneca might view it as a step down. It may well have been a step down—for them. But for most who could realistically aspire to such a trade, it would be an improvement in social position.[1403] And since there would always be far more people in a position to pay and work their way into such occupations, than those who

1401. Lucian, *On the Dream* or *Lucian's Career* 3–5 (in 12 he notes that even Socrates was a stoneworker, cf. Diogenes Laertius, *Lives and Opinions of Eminent Philosophers* 2.5.18–21).

1402. Lucian, *On the Dream* or *Lucian's Career* 6–18.

1403. As argued for the medical profession in Nutton 2013: 261–71 and Horstmanshoff 1990: 187–96; the same would hold true for engineers and astrologers (there would also be the same distinctions within such trades, in levels of education and accomplishment, as held for doctors).

would be rich enough to pursue them merely at their own leisure, we should *expect* ancient scientists to have arisen mainly from the middle class. All that would then be required of aristocrats is not that they wish to become scientists, but that they see the value in hiring, supporting, or appreciating those who do. As we have amply seen by now, the Roman elite certainly embraced such values, and many of the middle class responded in kind. And as far as the support and advancement of science is concerned, no other social groups mattered.

Such is the theory, and the facts bear it out. As it happens, "a significant number of ancient scientists about whom we have biographical information—including some of the most eminent—are sons of craftsmen or tradesmen."[1404] Even for the Roman period, where we have hardly any biographical information about scientists, other than Galen and Vitruvius, the indications available support this conclusion. And it's certainly the case for Galen and Vitruvius. Vitruvius expressed gratitude to his parents for funding his education as an engineer, which indicates his family had no better prospects to offer him; and Galen was the son of an engineer.[1405] Likewise, in the Roman period most of the authors writing avidly on scientific subjects were of the equestrian class, as exemplified not only by Pliny the Elder himself, but also his sources, and (as we saw in chapter 3.8.I) his own intended audience.[1406] As Mary Beagon explains, the early Roman empire "was a period in which the production of practical and technical manuals flourished," whose "authors were predominately members of the equestrian order, like Pliny, or 'new men' in the senate" who "lacked the prejudice of more aristocratic families against practical expertise," and their "active curiosity on such matters as architecture, agriculture, geography, and natural sciences was shared by their class as a whole."[1407] As Seneca's case

1404. Rihll 2002: 20.

1405. A summary of what we know of Vitruvius and a summary and bibliography on what we know of Galen are both available in Carrier 2016: 111–19.

1406. As argued, for example, in Syme 1969: 219–25. On Pliny the Elder's life and career see *DSB* 11.38–40, *NDSB* 6.116–21, *EANS* 671–72, *OCD* 1162, Syme 1969, French & Greenaway 1986: 1–10, and Beagon 1992: 1–25. For a discussion of his research habits, and a list of his writings (almost all lost), see Pliny the Younger, *Letters* 3.5.

1407. Beagon 1992: 5 and 6 (cf. 5–11).

exemplifies. Though increasingly wealthy members of the equestrian order could sometimes gravitate toward the greater snobbery of the aristocratic elite as they rose to join them, they could also acquire an immense passion for science and discovery, and show no disrespect for empirical methods of achieving it. We have already seen, even beyond Seneca's case, that such aristocratic snobbery was rarely in fact hostile to the physical requirements of sound science.

For the Roman period I already mentioned considerable scholarship establishing the social status of doctors and engineers, who constituted the principal pool of scientists in antiquity.[1408] All recent work establishes that they predominately came from the middle class, more or less as I just defined.[1409] This has been illustrated for the medical profession most recently by Vivian Nutton, who rightly identifies a huge variability in social origin and status, but the most accomplished usually came from the upper middle class, not the aristocracy.[1410] Nutton also argues that throughout his writings Galen promotes something like an ideal of middle-class respectability, representing it as superior to both the effete snobbery of the upper class and the uncouth ignorance of the lower class.[1411] The same has been shown for astronomers and engineers. For example, we know they either made their own instruments and machines or must have worked very closely with professional craftsmen on their manufacture.

We noted several examples of Ptolemy's competence and familiarity with building instruments and machinery in chapter 3.4, adding the particular example of armillary spheres in chapter 3.3. Even his work in geography securely links him to the middle-class profession of surveying.[1412] But the clearest example is Hero, which I also discussed in chapter 3.4. Serafina

1408. See relevant notes and remarks in chapter 1.3.

1409. In addition to the works cited in chapter 1.3, the situation is summarized in Rawson 1985: 84–86 (for doctors) and 86–88 (for architects and engineers).

1410. Nutton 2013: 257–71. Most of the examples he considers from lower levels of society would not have been scientifically educated doctors, much less engaged in scientific research, so actual 'scientists' should be counted among the higher ranking end of the curve.

1411. Nutton 2013: 243. For example: Galen, *Exhortation to Study the Arts* 5–9 (= Kühn 1.7–22).

1412. Cuomo 2001: 169–75.

Cuomo has documented his extensive experience and practical skills, and yet also his mastery of erudite science and mathematics, establishing that he was both a craftsman and a member of the learned elite, a combination most likely for a member of the upper middle class.[1413]

Evidence of Hero's considerable practical experience can be found, for example, in his discussion of friction as a factor in both physics and machine design.[1414] Hence A.G. Drachmann came to the same conclusion, observing that the content of Hero's *Mechanics* makes quite clear all throughout that in Hero's day any "student of the science of mechanics is at once the man who designs the building, the engineer who understands the working of the engines, and the overseer who directs the work," and, Drachmann adds, many passages confirm the fact that he also often cuts, fashions, and builds components of machines himself, and yet at the same time "must evidently be well founded in mathematics," mastering both the craft and the science behind such diverse technologies as cranes, pumps, and presses.[1415] And as we saw in chapter 2.7, Hero and members of his school explicitly said exactly that. A Roman engineer had to be both highly educated in the sciences *and* a practiced craftsman of considerable talent and experience.

In the very same fashion, ancient doctors were both the theorists and the practitioners, cultivating and manufacturing their own medicines and performing their own surgeries. The situation that Vesalius claimed had obtained before the Renaissance, of 'doctors' only touching books and leaving the practice of surgery to uneducated butchers and barbers, and of pharmacy to mere merchants, if at all factual, was a *medieval* development.[1416] As even Benjamin Farrington had to admit, "the ancient prejudice against the *cheirourgos*, the surgeon or manual operator" did not "become fully operative until after the fall of the Western Empire."[1417] In fact there is no

1413. Cuomo 2001: 161–68.

1414. Hero, *Mechanics* 1.20–23 and 2.32. The Arabic text of his *Baroulkos* (in *Mechanics* 1.1) also shows he included a calculation for expected friction in its design (cf. Drachmann 1963: 22–32). Hero also discussed the stress limits of machinery. See Schneider 1992: 212–15.

1415. Drachmann 1963: 49. See Schiefsky 2008.

1416. A development identified and attacked by Vesalius in the preface to his work *On the Fabric of the Human Body* (first published in 1543).

1417. Farrington 1946: 29.

evidence of such a prejudice *at all* in antiquity. Quite the contrary, there was a pervasive expectation that a doctor should be able to know and do everything required of his art. This held even for specialists, hence an eye doctor typically had to be both pharmacist and surgeon in every respect necessary for eye care. Consequently, there was no split between theory and practice in the field of medical research. Even Farrington could "find no clear proof that the prejudice against manual labor did, in fact, operate in Greek society to check the progress of the science of anatomy," to the contrary, "the names of great anatomists are too numerous" and "the progress, if spasmodic, still too remarkable to warrant the assertion that the science of anatomy, before the time of Galen, suffered from the prejudice engendered by the social structure of ancient society." Instead, "as Vesalius reminds us, Galen frequently expresses his pride in his manual skill."[1418]

Accordingly, ancient scientists often argued the vital importance of combining theoretical knowledge with practical experience and skill.[1419] Vitruvius captured this ideal with his remark that "engineers who work without a formal education and only aim at manual skill cannot gain the authority worthy of their labors, while those who trust only theory and books obviously follow a shadow and not reality."[1420] Much the same was said of medical scientists by Celsus half a century later.[1421] So when seeking to extend these values to the art of agriculture, and thus elevating its study to a science (at least by ancient standards), Columella borrowed the same ideas, arguing that to increase the productivity of our agricultural operations we need to combine practical hands-on experience with theoretical understanding, requiring a ready turn to books as well as experiments.[1422] Likewise, as Karin Tybjerg argues, Hero gives detailed practical instruction on the construction of scientific demonstration apparatuses specifically to show the importance

1418. Farrington 1946: 34.

1419. See Cuomo 2001: 201–05 and my discussion of developments in ancient scientific method introducing chapter 3.7 and concluding chapter 3.1.

1420. Vitruvius, *On Architecture* 1.1.2 (within the whole context of 1.1).

1421. Celsus, *On Medicine* pr.39. See related discussion in chapter seven of Carrier 2016.

1422. Columella, *On Agricultural Matters* 1.1.16–17 (he then surveys examples of needed areas of research and experimentation in 1.pr.22–28).

of practical craftsmanship to the discovery of the truth.[1423] She also argues that Hero saw technology as a means of 'doing philosophy', to the extent that "mechanical devices and skills play a central role in demonstrating physical theories," just as in his *Pneumatics* he argues that mere philosophers only have arguments, but engineers can demonstrate results perceptibly with machines, which is always more decisive—for the obvious reason that an apparatus can visibly demonstrate scientific principles in a way armchair philosophy never can.[1424]

Galen makes the same point not only for medicine but all other sciences as well, even using engineering and astronomy as examples.[1425] In fact, Galen and Hero both take pains to explain the practical hands-on procedures needed to conduct scientific research *and* to implement scientific knowledge in practical applications.[1426] Hence Galen used anatomical experiments and demonstrations to the very same end as Hero with his machines: to trounce philosophers in the public arena and to prove his own theories against them. As Tybjerg explains, "Both Hero and Galen use the generation of wonder, the exhibition of technical skill, hands-on descriptions, and references to observables to create an image of expertise that supports their physical or anatomical theories."[1427]

Galen's emphasis on visible progress and the superiority of empirical and practical knowledge has already been demonstrated.[1428] But as a prominent example, Galen's treatise *On the Doctrines of Hippocrates and Plato* is largely devoted to using anatomical dissection and experimentation to decide fundamental questions in natural philosophy, in the process effectively ridiculing those who argue from the armchair without actually 'getting their hands dirty' checking the facts. Though the primary aim of this

1423. Tybjerg 2005. Tybjerg 2003: 446–51 further argues that Hero uses this and other tactics to argue against those still clinging to anti-banausic attitudes (though our analysis in 4.6.I already qualifies the degree and popularity of such attitudes).

1424. Tybjerg 2005: 213–18 (quote from 215); Hero, *Pneumatics* 2.4–7 (see related discussion in chapter 2.7). See also Tybjerg 2003: 451–57 and 446–51.

1425. See chapter 3.6.VI (and related discussion in chapter 3.9.II and sections 4.1 above and 4.6.III below).

1426. For Galen see Nutton 2013: 237 and relevant discussion and notes below.

1427. Tybjerg 2003: 455.

1428. See, for example, von Staden 1995.

treatise was to establish the anatomical location of the deliberative soul (i.e. the mind) and thus confirm the theories of Plato and Hippocrates against such authorities as the Stoic sage Chrysippus, a secondary aim is clear throughout: to prove by example that empirical research makes progress in philosophy where arguments only stagnate—in other words, exactly what Hero said.[1429]

Hence Galen emphasizes the methodological point that "if anyone wishes to observe the works of nature, he should put his trust not in books on anatomy but in his own eyes" and seek empirical demonstrations from a hands-on expert like himself or "alone by himself industriously practice exercises in dissection," rather than merely reading what others say.[1430] For example, Galen relates how he at first thought skinning the apes he dissected was too trivial to do himself, so he had an assistant do it for him—until he discovered this procedure was destroying important data, so he began carefully skinning the apes himself.[1431] He tells this story to chastise other anatomists who do not dissect as carefully as this, but more haphazardly, and who then substitute arguments for observations. Instead, Galen insisted on more hard work and more attention to empirical fact. Though some shrank from such work because it was distasteful, clearly many did not. Aristotle had already said of zoological dissection that even the ugliest and lowest of animals are worth scientific study because of the intellectual pleasures such research induces, and in fact, Aristotle says, recoiling from this because the work is dirty or disgusting is childish.[1432] Accordingly, Galen was even willing to personally handle and inspect human feces and urine, considering such 'lab work' essential to diagnosis, even though it

1429. Confirming the analysis concluding chapter 3.9.II.

1430. Galen, *On the Uses of the Parts* 2.3 (= M.T. May 1968: 118–19).

1431. Galen, *On Conducting Anatomical Investigations* 1.3 (= Kühn 2.233–34). C.J. Singer 1956: 8 translates Galen as saying he was "avoiding the task myself as beneath my dignity" but neither the word 'beneath' nor 'dignity' is in the text, which reads: *men oun k'âimoi tôn hupêretôn tis exedere tous pithêkous, oknounti dêlonoti kai mikroteron ê kat'eme nomizonti t'ourgon*, "And so I too had one of my assistants skin the apes, which is to say I avoided it and thought the work too trivial for me to do," i.e. he though it was not worth the trouble, not that it was undignified.

1432. Aristotle, *On the Parts of Animals* 1.5.645a.

must have been repugnant.[1433] Likewise, "Galen believed that a good doctor should also be a good cook," hence he gives recipes and advice on diet and food preparation without apology, even though the cooking *profession* was among those most frequently despised as vulgar (thus demonstrating that it was not the activity itself that mattered but, as Aristotle said, how and why it was pursued).[1434]

Not only were Roman scientists craftsmen themselves, and not only did they insist upon the acquisition of the skills and experience of a craftsman, but they regularly associated with craftsmen. It is more than obvious that mechanics, carpenters, and blacksmiths must have worked closely with scientifically educated engineers in the construction of cranes and machinery, as well as on aqueducts and other complex projects.[1435] Mechanics, carpenters, and blacksmiths must also have worked closely with scientifically educated doctors in the construction of medical instruments and machinery, and also with astronomers in the construction of diopters, waterclocks, sundials, and a large array of other astronomical instruments.[1436] As G.J. Toomer observes, "great skill and precision in making and fitting the parts and graduating the arcs must have been demanded of the craftsmen" who made the astronomical instruments known to have been in use, simply "in order to attain the accuracy" shown in ancient astronomical texts, and in fact "extant artefacts from the Hellenistic period demonstrate that a high level of craftsmanship was attained," and thus evidently enjoyed by scientists—which entails they were often making these instruments themselves, and with consummate skill, or were in constant and detailed communication with those who did.[1437] We have a clear example: from the material evidence we have identified the workshop of a craftsman in Pompeii named Verus who was evidently engaged in the construction of mathematical instruments, including a groma, a portable sundial, and the parts of various uncompleted projects, and we can assume he was the go-to guy for more complex instruments like diopters—and probably medical

1433. See Nutton 2013: 243.

1434. Mark Grant 2000: 11.

1435. As argued, for example, in Di Pasquale 2002.

1436. See evidence in chapter 3.6.

1437. *OCD* 188–89 (s.v. "astronomical instruments"); and see examples in chapters 3.3 and 3.6.

instruments, as several sophisticated examples have been found in various buildings at Pompeii.[1438]

Vivian Nutton has already demonstrated that ancient doctors more routinely associated with accomplished artisans, and most frequently came from such a background.[1439] The same was surely true of ancient engineers, who in turn constituted the main pool of astronomers. Likewise, Roger French argues that Pliny the Elder got a great deal of information from tradesmen and artisans and was not only writing for members of that class, but was in regular communication with them, and it is evident the same was true of Aristotle and Theophrastus.[1440] For example, Pliny relates how he went to observe a metalworker in his shop while planning and constructing a monumental bronze statue for the emperor Nero, making this visit not only out of curiosity but also to learn and report something of the artisan's techniques.[1441] Even Seneca cited personal experience from his own agricultural excavations that rainwater does not wet the ground on his estate to a depth greater than ten feet, and described with evident respect the mechanical procedures he had observed for transplanting vines and trees.[1442] Though the aristocratic Seneca is unlikely ever to have picked up a shovel, he still must have observed and supervised any slaves he or his colleagues put to the task. Hence even when relying on others to perform menial labor, an aristocrat was not necessarily isolated from the empirical benefits of observing and supervising the work. Channels of communication were thus not closed by differences in social station.

As usual, our best information comes from our chattiest and most extant author, Galen, and though speaking for his own profession, the same surely held for engineers. For example, Galen says doctors should treat the poor to gain experience, travel to learn how climates and regions differ, test what they have read against their own experience, and admire hard

1438. See Cuomo 2001: 153–55, Dilke 1971: 71–73, and Di Pasquale 2002.

1439. Nutton 1995: 10–13 (and throughout).

1440. French 1994: 238–39 and 252–53. See chapter 3.1 for Aristotle and 3.5 for Theophrastus.

1441. Pliny the Elder, *Natural History* 34.18.45–47.

1442. Seneca, *Natural Questions* 3.7.1 and 86.14–21.

work rather than money and luxury.[1443] Accordingly, Vivian Nutton found Galen frequently interacting with the lower classes for his research.[1444] He was eager, for example, to converse with sailors about their cargoes and astronomical navigation techniques.[1445]

Painting the picture even more broadly, Nutton reports:

> [Galen's] inquisitiveness took him down the mines of Cyprus, to the shores of the Dead Sea, and, so he alleged, to the backwoods of Paphlagonia to learn the secrets of a herbalist-cum-poisoner. He obtained drugs from a camel caravan in Palestine, as well as from a search in the basements of the royal stores in Rome. He reports his own experiments with drugs, rather like his father's earlier tests on plants and wines, and he refused to write up a section on the mineral drug *terra Lemnia* until he had had personal experience of its production—and that took thirty years.[1446]

Likewise, in Galen's writings on regular and medicinal diets, "conversations with teachers and students are recorded, alongside impromptu meals with peasants in the countryside, comments by farmers" and much else besides.[1447] Galen also reports that when an elephant's heart "was removed by the emperor's cooks, I sent one of my colleagues experienced in such things to beg the cooks to allow him to extract the bone from it. This was done and I have it to this day," thus advertising the value of interacting with the lower classes for the benefit of science, as he had also done in his discovery of the osmotic properties of earthenware and grain.[1448]

Galen's kitchen story connects to another told by Plutarch, which

---

1443. Galen, *That the Best Doctor Is Also a Philosopher* 3 (= Kühn 1.58–59).

1444. Nutton 2013: 251–52.

1445. Nutton 2013: 223. See Galen, *Commentary on Hippocrates' Airs, Waters, Places*.

1446. Nutton 2000a: 962.

1447. Mark Grant 2000: 12.

1448. Elephant: Galen, *On Conducting Anatomical Investigations* 7.10 (= Kühn 2.620). It is not literally a bone that is meant but a bony tissue (cf. Kühn 2.618–19), since more specific tissue identification was not possible in antiquity without a microscope. Notably, Galen's 'colleague' was not a slave but *tina tôn gegumnasmenôn hetairôn*, "one of my educated friends," i.e. a social equal of Galen, who evidently had more experience dealing with kitchen staff. Grain osmosis: Galen, *On the Natural Faculties* 1.14 (= Kühn 2.55–56), discussed at the end of chapter 3.2.

more generally illustrates the world-straddling social position of ancient scientists.[1449] Plutarch reports that the physician Philotas, a friend of Plutarch's grandfather, attended medical school in Alexandria yet had middle class friends such as the royal chef. Philotas was eager to discuss and tour the kitchen operations in the morning with his friend, while at the same time being accepted as a regular dinner guest of Antony and Cleopatra at night. Here we have a scientist getting along just as easily with the lowest ranking craftsmen and the highest ranking elite, even on the very same day. He clearly did not consider touring the kitchens and conversing with chefs as beneath his dignity. To the contrary, he was excited by the opportunity, driven by curiosity to learn something about a large kitchen operation, both by direct observation and by conversing with experts. There are other examples of this. For instance, in his directed effort to make the water supply of Rome more efficient, Julius Frontinus specifically argues it would be disgraceful for the governing elite to run any operation without consulting hands-on experts in enough detail to know how things work—accordingly, he says he wrote *several* treatises following this method, recording as much useful practical experience on each subject as he could.[1450] In fact there were many more technical handbooks on the crafts than survive, yet their existence at the time entails greater collaboration between craftsmen and scholars, and an increasing interest in such fields among ancient readers. As just one example, not only is it remarkable that several formal treatises were written on the fishing industry, but it is even more remarkable that an elite writer like Athenaeus knew them, read them, and discussed them, and did not think it strange or in any need of apology or explanation that his imagined dinner guests—scholars, philosophers, and doctors—would have read and discussed them, too.[1451]

1449. Plutarch, *Antony* 28. See Scarborough 2012: 10–14.

1450. Frontinus, *On the Aqueducts of Rome* 1–2. Of his 'other' treatises only two are known (but not extant), his *Art of War* and one or more books on the science of surveying (some fragments of which survive). On this man and his extant writings see the end of chapter 3.4.

1451. Athenaeus, *The Dinnersages* 1.13c (citing prose treatises by Seleucus of Tarsus, Agathocles of Atrax, and Leonidas of Byzantium, about whom little or nothing is now known). See also chapter 3.5 and, to a lesser extent, Corcoran 1964 (though his generalizations about Roman interests are obsolete or untenable and his sourcing is poor).

III. Conclusion

Such was the reality, more nuanced than many scholars have imagined or claimed. But old notions die hard. When John Humphrey, John Oleson, and Andrew Sherwood (hereafter H.O.S.) compiled a recent sourcebook on ancient technology, they concluded it with a chapter entitled "Attitudes towards Labour, Innovation, and Technology" which is a typical example of what Ludwig Edelstein had complained about: an arbitrary collection of cherry-picked passages without regard for differences of time, place, author, or context, ignoring contrary evidence, or sometimes even the actual meaning of the texts they quote.[1452] A thorough analysis is unnecessary to demonstrate this point. The stereotypes and impressions they generate are already refuted by everything we have presented in this section, as well as material in chapter three. But I shall illustrate their folly with a few choice examples that exhibit the defects of the remainder.

They cite Lucian's preference for oratory over sculpture as evidence of a "prejudice against physical work," as if we could not find similar sentiments in modern parodies of a career in fast food, or on the balance sheet of any ambitious teen deciding between college or a life of factory work in a 'hick' town.[1453] My analysis of Lucian's account finds results very different from what is implied by the arrangement of H.O.S.[1454] Similarly, compare their treatment of Frontinus with mine (in chapter 3.8.I), or their naive use of Plutarch's fantasy about Archimedes (see section 4.6.I) or Seneca's atypical ludditism (see chapter 3.8.I and section 4.6) or those oft-abused tales about Tiberius and Vespasian (see chapter 3.8.II).[1455] But among their worst blunders is an entry entitled "Writing about the Banausic Crafts is Degrading," in which they cite as their only evidence a passage from Aristotle that in fact says no such thing, but almost entirely the opposite.[1456]

1452. Humphrey et al. 1998: 579–99 vs. Edelstein 1967: xiv-xv (which, ironically, they cite in their bibliography).

1453. Humphrey et al. 1998: 587.

1454. See chapter ten of Carrier 2016 and chapter 4.6.II here.

1455. Humphrey et al. 1998: 593 (Frontinus), 589–90 (Archimedes), 594–95 (Seneca), 595 (Tiberius and Vespasian, though their treatment of the latter is better than most).

1456. Humphrey et al. 1998: 583–84, quoting Aristotle, *Politics* 1.4.1258b-59.

To begin with, Aristotle says there that *many* books were written on the banausic crafts, which already entails few thought it degrading to write on such subjects. He also says these books are useful, and actually recommends them to those who are interested, which also does not sound like someone who thought writing them was degrading. But more importantly, H.O.S. translate Aristotle as saying "a general account of each of these industries has been given above, but—while it would be useful for their practical application to give an account of each one in detail—it would be vulgar to spend much time on them," and so Aristotle refers the reader to all those other books about them instead. Already the word 'vulgar' is not the same as 'degrading', which should have raised a red flag. But 'vulgar' is not even a plausible translation here, given that the actual word is 'burdensome' (*phortikon*), which in the given context clearly means just that: it would be too much bother for him to give a detailed account of what all those books say, in what is otherwise a mere digression in a book about politics.[1457] In fact, as far as I know, *no one* in antiquity ever said it was degrading to write about banausic industries. Aristotle can only be made to say so by ignoring the context and distorting what he said.

Such disastrous inattention to context is not unusual. In order to produce an "origin" of a "Greek prejudice against craftsmen," for instance, H.O.S. must first show evidence of a greater respect for craftsmen in the Archaic period. Yet to accomplish this they quote a single passage from Homer, and another from Hesiod, and nothing else.[1458] Worse, neither passage confirms what they claim, as both have parallels in later periods, thus eliminating any evidence of a change. And worse still, both passages include evidence refuting their thesis, by proving the existence of an aristocratic ethos even in the time *they* were written. The passage they quote from Hesiod, for example, only attacks idleness and sloth, as did many Greek and Roman authors thereafter—in fact, outside of comedy, one would have to work hard to find any contrary view at any point in history.[1459] And though Hesiod is clearly writing to the educated farmer, not the pampered aristocrat—and thus displays a healthy respect for chopping wood and plowing and other labors

1457. *LSG* 1952 (s.v. "*phortikos*"): 1. "of the nature of a burden, tiresome, wearisome, onerous" and only after that 2. "coarse, vulgar, common, low."

1458. Humphrey et al. 1998: 579–80.

1459. Hesiod, *Works and Days* 303–11, 410–13.

(though he considers the need for them a sign of social decline)—one will find some of the same in the *Georgics* of Virgil, leaving the obvious question of whether Hesiod, like Virgil, is really only playing the poet.[1460] Moreover, Hesiod actually attests to the existence of an aristocratic ignorance of the details of the work they command, essentially the attitude H.O.S. want us to believe only developed later.[1461]

Likewise, the passage they cite from Homer only says good hosts do not feed or house beggars but only itinerants who can ply some useful trade in exchange for their keep, which is a sentiment that probably any Greek or Roman would have agreed with, in any era.[1462] Moreover, again in direct contradiction to their thesis, Homer attests exactly the prejudice they are claiming evolved later, since the passage they quote is actually the response to an aristocrat scoffing at taking in a scruffy guest, a rather crucial context to omit. Moreover, at no time is this guest (actually Odysseus in disguise) ever treated as anything other than a servant, even once he is recognized as a storyteller (and thus a 'craftsman' according to H.O.S.). In every respect he is represented as socially beneath the aristocrats he performs to, who obviously would never have deigned to become a wandering storyteller themselves. In other words, this passage proves nothing that H.O.S. claim.

Hesiod and Homer also remained among the most widely read and revered poets throughout antiquity—despite advocating values that H.O.S. want us to believe were no longer respectable. Though many Greek and Roman authors did attack the values in Homer as immoral in their own time (such as in Plato's *Republic* or Plutarch's *How the Young Man Should Study Poetry*), an evident appreciation for manual labor is never among the values criticized in these assaults. It is thus hard to maintain that anything had changed much with respect to those values. There had always been an aristocratic ethos, there had always been a successful, hard-working middle class, and there had always been a complex variety of attitudes between and among them. Hence the reality was far more nuanced, as even a proper reading of Homer and Hesiod reveals, and as we have demonstrated

---

1460. That the need of labor was a sign of social decline: Hesiod, *Works and Days* 174–77.

1461. Hesiod, *Works and Days* 455–57 (repeating a proverb to the effect that the rich and presumptuous do not know what is involved in constructing a wagon).

1462. Homer, *Odyssey* 17.382–86.

throughout. In other words, this entire chapter in H.O.S.'s book is all but worthless, essentially fabricating an alternate reality with a textbook case of bad scholarship, rather than illuminating the actual facts of history. As this does nothing but mislead the reader and misrepresent the ancient world, it would do more harm to read it than to burn it. Let us hope they produce a more adequate second edition.[1463]

With regard to ancient science and natural philosophy, even the mildest form of their thesis that could still be maintained means little. Any art could gain respect among the elite if given a sound theoretical foundation, which only required that art to be useful, moral, rationally articulated, and organized in an erudite way. Insofar as there was any elite prejudice against the manual arts, that is exactly how it could be overcome, and was overcome in several cases. In fact, it can reasonably be said that such elite expectations actually resulted in the arts becoming *more* scientific, as the histories of medicine and engineering suggest. There were also far more people who could improve their social position by becoming a shop-tinkering scientist than those who would have anything to lose by it. Consequently, any aristocratic disdain there may have been (for becoming a scientist who either hobnobbed with craftsmen or practiced a craft themselves), would have had little effect on the advancement of science (or, for that matter, technology). As long as enough aristocrats supported these scientists by hiring them, praising them, patronizing them, dining with them, or simply treating them with more evident respect than they would have otherwise received, then the social status of empirical science in antiquity can be considered high. One might argue the situation in the early Roman Empire was comparable to the early British Empire in exactly this respect, when few aristocrats would deign to become a doctor or engineer, and would consider most doctors and engineers as socially beneath them, yet would eagerly employ them, converse with them, respect them, read their books, and elevate their wealth and status above the broader working class.

To see how little any actual aristocratic attitude hindered the methods or social position of natural philosophers, we can turn again to Galen, who tells a pertinent story:

1463. Oleson at least may have revised his views (judging from, e.g., Oleson 2004) and has overseen the production of a far more sophisticated *Oxford Handbook of Engineering and Technology in the Classical World* (Oleson 2008).

> I was recently present at a dispute between two philosophers. One claimed that water was heavier than wood, the other that wood was heavier than water. Both produced very long arguments, considering the matter from every possible angle. The chief point of the one philosopher was that any compressed substance—such as wood—is heavier. The other staked his claim on the notion that water has less of the void in it. And they proceeded in this way for a considerable length of time, producing arguments to reinforce their own plausibility, but without any proof—as if it were a matter incapable of being decided by observations of the kind which (as you know) I perform. The philosophers, who wished to continue this discussion, asked an engineer in what manner it could be clearly demonstrated which of the objects were heavier. It could not, they said, be done with a pair of scales, nor by means of a filled vessel, for it would be possible to set up the piece of wood, but not to fill the vessel with it, though it would be possible to fill it with water.
>
> As they continued in their usual fashion, the engineer laughed and said: "That's you through and through, you know-it-alls. You all reckon to understand what happens beyond the universe—a subject which admits of conjecture, but in which there can be no scientific knowledge. But when it comes to these kinds of questions—questions which are quite frequently understood by the man in the street—you are utterly at a loss. And so it is with the matter before us now, how to make a relative measurement of water and wood." And so everyone present begged the engineer to tell them how the weight of the wood could be scientifically and reliably measured against that of water, and he explained the matter succinctly and clearly, in such a way that it was understood by all except the two philosophers. He was in fact constrained to repeat the explanation a second and even a third time, and finally, with great difficulty, they managed to understand it.[1464]

Galen and his engineering friend later have a good laugh at their expense, but what is significant is that even these philosophers thought it appropriate and worthwhile to consult an experienced technician to resolve a question in natural philosophy. Of course it is also significant that the doctor Galen had an engineer as a friend, and both of them thought so

1464. Galen, *On the Affections and Errors of the Soul* 2.7 (= Kühn 5.99–100). Translation adapted from P.N. Singer 1997: 147–48. The method the engineer explained must have involved the hydrostatics of Archimedes and Menelaus (the only practical way to compare weights by volumes: e.g. weigh the wood on a scale, then immerse the wood in water and then weigh on a scale the water it displaces, and their weights by equal volume will be determined).

highly of the experimental method in natural philosophy that they could have a snicker at other philosophers who carried on arguing without it. And yet even *those* philosophers eventually decided to resolve their argument by conducting an experiment, and sought out an engineer to advise them how.

4.7 Lack of Institutional Support?

So there was no 'unbridgeable divide' between the bookish scientists and the men in the field who actually knew how things worked. Which leaves the other seemingly pervasive 'indicator' of a Roman disinterest in science: the absence of directed and subsidized research. Given all the above, as well as all the evidence of chapters two and three, why was there no institutional support for scientific research? Why did neither emperors, nor cities, nor private benefactors establish research institutes, like the Alexandrian Museum had once been in its Ptolemaic heyday? Even scientists themselves, who frequently interacted with each other and had their own dining clubs, did not pool resources to fund such a project.[1465]

I have already discussed the issue of state support for *education* in *Science Education in the Early Roman Empire*, and found that even the most popular avenues of educational advancement received only some support from state or private benefactors.[1466] So the lack of the same for science education does not indicate any special disinterest. Though it was always far less popular than an education in literature or oratory, at the same time there were many private or state funded Museums throughout the Roman world, which typically provided libraries or even free meals to members, such as the Aristotelian Society at Alexandria. Though Museums often admitted scholars in the humanities, all admitted scholars in the sciences, especially doctors. These societies would obviously have provided good networking opportunities for procuring patrons, accessing books, and exchanging ideas and resources with fellow scientists. So it cannot be said there was *no* institutional support for science in antiquity. But what was lacking were

1465. See discussion in Edelstein 1952: 596–602.

1466. In chapter 8 of Carrier 2016, which also covers every point to follow, including the role of libraries, science clubs, and the equivalent of universities in the Roman era. To which should be added Houston 2014 (on libraries).

salaries or stipends for scientific research or funded associations specifically devoted to collaborative research projects.

It is not as if the ancients were unaware of the utility of such a project. Plato (in the guise of Socrates) actually argues for state funded scientific research, for the obvious reason that it would produce more rapid progress, which seems essentially to have been what the Hellenistic kings more or less attempted. As Plato's Socrates makes the argument, in the context of mathematics and astronomy (and by implication harmonics as well):

> Since no city holds them in honor, these inquiries are barely pursued owing to their difficulty. The researchers also need a director, without whom no discovery would be made. And yet, in the first place, such a director will not be easy to find, and then if found, as things are now, researchers in these fields would be too arrogant to listen to him. But if the state as a whole should join in superintending these studies and treat them honorably, then they would accept advice, and intense and comprehensive investigation would bring out the truth. Since even now, lightly esteemed as they are by the multitude, and hampered by researchers not knowing what their work is good for, nevertheless progress is being made against all these obstacles, all due to the shear pleasure of it, so it will not be amazing if discoveries are still brought to light.[1467]

Xenophon, another pupil of Socrates, attributed a similar suggestion to the poet Simonides. Though almost certainly a fiction, Xenophon clearly means to endorse the ideas he credits to him. After bringing up the advantages of rewarding soldiers for their achievements in both readiness and battle, Xenophon's Simonides extends the same idea to the whole sphere of civic utility:

> Even agriculture itself, most useful of all occupations, but just the one in which the spirit of competition is conspicuous by its absence, would make great progress if prizes were offered for the farm or village that can show the best cultivation, and many good results would follow for those citizens who threw themselves vigorously into this occupation. For apart from the consequent increase in the revenues, sobriety far more commonly goes with industry; and remember, vices rarely flourish among the fully employed. If commerce also brings gain to a city, the award of honors

1467. Plato, *Republic* 7.528b-e (of math and astronomy, but implies also harmonics at 7.530d-e).

for diligence in business would attract a larger number to a commercial career. And were it made clear that the discovery of some way of raising revenue without hurting anyone will also be rewarded, this field of research too would not be unoccupied. In a word, once it becomes clear in every department that any good suggestion will not go unrewarded, many will be encouraged by that knowledge to apply themselves to some promising form of investigation. And when there is a widespread interest in useful subjects, an increase of discovery and achievement is bound to come. In case you fear, Hiero, that the cost of offering prizes for many subjects may prove heavy, you should reflect that no commodities are cheaper than those that are bought for a prize. Think of the large sums that men are induced to spend on horse-races, gymnastic and choral competitions, and the long course of training and practice they undergo, all for the sake of a paltry prize.[1468]

As I noted in chapter 3.1, a century or so after Plato and Xenophon wrote, several Hellenistic royal courts did essentially what they asked for. Though we do not know the precise details, they gathered a staff of well-paid experts in various fields and, as far as we can tell, either directed or expected them to produce results. We only have definite evidence of this in the field of military technology (particularly with respect to naval and artillery development), but there is indirect evidence of results in other fields. Philo of Byzantium argued (from having seen it himself) that progress in war technology is made by hands-on experimentation, consultation with craftsmen, and adequate funding of researchers, and it is hard to imagine the same would not have been obvious for any other field.[1469] As we have seen, arguments from silence are rather ineffective here. Just as evidence of an acceleration in innovative naval technology is undeniable during this same period, even though we have no treatises on the subject and no idea

1468. Xenophon, *Hiero* 9.7–11. Although this is not the same point as made in Aristotle, *Politics* 2.8.1268a, who says the architect Hippodamus, as part of his ideal constitution, "proposed a law that those who discovered anything of advantage to the state should receive honors," for a little later Aristotle makes quite clear that technologies were not meant (nor even advances in the arts), but the exposure of secrets and conspiracies and the proposing of new laws.

1469. Philo of Byzantium, *Siegecraft* 4.1–5 (= Marsden 1971: 107–09 = DeVoto 1996: 5–9). Zilsel's claim that Philo was "not a scholar but a military engineer," a sort who "were never admitted as fellows to the Alexandrian Museum," is simply and entirely false (Zilsel 1945: 328–29).

who was directing the relevant research (and yet someone must have been and it is almost inconceivable that none wrote of it), the same can be said of mechanics and pneumatics and hydrostatics and similar subjects.[1470]

In other words, the Ptolemies and other royal courts recognized the advantages enough to fund something like the institute Plato recommended, and it had something like the effects Xenophon predicted, though not quite as marvelously as either expected. The Sicilian tyrant Dionysius, for example, engaged a similar project to fund research in military weaponry, luring artisans with huge salaries and prizes for achievement, mingling with them to encourage their work as he would his soldiers, and inviting winners to dine with him.[1471] Why, one might ask, was that model abandoned?

In truth, the Roman emperors continued to support the same Museums, and thus did not abandon the basic model established in the Hellenistic period, though they perhaps took less personal interest in selecting their members or directing their activities. As argued in chapter 3.8.I (and elsewhere), this is a neglect that several Roman intellectuals were perhaps delicately attempting to protest and reverse, though none as far as we know proposed anything as bold as Plato or Xenophon had suggested, or as clear and well-thought-out as Francis Bacon would eventually suggest himself (though he had the advantage of witnessing the rise of the Scientific Revolution to inspire him). But Plato and Xenophon were still being read in the Roman empire, in fact rather widely, so their ideas would not have been forgotten, and similar notions were hinted at by scientists of the time.

Pliny, for instance, laments that "we" do not give any thanks to astronomers, but prefer to compose annals of blood and slaughter, "so that the crimes of men may be well known to those who are ignorant of the very universe," a gentle suggestion that astronomers should receive more honors than they do.[1472] More boldly, perhaps—for a treatise addressed to the reigning emperor, who also happened to be Pliny's personal friend—is the following tale:

1470. On rapid advances in naval war technology in this period see Foley & Soedel 1981 (with scholarship cited in the notes on naval technology in chapter 3.6.III and 3.8.IV).

1471. Diodorus Siculus, *Historical Library* 14.41.1–43.4. See discussion in Rihll 2007: 26–30 and D. Campbell 2011.

1472. Pliny the Elder, *Natural History* 2.6.43.

> When King Alexander the Great was fired up with a passion to know the natures of animals, and delegated this research to Aristotle, a man of supreme eminence in every field, some thousands of people all throughout Greece and Asia Minor were ordered to make themselves available to him—including all those who make a living by hunting, fowling, and fishing, and those who were in charge of warrens, herds, beehives, fishponds, and aviaries—so not any living thing would be unknown to him. By thus making his inquiries to them, Aristotle composed his famous books on zoology in roughly fifty volumes...[thus fulfilling] the central desire of the most glorious of all kings.[1473]

It does not matter for our purposes how much of this is truth or legend.[1474] We know Aristotle did consult with the very people Pliny describes, who had a first-hand working knowledge of animals, and acquired much of his information (and specimens) from them, as from others among the working classes, whether Alexander the Great had anything to do with this or even any particular interest in the matter. But what is important here is that Pliny not only *believed* (or wanted his readers to believe) that Alexander took such an active interest in scientific research, but he portrays this as the mark of a great king, a praiseworthy action. We also see no hint of any disdain or apology for the prospect of communicating extensively with the working classes for the advancement of science. In fact, the way Pliny tells this story entails that he expected his audience—which included most conspicuously the emperor himself—would not hold any such revulsion but in fact would agree that this behavior made Aristotle a more reliable authority and Alexander a more glorious king. Pliny may have hoped Titus would get the hint. Perhaps it is unfortunate that Titus did not live more than two years after receiving Pliny's book, and then was quickly replaced by his tyrannical brother Domitian.

Vitruvius *also* called for public support of the arts and sciences, *also* in a book written to a Roman emperor (in this case Augustus), *also* by a man who had personally served with him in the field (and with his predecessor Julius Caesar). Vitruvius complains that pensions, honors, and parades are given for athletes but not writers or inventors, declaring:

1473. Pliny the Elder, *Natural History* 8.17.44.

1474. It is still debated (see, for example, Healy 1999: 72–73).

> I am amazed that the same honors—or honors greater still—are not bestowed on writers who provide every nation with endless utility for everlasting ages. For this would have been a much more worthy institution to have set up, because athletes only make their own bodies stronger by exercising, but writers strengthen not only their own wits, but indeed everyone's, by preparing books for learning and the sharpening of minds.[1475]

So they should receive even greater rewards than athletes, possibly even triumphs and deifications. He goes on to list examples of "the discoveries" of the kind of honor-worthy writers he has in mind—discoveries "that have been useful for people in improving their lives"—and almost all of them are advances in mathematics, technology, and engineering science, including Archimedes' discovery of the laws of hydrostatics.[1476]

Moses Finley once claimed that "Vitruvius saw neither a virtue nor a possibility in the continued progress of technology through sustained, systematic inquiry," a statement that could only be true when we retain the crucial words "sustained" and "systematic," as otherwise Vitruvius was well aware of the value and possibility of technological progress (as we saw in chapter 3.9.II).[1477] But even Finley's claim that he saw "neither a virtue nor a possibility" in "sustained" and "systematic" inquiry is dubious, as a century later Hero explicitly argued for both in his *Siegecraft*, while Vitruvius never denies the value of either and surely would have supported both—and seems to have hoped his emperor would agree.

Even during the Scientific Revolution state interests failed to fund scientific research, which had to be done on individual initiative with private resources, whether from personal wealth or the largesse of a patron, and usually outside the university system, which was not organized to support much more than teaching, nor geared to promote innovation. In other

---

1475. Vitruvius, *On Architecture* 9.pr.1–2. This is similar to Aristotle, *Rhetoric* 3.14.2.1414b, which says "Isocrates disparaged the Athenians because they rewarded bodily excellences, but instituted no prize for men of wisdom" (i.e. Isocrates, *Panegyric* 1, though in *Panegyric* 45 he says Athens *did* have prize-winning contests in "eloquence and wisdom and all the other arts").

1476. Vitruvius, *On Architecture* 9.pr.3–17.

1477. Finley 1985: 146.

words, the situation was not very different from antiquity. The founding of the first institutions ever *explicitly* devoted to scientific progress—the Royal Society of London for the Improvement of Natural Knowledge, followed by the Académie des Sciences shortly thereafter—effectively marked the *conclusion* of the Scientific Revolution, for such institutes were concrete manifestations of the triumph of its revolutionary ideals. Before that, the invention of the patent had become something of a compromise between doing nothing and actively funding research—since awarding and enforcing patents required minimal risk of expense on the part of the state, yet motivated advances in technology all the same. But patents aided science only indirectly, and in no coordinated way.[1478] Direct state support, such as in the form of honors and prizes or salaried research positions, would have been of greater benefit to science (and indeed, even to technology as well, as the Russian and American space programs can attest), rather than letting patents encourage only random technological advances (although the latter certainly aided the former, so directly promoting both would have been ideal). But it essentially took a whole Scientific Revolution to finally convince any nation of this.

The natural inference is that before any kind of direct support could become a reality, the wealthy or governing elite had to be convinced, first, what research method to support (and ancient scientists had not yet reached a consensus on that point), and second, that supporting that method would produce results worthy of the expense (and the slow pace of progress in antiquity might not have been encouraging in that regard). Accordingly, though there may be several reasons why the idea of an 'institute for scientific progress' was only truly successful in the latter half of the 17th century, two reasons that seem most likely, which may have operated in conjunction, are, first, the sudden impact of a series of momentous new technologies and discoveries in relatively tight succession and, second, a growing consensus among scientists as to the methods such an institute should employ and promote, especially against their religious and philosophical opponents at the time. Such a consensus had not yet been achieved by the Roman era, at least not sufficiently to win the attention and approval (and eventual support) of a Roman emperor. It is possible this may have happened had things been allowed to continue as they were, and this would be even more

1478. Consider the remarks of Meissner 1999: 25–28, 450.

likely if any comparable series of discoveries had arisen to inspire and support such a development.[1479]

However, though formal support did not arrive, ancient scientists were not without *informal* support (as already noted in chapter 3.1). In fact, sometime after Vitruvius, an Aristotelian writing in Aristotle's name 'responded' to his complaint by arguing that the reason prizes are awarded to athletes and not philosophers is that for athletes the prize is more valuable than the competition, whereas studying philosophy was itself worth more than any prize.[1480] He meant moral and scientific knowledge are intrinsically valuable and rewarding, and in fact more so than any award of money (and as we have seen, this was a widespread sentiment). But scientists were also rewarded in other ways as well. They could receive the prizes of eternal glory in history books, acceptance among the elite, or actual gifts or pensions from patrons.[1481] Though such elevation in status and security attended most learned men, science and natural philosophy were among those recognized paths to distinction, open to all men of means who cultivated a passion for it.

Such a value for a scientific education in both practical and intrinsic terms is reflected by a story that both Vitruvius and Galen tell. When the philosopher Aristippus washed ashore after a shipwreck, he had no trouble finding hosts to take him in and restore his losses, precisely because of their respect for his manners and erudition, and thus he announced it was better to have "the sort of wealth and possessions that can survive a shipwreck." In other words, an education is like a coin that cannot be lost or stolen, yet can be cashed everywhere. As Theophrastus once said, according to Vitruvius, "an educated person is the only one who is never a stranger in a foreign land, nor at a loss for friends even when bereft of household and intimates," but to the contrary, "he is a citizen in every country, and may look down without fear on the difficult turns of fortune."[1482] This is essentially what Lucian had

1479. All this I have already suggested and discussed in chapter 3.6.VI, 3.9.I, and 3.10.

1480. Pseudo-Aristotle, *Problems* 30.11.

1481. On this ideal of historical glory see discussion of Diodorus in chapter 3.9.II and Apuleius in chapter 4.2 and examples in chapter 4.5. On salaries and pensions see chapter 3.1.

1482. Vitruvius, *On Architecture* 6.pr.1–3; Galen, *Exhortation to Study the Arts* 5 (=

identified among his reasons for preferring life as an orator. And as both Galen and Vitruvius argue, a scientist could expect the same.

Hence I think Ludwig Edelstein exaggerates the lack of social support for the scientific enterprise.[1483] Though scientific research had not won direct support, the scientists who were able and interested could find their own support in Museums and patronage and lucrative careers, as well as in the enjoyment of more intangible rewards. In fact, as Edelstein concludes, that so many individuals chose to pursue scientific research century after century, despite the near total lack of public support for their research, and instead undertaking the entire motivation, labor, and expense on their own, or securing it all on their own initiative, indicates that the social value of natural philosophy was not negligible—however low it may have been, it was high enough to continually attract the passionate attention of many intellectuals in every generation.[1484] It may only have been a matter of time before they saw the brighter light, and assembled a larger enterprise to advance the same goals.

4.8 Evidence of Non-Christian Hostility to Science

Values were not uniform in antiquity, even within any given social class, but more so from one class to the next. The evidence of praise and admiration surveyed in chapter 3.5 and above is certainly broad enough to represent a significant swath of elite culture, and their frequently unapologetic assumption of reader agreement entails these attitudes were not unusual among their peers. Nevertheless, there were certainly members of the elite who had no interest in the subject, or who had philosophical commitments not always conducive to supporting or recognizing good science, or who may have been more attached to their religious superstitions than the rationalizing trends of science and philosophy.

We should also not expect to find anything much better than indifference among the vast majority of the population, who were either illiterate or

Kühn 1.8–9).

1483. Edelstein 1963: 27, 28.

1484. Edelstein 1952: 598–99.

poorly educated, and typically even more religious and superstitious—certainly not commonly fans of rational philosophy, much less science. In truth, many probably had no opinion of natural philosophers at all, or did not even know what one was, much less did, or only knew them in other roles, having interacted with doctors (as healers) or architects (as employers). But many will have gained some familiarity with natural philosophy from public orations and popular hearsay, or by conflating scientists with philosophers in general. And any general resentment or hostility against the privileged classes, or against the perceived arrogance or hypocrisy of various philosophers, would have operated as much against natural philosophers as anyone else, as the public would not normally have been informed enough even to know there was a difference, much less what that difference might have been—and knowing the difference might not have mattered anyway.

There is nothing from extant pagan sources comparable to the Christian level of hostility that we shall see in chapter five, though there may have been a lot of unvoiced discontent among the masses that Christianity later tapped into. Scattered examples of a minority undercurrent of hostility to science among the pagan population (elite and non-elite) have already appeared in chapters 1.3, 2.6, and 2.8, but as is evident throughout, such material is often limited and vague.[1485] In fact all the evidence of more negative pagan attitudes, insofar as there is any, is indirect and often highly qualified or unclear. For instance, the fate of the councilman who 'profaned' the 'Triton of Tanagra' with a science experiment (as discussed at the end of section 4.5) says something about a negative attitude among certain segments of the non-elite, but exactly what is hard to determine. Yet that is the only example of its kind I could find, and about as clear as the evidence ever gets (which is to say, not much). Since the majority of the population, undereducated as it was, produced little writing (and none on this topic) we can only attempt to read between the lines of what the elite said about them (which is, again, very little on this topic), drawing on what we know of the social and psychological context of ancient Rome specifically, and class-torn societies generally. But that still does not get us very far.

For example, there is abundant evidence of popular reactions to lunar

1485. Likewise indications of indifference amidst praise documented throughout Carrier 2016.

eclipses in the Roman period, which demonstrates widespread scientific illiteracy and superstition.[1486] But though demonstrating ignorance, this indicates nothing discernible about attitudes, whether toward astronomy or astronomical research, or natural philosophy generally. Likewise, there is evidence of some popular hostility to scientific medicine, and of a converse preference for magic and miracles, but that doesn't translate well into evidence of hostility to natural philosophy or even scientific research in any general sense, since medical practice is not identical with it. The issue of ancient hostility to medicine is actually a very complex multi-faceted question that would easily take a considerable dissertation to address in its own right. For instance, much of it was directed against the actual or perceived moral failings of practitioners, rather than the utility of the science itself (see related evidence in chapter 3.2, 3.3, and 3.5).

So, too, for elite philosophical responses to natural philosophy. Apart from Cynicism (which was never influential and generated next to no extant texts, and none relating to this question), and the skeptical sects (which nevertheless embraced a qualified science, in fact everything visibly useful), natural philosophy was embraced by all schools, which only vented hostility (if any) toward certain approaches or conclusions *within* natural philosophy, which is a phenomenon not very pertinent to the present thesis (regarding the level of support for science, scientific progress, and natural knowledge generally), and, like attitudes toward medical practice, this is also a complex multi-faceted question that, again, would easily take a considerable dissertation to address in its own right. I could only briefly survey some of the sectarian differences regarding natural philosophy in my previous work on Roman science education (Carrier 2016; and somewhat more here in section 4.1), and even those differences were significantly diluted by the popularity of eclecticism (especially among scientists themselves).

Apart from some Platonist disdain for hands-on empirical work (as we saw from Plutarch in section 4.6), an attitude wholly unique to that sect, the only other stark example of what one might identify as significant philosophical hostility toward the natural philosopher would be the attitude of the Pyrrhonists. Yet they were not in fact hostile to science or scientific research, only unempirical speculation or unwarranted certainty. Otherwise

1486. For example see: Quintus Curtius Rufus, *History of Alexander* 4.10.1–7; Seneca, *Natural Questions* 7.1.2; Tacitus, *Annals* 1.28.

(as noted in section 4.1) they accepted what we would call predictively successful medical and astronomical knowledge (and no doubt mechanical knowledge as well), and thus supported all research toward advancing such knowledge (whether they called it 'knowledge' or not).[1487] Even in Sextus Empiricus's treatise *Against the Natural Philosophers*, the only book of any such title from our period of interest, where surely of all places we should expect to find hostility toward the natural philosopher, instead all he attacks is theology and metaphysics, not empirically pragmatic science.[1488] And though in this manner he challenges the underlying assumptions of science, in each case he only arrives at suspended judgment, not refutation or rejection, which leaves him affirming what 'seems' to work in practice. In this respect Sextus does not differ even from modern scientists who denounce speculative philosophy while praising empirical 'philosophy', i.e. science.

And yet, as philosophers even today point out, such sentiments are themselves taking a philosophical position. For instance, Sextus argues that natural philosophers cannot prove anything moves, yet he accepts our observation of movement as irrefutable. In other words, though he 'withholds judgment' on whether these observed motions are produced by any real movement, he nevertheless conducts himself as a doctor (for instance) by following the appearances, without deciding what causes those appearances, or what they *truly* signify (as opposed to what they *appear* to signify, which he accepts).[1489] What we have here is in practice just another variety of debating what the content and methods of natural philosophy should be. Otherwise the Pyrrhonists accept a considerable amount of scientific knowledge, everything necessary for observable success in practice,

1487. See also Carrier 2016: 100. Adding to the resources cited there: Bett 2011 and *OCD* 1324–25 (s.v. "Sceptics").

1488. Sextus Empiricus, *Against the Natural Philosophers* 1.12–194 (whether anything can be said about the existence and nature of any gods), 1.195–2.351 (whether anything can be said about the metaphysics of causation, the structure of matter, the nature of space, motion, time, abstract numbers, and the metaphysics of generation and decay), almost all of which consists (in practice) of disputing semantics rather than pragmatic beliefs.

1489. Sextus Empiricus, *Against the Natural Philosophers* 2.66–69 and 2.168, with *Outlines of Pyrrhonism* 1.236–41. On this allowance of science as a pragmatic acceptance of the appearances see the analysis of J. Barnes 1988.

and they only cast doubts on the rest—without vilifying it or expressing any moral alarm over it. And this represents a minority sect. Other philosophies, which were even more favorable to science and natural philosophy, were far more popular among the elite, and yet even in Sextus we do not find any real hostility—or even indifference, as he clearly studied the subjects he discusses in meticulous and learned detail, and never attempts to persuade his readers to avoid doing the same.

Such is the philosophical situation. More problematic are attempts to find evidence of hostility among the elite in general, since such evidence generally ends up dissolving on closer analysis, such as the supposed evidence of hostility or indifference to scientific *progress* surveyed in chapter 3.2, or the alleged hostility to *hands-on* research examined in section 4.6. Similarly, occasional expulsions of philosophers from Rome (most notably by Vespasian and Domitian) had nothing to do with science or natural philosophy, but were only aimed at removing moral and political agitators from the capital, which at worst would have relocated scientific activity to other cities (like Alexandria) where it was already more commonly pursued—and hardly even that would result, as such expulsions were always temporary and brief (as abundant sources confirm, Rome rarely lacked philosophers), and might not have affected scientists at all (by no doubt exempting doctors and engineers, who were always needed in the city despite their devotion to natural philosophy).[1490] At most a good case can be made that elite indifference to natural philosophy was not uncommon, judging from its limited place in ancient education (as shown in Carrier 2016), the lack of institutional funding (examined in section 4.7), and the

---

1490. See Breebaart 1976 and Toynbee 1944. These were also exceptional events. Domitian expelled philosophers not just from Rome but from the whole of Italy (according to Suetonius, *Domitian* 10 and Aulus Gellius, *Attic Nights* 15.11.3–5), but this is listed among his criminal acts (which were abolished by the Senate after his death, cf. *OCD* 411, s.v. "*damnatio memoriae*"). Suetonius does not mention Vespasian having done anything like this, and even implies the contrary (Suetonius, *Vespasian* 13–15). We hear only a confused account from John Xiphilin's epitome (of Cassius Dio, *Roman History* 66.13) that moral and political agitation by certain Cynic and Stoic philosophers led Vespasian to expel "all the philosophers," but the context of the occasion suggests only certain philosophers were meant (e.g. "all the [offending] philosophers" not "all philosophers"). This is supported by Gellius, who lists Domitian's act as the only occasion of such a general expulsion under an emperor.

demonstrated ignorance of this or that author (such as the two philosophers in chapter 4.6.III whom Galen poked fun at because they didn't know even basic hydrostatics), but this detracts little from the evidence of widespread admiration and praise (especially since the same ignorance and indifference to science can be found even in modern populations, and has always been the case in every human century).

My last example of the difficulty in locating actual hostility among the non-Christian population is the Roman recension of the *Life of Aesop*, which as a critique of the follies of elite society might reflect lower or middle class values, or, like the satires of Juvenal or Lucian, it might reflect instead elite criticism of their own class. Or any combination thereof. Thus already we cannot identify clearly whose values this document represents. And even what it says is problematically vague. For example, the philosopher Xanthus, when asked for advice by a gardener, replies, "How can anything I say help you as a gardener? I'm no craftsman or smith to make you a hoe or a leek slicer. I'm a philosopher!" This seems to reflect a popular view of philosophers as useless. If Xanthus is meant to represent philosophers as a class, then the rest of the *Life* depicts them as hypocrites and fools, who take large fees and give fancy speeches, but are worthless and deplorable in comparison with simple common folk.[1491] Yet this contrasts so sharply with the evidence of section 4.6.II that we can only conclude that Xanthus and his cronies represent the *mere* philosopher (or even the mere *natural* philosopher, like the armchair hacks repeatedly attacked by Galen), and not philosophers who practice a useful art like medicine, engineering, or astronomy. Whether the general public could make this distinction may be an open question, but even if they could not, then their error would lie in failing to realize that doctors, engineers, and astronomers were *also* natural philosophers, but not in deriding these professions as equally useless. For clearly Xanthus neither has nor practices any of the skills ancient scientists had, and is made the butt of every joke for that very reason, yet doctors, engineers, and astronomers were not invisible to the public, nor were the products of their skill.[1492] Thus even here we cannot really ascertain

---

1491. *Lives of Aesop*, vita g § 35 (= e cod. 397 bibliothecae pierponti morgan, recensio 3). See discussion of this work in chapter 2.8.

1492. In case it is not obvious, the social visibility of astronomers would result from, among other things, their production of public sundials (cf. Cuomo 2001:

any attitudes toward ancient scientists, since nothing is said about them specifically, and what is said excludes them implicitly.

The above sample, though not exhaustive, is representative of all the evidence I could find of negative attitudes toward natural philosophy among non-Christians. In contrast to all this, which is vague, highly qualified, and often hard to find, evidence of pagan *regard* is (as we have seen) quite clear, unambiguous, and easily found—as is evidence of Christian *hostility*.

4.9 The Path to Christian Values

I showed the diversity and sometimes ambiguity of Jewish values in the context of ancient education in chapter nine of *Science Education in the Early Roman Empire* (Carrier 2016), and I have already said something of Christian values there, and in chapters 2.6 and 3.8.IV here (and we will devote most of our attention to that subject in the next chapter). As in education, so in science, Hellenized Jews often embraced Greek values. For instance, when Josephus retells an old Jewish legend about Ptolemy Philadelphus interrogating the Jewish elders, he replaces the original subject of discussion (ethics) with natural philosophy, so now it is *this* subject on which their "precise explanations concerning every single problem suggested to them for discussion" produced "delight" in the king and confirmed the wisdom of the Jews.[1493] Such an emendation suggests an increased prestige for the knowledge of a *physicus*, at least within the tradition history of this story.

Another example is Philo. It is impossible to know how typical this Jewish philosopher was, especially as his interests were heavily Platonic, but he was renowned, we have no comparable material from any other Jewish intellectual in antiquity, and he is the only Jewish author who says anything

151–153) and *parapegmata*, astronomically-based agricultural calendars erected and maintained in most ancient towns and cities for the benefit of farmers and merchants (cf. Taub 2003: 20–37, 41–43, 173–76; Lehoux 2007b; Hannah 2009: 49–49). Meanwhile the visible products of doctors and engineers is more than obvious.

1493. Josephus, *Jewish Antiquities* 12.12.99. For the original context, see the earliest version still extant in (Pseudo?)-Aristaeus, *Letter to Philocrates* 187–292 (which includes too much detail to be in any doubt about the subject of discussion).

directly on the value of natural philosophy, for which he brought both praise and some censure.

Philo wrote:

> Mind fathered both the cruder and the finer arts, and was also the parent of philosophy, the greatest of blessings, employing each part of philosophy to benefit human life—the logical, to produce absolute exactitude of language; the ethical, for the amelioration of character; and the physical ('natural philosophy'), to give knowledge of heaven and the universe.[1494]

And besides the bounty of the natural world that God has given us:

> We must also mention the higher, nobler wealth, which does not belong to all, but to truly noble and divinely gifted men. This wealth is bestowed by wisdom through the doctrines and principles of ethics, logic and physics ['natural philosophy'], and from these spring the virtues, which rid the soul of its proneness to extravagance, and engender the love of contentment and frugality, assimilating the soul to God.[1495]

Thus, for Philo, the study of natural philosophy is among the 'greatest of blessings' and the mark of a 'truly noble and divinely gifted' mind—so long as it is employed to 'benefit human life' and to reforming the student's character toward a more virtuous and godly nature.

In every respect it appears Philo adapted to a Jewish context the ideals of his pagan contemporaries (as seen for example in sections 4.1 and 4.2 above). Still, this is among the most we ever hear from him on the subject, and it is hard to find in his vast number of extant works any of the praises of curiosity, empiricism, or progress that we have seen from pagan authors. So it is not clear exactly what he means by 'physics' or how he thinks it should be pursued. What often passed for science in Jewish texts was inferior to the cutting edge knowledge of pagans, much distorted by superstition, and subordinated to religion.[1496] Likewise, Philo still thought a life of God, devoted to revelation and thus no longer pursuing scientific study or research, was better than a life of science, which was only better than a life

---

1494. Philo of Alexandria, *On the Special Laws* 1.336.

1495. Philo of Alexandria, *On the Virtues* 8.

1496. Ben-Dov & Sanders 2014; Newmyer 1996.

of vice.[1497] So his enthusiasm for natural philosophy is somewhat muted and qualified.

For instance, Philo praises as supremely admirable the life and philosophy of the Essenes, who had rejected essentially all interest in science and natural philosophy. As Philo explains:

> Of philosophy the Essenes have left the logical part to word-catchers, as being unnecessary to the attainment of virtue, and the physical part to star-gazers, as too high for human nature, except so much of it as is made a study concerning the existence of God and the creation of the universe, but the ethical branch they study very elaborately.[1498]

This might not mean Philo or even the Essenes despised *others* who undertook such studies, though it does reflect a Jewish attitude that walked a fine line between the early Christian disdain for science and a limited or qualified respect for it. Such attitudes could easily fall on the negative side, as we shall see for the Jewish authors of the Wisdom of Solomon (in chapter 5.6.II). So it is not surprising that the Christian Eusebius (whose views we will examine in chapter 5.4) finds the root of his own hostility in the same Jewish Essene origins, quoting with admiration Philo's account of them as embracing only the ethical part of philosophy, leaving the rest alone as inferior, or actually useless, or even morally dubious.[1499]

Very few would reverse this order of values. In fact, there is only one Christian text in our period that offers clear praise for natural philosophy: the *Panegyric* of Gregory Thaumaturgus, who praises his teacher—the heretic Origen—for having insisted upon a thorough study of philosophy. He first emphasized logic, later ethics, but in between:

> He also sought after the humble part of the soul of those amazed by the magnificence and wonder and the intricate and infinitely wise construction of the cosmos, of those who marvel beyond reason and cower in terror, not knowing at all what to conclude, just like irrational animals. He awakened this, too, and set it straight with other studies in natural philosophy, clarifying each kind of being and very skillfully reducing them

1497. See discussion and evidence in Carrier 2016: 141–43.

1498. Philo, *That Every Good Man Is Free* 12.80.

1499. Eusebius, *Preparation for the Gospel* 8.12.9.

> to their most fundamental elements, weaving them into his discourse and ranging over the nature of the whole and each part, as well as the complex revolution and transformation of what is in the cosmos, until he girded our souls with a rational wonder, in place of an irrational one, carrying our souls along with the clarity of his teaching and reasoning, imparting everything he had both learned and discovered about the divine operation of the whole and its faultless nature.
>
> This is the lofty and god-filled education taught by the study of nature that is most attractive to all. And what need is there to speak of the subjects of divine mathematics—such as geometry, which is beautiful and indisputable to everyone, and astronomy, which soars above? Each of these he imprinted on our souls, teaching us or reminding us, or something, I do not know how to describe it. One he singled out, setting it up as the foundation of them all, unshaken, and that was geometry, an unfailing groundwork, and that also leading up to what is high above, through astronomy, making even heaven reachable to us, as if each of these studies were a ladder shooting up to heaven.[1500]

There is no explicit mention of the 'natural philosopher' here, but the implications are clear: Gregory thought very highly of the study of natural philosophy, especially astronomy, and assumed his audience would agree—indeed, Gregory calls natural philosophy a "lofty and god-filled" study "most attractive to all."

Still, as with Philo, it is unclear whether Gregory is actually praising the scientific values of curiosity, empiricism, and progress, or only a scripturally suitable dogma. For Gregory goes on to praise Origen for placing scripture and revelation above all else, and teaching that natural philosophy was merely a stepping stone to a study of scripture. In this he sounds more like Philo, walking that same delicate line. Hence Gregory is *also* proud to declare how Origen did not let his students read any philosopher who excluded God's role as creator and director of the universe, believing such

---

1500. Gregory Thaumaturgus, *Panegyric Oration on Origen* 8. There is some dispute as to the actual identity of the author, but his identity is supported by Eusebius (*History of the Church* 6.30), who was using Origen's library at the time and thus would be in a good position to know, while arguments against the attribution are not very persuasive. Whatever his name, the author was certainly a student of Origen writing in the middle of the 3rd century. See Trigg 1998: 36–37 and 249 (n. 6); and Crouzel 1979 and 1969; with *OCD* 636 (s.v. "Gregory (4) Thaumaturgus") and *ODCC* 713–14 (s.v. "Gregory Thaumaturgus, St.").

reading was worthless and would lead the faithful astray.[1501] So there were limits, even for Origen and Gregory. But apart from all this, some of the same appreciation for the value of studying nature that we have seen among pagans was expressed here, too: such learning allays fear by increasing our understanding, it draws our mind up to higher things, and can even generate greater appreciation for the genius of God. But no other Christian text survives from our period with anything like this high opinion of natural philosophy. As we shall see later, all other extant Christian discussions are decidedly hostile or considerably more ambivalent.

Hence, for example, when Eusebius expresses respect for the knowledge of nature exhibited in the Bible, he specifically points out that it is valuable and authoritative because it was directly revealed by God.[1502] He does not express any value for human inquiry or the actual study of nature, much less for any rational, empirical approach to this study. Instead, it is only God who is supposed to know everything, and if he wanted us to know, he would have told us (through revelation, inspiration, or scripture, as we'll see in chapter five). The most curious example of this notion appears in a surviving Arabic translation of the *Infancy Gospel of Jesus*, believed to derive from a 2nd or 3rd century Greek original. This specifically depicts Jesus as omniscient in the subjects of astronomy, medicine, and natural philosophy—not through any education, and not himself advocating any education, but simply because his was the mind of God, which he displayed merely to show off his uniquely divine wisdom. Indeed, in the extant Greek version of this Gospel, Jesus is such a clever child he is sent to school at an early age, but in class he petulantly insults, demeans, and terrifies his teacher, impatiently insisting that he does not need to be taught because he already knows everything.[1503]

Similarly, in the Arabic version of that Gospel, scientific knowledge is depicted as miraculous and in many ways beyond the reach of mortals. Here we are told that astronomers and natural philosophers were among the "teachers and elders and learned men of the sons of Israel" whom Jesus

---

1501. See Carrier 2016: 149–55; and related discussion in chapter 5.9 here.

1502. Eusebius, *Preparation for the Gospel* 11.7.

1503. *Infancy Gospel of Thomas* 6:1–8:2, which contains a detailed account of this classroom nightmare that is astonishing to read. In fact, this entire Gospel reads like a proposed screenplay for *The Omen* that the studio rejected because it was too scary.

debated in the temple when he was twelve.[1504] But as this is fiction aimed at glorifying the miraculous knowledge of Jesus, and written long after the temple had ceased to exist, we cannot know how realistic this picture is. Nevertheless, while Jesus "explained the books, and the law, and the precepts, and the statutes, and the mysteries contained in the books of the prophets—things which the understanding of no creature attains to," which is all the passage this is based on in the Gospel of Luke ever implied (and which the Infancy Gospel builds on considerably), the Arabic *Gospel of the Infancy* adds that when "a philosopher who was there present, a skilled astronomer, asked the Lord Jesus whether he had studied astronomy," Jesus answered him in meticulous detail. And to prove it, this Gospel's author surveys a lot of the astronomical basics of his time; but then, we're told, Jesus added "other things beyond the reach of reason" (which, conveniently, the author of this Gospel neglects to write down).

After that exchange, we're told "there was also among those philosophers one very skilled in treating of natural science, and he asked the Lord Jesus whether he had studied medicine," and again Jesus answered and "explained to him physics and metaphysics, the nature of things above and the nature of things below, and the powers and humors of the body, and the effects of these," and other details of anatomy and physiology that the author felt free to list. And then, of course, the young Jesus went on to explain "other things beyond the reach of any created intellect" (which again the author conveniently cannot think to record). Naturally, after witnessing all this, "the philosopher rose up, and adored the Lord Jesus, and said: Oh Lord, from this time I will be your disciple and slave." Hence despite being the only Gospel to exhibit knowledge of ancient astronomy and medicine in more than trivial detail, there is no real respect for scientific knowledge in this document—instead, the moral of the story is that only God knows everything, that any new science is beyond human reason or comprehension, and that one should simply submit and worship the Lord. This is a radically different attitude than any we have seen from the pagan elite.

1504. *Arabic Gospel of the Infancy of the Savior* 50–53 (greatly expanding on Luke 2:42–47), quoting the English translation of Roberts & Donaldson 1896. The extant Greek text (cf. 19:1–5) does not expand on this passage from Luke anywhere near this extensively nor in the same way, but there is evidence the Arabic text might be closer to the original Greek text (cf. Klauck 2003: 73–78 and Schneemelcher & Wilson 1991: 1.414–69).

4.10 Summary & Conclusion

Joseph Ben-David claims the ancient scientific community was not "capable of convincing society in general about the importance of its enterprise."[1505] That was probably true if by "in general" we mean large numbers of the general population or by "importance" we mean economically. Only the elite, however—the upper and upper middle class—had any power to affect the fortunes of science and scientists, and therefore it was only the elite that the ancient scientific community had to convince. It is quite clear by now that many among the elite were indeed convinced. They were not all yet convinced of its economic value, but most were persuaded it had moral and practical value, and many agreed its value in these regards was considerable.

As Elizabeth Rawson says, "there was enough interest in *physica*, natural science, which still counted usually as a branch of philosophy, to make the common generalization about the Romans' philosophic interests being confined to ethics not wholly fair," or indeed fair at all.[1506] Even Seneca, whose preoccupation with ethics was far greater than average, and at least as great as any extant Roman author, was in love with science and natural philosophy and wrote considerably on the subject. Even from the start, three of the earliest works to bring Greek philosophy into the Latin language through the popular beauty of verse emphasized natural philosophy: Lucretius *On the Nature of Things*, the *Empedoclea* of Sallustius (lost but most likely a translation of Empedocles' famous treatise *On Nature*), and yet another poem *On the Nature of Things* by Egnatius (also lost and exact content unknown), all composed in the 1st century B.C.[1507] A more notable example is the role the Roman elite played in recovering and promoting Aristotelian studies.[1508] These only add to the abundant evidence provided throughout the present chapter, furthering the same conclusion, that natural philosophy was well respected and enjoyed an elevated status in the early Roman empire.

1505. Ben-David 1984: 42.

1506. Rawson 1985: 282.

1507. Rawson 1985: 284–85.

1508. See Rawson 1985: 289–91 and Gottschalk 1987 (who rightly explains that Aristotle's works were never actually 'lost' as Roman legends liked to claim, though there was a revival of interest in them, and a new critical edition was produced).

Though "profound differences between notable Romans make generalizations about their characteristics and attitudes ridiculous," and indeed the early Roman empire is characterized by a widespread diversity of values and perspectives with open and active debate among them, nevertheless there is more than enough evidence to demonstrate that a distinctly positive attitude toward the scientific enterprise was commonly voiced and accepted among the educated elite.[1509] Though very different, even contrary attitudes could be found, these voices were evidently not numerous or respected or influential enough to matter—unlike today, when real threats to the scientific enterprise are being mounted by many respected and influential people, a situation with no parallel in the Roman empire until its decline. And though aristocratic snobbery was certainly more pronounced then than today, it was also more pronounced two hundred years ago, or even one hundred years ago, in either case at the height of the Industrial Revolution and right before, during, and still after the Scientific Revolution—yet this was no barrier to either.

Of course, widespread science illiteracy, which pervades even modern society, was far more severe in antiquity, for the obvious reason that science did not command as large a place in public culture and education as it does now. This may have limited the number of scientists, which in turn may have slowed scientific progress—through all periods, Greek and Roman. But it did not cause scientific stagnation or decline, until Roman society itself declined as a whole, into a degenerative, superstitious fascism. Though science did not carry the prestige in antiquity that it does today (though it does not hold quite as much prestige today as some historians tend to claim), it was sufficiently prestigious then to inspire and support a fair number of creative scientists in every generation, who predominately came from what we can call the upper middle class. As prosperous professionals they may have received the condescension of the aristocratic elite, but were nevertheless relied upon, paid well, and regarded as better company than the laboring masses. In fact, whatever other opinions they held, most among the Roman elite considered doctors, engineers, and astronomers to be useful and respectable, and frequently elevated the economic and social status of anyone who mastered and pursued such fields. So it cannot be said the Roman attitude toward scientists was hostile or negative or even

1509. Quote from Greene 1994: 25.

dismissive. Even when those in power might have exhibited ignorance or distaste for science or the people who pursued it, they nevertheless put them on a loftier pedestal than the vast majority of the population. The Christians, however, saw things very differently.

5. Christian Rejection of the Scientist

Almost all explicit discussions of the natural philosopher and his activity by Christians in this period are hostile or unsupportive. There is no group in antiquity who more clearly and ardently took such a position. Even the most hostile of pagan philosophers had mixed impressions, and at least supported some aspects of the pursuit of natural studies, or at most debated what its methods and aims should be. But most Christians were not at all interested, or were even a little disgusted by the whole enterprise, frequently dismissing it as arrogant and useless, or even an example of everything they became a Christian to reject and oppose. This did not prevent Christians from using the work of natural philosophers to support their own arguments—as for example in their many treatises on the nature and possibility of resurrection.[1510] But Christians who employed this knowledge of the natural world treated it as little more than a scholastic tradition, to be believed or attacked, but not empirically questioned or improved upon. When using such knowledge, Christian authors show little or no awareness of the discipline and effort needed to acquire it, and often attack or ignore the very methods and assumptions that were required for its advancement, or even deride the motivations of those who seek it.

Marshall Clagett argues that three factors combined to produce a significant decline in the sciences after the 3rd century A.D. During and after that century, there was an observable increase in the embrace of

1510. Bynum 1995: 19–58.

supernaturalism that eventually began to overtake and eclipse the best traditions of Greco-Roman rationalism.[1511] Partly as a result of this, Christianity became increasingly attractive and eventually dominated the culture, leading to an increased diversion of intellectual activity away from genuine scientific research and toward scriptural and theological studies. This triumph of Christianity then brought with it a reversal of epistemic values that did substantial harm to the scientific spirit.

Clagett's three-stage theory seems undeniably correct. The broad popularity of supernaturalism was always a fact in ancient society, but it does become more obvious and dominant from the 3rd century on, certainly among the elite—anyone who compares the literature before and after that century can hardly deny it. There is an obviously rising preference for the fabulous and mystical over the empirical and cautious, and more and more errors begin to creep into the rehashing of past scientific theories and findings.[1512] It is likewise true that beginning in the 3rd century, the Christian Church "began to attract in fairly significant numbers men who might have gone into philosophy or science but now undertook the writing of Christian apologetic and dogmatic literature," so in a sense the Church had "siphoned off men who might well have pursued natural philosophy or science" and inspired them instead to devote their lives to scripture and theology (or the abandonment of worldly life altogether for a life of God).[1513] The result could only have been a marked decline in the number of scientists—whose number was already small—and this does appear to be the case, as original scientific writers become almost impossible to find in extant sources after the 3rd century, yet there is no notable decline in the production of intellectual literature. Finally, as I myself shall demonstrate shortly, Clagett is entirely correct to emphasize the Christian elevation of prophecy, revelation, and scripture over genuine scientific values, which "constitutes a fundamental alteration from the spirit of Greek rationalism," which does not mean a deviation from *rationality* (broadly defined), but from the particular rationalist values of the Greeks and Romans that had been producing successful scientific inquiry for the previous six hundred years.[1514]

1511. Clagett 1955: 118–29.

1512. Clagett 1955: 146–82.

1513. Clagett 1955: 120, 130.

1514. Clagett 1955: 133.

Clagett himself shows that the result of this upheaval in epistemic values was, first, a varying degree of hostility toward any effort to approach scientific knowledge, apart from scripture or agreement with 'correct' theological doctrines, and, second, a denigration of scientific ambition and curiosity, in line with the belief that only pursuing knowledge of god and his will is worthwhile or even morally appropriate.[1515] Though philosophy could serve as a "handmaiden" to this pursuit, it was only acceptable in that role insofar as it furthered and secured that end, which is the exact opposite of the scientific spirit, in which the knowledge of nature and its practical uses are a driving aim, and the curiosity and ambition to solve questions in natural philosophy are not regarded as vain or morally suspect.[1516] In the words of G.E.M. de Ste. Croix, "the anti-scientific attitude of so many pagan thinkers was taken over in an intensified form by early Christianity," and despite his seeing attempts by modern apologists to dispute it, "in the first five Christian centuries" he had "found in the Fathers of the Church attitudes to Greek science ranging only from indifference to hatred, with sometimes a note of fear."[1517]

We will see in coming sections how correct these conclusions are, from a thorough examination of the primary evidence from the first three centuries of Christianity. Nevertheless, it was not solely Christians who embraced these hostile or negative views of scientific endeavor—there had always been fringe elements of this opposition or indifference among the pagan population. Kudlien, for example, finds a case where a doctor

1515. Clagett 1955: 130–45, in which he further documents that Christian attitudes toward science and natural philosophy were never very positive. R.M. Grant 1952: 87–126 provides a representative survey of the poor quality of scientific knowledge in early Christian writers, the often unscientific way in which such knowledge was used, and their generally hostile attitude toward scientific values (and though he cites Christian sources in adequate abundance, his assessment of the state of Roman science on p. 120 is unsourced and wholly incorrect—he clearly could not have read the works of Galen, Hero, or Ptolemy).

1516. The idea of philosophy as the "handmaiden" (i.e. slave) to theology and faith actually derives (word and all) from the Jewish philosopher Philo of Alexandria in the 1st century A.D. (see chapter 4.8 here and chapter nine of Carrier 2016), only later picked up by Christian intellectuals like Clement of Alexandria (Clagett 1955: 134–35). See related note in section 5.6.III below.

1517. In Crombie 1963: 87 (with: 79–87).

named Thessalus abandoned scientific medicine and started getting his only "reliable" knowledge through direct revelations from the god Asclepius, who in a dream "explicitly called attention to the fact that a self-acquired medical knowledge was worthless in comparison to a medical revelation given by a god to an elected person."[1518] This represents a kind of thinking that is quite the reverse of scientific reasoning.[1519] Thessalus' date is unknown, but though such patterns of thought always existed in all eras of pagan antiquity, they did not come to predominate among the literate elite until the 3rd or 4th century A.D. In many ways, Christianity's success was a symptom of this overall trend, not its cause. But the eventual effect was closer to what one might expect if the Cynics had conquered Rome.

G.E.R. Lloyd concurs with Clagett's analysis, explaining that:

> Greek science coexisted with magic, superstition and irrationalism of various kinds from the very beginning. What marks Christianity out is not the particular doctrines associated with it, so much as the fact that those doctrines eventually received unprecedented state approval and support.[1520]

Among the changes this brought (right from the highest levels of power and prestige all the way down), faith in a sanctioned "revelation" came to be seen as the best and only secure source of knowledge, so much so that "for the faithful, empirical inquiry is unnecessary, a distraction from the practice of his religion and possibly a source of dangerous heresy," a view that "if rigidly adhered to, meant the end of scientific research." Thus, "with the Christian Church, religion became established institutionally in ancient society in a way in which science never did," and as a result, most intellectuals after the 4th century A.D. "preferred revelation to reason and sensation and put faith above knowledge."[1521] This does not mean Christian intellectuals rejected reason, sensation, or knowledge, only that these things took a subordinate place, in effect flipping scientific values upside down—instead

1518. Kudlien 1970: 24–25.

1519. On this kind of thinking throughout antiquity see Dodds 1951.

1520. Lloyd 1973: 167 (examples of the prevalence of this Christian hostility: 167–71).

1521. Lloyd 1973: 169–171.

of subordinating beliefs to experience, experience was now subordinated to beliefs. A respectable measure of logic and reason and knowledge, even common sense, might survive in such conditions—but science cannot.[1522]

To demonstrate this characteristic of the early Christian mindset, we will first look at all the passages in Christian literature before the reign of Constantine that specifically discuss attitudes toward the natural philosopher and their activity.[1523] Then we will examine the underlying sentiments producing such views, by looking at how early Christian literature (including the New Testament) portrays academics and the pursuit of knowledge generally. I have already covered the role of natural philosophy as limited and even opposed in Christian education in chapter nine of my previous book, *Science Education in the Early Roman Empire*. And though only four Church Fathers explicitly discuss the natural philosopher and their activity before 313 A.D., they are among the grandest intellectuals of the early Christian era: Clement of Alexandria, Tertullian, Lactantius, and Eusebius.

5.1 Clement of Alexandria (c. 200 A.D.)

Among the 'orthodox', Clement of Alexandria comes closest to voicing some respect for natural philosophy.[1524] But his view is qualified to the point of ambivalence regarding the actual study of nature, and his opinion of the natural philosopher is not very high. Clement does criticize his Christian peers for what we might call "throwing the baby out with the bathwater." He complains that "the multitude are frightened by Greek philosophy, as children are at masks, being afraid lest it lead them astray" and as a result "some do not wish to touch either philosophy or logic, and all the more, they do not wish to learn natural philosophy" but instead "they demand bare

1522. On this "flipping" of epistemic values upside down, see discussions in Walzer 1949.

1523. This adds to several other examples of Christian attitudes already discussed in chapters 2.6, 3.8.IV, and 4.8 here, and chapter nine of Carrier 2016, which supplement the evidence in the present chapter.

1524. For background on Clement of Alexandria see *OCD* 331 (s.v. "Clement of Alexandria (Titus Flavius Clemens)") and *ODCC* 364–65 (s.v. "Clement of Alexandria").

faith alone."[1525] Some Christians even argued that philosophy came from the Devil.[1526] Thus Clement recognized a pervasive Christian hostility toward natural philosophy. As we shall see, this was an attitude expressed even by leading Christian intellectuals, so it was not an opinion held only by the Christian "masses." Nevertheless, Clement at least thought they were going a little too far. "I call him truly learned who brings everything to bear on the truth," he argues, "so that, from geometry, and music, and grammar, and philosophy itself, culling what is useful, he guards the faith against assault."[1527] Hence Clement declares "how necessary it is for him who longs to receive of the power of God, to philosophize about intellectual subjects!"[1528]

However, Clement does not mean these pursuits were necessary in order to understand nature or natural causes. To the contrary, philosophy was necessary to understand *scripture*, through its talent for logic and interpretation, and its utility in combating heresy.[1529] In all other respects philosophy went too far and soon wore out its welcome. Hence Clement warns against making philosophy an occupation rather than a tool in the service of faith:

> For "the wisdom of the world is foolishness with God," and of those who are "the wise the Lord knows their thoughts are vain." Let no man therefore glory on account of preeminence in human thought. For it is written well in Jeremiah, "Let not the wise man glory in his wisdom...but let him that glories glory in this, that he knows and understands that I am the Lord."[1530]

Accordingly, Clement argues that we should reject all philosophy as foolish which does not acknowledge God as the creator and first principle of everything, because, first, otherwise philosophy will lead us astray into atheism, and, second, once we have found God's truth, we no longer need philosophical inquiry.

1525. Clement of Alexandria, *Stromata* 6.10:80.5–81.1 and 1.9:43.1.

1526. Clement of Alexandria, *Stromata* 6.8:66.1.

1527. Clement of Alexandria, *Stromata* 1.9:43.4.

1528. Clement of Alexandria, *Stromata* 1.9:44.3.

1529. See Clement of Alexandria, *Stromata* 1.10.

1530. Clement of Alexandria, *Stromata* 1.11:50.1–50.3 (quoting 1 Corinthians 3:18–20).

Clement quotes several passages in the New Testament in support of this, which we will examine further in section 5.6. But as an example, Clement explains:

> "Seek, and you shall find." But seeking ends in finding, driving out the empty and trifling, and approving of the contemplation which confirms our faith. "And this I say, lest any man beguile you with enticing words," says the Apostle [Paul], clearly to those who had already learned to distinguish what was said by him and what had been taught subsequent to that. "As you have therefore received Christ Jesus the Lord, so walk in Him, being rooted and built up in Him, and established in the faith," and as persuasion is being established in the faith, "Beware lest any man steal you away from faith in Christ with philosophy and vain deceit," which does away with providence, "following the tradition of men," for the philosophy which is in accordance with divine tradition establishes and confirms providence, which, being done away with, the governance of the Savior appears a myth, and we are then influenced "according to the elements of the world rather than Christ." For the teaching which is agreeable to Christ deifies the Creator, and traces providence in particular events, and knows the nature of the elements to be capable of creation and change, and teaches that we ought to pledge ourselves to the power near to God, and submit to his governance as Lord over all education.[1531]

Therefore, when Clement does speak favorably of "philosophizing" he is not talking about natural philosophy in the sense embraced by actual natural philosophers, but of a very different endeavor, that of identifying the role of God in nature.[1532]

When it comes to natural philosophy as natural *inquiry*, Clement associates it with faithlessness, folly, and childish questions, and considers it wholly unnecessary or even a dangerous distraction. For example, Clement declares that "philosophers are children, unless they have been made men by Christ," so when they ignore the role of God in natural events, and instead seek answers in natural causes, they end up asking only "juvenile

1531. Clement of Alexandria, *Stromata* 1.11:51.4–52.3. His anchor quotes come from Colossians 2. I'll discuss later where he is getting this material from in the Bible.

1532. Clement further explains the limited uses of philosophy in *Stromata* 1.17 and 1.20.

questions."[1533] Such philosophers tend to end up with preconceptions that "incline them to disbelieve" and from this they are "proved to be without understanding, unbelievers, and fools."[1534] In contrast, Clement says "the true philosophy has been communicated by the Son."[1535] Therefore, even when it appears Clement has something good to say about natural philosophy, he is not approving of the activity of a scientist, but the mere utility of "philosophizing" about nature when "interpreting" scripture and combating heretics. Clement emphasizes this distinction when he notes that:

> For in the Epistles Paul plainly does not disparage philosophy, but merely deems it unworthy of the man who has attained true knowledge to go back to Greek philosophy anymore, figuratively calling it "the rudiments of this world," as being most rudimentary, and merely a preparatory training for the truth.[1536]

Therefore philosophy has only one valid aim: to be a stepping stone to Christian salvation. There is little room here for any of the scientific aims of natural philosophy.

Clement belittles philosophy along the same lines when he says Paul warned against those who would "entice believers again to return to philosophy, the elementary doctrine," because:

> This fragmentary philosophy is very elementary, while truly perfect knowledge deals with intellectual objects, which are beyond the sphere of the world, and with objects still more spiritual than those which "eye saw not, and ear heard not, nor did it enter into the heart of men," until the Teacher taught them to us.[1537]

1533. Clement of Alexandria, *Stromata* 1.11:53.2 and 51.2.

1534. Clement of Alexandria, *Stromata* 1.18:88.4–88.8.

1535. Clement of Alexandria, *Stromata* 1.18:90.1–90.2. What Clement considers to be "true philosophy" is scriptural truth, as articulated in *Stromata* 6.7 and 6.8.

1536. Clement of Alexandria, *Stromata* 6.8:62.1 ("true knowledge," *gnôsis*, is only attained by receiving and understanding prophetic scripture, as Clement argued in *Stromata* 6.7).

1537. Clement of Alexandria, *Stromata* 6.8:62.3–62.4 and 6.8:68.1.

Thus, knowledge of nature did not hold a very high place in Clement's hierarchy of wisdom. Though "knowledge is the principal thing," what he means is that a studious Christian "applies himself to the subjects that are a *training* for knowledge, taking from each branch of study its contribution to the truth."[1538] True knowledge, he thus explains, is knowledge of Christ and God, and philosophy is merely a step toward that, and not an end in itself.

Clement says there are useful things to learn from logic (dialectics) as well as the quadrivium of musical theory, arithmetic, geometry—and astronomy, the only real 'science' that makes his list of worthwhile pursuits. This was a preference that would come to characterize the Church for centuries. Like some later Christians, Clement embraces astronomy because one who studies it is "raised from the earth in his mind, he is elevated along with heaven, and will revolve with its revolution, studying ever divine things, and their harmony with each other," which ultimately leads "to the knowledge of Him who created them," a benefit similar to that which even pagans saw in natural philosophy, but with a different means and a somewhat different end in mind.[1539]

Instead, for Clement, knowledge of God is the proper aim of philosophy, rather than knowledge of the natural world, since "it is necessary to avoid the great futility that is wholly occupied in irrelevant matters." So instead the knowledgeable Christian:

> Avails himself of the branches of learning as *auxiliary preparatory exercises*, in order for the accurate communication of the truth, *as far as attainable* and with *as little distraction as possible*, and for *defense against evil arguments* aimed at destroying the truth. He will then not be deficient in what contributes to proficiency in the curriculum of studies and in Greek philosophy—*but not principally, only necessarily, secondarily, and as a matter of mere circumstance.* For what those laboring in heresies use wickedly, the knowledgeable will use rightly.[1540]

1538. Clement of Alexandria, *Stromata* 6.10:80.1–80.2.

1539. Clement of Alexandria, *Stromata* 6.10:80.3–80.4 (referencing the same legend in Philo that Abraham began an astronomer and ended a man of God: see discussion in chapter 4.8 here and Carrier 2016: 142).

1540. Clement of Alexandria, *Stromata* 6.10:82.4–83.1. On the "curriculum of studies" (here *mathêseis tas egkuklious*) see chapter five of Carrier 2016. For philosophy as a mere "preparatory" study that is largely obsolete, and entirely subservient to theology, see Clement of Alexandria, *Stromata* 1.5.

Clement's qualifications here (which I have marked with italics) allude to a wider Christian attitude toward learning that we shall see more clearly elsewhere: studying anything other than scripture is auxiliary, not primary; a necessary evil rather than something to enjoy; a preparation for studying scripture, not a preparation for studying nature; nor is it acceptable as an end in itself, but only when it does not distract us from more pious pursuits. Ultimately, its utility lies not in discovering the truth, but in defending the fixed truth of a dogma already accepted by faith.

"The only wisdom," Clement concludes "is the God-taught wisdom we possess, on which depend all the sources of wisdom, which only make conjectures at the truth" because human reason is unreliable.[1541] In this respect Clement had something different in mind than what the *physici* advocated.

> The most ancient of the philosophers were not carried away to disputing and doubting, much less are we, who are attached to the genuinely true philosophy, which the Scripture openly commands us to discover by examining and searching. For it is only the more recent of the Greek philosophers who, by empty and futile love of fame, are led into useless babbling in refuting and wrangling.[1542]

But Christians, Clement argues, can find in scripture the answer to any question, since scripture is the unassailable word of God and therefore free of error, for "to those who thus ask questions, in the scriptures, there is given from God" what they are seeking, "the gift of the God-given knowledge, by way of comprehension, through the true illumination of logical investigation."[1543] So when he says, in this context, that "it is incumbent, in applying ourselves not only to the divine scriptures, but also to common notions, to institute investigations, with discovery ceasing at some useful end," by "investigations" he means logical analysis, and by "useful end" he means an understanding of God's word.[1544] And then inquiry should cease.

1541. Clement of Alexandria, *Stromata* 6.18:166.4–166.5 (concluding the point argued throughout 6.18).

1542. Clement of Alexandria, *Stromata* 8.1:1.1–1.2 (possibly part of a separate lost work on logic).

1543. Clement of Alexandria, *Stromata* 8.1:2.1–2.2.

1544. Clement of Alexandria, *Stromata* 8.1:2.4–2.5.

The above survey has prepared us for the most direct statements of Clement's attitude toward the activity of the scientist in antiquity, where we can now see how Clement's opposition is closer to that found from early pagan philosophers like Plato. But he bridges the gap between their critiques and the more thorough Christian abandonment of the scientific enterprise that we shall see in other authors.

Two passages from Clement serve to illustrate:

> Plato says "Do not imagine philosophy is spending one's life stooped over practical skills or in the pursuit of wide erudition. No. That is in my view a scandal." I suppose he knew with Heraclitus the truth: "Much learning does not teach intelligence." In the fifth book of the *Republic* he says:
>
> "Are we to rank as philosophers all these and others engaged in similar studies and those concerned with minor practical skills? No...only caricatures of philosophers."
>
> "Whom do you call the true philosophers?"
>
> "Those whose joy is in the contemplation of truth."
>
> For philosophy does not consist in geometry with its postulates and hypotheses, or in music, which operates by approximation, or in astronomy, which is stuffed full of arguments having to do with physical nature, arguments that are slippery and depend on probability. Philosophy operates through knowledge of the good in its own being, and through the truth. These are not identical with the good, but more like paths to it.[1545]

Hence Clement proposes instead:

> By faith alone is it possible to arrive at the first principle of the universe, as this is the subject of teaching, which is taught on the basis of previous knowledge. But the Greeks had no previous knowledge of the first principle of the universe, whether Thales, who took water as the primary cause, or any of the other natural philosophers who succeeded him. Anaxagoras may have been the first to establish Mind in charge of material

---

1545. Clement of Alexandria, *Stromata* 1.19:93.1–93.4, citing Plato, *Lovers* 137b (also known as *Demodicus*) and *Republic* 5.475d-e. In retreating from empiricism into mystical armchair speculation, Clement's 'Christianized' philosophy was moving in very much the same direction as pagan Neoplatonism, at nearly the same time (and notably both dominated the world after the 3rd century, Christendom winning out only by being the more ruthless): cf. Remes 2008 with *OCD* 1007, and 722, 1163–64, 1190–91 (s.v. "Neoplatonism," and "Iamblichus (2)," "Plotinus," and "Porphyry").

> objects, yet not even he noted the cause of creation, sketching in irrational vortices, leaving Mind mindlessly inactive. So the Word says, "Do not call anyone your teacher on earth." For knowledge is a state resulting from demonstration. But faith is a grace which helps its possessor to climb from things which cannot be demonstrated to the ultimate simplicity, which is not matter, has no connection with matter, and is not subject to matter. But those who believe not, as to be expected, drag all down from heaven, and the region of the invisible, to earth, "literally grasping with their hands rocks and oaks," according to Plato. For, clinging to all such things, they insist that that alone exists which can be touched and handled, defining body and essence to be identical.[1546]

Clement translates the Platonic disdain for opinion (which is derived from the senses) and corresponding elevation of reason (which alone can, through contemplation, arrive at true knowledge) into a Christian scheme dividing the futile knowledge of natural philosophers, which only limit themselves to studying what they can see (or what Aristotle called "nature," the sum of all things capable of motion or change), from the knowledge provided by *faith*, which is the only *true* knowledge, for "faith, advancing over the pathway of the objects of sense, leaves 'opinion' behind, and speeds to things free of deception, and reposes in the truth."[1547]

Clement ultimately develops an entire epistemology wherein Christian 'faith' is 'infallible' and therefore constitutes the real 'criterion of truth', such that "if you will not have faith, neither will you understand."[1548] Yet, he says, all the evidence a man of faith needs is scripture—the Word of God. In contrast, "conjecture is only a feeble supposition, a counterfeit of faith."[1549]

1546. Clement of Alexandria, *Stromata* 2.4:14.1–15.1. Clement here paraphrases and 'interprets' the Biblical Jesus, who says to his disciples: "Do not be called 'Rabbi', for one is your Teacher (*didaskalos*) and you are all brothers; nor call [anyone] on earth your father, for one is your Father: the one in heaven; nor be called tutor (*kathêgêtês*), because one is your Tutor: Christ" (Matthew 23:8–10).

1547. Clement of Alexandria, *Stromata* 2.4:13.3–13.4. On defining 'nature' see chapter 2.1–2.2.

1548. Clement of Alexandria, *Stromata* 2.4:17.4 (quoting Isaiah 7:9), faith is declared "the infallible criterion" in 2.4:12.1. See analysis in Osborn 2005: 155–212.

1549. Clement of Alexandria, *Stromata* 2.4:16.1. Clement's epistemology is the whole subject of *Stromata* 2.4 (and furthered in 2.11), which throughout argues the role of obedience in the order of knowledge: first believe, then you will know.

So instead of examining nature and studying past research on the natural world, faith is defined by obedience to scripture: believe *first*, then you will understand—nothing else is needed, except as an aid in grasping the meaning of scripture and the nature of God. This is quite the opposite of any kind of scientific epistemology. For Clement, one must simply choose to believe, and this alone will 'cause' the only truth to be 'revealed' to you. Everything else is pointless or unreliable, or merely a means to the same unscientific end. Hence Clement argues that the only natural philosophy a Christian need concern themselves with is that of Genesis and its elucidation, with the opinions of natural philosophers otherwise rejected.[1550]

5.2 Tertullian (c. 200 A.D.)

Tertullian takes this Christian attitude to its natural conclusion.[1551] He sees no use for scientific research, for "What concern have I with the conceits of natural philosophy? It were better for one's mind to ascend above the state of the world, not to stoop down to uncertain speculations."[1552] He mentions debates about the size and shape of the earth and sun, and replies:

> Now, pray tell me, what wisdom is there in this hankering after conjectural speculations? What proof is afforded to us, notwithstanding the strong confidence of its assertions, through the useless affectation of a scrupulous curiosity tricked out with an artful show of language?[1553]

Clement also draws on Hebrews 11 here, which we will discuss in section 5.6. For his complete discussion of faith as a method and source of knowledge see *Stromata* 2.1–2.4 and 5.1.

1550. Argued in Clement of Alexandria, *Stromata* 4.1. Also compare his statement that he will 'next' show what sort of natural philosopher a Christian should be (*Stromata* 6.18:168.4), and the content of what actually follows (*Stromata* 7).

1551. For background on Tertullian see *OCD* 1444–45 (s.v. "Tertullian (Quintus Septimius Florens Tertullianus)") and *ODCC* 1591–92 (s.v. "Tertullian, Quintus Septimius Florens").

1552. Tertullian, *To the Nations* 2.4.47.

1553. Tertullian, *To the Nations* 2.4.47.

This inspires him to mock the legend that Thales was so engrossed in studying the stars one night that he stumbled into a well. To pagans, this tale often represented his single-minded devotion to higher concerns.[1554] But to Tertullian it symbolized his damnation: "His fall," Tertullian says, "is a figurative picture of the philosophers—of those, I mean, who persist in applying their studies to a vain purpose, since they indulge a stupid curiosity on natural objects, which they ought rather have for their Creator and Governor."[1555]

Like Clement and other critics of natural philosophy, Tertullian also condemns it for the fact that there is so much disagreement—though unlike Cicero, who took this as a reason to adjust the epistemic status of science and treat it as probable rather than certain knowledge, yet useful and worthwhile all the same, Tertullian takes the same fact as grounds to jettison the whole thing as vain hubris. As he puts it, "philosophers have ingeniously composed their naturalist views out of their own conjectures" but "these are only a doubtful conception," for "all things with the philosophers are uncertain, because of their variation."[1556] And so, though the pagans find the wisdom of their philosophers to be the best, it has a fatal weakness: it is a mere "opinion that proceeds from an ignorance of the truth." For only "God is the Father and Lord of wisdom and truth," and as Solomon wrote, "the fear of the Lord is the beginning of wisdom."[1557] But, Tertullian explains, the philosophers do not know God, and you cannot fear what you know nothing about. Since the philosophers do not fear the Lord, they cannot even make a beginning of wisdom.

This is essentially what Clement argued: you must choose to believe and thus recognize God *first*, and only *then* can wisdom follow. But Tertullian is even more explicit: "he who shall have the fear of God, even if he be ignorant of everything else, if he has attained to the knowledge and truth of God, will possess full and perfect wisdom," needing no other—and since philosophers "have not clearly realized" this, their 'wisdom' is useless.[1558]

1554. Articulated at length in Plato, *Theaetetus* 173e-176a (more briefly in Diogenes Laertius, *Lives and Opinions of Eminent Philosophers* 1.34).

1555. Tertullian, *To the Nations* 2.4.47.

1556. Tertullian, *To the Nations* 2.1.41.

1557. Tertullian, *To the Nations* 2.2.42.

1558. Tertullian, *To the Nations* 2.2.42.

Natural philosophy is therefore useless. Knowledge of God is all anyone needs. Indeed, "vain are those supports of human learning, which, by their artful method of weaving conjectures, belie both wisdom and truth."[1559] Thus, unlike Clement, Tertullian is ready to throw out the whole thing.

Worse, Tertullian says, is the fact that, as he sees it, "according to the naturalistic way of thinking, he who has spoken the best is supposed to have spoken most truly, instead of him who has spoken the truth being held to have spoken the best," thus showing his disdain for eloquence and well-constructed logical arguments.[1560] In the end, Tertullian declares, though everyone is fallible, it is better our errors be simple ones, than that they derive from those complex speculations "of the natural philosophers." Natural philosophy is therefore the way of error and damnation, fundamentally ungodly, and best swept away with the rest of the rubbish.

Even worse than that, Tertullian argues, physics also leads to atheism. To make the point, he gives numerous examples of how everyone knows to look for the author of anything that happens—for example, no one blames the sword for the wound, but the one who wielded it. So, he says to the pagans:

> On all other occasions, your conduct is right enough, because you consider the author, but in physical phenomena your rule is opposed to that natural principle which prompts you to a wise judgment in all other cases, removing out of sight as you do the supreme position of the author, and considering rather the things that happen, than him by whom they happen. Thus it comes to pass that you suppose the power and the dominion belong to the elements, which are but the slaves and functionaries.[1561]

In other words, the very objective of science—to discover the natural principles and causes of things, indeed to study the phenomena of nature in any way at all, rather than God—is damnable. God is all we ought to be studying. This is actually a return to a prescientific way of thinking, reducing explaining natural phenomena to the task of inferring the intentions of the supernatural agents causing it, rather than the mechanisms producing it. Likewise, Tertullian despises the tendency of philosophers to 'reinterpret' religion in naturalistic terms and to argue that the names of the gods really

1559. Tertullian, *To the Nations* 2.6.50.

1560. Tertullian, *To the Nations* 2.6.50.

1561. Tertullian, *To the Nations* 2.5.48–49.

only refer figuratively to natural properties of the world.[1562] For all these reasons, natural philosophers have "wandered away" from "the beginning of wisdom, that is, the fear of God" because "proofs are not wanting that among the philosophers there was not only an ignorance, but actual doubt, about the divinity." He quotes several early philosophers stating various states of agnosticism about the divinity as proof of his point.[1563] And this, he claims, leads to moral corruption.

Tertullian's disdain for philosophical enquiry is quite thorough. "Philosophers," Tertullian says, are not worthy of respect, for they are the "mockers and corrupters" of the truth "with hostile ends" who "merely affect to hold, and in doing so deprave" the truth, "caring for naught but glory," while the "Christians both intensely and intimately long for and maintain with integrity" the truth, since they are the ones "who have a real concern about their salvation." And for precisely this reason, Christians and philosophers "are like each other in neither their knowledge nor their ways." He exemplifies his point with the report that Thales, "the first of natural philosophers," professed ignorance and disinterest in God, and then he catalogues the various alleged crimes and immoralities of later philosophers, in contrast to the allegedly superior moral virtues of Christians.[1564] Hence he concludes only ethical knowledge is worthwhile, and only if it actually results in moral action.

Thus, since philosophy—especially in physics and logic—is not about moral reform, it is worthless. As Tertullian puts it, in a sweeping condemnation of all philosophy:

> Where is there any likeness between the Christian and the philosopher? Between the disciple of Greece and of Heaven? Between the man whose object is fame, and whose object is life? Between the talker and the doer? Between the man who builds up and the man who pulls down? Between the friend and the foe of error? Between one who corrupts the truth, and one who restores and teaches it? Between its thief and its custodian?[1565]

1562. Tertullian, *Against Marcion* 1.13. For more evidence of this view see chapter 2.5.

1563. Tertullian, *To the Nations* 2.2. For more evidence of this view see chapter 2.6.

1564. Quotes and paraphrases from Tertullian, *Apology* 46.

1565. From Tertullian, *Apology* 46.

The philosopher is therefore the antithesis of the Christian, and is everything the Christian should not be. Hence Tertullian dismisses all philosophy with the infamous quip, "What has Athens to do with Jerusalem?" The answer, for Tertullian, was 'nothing'.[1566]

> After Jesus Christ we have no need of curiosity, after the Gospel no need of research. When we believe, we have no wish to believe anything else. For this we believe first of all: that there is nothing else we have to believe.[1567]

In fact we do not need to believe anything other than Christian doctrine, he argues, because *all* philosophy leads to heresy of one form or another, and all inquiry beyond the faith is useless, even immoral, for the simple reason that if God did not tell us, we do not need to know it.[1568]

All this is exemplified in Tertullian's *Treatise on the Soul*, which attacks the uncertainties and disagreements among philosophers and scientists about the nature of the soul, and then concludes with why the Christian approach (of relying on scripture and inspiration) is superior:

> It is better not to know what God has not revealed, than to know it from man, because man is presumptuous. ... Hence for a Christian only a few words are needed to have knowledge of this subject. For there is always certainty in those few words. Man is not allowed to investigate any further than what he is allowed to discover anyway. For the Apostle [Paul] forbids "endless questions." Again, man is not permitted to discover any more than what is learned from God. For that which is learned from God is all there is to know.[1569]

This, in a nutshell, expresses how and why Christianity was often diametrically opposed to the methods, aims, and assumptions of even the best natural philosophers. Thus, "like most patristic authors," Heinrich von Staden observes, "Tertullian was strongly opposed to the scientific research

1566. Tertullian, *Prescription against Heretics* 7.

1567. Ibid., using the words *curiositas* (curiosity) and *inquisitio* (research; more precisely, "critical inquiry").

1568. Tertullian, *Prescription against Heretics* 11–14 (in fact, because Church doctrine comes from God, everything that disagrees with Church doctrine is by definition false: 21–28).

1569. Quotation from Tertullian, *On the Soul* 1 (first line) and 2 (remainder).

of pagan scientists" and "did not refrain from disparaging their work," an observation more restrained than G.E.R. Lloyd's hyperbolic remark that "Tertullian was totally opposed to the scientific investigations of pagan researchers and did everything he could to defame them and their work," which perhaps overstates the matter, though not by much.[1570] It is still clearly the case, in the words of Henri Crouzel, that Tertullian "offers no positive reflections on the use of philosophy by Christians."[1571] And since science was a prominent department of philosophy, it, too, had to go.

This is exemplified by Tertullian's attitude to neurophysiological research. He prefaced his treatise *On the Soul* with an extended argument that the truth in every subject, when not already obvious, can only be learned from God, and in fact only by Christians, through Scripture and the Holy Spirit, since all other spirits and gods are false, and all human sources of knowledge are unreliable. Hence we don't need science if we have God, and in fact if God wanted us to know anything more about the world, he would already have told us. This was not an uncommon view in the early medieval period. Again according to Heinrich von Staden, "a theme common to many Church Fathers" is that "what God has hidden is not intended for human eyes and therefore should not be unveiled artificially," as for example by practicing dissection.[1572]

The consequence of this for the advancement of science is revealed by contrasting Tertullian's treatise *On the Soul*, and Galen's comparable treatise *On the Doctrines of Hippocrates and Plato*, written around the same time.[1573] As Galen demonstrates, scientists had already decisively and empirically proved that all mental functions reside in the brain. They had even proved that different functions reside in different parts of the brain by selectively

1570. von Staden 1989: 142; Lloyd 1973: 76.

1571. Crouzel 1989: 156.

1572. von Staden 1989: 143. This idea originates with Socrates (see relevant notes and discussion in chapter 4.2 and section 5.4 here, and Carrier 2016: 100–01), but was rejected by every subsequent school of thought except the Cynics (who were socially uninfluential) and the Christians (who only became influential after the decline of the empire).

1573. Tieleman 1996 establishes the latter as a treatise using empirical and experimental science to answer fundamental questions about the nature of the soul.

resecting the brains of live animals, and they traced all the motor and sensory nerves to their end-points in the brain, leaving no rational doubt that all mental function resided there. Galen added to this research with his own extensive investigations into the physiological origins of human speech. So the antiquated idea that the mind resided in, say, the heart, had been scientifically refuted. How does Tertullian react to this fact? In his own treatise on the subject he dismisses all this scientific evidence and declares the whole debate unresolvable, a demonstration of the folly of human reason, an attempt to go beyond what God saw fit to reveal. Instead, Christians have no need of any scientific research. All they need is Scripture, with perhaps some help from divinely 'inspired' visions and armchair reasoning.[1574]

5.3 Lactantius (c. 300 A.D.)

A century later we find little has changed. At the dawn of the 4th century, between the Great Persecution of 303 A.D. and Constantine's negotiated Edict of Toleration of 311 A.D., Lactantius wrote the *Divine Institutes*, a monumental tract 'refuting' heathen wisdom and building up Christian religion as the only true course for man.[1575] At the time Lactantius was Constantine the Great's advisor and tutor of his eldest son, and would soon become an admired author in the Christian tradition. So Lactantius represents an important point of contact between the development of Christian attitudes towards scientists and the entire future of the Roman World.

In book 3 of his *Institutes*, Lactantius attacks pagan philosophy in its entirety as nothing but vanity and confusion, which detracts from the only ultimate good: knowledge of God. He reflects the trend we have already seen, and captures it in such eloquent detail that we will analyze large portions of his attack. He argued that knowledge is never sought for its

1574. Tertullian, *On the Soul* 1–4, 7, 26 (scripture is authoritative); 5–6, 8, 10–25, 27–58 (armchair reasoning); 9 (visions received by a church lady counted as evidence).

1575. For background on Lactantius see *OCD* 789 (s.v. "Lactantius (Lucius Caelius (Caecilius ?) Firmianus also called Lactantius)") and *ODCC* 942 (s.v. "Lactantius"). For another example of his attitude, which supports and informs the following, see Carrier 2016: 160–63.

own sake, but for some other end, and most knowledge, whether practical or theoretical, is sought for "subsistence, glory, or pleasure."[1576] But even beasts desire these, he says, so they cannot be ends peculiar to man, and "if those things which knowledge produces are common to man and other animals, it follows that knowledge is not the chief good."[1577] In other words, the pleasure at discovery praised by Cicero, the value of the glory accorded to scientists by Apuleius, the practical applications of scientific theory lauded by Galen, are thus all rejected by Lactantius as beastly aims, unworthy of men.

So Lactantius asks why we seek knowledge at all:

> If regarding the causes of natural things, what happiness will be proposed to me, if I shall know the sources of the Nile, or the vain dreams of the natural philosophers respecting the heaven? Why should I mention that on these subjects there is no knowledge, but mere conjecture, which varies according to the abilities of men? It only remains that knowledge of what is good and evil is the chief good.[1578]

And so ethics, particularly Christian ethics, informed by a recognition and contemplation of God and his Word, is the only aim peculiar to man and therefore the only aim worthy of men who would not be beasts. He specifically names natural philosophers and their work, the very object of their study and interest, as useless to human happiness. Astronomy, physics, geography, are a waste of time. And nothing can really be known about them anyway. Hence Lactantius concludes his entire argument with a predictable exhortation: "Let him who wishes to be wise and happy hear the voice of God, learn righteousness, understand the mystery of his birth, despise human affairs, and embrace divine things," for only then may he "gain that chief good to which he was born."[1579] Acknowledge God, study God. Despise human affairs, and embrace the divine in their place. There is no place here for science.

However, Lactantius is not entirely opposed to the arts. There are some

---

1576. Lactantius, *Divine Institutes* 3.8.25.

1577. Lactantius, *Divine Institutes* 3.8.27.

1578. Lactantius, *Divine Institutes* 3.8.29–30.

1579. Lactantius, *Divine Institutes* 3.30.8.

things we can know, and should pursue—whatever is necessary for life, and whatever is obvious to everyone:

> There are many things which nature itself, and frequent use, and the necessity of life, compel us to know. Accordingly you must perish, unless you know which things are useful for life (in order that you may seek them) and which are dangerous (that you may shun and avoid them). Moreover, there are many things that experience finds out. For the various courses of the sun and moon, and the motions of the stars, and the computation of times, have been discovered, and the nature of bodies and the power of herbs by students of medicine, and the nature of soils by the cultivators of the land, and signs of future rains and tempests have been collected. In short, there is no art which is not dependent on knowledge.[1580]

But, he says, only these practical arts thus have a place—theory he seems to regard as having no benefit to practice. All theoretical science is therefore cast out. The knowledge of causes, of hidden structure or principles, is rejected as vain and unattainable. Hence Lactantius outright declares, "the common people sometimes have more wisdom, because they are only so far wise as is necessary."[1581] According to such a view, too much knowledge, more than we need, is the antithesis of wisdom. And just about everything we would count as science today is more than Lactantius thinks we need.

Lactantius argues that "philosophy appears to consist of two subjects, knowledge and conjecture, and of nothing more" and yet "knowledge was taken away by Socrates, and conjecture by Zeno."[1582] He argues there is no knowledge because philosophers cannot agree, while conjecture belongs solely to the "rash and foolish man."[1583] Lactantius attacks all philosophy with this line of reasoning, but natural philosophy especially:

> For to investigate or to wish to know the causes of natural things—whether the sun is as great as it appears to be, or is many times greater than the whole of this earth; whether the moon be spherical or concave; whether the stars are fixed to the heaven, or are born freely through the air; of what

1580. Lactantius, *Divine Institutes* 3.5.1–2.

1581. Lactantius, *Divine Institutes* 3.5.4.

1582. Lactantius, *Divine Institutes* 3.3.1 and 3.4.2 (united by the argument of 3.3–4).

1583. Lactantius, *Divine Institutes* 3.4.1.

> magnitude the heaven itself is, of what material it is composed; whether it is at rest and immovable, or is turned round with incredible swiftness; how great is the thickness of the earth, or on what foundations it is poised and suspended—to wish to comprehend these things, I say, by disputation and conjecture, is as though we should wish to discuss what we may suppose to be the character of a city in some very remote country, which we have never seen, and of which we have heard nothing more than the name. If we should claim to ourselves knowledge in a matter of this kind, which cannot be known, should we not appear to be mad, in venturing to affirm something on which we may be refuted? How much more are they to be judged mad and senseless, who imagine that they know natural things, which cannot be known by man![1584]

Scientists are thus mad and senseless, and their activity is as vain as declaring to know things about a city they have never seen nor heard a thing about.

Lactantius was either ignorant of or unimpressed by the fact that empirical data was used by scientists to discover some of these very things (such as the diameter of the earth by Eratosthenes). Whether he does not know or does not care, it is all a waste of time to him. So:

> It remains that there is in philosophy only conjecture; for that from which knowledge is absent, is entirely occupied by conjecture. For everyone conjectures that of which he is ignorant. But they who discuss natural subjects, conjecture that they really are as they discuss them. Therefore they do not know the truth, because knowledge is concerned with that which is certain, conjecture with the uncertain. … Yet this is the very thing which philosophers do, who discuss what is taking place in heaven.[1585]

Predictably, Lactantius asks, "Does wisdom therefore nowhere exist?" He answers in the affirmative—although "it was amongst them, but no one saw it."[1586] For knowledge of the hidden truths of nature can only be learned from scripture and revelation, because the truth "God has revealed to us," hence "we do not arrive at it by conjectures but by a communication from heaven," unlike philosophers, who can never discover the truth because

---

1584. Lactantius, *Divine Institutes* 3.3.4–7.

1585. Lactantius, *Divine Institutes* 3.3.7–15.

1586. Lactantius, *Divine Institutes* 3.6.1.

"man cannot arrive at the truth by thinking and discussion, but only by learning and hearing from Him who alone is able to know and teach," which is why all philosophy is false, natural philosophy included, because theories "which have no support from prophesies or divine voices cannot have any foundation or stability."[1587]

Lactantius thus censures the natural philosophers for claiming they could one day know everything, hence rejecting the very idea of scientific progress, and then he censures the skeptics for claiming nothing could ever be known. However, unlike Cicero or even Sextus Empiricus, who argue from the latter observation to a moderate epistemology of probable knowledge—a principle central to the development and success of modern science—Lactantius argues that all natural philosophy is "obscure" and "incomprehensible" and only what is manifestly obvious to everyone is worth acknowledging, and for that reason there was nothing left to study, physics having long ago "attained to its greatest increase" and already "necessarily growing old and perishing" by the time of Arcesilas (founder of the Middle Academy in the early 3rd century B.C.), who "rightly saw that they are arrogant, or rather foolish, who imagine that the knowledge of the truth can be arrived at by conjecture."[1588] And so, though the skepticism of Arcesilas went too far, Lactantius says it was on the right track:

> How much more wisely and truly he would act, if he should make an exception, and say that the causes and systems of heavenly things only, or natural things, because they are hidden, cannot be known, for there is no one to teach them, and they ought not to be inquired into, for they cannot be found out by inquiry! For if he had brought forward this exception, he would both have admonished the natural philosophers not to search into those things which exceed the limit of human reflection, and he would

1587. From Lactantius, *Divine Institutes* 7.1.11, 7.2.9, and 7.2.11 (in the context of the whole of 7.1–2 the sentiment is clearly meant to be universalized to all doctrines about nature) and 7.2.9 and 7.2.11 (here again universalized; see also the arguments of 1.1, 3.1, 3.30, and 6.18.1). Hence Lactantius outlines a Christian 'version' of natural philosophy in 7.3–14, supposedly based on scripture though in fact almost entirely conjecturing from fundamental Christian dogmas (which he elaborates in his separate work *On the Craftsmanship of God*), while in 7.15–26 he lays out what scripture supposedly proves about the coming end of the world.

1588. Lactantius, *Divine Institutes* 3.6.5 and 3.6.9. On Arcesilas and the Middle Academy see *OCD* 2 and 136 (s.v. "Academy" and "Arcesilaus (1) or Arcesilas").

> have freed himself from the ill will arising from calumny, and would certainly have left us something to follow.[1589]

There can be no question that Lactantius had a decidedly negative attitude toward the scientist and his activity. Not only theorizing, but scientific inquiry itself could produce no results, so the only wise move is to abandon the enterprise altogether, for "natural philosophy is superfluous," and ultimately "useless and inane."[1590]

Not only is all philosophy thus "useless and inane," Lactantius further concludes that "if all things cannot be known, as the natural philosophers thought, nor nothing, as the Academics taught, philosophy is altogether extinguished."[1591] So, apart from the purely practical arts, knowledge of God alone is the only proper object of study. And for this reason, once again, natural philosophy actually contributes to atheism: for "those philosophers who wish to free the mind from all fear, take away even religion," alluding to the same point made by Tertullian, that all wisdom begins with fearing the Lord.[1592] Thus:

> We must look up to the heaven, to which the nature of the body calls us. But if it is admitted that this must be done, it must either be done with this view, that we may devote ourselves to religion, or that we may know the nature of the heavenly objects. But we cannot by any means know the nature of the heavenly objects, because nothing of that kind can be found out by reflection, as I have before shown. We must therefore devote ourselves to religion, and he who does not undertake this prostrates himself to the ground, and, imitating the life of beasts, abdicates the office of man. Therefore the ignorant are more wise, for even if they err in choosing religion, they still remember their own nature and condition.[1593]

1589. Lactantius, *Divine Institutes* 3.6.16–17. This is a sentiment extending all the way into Jewish literature (e.g. 1 Enoch), whereby knowledge not revealed by God is meant not to be known; and thus when revealed, must be demonic.

1590. Lactantius, *Epitome of the Divine Institutes* 30.6 (= 35.6 in a variant numeration): using *inanis*, "empty, groundless, worthless, inane," and *inutilis*, "useless, unprofitable, harmful." This epitome was written by Lactantius himself, as explained in its preface.

1591. Lactantius, *Divine Institutes* 3.6.20.

1592. Lactantius, *Divine Institutes* 3.10.9.

1593. Lactantius, *Divine Institutes* 3.10.12–15.

Hence the scientist is a beast who abdicates his status as a human being, and the uneducated masses are wise, because they are pious, and know their place—for they do not aspire to know anything but God.

As Tertullian exemplified this attitude in neurophysiology, so Lactantius exemplified it in astrophysics. For he pauses to ridicule at some length the very idea (otherwise scientifically well-founded even in his own day) that the earth is a sphere.[1594] "Is any one so stupid," Lactantius asks, "that they believe there are men whose footsteps are higher than their heads? Or that...crops and trees grow downwards, or rain, snow, and hail fall upwards!?"[1595] The sphericity of the earth was often denied by the ignorant masses on similar grounds of apparent absurdity.[1596] But Lactantius is no illiterate—he is clearly a well-educated man. Just as Tertullian was thoroughly unimpressed by an extensive body of scientific evidence confirming the role of the brain in producing human thought, Lactantius simply scoffs at the empirical evidence astronomers had collected that the earth was a sphere. He attacks only one astronomical argument, in fact the weakest (the evident rotation of the stars), but whether he knew any others is unclear, since he says he offered only one example of many, which he also promises he could dispatch as easily. Yet ancient astronomers had amassed at least six persuasive empirical proofs that the earth is a sphere, none of which Lactantius mentions:

1. As ships approach or depart land, mountains rise and sink along the horizon, and do so in every direction. Ships likewise rise and sink to a viewer on shore.
2. The rising and setting times (and in some cases the outright visibility) of northern and southern stars vary relative to an observer's latitude.
3. Noonday shadows vary in length in proportion to an observer's latitude.

1594. Lactantius, *Divine Institutes* 3.24.

1595. Lactantius, *Divine Institutes* 3.24.1 (as the full context of 3.24 makes clear, Lactantius did not even accept that the heavens revolved beneath the earth).

1596. That many ignorant people insisted the earth must surely be flat (and some for the same reasons as Lactantius) is attested in Pliny the Elder, *Natural History* 2.65.161.

4. A lunar eclipse is seen to begin at different hours of the night according to an observer's longitude (equivalent to the modern argument from time zones).
5. The duration of lunar eclipses is always the same regardless of the time of night or the moon's location in the sky (which geometrically requires the earth to be a sphere).
6. The earth's shadow on the moon is always round regardless of when, or where in the sky, the moon is in eclipse (which also geometrically requires the earth to be a sphere).

The observations and geometry in all five cases had been thoroughly worked out even before the Roman era, and were rock solid and in mutual agreement.[1597] But Lactantius stubbornly refused to acknowledge any of it. The earth was flat. End of story. But whether from ignorance or arrogance, the result is the same: here we see again, science is entirely abandoned, and even its best conclusions declared stupid and contrary to all reason.

5.4 Eusebius (c. 300 A.D.)

Finally, Lactantius's contemporary, Eusebius, presents essentially the same view in the *Preparation for the Gospel*. Eusebius was also connected to the Constantine family.[1598] And he expands on Lactantius's claim that the truth was among the philosophers, yet they did not see it, by arguing at excruciating length that everything 'good' about pagan philosophy was actually stolen from the Jewish 'scriptures', while everything *else* was rubbish.[1599] By and large, the good stuff was select ethical and practical

1597. Aristotle, *On the Heavens* 2.14.297a-298a; Strabo, *Geography* 1.1.20; Pliny the Elder, *Natural History* 2.65.161–166 and 2.71.177–77.187; Cleomedes' *On the Heavens* 1.5; Theon of Smyrna, *Aspects of Mathematics Useful for Reading Plato* 3.1.120–4.124; Ptolemy, *Almagest* 1.4; plus various passages throughout Ptolemy's *Geography* and Geminus' *Introduction to Astronomy*.

1598. For background on Eusebius see *OCD* 555–56 (s.v. "Eusebius") and *ODCC* 574 (s.v. id.).

1599. The same argument was previously made by Tertullian (*To the Nations* 2.2, *Apology* 47, *Testimony of the Soul* 5) and Clement of Alexandria (*Stromata* 1.15, 1.17,

advice, and the bad stuff was everything scientific or theoretical—except whatever truths about nature God had already revealed to us in scripture. Though the *Preparation* was probably written after 313 A.D. (which was about the time he received the bishopric of Caesarea and gained access to Origen's famous library there), it is a crucial document for understanding the early Christian worldview, as it probably represents opinions Eusebius held before his patron Constantine became Emperor of Rome.[1600]

Eusebius finds the root of his attitude in its Jewish Essene origins, quoting with admiration Philo's account of the Essenes as embracing only the ethical part of philosophy, leaving logic to "word-catchers," being "unnecessary to the attainment of virtue," and leaving the study of nature to "star-gazers," being "too lofty for human nature," though accepting that they studied what related to the existence of God and the creation of the universe.[1601] As I noted before, this did not entail Philo or even the Essenes despised the taking up of these studies by *others*, though it did reflect a Jewish attitude that walked a fine line between the early Christian disdain for science and a certain respect for it, provided it was conducive to virtue.[1602] Whereas as a Christian Eusebius goes much further, calling the disputes of scientists "blasphemies," especially those who exclude any role for God or a Divine Mind in the order of things.[1603]

In fact, Eusebius asks, since philosophy is all just hypothesizing from circumstantial evidence, "what of any use can we expect to learn from the teachings of philosophers?"[1604] The answer for Eusebius was 'nothing'. Since all their learning did them no good, while many among the unlearned were paragons of piety and virtue, philosophy was clearly of little use.[1605] He

---

2.5, 2.18, 5.14, 6.3).

1600. Carriker 2003 surveys the known contents of Origen's library available to Eusebius.

1601. Eusebius, *Preparation for the Gospel* 8.12.9, quoting Philo, *That Every Good Man Is Free* 12(80). See also chapter 4.9 and Philo *On Agriculture* 16 and *On the Changing of Names* 74.

1602. As discussed in chapter 4.9 here and chapter nine of Carrier 2016.

1603. Eusebius, *Preparation for the Gospel* 14.16.11. See also chapter 2.6 here.

1604. Eusebius, *Preparation for the Gospel* 14.10.7.

1605. Eusebius, *Preparation for the Gospel* 14.10.11.

declares that all the natural philosophers before and after Plato, "stand in opposition alike to the doctrines of the Hebrews and of Plato and to the truth itself" and so thoroughly contradict each other as to refute themselves.[1606] Thus, like previous Christian authors, the reproach that philosophers cannot agree about anything is taken not as a call for better methods or a more humble epistemological standard, but for the complete rejection of the entire enterprise—as is clear even from his chapter titles, like "How much there is disagreement among the natural philosophers concerning first things and so we also advisedly put them aside."[1607]

Eusebius tells us he "shall drag out to light both the discrepancies of their doctrines and the futility of their eager studies," though he insists he does not do this "as a hater of the Greeks or of reason, far from it, but to remove all cause of slanderous accusation, that we have preferred the Hebrew oracles because we are very little acquainted with Greek culture,"[1608] and to prove that "we have not without right judgment neglected the useless learning of such subjects as these."[1609] Eusebius then expands on the hostile attitude of Plato in support of his own rejection of science, just as Clement had, but now even more sternly, quoting both Plato and Xenophon explaining how Socrates condemned the studies of natural philosophers as foolish madness—being at worst impious, and at best a futile waste of time.[1610]

We should only bother learning as much astronomy, for example, as is already known and used by hunters and pilots, much as Lactantius argued. Anything more theoretical, anything that has no common practical

1606. Eusebius, *Preparation for the Gospel* 15.1.10.

1607. Eusebius, *Preparation for the Gospel* 1.pr.1 (and also 1.8.1). Lactantius deployed the same argument, in *Divine Institutes* 1.1, 3.1, 3.4, 5.3, etc., and *Epitome* 32 (= 27). Likewise the Christian author Hermias, who may be a predecessor to Eusebius (or an incompetent successor), composed the mediocre but viciously hostile *Mockery of the Profane Philosophers*, satirizing pagan disagreements on questions of natural philosophy (dated variously from the 2nd to the 6th century A.D.): cf. *OCD* 670 (s.v. "Hermias (3)") and *ODCC* 761 (s.v. "Hermias").

1608. Eusebius, *Preparation for the Gospel* 14.2.7 (cf. 14.3.6).

1609. Eusebius, *Preparation for the Gospel* 14.13.9 (repeated in 15.61.11, etc.).

1610. Eusebius, *Preparation for the Gospel* 14.13.9 (quoting Plato, *The Republic* 530e-531c); Eusebius, *Preparation for the Gospel* 1.8.14–19 (quoting Plato, *Phaedo* 96a and Xenophon, *Memorabilia* 1.1.11–13). On Socrates see related notes in section 5.2 and chapter 4.2 here and in Carrier 2016 (index, "Socrates").

application—like learning the distance from the earth to the moon, the causes of planetary motion, or the periods of comets—is of no use, and even takes valuable time away from proper pursuits. The most obscure subjects, like cosmology, were to be avoided altogether, because humans simply cannot discover such things, and God would probably not be pleased with anyone who tried to discover what He had not saw fit to reveal.[1611] In fact Eusebius had no patience for *any* scientific inquiry, ultimately rejecting all scientific questions as "unprofitable, meddlesome, and pointless," declaring with pride that Christians waste no time on them, "for we see no use in them, nor any tendency to help or benefit mankind." All Christians need do is worship God and live righteously. As science aids in neither, it is happily discarded.[1612]

In contrast, Eusebius praises Hebrew scripture as containing perfect wisdom passed down unerringly for ages, considering it vastly better than the inquiries of scientists, who have only "tossed about in shallow waters." Though they "wandered over the wide earth, and set the highest value on the discovery of the truth, and were familiar with the opinions of all the ancients, and carefully studied" them, even the ideas of other cultures, everything they said was nevertheless both "false and contradictory."[1613] But the knowledge of nature revealed in scripture has been communicated directly by God, rather than hypothesized by men, and is therefore superior to anything philosophers have to offer.[1614] Likewise, though the study of moral philosophy is "practicable and useful," any "discussions about nature are quite the contrary, neither being comprehensible, nor having any use even if they are clearly understood."[1615] Since the study of nature will not make us "wiser, or more just or brave or temperate, or even stronger or

---

1611. This is the argument of Eusebius, *Preparation for the Gospel* 14.10–11 (eventually quoting Xenophon, *Memorabilia* 4.7).

1612. Eusebius, *Preparation for the Gospel* 15.61.11. This is the argument of the entire fifteenth book of his *Preparation for the Gospel* (cf. 15.1 and 15.61–62), which surveys a wide array of questions in science and natural philosophy (including astronomy, meteorology, and human physiology), and declares them all void of any value or solution.

1613. Eusebius, *Preparation for the Gospel* 14.9.4.

1614. See, for example: Eusebius, *Preparation for the Gospel* 11.7 and 14.10.

1615. Eusebius, *Preparation for the Gospel* 15.62.7–8.

prettier or richer," we should give no care to it.[1616]

Finally, like all the Christian authors before him, Eusebius suspects that the study of natural philosophy leads to atheism, for most natural philosophers "did not assume any creator or maker of the universe, nay, they made no mention of God at all, but referred the cause of the All solely to irrational impulse and spontaneous motion,"[1617] and "such too is their opinion concerning first principles," where they allowed "no god, no maker, no artificer, nor any cause of the universe, nor yet gods, nor incorporeal powers, no intelligent natures, no rational essences, nor anything at all beyond the reach of the senses."[1618] Consequently, the study of nature borders on being impious or even immoral.[1619] Eusebius, of course, insists nothing can really be known about such things, because "physics is beyond us," as the constant disagreements among natural philosophers has proved (at least to him, and many of his Christian peers).[1620] There is also a disdain here for the very notion that nature could be explicable without God as the explanation of it. Which is a complete reversal of the cognitive revolution that started science to begin with: when the Presocratics started looking for natural causes rather than divine in all of nature's ways. For all such reasons, we see in these remarks of Eusebius an ever-present fear that godlessness is the inevitable end of any free pursuit of science.

5.5 Christian Anti-intellectualism?

We have examined the only explicit discussions of the natural philosopher in early Christian literature. And from what these authors said, it seems evident that the predominant Christian attitude toward the natural philosopher and his activity, through all three centuries of the early Roman Empire, was one of hostility or indifference—either way, decidedly negative. This might reflect a more general anti-intellectualism, not unlike that found

1616. Eusebius, *Preparation for the Gospel* 15.62.8–9.

1617. Eusebius, *Preparation for the Gospel* 1.8.13.

1618. Eusebius, *Preparation for the Gospel* 14.14.7; cf. also 15.62.

1619. Eusebius, *Preparation for the Gospel* 15.62.11–15.

1620. Eusebius, *Preparation for the Gospel* 15.62.9–10.

in modern American society.[1621] Like today, in antiquity some Christians may have developed a negative opinion from a certain disdain for the elite and learned, matched by a superior respect for the pious but uneducated common folk.[1622] It is well known that Christianity began with and was largely driven by, or sold to, the non-elite, and as a result we can expect that it absorbed, or promoted, attitudes shared by that group. As outsiders looking in, many Christians apparently saw natural philosophy as a bulwark of the enemy, the tool of heretics and heathens, and thus as something to be jettisoned rather than embraced.

A central element of such anti-intellectualism may be reflected in the Christian longing for simple rather than complex answers, and for the comfort of conviction rather than the anxiety of doubt and uncertainty. Being certain of the truth is often seen as better than having questions. From the one comes comfort and safety; from the other, danger and disquiet. A recent psychological study found that modern American conservatism, which is still strongly correlated with the more traditional modes of Christian belief, is highly correlated with several characteristic features of anti-intellectualism.[1623] Among the psychological variables that were found to predict political conservatism were a relative excess of "death anxiety," "dogmatism" and "intolerance of ambiguity" and of "uncertainty." Ancient Christianity would have appealed to the same psychological mindset, by resolving fears of death, and answering any intolerance of uncertainty and ambiguity by replacing arguments and questions with dogma and faith. We have already seen hints that the complexities and uncertainties of natural theory and inquiry were regarded with horror, leading many devout Christians to find their only safe harbor in the simplicity and absolute certainty of faith.

As Robin Lane Fox puts it, while others "surveyed the philosophers' disagreements with a certain detachment, Christianity captured" hearts and minds "by giving them a firm dogma" as a "powerful counter to anxiety."

1621. See Hofstadter 1962, Pigliucci 2002, Noll 2008, Jacoby 2008, and Pierce 2010.

1622. See Alexander 2002; with chapters nine and ten of Carrier 2016.

1623. Jost et al. 2003. Hunsberger & Altemeyer 2006 present evidence and scholarship on the association of these personality traits with modern conservative Christianity.

Hence Christian writers repeatedly "emphasize the general certainty which they found in their new faith."[1624] Simplicity was likewise recognized in antiquity as a lure for the naive who did not approve of things difficult to learn—as Galen and Ptolemy noted, many people tend to slander, attack, or reject whatever requires considerable study to understand.[1625] Ambiguity, complexity, debate, and uncertainty are fundamental to science and natural philosophy, but anathema to what many people wanted, so Christianity offered the opposite: the certainty, clarity, and simplicity of a dogmatic faith. That put Christianity at odds with any sound natural philosophy. And that may explain the ambivalence and hostility we have seen already, and will see more of below. We will also see more evidence in following sections that these underlying psychological factors were indeed a motivating force within Christianity.

This Christian hostility also seems to have stemmed from a conviction that turning one's attention to any question other than how we ought to serve God was an invitation to atheism, and thence to immorality and damnation. Hence Gregory Thaumaturgus informs us that even his teacher Origen, who was perhaps unique in insisting his students read all the philosophy they could, refused to let his students read anything written by "godless" philosophers, because such men "deny there is either a God or providence" and therefore "our soul, intended for piety, might accidentally be defiled" by merely "hearing their words," and, besides, books written by atheists "were not worth reading" anyway. Hence Origen told his students that nothing written by atheists was "at all worthy of being considered by men who embrace religion."[1626] As we have seen, and will continue to see, this seems to have been the common if not a universal sentiment among Christians of the time.

So at worst, the study of nature could be viewed as a temptation to

1624. Fox 1987: 330–33. The clearest examples of this sentiment are voiced in Justin Martyr, *Dialogue of Justin and Trypho the Jew* 2–8; Tatian, *Address to the Greeks* 29–30; and Theophilus, *To Autolycus* 1.14.

1625. Ptolemy, *Tetrabiblos* 1.1.2–3. For Galen see Carrier 2016: 161–63 (where I also show that Lactantius confirms the same observation).

1626. Gregory Thaumaturgus, *Panegyric Oration on Origen* 13. On the identity of this Gregory see note in chapter 4.9. For discussion of Origen's attitude and its consequences see section 5.9 (and related discussions in chapter 4.8 here, and Carrier 2016: 150–54).

atheism, and at the very best, with few exceptions, the study of nature was viewed as a waste of time, seeking after what we could never really understand, and for no worthy purpose, an occupation which only distracted us from our more proper and fitting pursuit: the contemplation of God and the moral life He commands us to lead. Apart from, at best, some allowance for non-theoretical improvements in the purely practical arts, and the use of past knowledge and philosophical reasoning as a stepping stone to a more secure faith reliant upon scripture, science had no place in Christian society. It was thought better to eliminate fear by eliminating doubt, complexity, and ambiguity; and faith in the Gospel promised to achieve that in a way science and philosophy never could. Such was the predominant Christian view in the first three centuries.

5.6 Evidence in the New Testament

Elsewhere in early Christian literature, though the scientist or natural philosopher is not explicitly discussed, we can nevertheless infer varying attitudes of hostility toward their methodological aims and assumptions in how Christians speak of knowledge and academics in general. Which further supports the conclusions proposed above, and thus explains why even elite Church fathers were so hostile to the scientific enterprise.[1627] We will address first the New Testament, then a sample of other Christian writers of the first three centuries. Though the Bible nowhere mentions the 'natural philosopher' as a class, or names any particular science, it has a lot to say about the proper aims and means of acquiring knowledge, and what it says is not very friendly to ancient science or philosophy.

I. 'Epistemic Values' in the New Testament

To get a complete picture, we must first look at the implied epistemology of the Christian Bible, and the methods and interests expressed there regarding the pursuit of knowledge and understanding. None of the authors of the New Testament seem very impressed by rational, historical, scientific,

1627. This connection with scripture is most explicit, for example, in Clement of Alexandria, *Stromata* 1.11, where many of the following passages are quoted or paraphrased.

or dialectical methods, so these get no significant mention there. Instead, what we find used and advocated are generally unempirical paths to "truth." For example, Paul always 'proves' what he says is true by appealing to the efficacy of apostolic miracle-working, to scripture, to private revelation, and to his upstanding behavior or 'suffering' as proof of his sincerity.[1628] Science, formal logic, empirical facts—these never seem relevant to him. Instead, Paul's epistemology is consistently mystical and supernatural. And the fact that he *only* argues in this mode entails his fellow Christians probably shared these epistemic values.

Miracles, for example, were seen as God's way of confirming the Gospel's truth. Paul's teachings are to be heeded, for "the signs of the apostle were accomplished among you in all endurance, by signs and marvels and powers," and "our gospel did not come to you in word alone but also in power and in the holy spirit and ample certainty, because you know as what sort of men we came to you."[1629] This is the same evidential standard used to persuade people of any other magic, oracle, god, or superstition, and deployed by many a temple, guru, and huckster throughout history.

In fact, the author of Hebrews says, we must heed what we are told by the apostles, for if what even angels have spoken has come true, then:

> How shall we escape if we neglect so great a salvation, which, having first been received when spoken through the Lord, was confirmed to us by those who heard, God having added to their testimony with signs and marvels and many different powers and distributions of the holy spirit, according to his will.[1630]

What exactly these "signs" and "marvels" and "powers" were is not exactly explained, but apart from the fanciful tales of Acts and the Gospels, the evidence from the epistles is substantially mundane.[1631] For example,

---

1628. For examples of the latter see 2 Corinthians 11:23–33, 12:7–10, and for a good discussion see Barnett 1997: 534–77. For context and background on Paul the Apostle see *OCD* 1095–96 (s.v. "Paul, St.") and *ODCC* 1234–38 (id.).

1629. 2 Corinthians 12:12 and 1 Thessalonians 1:5.

1630. Hebrews 2:3–4.

1631. Acts calls the miracles performed by Jesus "powers and marvels and signs" (Acts 2:22) and also calls the miracles performed by the Apostles "marvels and signs" (Acts 2:43, 4:30, 5:12, 14:3, 15:12) and "powers" (Acts 8:13, 19:11), but when

Paul discusses the various "distributions" of the holy spirit alluded to above, in 1 Corinthians 12:

> For to one of us God grants the word of wisdom through the Spirit; and to another, the word of knowledge according to the same Spirit; to another, faith in the same Spirit; and to another, gifts of healing in the one Spirit; and to another, workings of power; and to another, prophecy; and to another, the distinguishing of spirits; to another, different kinds of utterings; and to another, the interpretation of utterings.[1632]

What is evident here, and we shall see confirmed below, is that "wisdom" and "knowledge" and "belief" came to the Christian through the inspiration of the holy spirit, not from research or study or learning or observing the natural or human world. Indeed, the very fact that Christians exhibited such unlearned (or "innate") wisdom, knowledge, and strength of conviction was itself seen as evidence of the gospel's truth, just as were miraculous powers of psychosomatic healing, other undefined "works of power," and the ability to "prophecy," babble, and 'interpret' prophecy and babbling.

Here, "works of power" may mean the power to exorcise demons, by invoking the "power" of Christ's name against them.[1633] However, the Gospel According to Mark says the "signs" of an apostle would be healing, speaking in tongues, casting out demons, *and* immunity to poisons, which may mean "works of power" included not just exorcism but public shows of invulnerability to supposedly poisoned drinks and snakes.[1634] As for the

it's the apostles, these are generally quite mundane, e.g. inspired babbling (e.g. Acts 2:1–18, 2:33, 10:44–48), escapes from prisons (e.g. Acts 12:6–11; 16:23–30), and demonic exorcism and psychosomatic healing (e.g. Acts 3:1–16, 4:9–17, 8:6–7, 14:8–11), i.e. always curing only 'demonic possession', blindness, and paralysis, which have known psychological causes and cures (e.g. Shorter 1992; note even the alleged psychosomatic *causing* of blindness in 9:8–18 and 13:11–12), never any demonstrably real biological ailment like wounds, tumors, diarrhea or vomiting, or lost limbs or organs.

1632. 1 Corinthians 12:8–10 (for useful discussion see Fee 1987: 590–99).

1633. This might be implied in Ephesians 1:19–22 and 3:7–12, and is suggested by Mark 16:17 (cf. Acts 19:13–17; Luke 10:17) and Justin Martyr, *Dialogue of Justin and Trypho the Jew* 30. The same ambiguous phrase appears again in Galatians 3:5.

1634. Mark 16:17–18 (see extensive discussion in Kelhoffer 2000); e.g. Acts 28:3–6 depicts Paul's expected immunity to snake venom.

rest, Paul explains in detail that speaking in "tongues" was also a "sign" and consisted of unintelligible babbling that required a spiritual interpreter, while prophesy was an intelligible discourse inspired by the holy spirit, conveying messages from God.[1635] The power to "distinguish spirits" probably indicates a talent for telling the difference between the influence of good and evil spirits on inspired messages or events.[1636] In every case, useful knowledge pertains to the spiritual world and comes *from* the spiritual world. The natural world does not even make the list, as either an object of interest or a source of useful knowledge. And instead of a rational methodology at all, on this dimension their epistemology was grounded on the singular fallacy of 'if I can perform a faith healing act, everything else I say must be true'.

Next on the list of valued sources of knowledge for the Biblical Christian is scripture, which meant the 'divinely inspired' writings of the Jews, often in fact through finding by divine inspiration a hidden mystical "bible code" in a special arrangement and interpretation of passages.[1637] The Christian gospel itself was derived from scripture: "For I handed down to you, among the first things, what was also handed down to me: that *according to the scriptures* Christ died for our sins, and that he was buried and that *according to the scriptures* he was raised on the third day."[1638] Acts even claims that finding these facts in scripture was alone sufficient to convert people, and

---

1635. I Corinthians 14. References to "speaking in tongues" and to "prophesying" as an ongoing phenomenon in the Church (and as a proof of the Gospel's truth) can be found throughout the New Testament: Mark 16:17; Acts 2:3–4, 2:11, 10:46, 19:6; I Corinthians 12:28, 12:30, 13:1–2, 13:8, 14:1–28; and: Romans 12:6; I Thessalonians 5:20; Acts 21:9; I Corinthians 11:4–5, 13:9. The phenomenon is discussed in Fee 1987: 652–713. Prophesy was also associated with dreams, visions, and "revelations" (Acts 2:17–18; I Corinthians 14:26–33; 2 Corinthians 12:1; Revelation 10:11). The power of prophecy was thought to be conferred on converts by elders through the "laying on of hands" (I Timothy 4:14; 2 Timothy 1:6; Acts 8:18, 19:6). For the entire range of schizotypal behaviors revered within the early churches and their known scientific background see Carrier 2014a: 124–37.

1636. As suggested in I Thessalonians 5:21; I John 4:1–5:13; 2 Peter 1:19–2:22; Matthew 7:15–20, 24:11–12, 24:23–29.

1637. A method of extracting hidden claims in the scriptures invented by the Jews and called *pesher*: Carrier 2014a: 87–88.

1638. I Corinthians 15:3–4 (emphasis added). See also Carrier 2014a: 137–43.

those who were persuaded to convert by studying scripture were especially praised as "more noble."[1639]

Paul tells us they were finding "hidden" meanings in these sacred texts, which predicted, confirmed, and explained the Christian gospel:

> To the one who can support you according to my gospel and the proclamation of Jesus Christ, according to the revelation of the mystery which has been kept in silence through endless times, but is now made clear, and so is *made known through the prophetic scriptures*, according to the commandment of the eternal God, to all the nations, for submitting to the faith.[1640]

This reflects a general esteem for a scriptural epistemology, where the truth is to be found in scripture, with the assistance of divine "revelation" for understanding what a passage is actually saying. As Paul writes, "For whatever was written beforehand was written for our education, so that through patience and through the exhortation of the scriptures we may have hope," and Paul exhorts his followers "that among us you might learn not to go beyond what has been written."[1641]

As the author of 2 Peter likewise explains:

> We have the more reliable prophetic word, which you do well to cling to, as to a lamp shining in a squalid place, until the day shines through and the Morning Star rises in your hearts, knowing this first: that no prophecy of scripture comes from an individual's explanation, for prophecy is never brought by the wishes of man. Rather, men who are carried by the holy spirit speak from God.[1642]

This reliance on scripture is extended by the author of 2 Timothy even to education:

1639. Acts 8:27–39 and 17:11–12 (see commentary in Bruce 1988: 173–79, 326–28).

1640. Romans 16:25–26 (emphasis added); the same point is made in Ephesians 3:3–11 (a forgery that nevertheless represents what the winning faction of Christianity wanted Paul to have said); e.g. Acts depicts Paul "confounding" his opponents with his interpretations of scripture (Acts 18:28).

1641. Romans 15:4 and 1 Corinthians 4:6.

1642. 2 Peter 1:19–21.

> Remain in the things which you learned and were assured of, knowing from whom you learned them, and that from the time you were born you have known the sacred writings which can make you wise to salvation through trust in Jesus Christ. Every divinely-inspired scripture is useful for education, for refutation, for correction, for schooling in righteousness.[1643]

Thus he places the scriptures front and center as *the* school text, the one true source of wisdom and knowledge and the ultimate source for refuting and correcting the claims of others. This easily became antithetical to the aims and methods of natural science. It's certainly no method any real scientist can heed.

Scripture had to be interpreted according to ongoing "revelations" from God, too—rather than, for example, philosophical analysis or scientific evidence. As noted above, Paul claimed that wisdom and knowledge come "through the spirit" and the secret gospel hidden in scripture was "revealed" by God.[1644] Likewise, the gospel communicated to him had told him what happened "according to scripture," by which he did not mean a story or report handed down to him from witnesses or even other men, but a mystery directly delivered to him by a revelation from God.[1645] "For I make known to you," Paul explains, "that the Gospel that was preached by me is not according to a man, for neither did I receive it from a man nor was I taught it, except through a revelation of Jesus Christ," when God "revealed his son in me."[1646] Elsewhere we find the same idea that the "mystery" of the Christian gospel was "made known" to him "according to a revelation."[1647] In fact, it appears that Paul often received revelations from God, as he sometimes distinguishes this special knowledge from his own opinions.[1648]

1643. 2 Timothy 3:14–16.

1644. 1 Corinthians 12:8 and Romans 16:25–26.

1645. 1 Corinthians 15:3–4 and following note.

1646. Galatians 1:11–12 and 1:15–16.

1647. Ephesians 3:3 (written by someone passing themselves off as Paul).

1648. For example, Paul refers to "the outstanding quality of the revelations" God granted him (2 Corinthians 12:7), including an actual mystical conversation he had with Jesus (2 Corinthians 12:8–9); he made journeys "according to revelation" (Gal. 2:2; cf. Acts 13:2); hints at a specific example (2 Corinthians 12:1–10, cf.

As later Christians imagined Paul, he did not consider himself uniquely privileged in this, for though the secret meaning of scripture "was not made known to the sons of men in previous generations" it was "now revealed to his holy apostles and prophets in the spirit," thus connecting the role of prophecy in the church to revelation as a source of knowledge.[1649] Indeed Paul is imagined as having prayed that God would give his fellow Christians "a spirit of wisdom and revelation" and thus "illuminate the eyes of their heart," thus connecting the idea of revelations from God to his notion of spirit-granted knowledge and wisdom.[1650]

This reliance on "revelations" from God superseded all human learning, which necessarily included science and natural philosophy. This is a direct reversal of the methods of the scientist: to trust in tradition (scripture) and intuition (revelation), instead of independent observation and empirical evidence. Paul's epistemology bypasses every concern of a scientist or natural philosopher, skipping the need to discover and understand the causes of things, by directly accessing useful knowledge through the word of God. This is no aberration. Centuries later even Eusebius was still arguing both points—the supreme authority of scripture and the superior value of acquiring knowledge of the natural world from divine revelation rather than human reason.[1651]

As Paul himself explained:

> We speak wisdom among the perfected, but not wisdom of this age, nor of the rulers of this age, who are impotent. Rather, we speak God's wisdom in a mystery, wisdom that has been kept hidden, and which God preordained before the ages, for our glory, and which none of the rulers of this age has known. For if they had known, they would not have crucified the Lord of glory. But as it is written, "What eye has not seen and ear has not heard, and what has not entered the heart of man, that's what God made ready for those who love him." For God revealed it to us through the spirit. For the spirit searches out everything, even the deep things of God. For who among men knows a man's affairs except the man's spirit that is in him?

Barnett 1997: 556–77); and occasionally distinguished between his own opinions and instructions from God (1 Cor. 7:12, 7:25 vs. 14:37).

1649. Ephesians 3:5.

1650. Ephesians 1:16–18.

1651. Eusebius, *Preparation for the Gospel* 11.7 (see also section 5.4 above).

> In the same way, no one has known God's affairs except the spirit of God. And we did not receive the spirit of the world but the spirit that is from God, so we would know the things granted us by God, which we talk about not in words taught by human wisdom but those taught by the spirit, interpreting spiritual things with spiritual words.[1652]

Paul thus denigrates the human wisdom of "this age" and of "the world" and of "the rulers of this age," which would have included natural philosophy (certainly if he were referring to the Greco-Roman elite, but even if not), and he elevates in its place a "spiritual" wisdom known only from God, through "revelation," not inquiry. Examples of this method of acquiring knowledge can be found throughout the book of Acts.[1653]

Therefore, instead of defending and advocating the principles of analysis and inquiry trumpeted by the natural philosophers, Paul argued that "truth" had to be grasped spiritually, on faith—not learned from empirical investigation, or from human teachers, but directly from God, through revelation and scripture. For only Christ in heaven can be one's teacher and tutor, and revealed knowledge trumps all other knowledge.[1654] For "the spiritual man interrogates everything, but is himself interrogated by no one, for who knows the mind of the Lord that he may instruct him? Yet we have the mind of Christ" himself.[1655]

The sociological analyses of Rodney Stark, Bruce Malina, and others, indicates that in antiquity revealed knowledge tended to appeal to a wider segment of the population than rational or empirical approaches did.[1656]

1652. I Corinthians 2:6–13. Discussed in Fee 1987: 97–120.

1653. Acts 7:55–56, 10:1–7, 11:5–14, 12:6–11, 16:9–10, 22:17–21; see also 2 Corinthians 12:1–5 and, of course, the entire book of Revelation. Outside the New Testament, *Diognetus* 11–12 also articulates the nature and superiority of this 'holy spirit' epistemology (and cf., e.g., Lactantius, *Divine Institutes* 3.6; Tatian, *Address to the Greeks* 27; etc.).

1654. Matthew says Christ alone is our *didaskalos* (teacher, master, chorus trainer) and *kathêgêtês* (guide, tutor, professor) according to Matthew 23:8–10.

1655. I Corinthians 2:15–16.

1656. Stark 2001: 9–29; Malina 2001: 1–13, 129–31; Malina & Pilch 2000: 1–24, 41–44; Malina & Neyrey 1996: 212–18. See also: Pilch 2002; Segal 2004; Fales 1996a, 1996b, 1999. Further discussion relating to the early Christian context: Malina & Rohrbaugh 2003: 140, 369, 398–99 and 1998: 282–85. Further discussion relating

Thus, insofar as Christianity sought or found a broader appeal than rational or empirical philosophies, a Christian elevation of mysticism over empiricism should be expected. So when we catch glimpses of the actual methods that Christians respected, we find mysticism trumping empiricism every time. Consider Paul's moving appeal:

> When I came to you, brethren, I did not come with superiority of speech or of wisdom when I proclaimed to you the testimony of God. ... My message and my preaching were not in persuasive words of wisdom, but in a demonstration of the spirit and of power, that your faith should not rest on the wisdom of men, but on the power of god.[1657]

Thus, Paul openly disavows the established rhetorical principles of evidence and argument, and says instead that the miracles of the Holy Spirit are all he came with, and all that God wants Christians to trust as evidence. Miracles and revelations and an apostle's word are always sufficient in the discourse of the New Testament. No empirical inquiry or dialectical debate or even formal education was necessary. After all, "the Lord will give you understanding in everything."[1658]

This is a point we find repeated in the Gospels. The advice of the Roman educator Quintilian, of acquiring a strong background in academic knowledge (including natural science) and learning the forensic skills of a critical thinker, was essentially a waste of time. For all useful information will come to you through divine inspiration.[1659] That is why "we walk by faith and not by sight," not looking at the visible things, which are temporary, but at the invisible things, which are eternal.[1660] Hence the Christian is told

---

to the pagan context: Dodds 1951: 64–101, 102–34; Fox 1987: 102–67; P. Green 1990: 408–13, 594–95; I.M. Lewis 2003.

1657. I Corinthians 2:1–5.

1658. 2 Timothy 2:7.

1659. Mark 13:11; Luke 12:11–12, 21:13–15. On how this translated into the Christian view of education in general see chapter nine of Carrier 2016. On Quintilian's opposite view see chapters five and six therein. For the consequences of this, as played out between Galen and Lactantius, see chapter ten therein.

1660. 2 Corinthians 5:7 and 4:18, with relevant commentary in Barnett 1997: 245–77 (the context here is evidence that the brethren are aging and dying, which we are to discount in favor of the 'unseen' evidence that they will live again: see

"you have no need for anyone to teach you" because the anointing of Christ "teaches you about all things and he is true and not false, and just as he has taught you, you abide in him."[1661] Thus, Christians are basically exhorted to ignore the evidence of their senses, and trust instead in the invisible certainties of their heart, since that is where God speaks to them. They are then asked to rely on this inner faith to trust Scripture and the gospel, and thus asked to have faith in what God or his Scripture says, regardless of the evidence of their senses.[1662]

Paul further reveals the epistemic values of his fellow Christians when he declares "what use shall I be to you unless I speak to you either in revelation, or in *gnosis* [spiritual knowledge], or in prophecy, or in *didachê* [received doctrine]?"[1663] Notably, "evidence" and "logic" do not make his list. If a claim does not come by revelation, prophecy, inspiration (*gnôsis*), or tradition (*didachê*), it is "of no use" and not even worth mentioning. As we have seen, wisdom, knowledge, and faith all come from the Holy Spirit, not from research, nor from logical debate, nor from making inquiries or investigations or observations. In fact, prophecy and revelation are to be "tested" not by scientific, historical, or empirical investigation, but by whether the inspired message is moral and in agreement with prior dogma.[1664] So when Paul institutes a hierarchy of authority, his list goes: "first apostles, secondly prophets, thirdly instructors, then powers, then gifts of healing, then the ability to help, then to administer, then varieties of speaking in tongues."[1665] There is no place here for science or philosophy in

---

Carrier 2005b: 125, 139–41).

1661. 1 John 2:27.

1662. This is the entire argument of Hebrews 11, that "faith" (*pistis*), faith in Scripture and the gospel, is their "evidence" (*hupostasis, elegchos*), which they will be rewarded for (hence the context: Hebrews 10:19–39), cf. Bruce 1990: 276–331; and Carrier 2009: 236–40.

1663. 1 Corinthians 14:6.

1664. This is implied in Galatians 1:6–17; 1 Thessalonians 5:19–22; 1 Timothy 6:3–4, 6:20–21; 1 John 4:1–5:13; 2 Peter 1:19–2:22; and in Matthew 7:15–20 and 24:11–12, 24:23–29; James 3:13–4:17; 2 Thessalonians 2:1–12; 1 Timothy 4:1–7; 1 John 4:4–6. In contrast, Paul essentially condemns philosophical 'reasoning' (see section 5.6.II below).

1665. 1 Corinthians 12:28. See also Romans 12:6–8 and Ephesians 4:11.

the classical sense, or any or critical thinking, and certainly no place for a scientist to claim any respect or authority.

It is clear from all of this that the Christians did not share the same epistemic values or interests that inspired the natural philosophers, who would sooner seek to examine the natural causes of phenomena like healing or prophecy than to rely on them as evidence of a god's power or his wizard's honesty. Instead, all the evidence that matters to Christians is evidence of God's will, not evidence of the causes and operations of natural phenomena. Scientists are keen to understand nature, but the Christians are keen to understand the nature of God's plan.

A clear expression of this attitude appears, for example, in Tertullian's *Treatise on the Soul*, as we saw above, where Tertullian is alarmed by uncertainty, doubt, questions, complexity, which are all entailed by any causal and empirical inquiry, so he takes the solution to be the dogmatic certainty and simplicity of Christian ways of knowing: direct from God, with few questions and little research. This typified Christian thinking for centuries.[1666]

II. Against the 'Wisdom of the Wise'

The Bible's implied demotion of or disinterest in natural philosophy also appears to be an inevitable outcome of the Christian's apocalyptic worldview.[1667] For there would be little point in studying nature if the natural world was soon to be burned away:

> The day of the Lord will come like a thief, in which the heavens will pass away with a rushing sound, and the elements shall be dissolved, burned with an intense heat, and the earth and the works that are in it shall be found undone. Since all these things are thus to be undone, what sort of people must you be in holy behaviors and pieties, expecting and urging on the coming of the day of the Lord, through which the heavens will be dissolved by burning and the elements will be melted down by intense heat? But according to his promise, we expect a new heaven and a new earth, in which righteousness dwells.[1668]

1666. For further analysis of early Christian epistemology see Carrier 2009, chs. 7, 13, and 17.

1667. Carrier 2010: 408. For example: Hebrews 1:10–12; 1 Corinthians 1:28, 6:13, 7:31; and passages cited below.

1668. 2 Peter 3:10–13.

The message here is: do not waste time on the things of this world, devote all your time to moral piety, and urge on the destruction of the universe, so a new world can be made in its place, which will be morally superior to this one.

The author of 1 John carries the same logic further:

> Do not love the world or the things that are in the word. If anyone loves the world, the love of the Father is not in him. For all that is in the world, the desire of the flesh and the desire of the eyes and the false pretension of life, is not from the Father but is from the world. The world is passing away, along with its desire. But he who does the will of God remains to eternity.[1669]

And Paul said pretty much the same thing.[1670] Though this does not entail *hostility* toward the world, just an indifference to it, people who thought like this would still have had very little incentive to become natural philosophers or even to support their activity—indeed, they might be suspicious of anyone who had such an interest, as possibly "loving the world" or the "things" or "works" of the world, embracing an ungodly passion.

The author of James is even more adamant. "You adulteresses," he declares, "do you not know that friendship with the world is hatred of God? So anyone who decides to be a friend of the world makes himself an enemy of God."[1671] Though these condemnations were not aimed at scientists or natural philosophy specifically, it was not a far leap to conclude that any natural philosopher, by passionately studying and discussing the world, was indeed a friend of the world, and therefore an enemy of God. So we should not be surprised when we find (as we have) that this is the very attitude among later Christian intellectuals. The New Testament is full of remarks that exhort the Christian to despise the world and focus all his attention on higher things. Nature was not something to study, but something to escape from. As Paul says, before baptism we were children, and "we were imprisoned by the elements of the world," but in baptism the world dies to us and we die to the world, and in this baptismal "death" we escape from the

---

1669. I John 2:15–17.

1670. I Corinthians 7:29–31.

1671. James 4:4.

"elements of the world" and hence no longer live "in the world."[1672] Thus, while the scientist would pursue and study the 'elements of the world', the Christian seeks to escape them, and considers themselves already dead to them.

And so the Christian is admonished, "contemplate the things above, not the things on earth."[1673] Though the target here is immoral lust, there is no indication of an exception being made for the "lust" to study and understand the causal mechanisms behind the creation—such rationalizations would not be seen in Christian literature for many centuries. Instead, from the reasoning we find in the New Testament, anything other than the "supremacy of the knowledge" of Christ could readily be dismissed as "trash."[1674] Hence Paul exhorts his fellow Christians to "keep watch, lest someone come along and carry you off as a captive, through philosophy and empty deceit, according to the tradition of men, according to the elements of the cosmos, and not according to Christ."[1675] In effect this is an attack on philosophy, and natural philosophy in particular, as hostile to faith. The "traditions of men" certainly included the theories and schools of thought that comprised ancient science and philosophy, and to argue from "the elements of the cosmos" was the particular activity of the natural philosopher, which is here denounced as a vain fraud that will lead us away from Christ, since only a 'philosophy' that argues from Christ is valid.[1676]

Instead, "the man of life does not receive the things of the spirit of God, for to him they are foolishness, and he cannot know them, because they are

---

1672. Galatians 4:3, 4:8–9 and 6:14, and Colossians 2:8 and 2:20 (though the latter written by Pseudo-Paul), with relevant discussion in Fung 1988: 306–07 and Dunn 1996: 145–51, who also note that early Christians linked the natural elements (e.g. 2 Peter 3:10–12) with the governance of demons (e.g. Justin Martyr, *Apology* 2.5).

1673. Colossians 3:2.

1674. Philippians 3:8.

1675. Colossians 2:8. For the context of this hostility in Paul and Pseudo-Paul see Judge 1983.

1676. In part because the elements were believed to be operating under the control of demonic forces: Carrier 2014a: 180–93. For some discussion of these and other passages from the epistles hostile to philosophical inquiry see Judge 1983: 11–14.

discerned spiritually."[1677] The "foolishness" here that the man of the world cannot understand is the "story of the cross" which alone saves, but the "wise" are not privy to spiritual knowledge of this truth because:

> It is written, "I will destroy the wisdom of the wise. And I will reject as spurious the discernment of the discerning." Where is the wise man? Where is the scribe? Where is the investigator of this age? Has not God made foolish the wisdom of the world? For since through its wisdom the world did not know God, in God's wisdom He thought it was a good idea to save the believers through the foolishness of His teaching. Because the Jews ask for signs and the Greeks search for wisdom, but we teach Christ crucified—to Jews a stumbling block, to Gentiles foolishness, but to those who are called, both Jews and Gentiles, it is Christ the power of God and the wisdom of God. For the foolishness of God is wiser than men, and the weakness of God is stronger than men. Just look at your calling, brothers, that not many are wise with regard to the flesh, nor many powerful, nor many of noble birth. But God selected the foolish things of the world in order to shame the wise, and selected the weak things of the world in order to shame the strong.[1678]

The anti-elitism of Paul's message is clear: there is an obvious disdain here for the educated, the philosophical thinkers, the investigators of the world. Instead, the Christians see themselves as their opposite. It would be hard for a Christian to defend the merits of the activity of a natural philosopher when such "wisdom of the world" is the very thing God intends to "destroy" and "reject as spurious" and "put to shame."

Paul echoes the same sentiment later in the same letter: "Let no one con himself," Paul warns, "if anyone thinks he is wise among you in this age, let

---

1677. I Corinthians 2:14. A similar sentiment is voiced in I John 4:4–6; likewise, Jesus is made to have said: "I praise you, o father, lord of heaven and earth, that you did hide these things from the wise (*sophoi*) and intelligent (*synetoi*) and revealed them to children (*nêpioi*)" (Matthew 11:25; Luke 10:21).

1678. I Corinthians 1:19–27 (perhaps loosely quoting Isaiah 29:14; cf. Fee 1987: 66–88): *sophos*, "wise man," essentially means philosopher, while *suzêtês*, "co-investigator," means one who debates and discusses questions and theories with others, with the aim of discovering the truth (which often included appeals to evidence and scientific argument), or who conducts research with others (hence also describing scientific collaboration), cf. *LSG* 1670 (s.v. "*suzêteô*," "*suzêtêsis*," "*suzêtês*," "*suzêtêtikos*").

him become a fool, so he may become wise," because "the wisdom of this world is foolishness before God, for it is written, 'he catches the wise in their craftiness' and again 'the Lord knows the arguments of the wise that they are groundless.'"[1679] Hence for those who do not correctly know God's will, "knowledge produces arrogance," with the result that "if anyone thinks he knows something, he does not yet know as he ought to know."[1680] Instead, to become "perfected" (and thus "mature") we must seek "the unity of faith and of the knowledge of the son of God," and by thus "growing up" we will "no longer be children thrown into confusion and carried around by every wind of doctrine, in the dice-play of men, in the craftiness leading to the trickery of error," an allusion to pagan philosophers being children, an idea we already saw had influenced Clement.[1681] Hence Paul was imagined as exhorting his fellow Christians to "walk no longer as the Gentiles also walk, in the folly of their mind, darkened in their understanding."[1682]

A similar attitude is found in the Wisdom of Solomon, an early Jewish text whose thought is very similar to that of early Christianity and substantially influenced it:

> For what man will know the will of God? Or who will infer what the Lord wants? For the calculations of mortals are worthless and our conceptions are dubious. For a perishable body weighs down the soul, and our earthly tent burdens the mind that thinks countless thoughts. We scarcely figure

---

1679. 1 Corinthians 3:18–20 (cf. Fee 1987: 150–56). Paul's second quote comes from the Septuagint text of Psalms 94:11 (though he notably replaces *anthrôpôn*, "men," with the more specific *sophôn*, "wise men," i.e. philosophers), but his first quote conflates several verses into one, drawing on Job 5:12–14, where God "frustrates the plans of the crafty" (*panourgôn*), "catches the wise in their thought" (*phronêsis*), and "thwarts the advice of complex men," who are lost in darkness and grope around in the day as if it were night. Instead, Paul says God "catches the wise in their craftiness" (*panourgia*), thus equating all three types of men, and identifying the *sophos* with men who are lost in the dark (a metaphor we'll see Paul using elsewhere).

1680. 1 Corinthians 8:1–2.

1681. Ephesians 4:13–15. On Clement's use of this metaphor see section 5.1.

1682. Ephesians 4:17–18. The author pretending to be Paul here says "walk" using *peripatein*, which may imply an allusion to the Aristotelians, who were called the Peripatetics because they began by teaching in the Peripatos, the "walking area" of temples (just as the Stoics began teaching in the Stoa, the "Porches").

> out the things on earth, and we discover things close at hand only with labor—but who has traced out the things in the heavens? And who has known your counsel, unless you have given them wisdom and sent your holy spirit from on high? And in such a way the tribes on earth were set straight and men were taught what was acceptable to you, and were saved by wisdom.[1683]

All the same sentiments are here: that human wisdom is useless and a waste of time, that we cannot figure out the natural world, least of all the planetary system, that it is pointless to try and locate and understand the hidden causes of things, that we barely even understand what is right before our eyes, and the only way to learn what is useful and approved by God is to receive such knowledge directly, through the inspiration of his holy spirit. That is the real wisdom, the only wisdom that will save us.

This is an attitude quite antithetical to science—yet it reflects exactly what Paul says in several of his letters. Since scientists explained everything by appealing to natural causes rather than divine will, they were among those who do not recognize God's design in nature. All such people, Paul says, "became foolish in their reasonings and their senseless heart was darkened" such that "claiming to be wise, they became foolish."[1684] By describing such thinkers as having "darkened hearts" deprived of sensation, Paul directly paints natural philosophers and other intellectuals as the exact opposite of Christians, whom Paul prayed would have the "eye of their heart illuminated."[1685]

There the target of attack are those who do not find God among the causes of things. But elsewhere this target includes those who try to arrive at precise definitions, which the New Testament derides as a "wrangling over words" that is "useless," bringing only strife and ruin, or as nothing but "fruitless discussion" by those who "neither understand what they are saying nor grasp the matters about which they make confident assertions," and who eventually become "conceited" and "understand nothing" and "have a

1683. Wisdom of Solomon 9:13–18.

1684. Romans 1:19–25 (cf. Moo 1996: 95–108). Since it was the natural philosophers who catalogued and studied the evidence Paul refers to here (the empirical evidence of God's creation), and yet supported paganism or pantheism or atheism instead of the 'truth', they are clearly among Paul's targets in this passage.

1685. Ephesians 1:18.

morbid interest in controversial questions and disputes about words," from which arises every evil.[1686] Thus even formal logic and the Socratic method get a sound beating here. As Clare Drury describes it, "Clearly, acceptance of sound doctrine means not asking questions or questioning definitions," for, "a clear exposition of accepted doctrine was the only proper method of teaching" and "discussion could only lead to dispute, and so must be avoided." After all, the thinking was, "If the teacher is above [moral] reproach, then opponents have no grounds for raising questions" in the first place.[1687]

In contrast, when the Christian has questions, he is told to ask *God* for the required wisdom, and not only that, but to "ask in faith without any doubting, for the one who doubts is like the surf of the sea, driven and tossed by the wind" and "such a man cannot expect to receive anything from the Lord, since he is a man of two minds, unstable in all his ways."[1688] Ask in faith—without doubting. This is quite the opposite of a passion for debate and inquiry essential to scientific progress. The man who doubts is aimless and unstable and worthy of no help from God. Hence rather than making heroes of those who question and inquire and explore the world, Christian ideology practically made villains of them. Indeed, Christians preached that God had punished Zacharias, by striking him mute, merely for *requesting* evidence.[1689] The epistemic values here are again quite the opposite of those most widely held by the natural philosophers of the era, while the critical minds and skepticism embraced by natural philosophers is attacked as ungodly and unstable. Human "reasoning" must be "cast down" along with everything else that is held up against "the knowledge of God" and in

---

1686. 2 Timothy 2:14, 2:16 and 2:23, and 1 Timothy 1:6–7 and 6:3–4. These passages are discussed in Towner 2006: 104–21, 392–405 (most relevantly), and 514–51.

1687. Drury 2001: 1229.

1688. James 1:5–8 (cf. Adamson 1976: 55–61 and 1989: 267–75). The context is the pursuit of wisdom (*sophia*) through prayer, and though moral wisdom is here (as always) the Christian's principle concern, the argument nevertheless conveys a general belief that *all* doubters are unstable (since the form of the argument states this as a general principle and then applies it to the particular case). Note also how this approach to acquiring moral wisdom contrasts epistemologically with that of the philosophers, who uniformly argued this could only be gained by applying human reason to a valid knowledge of nature (see chapter 4.1 and 4.2).

1689. Luke 1:18–20.

their place we must "bring every thought into captivity to the obedience of Christ."[1690]

III. Conclusion

Though there is no explicit condemnation of the scientist or their activity in the Bible, we have seen that the New Testament consistently advocates an epistemology that is unfavorable or even antithetical to them, expressing completely different interests and methods. In many ways, the early Christian worldview resembled that of the Cynics, rejecting natural philosophy and even dialectics as a waste of time, or even a source of folly, while embracing ethics as the only truly valid branch of study, and assuming wisdom can *only* be demonstrated by moral action.[1691] But unlike the Cynics, the Christian approach to ethics was even more un-empirical: knowledge of moral truth, which meant knowledge of God's will, was not known philosophically, it was not reasoned out from the evidence of the natural world or of human nature, but was learned either directly from God through the divine inspiration of the holy spirit, or through intermediary "authorities" (like the prophets and apostles) who learned it in such a way.

According to recent sociological research on antiquity, this involved what we would today call "altered states of consciousness" (real or pretended) in which "revelations" and "inspired statements" would come to an individual, providing 'direct' information or revealing the 'true' meaning of authorized scriptures, which were assumed to contain the genuine word of God.[1692] Thus, the Christian approach to knowledge altogether bypassed nature, formal logic, and empirical observation, while their proper objective was not to understand the operation of nature but to know the will of God.

1690. 2 Corinthians 10:4–5 (cf. Barnett 1997: 464–66). The war metaphor that Paul employs here echoes the idea of making philosophy the 'handmaiden' (i.e. slave) of faith. The connection is even more explicit in Origen, who references scripture on subjugating captive women taken in war as a metaphor for the proper Christian approach to philosophy (as discussed in Carrier 2016: 152).

1691. For example: James 3:13. This attitude has already been seen in previous sections among leading patristic authors (in sections 5.1 through 5.4).

1692. Carrier 2014a: 124–37. See references to Stark, Malina, etc., in the relevant note in section 5.6.1, where the concept of altered states of consciousness is also discussed (from both the modern and ancient perspectives), as well as the value of merely *claiming* to have had such experiences.

Even in practical affairs, a trust in the miraculous replaced any interest in an accurate understanding of nature. Hence James advises those who are ill to seek the prayers of Christian elders, which would be sure to heal them.[1693] The idea of going to a doctor never comes up, and the value of acquiring, advancing, or even employing medical knowledge is not even mentioned.

5.7 Evidence from Christian Writers

We see all the same epistemic values reflected in the earliest Christian apologetic works of the 2nd century, when the New Testament documents were still being selected and assembled, further supporting our previous conclusions. Justin Martyr's *Apology*, for example, rests almost entirely on the argument 'scripture says so, therefore it's true'.[1694] Most explicitly, he declares "this should now be obvious to you—that whatever we assert, in conformity with what has been taught to us by Christ, and by the prophets who preceded Him, is alone true, and older than all the writers who have existed," and therefore this form of knowledge takes precedence over all others.[1695] In fact, just about every single thing Justin believed about Jesus he apparently could learn from scripture.[1696]

Justin tells us he tried to study every philosophy, but reports with regret either that faith in God was devalued by the philosophical schools, or they demanded money, or they required him to study the sciences (he names astronomy, geometry, and musical theory in particular), which he says he had no time for.[1697] Though Justin singles out the time-consuming need to study the sciences in order to advance in philosophy, his story suggests he also disliked the need to study and debate numerous rival theories, reflecting that same Christian distaste for philosophical disagreement and dispute

1693. James 5:13–18 (cf. Adamson 1976: 196–202). Later Christian responses to scientific medicine are discussed in section 5.7.

1694. For background on Justin and his apologetic see *OCD* 782 (s.v. "Justin Martyr") and *ODCC* 915 (s.v. "Justin Martyr, St.").

1695. Justin Martyr, *Apology* 1.23.

1696. Justin Martyr, *Apology* 1.31.

1697. I discuss Justin's disdain for science education in more detail in Carrier 2016: 148–51.

we observed before. As Cicero said, "one could set out all the systems of the natural philosophers, but it would be a long story," and that fact in itself may have been a major turn-off for Justin and many Christians like him.[1698] However that may be, after sampling several major philosophies, Justin says he ended up a Platonist only because Platonism agreed with his prior assumption of a mystical, unempirical approach to knowledge. And then from there he reasoned his way to Christianity by conversing either with himself or a Christian elder (depending on how you interpret his story). Thus, Justin is telling us he chose Christianity because it was the only philosophy that placed God first, taught its doctrines for free, and did not require any research or advanced study. He adds, as the final blow that converted him, the fact that Christianity was based on the oldest and thus most venerable of prophetic books.[1699]

It seems quite clear, then, that Justin was attracted to Christianity precisely because it abandoned the interests and methods distinctive of natural philosophers. Besides "scripture," the only evidence he says helped persuade him is the present efficacy of Christian miracle-working, the same "signs" Paul listed as demonstrating God's approval of the Christian faith—in Justin's case this meant, first, the ability to cast out demons, and then all the "gifts" of the holy spirit, which Justin lists as "the spirit of understanding, of counsel, of strength, of healing, of foreknowledge, of teaching, and of the fear of God." Justin concludes from this that "we who have received gifts from Christ, who has ascended on high, prove from the words of prophecy that you, 'the wise in yourselves, and the men of understanding in your own eyes', are foolish."[1700] Thus not only does Justin's list complete the epistemic values we identified in the bible (scripture, revelation, miracles, and the superior morality of believers), but he uses this to beat up on all other "wise men" and "men of understanding" (which would include scientists) as obviously "fools," whereas the Christians are 'truly wise'. This is the very same attitude we observed in the Bible, thus confirming the effect of those very ideas on its adherents.

1698. Cicero, *Prior Academics* (= *Lucullus*) 2.36.117.

1699. Justin Martyr, *Dialogue of Justin and Trypho the Jew* 2 (for the story of his conversion: 3–8; on how the venerability of scripture convinced him: 7–8).

1700. Justin Martyr, *Dialogue of Justin and Trypho the Jew* 30 (exorcism) and 39 (quotation and gifts of the spirit).

In the same way, Justin's contemporary, Athenagoras, compared 'prophets' to 'philosophers' and found the former to be the only reliable source of truth, while the latter were "not competent" to discover the truth, because they argue only by conjecture, disagree amongst themselves, and rely on human thought and testimony. In contrast, 'prophets' are inspired by the Holy Spirit of God and therefore have direct access to the unassailable truth, which God reveals to them. Hence "it would be irrational for us to give heed to mere human opinions, and cease to believe in the Spirit from God, who moved the mouths of the prophets like musical instruments."[1701] This attitude would again have jettisoned natural philosophy, and thus all scientific theory and inquiry, along with the rest of philosophy.

Justin's predecessor, Aristides, also took the same path as Justin: he surveyed all the alternatives and found them illogical, then he observed the Christian lifestyle and found it godly, then he read the Gospels and was convinced.[1702] Justin's pupil Tatian tells a similar story: the alternatives offered by pagan religion and philosophy morally disgusted him, while the antiquity and prescience of the scriptures impressed him.[1703] And then "my soul being taught of God, I discerned that the former class of writings," meaning the philosophical books of the Greeks (which would include the works of scientists) "lead to condemnation, but that these," meaning the scriptures, "put an end to the slavery that is in the world" by revealing what others have been "prevented by their error" from understanding.[1704] Scripture, moral standards, inspiration from God—these he values, while he denigrates the methods, aims, and conclusions of philosophers.

Tatian is quite blunt:

1701. Athenagoras, *A Plea for the Christians* 7 (for the observation that scripture was his only trusted source of 'evidence': 9–10). For background see *OCD* 195 (s.v. "Athenagoras") and *ODCC* 121 (s.v. id.).

1702. Aristides, *Apology* (esp. 2 and 16 in the Syriac edition, which represents the earliest known form of the text; the Greek version was heavily redacted in later centuries and thus is not authentic). For background see *ODCC* 101 (s.v. "Aristides").

1703. For background on Tatian see *OCD* 1433–34 (s.v. "Tatian") and *ODCC* 1579 (s.v. id.).

1704. Tatian, *Address to the Greeks* 29 (on scripture as the superior source of knowledge: 31–32; on his studying all the philosophical schools and finding them wanting: 35).

> I could also laugh at those who in the present day adhere to Aristotle's teachings, people who say that sublunary things are not under the care of providence, and so, being nearer the earth than the moon, and below its orbit, they themselves look after what is thus left uncared for.[1705]

In so saying, Tatian directly attacks the very interests and activity of the natural philosophers, who study nature and seek out the natural (rather than divine) causes of things. "While staring all agape at the sky," Tatian scorns, "you stumble into pitfalls," echoing Tertullian's use of the legend of Thales falling down a well as a symbol of his damnation. Because none of the claims of philosophers agree, "they are all worth nothing" and because philosophers are no better behaved than anyone else, "we have abandoned you, and no longer concern ourselves with your tenets, but follow the word of God." So much for natural philosophy.[1706] To Christians like Tatian, the activity of scientists was worthless and of no concern. Only the "Word of God" deserved our attention.

Tatian is most hostile to scientific medicine, one of the most popular of sciences, condemning the whole of pharmacology, for example, as demonically evil.[1707] Though he may have been more permissive of drugless medical care, he offers no positive support even for that, much less for medical research or its required values. Instead, Tatian taunts the medical patient for lacking faith, rhetorically asking, "Why is he who trusts in the system of matter not willing to trust in God? For what reason do you not approach the more powerful Lord, but rather seek to cure yourself?" Though the context is that of taking palliative or curative drugs, his argument could only be valid if *all* non-divine treatments entailed the same conclusion. In other words, for Tatian, since God will heal you if you have faith (if it is His will that you be healed at all), trusting in doctors is tantamount to doubting

---

1705. Tatian, *Address to the Greeks* 2 (sections 1–3 and 25–26 also argue at length that philosophers have never produced anything great or noble but have all been immoral and vainglorious and have only contradicted each other, while ignoring God).

1706. Tatian, *Address to the Greeks* 26 (compare with Tertullian, *To the Nations* 2.4, discussed in section 5.2 above).

1707. For his entire assault on medicine see Tatian, *Address to the Greeks* 18. For a complete analysis see Amundsen 1995 and 1996: 158–74.

God's power or even His will. Though this was not an uncommon view among early Christians, many ignored such condemnations, and eventually most came to accept an ossified version of scientific medicine. But many Christians like Tatian offered little support for the Galenic ideal that the study of natural philosophy, and a constant empirical pursuit of scientific discovery and improvement, were essential to the art of medicine.[1708]

The same attitudes can be found in Hippolytus almost a century later, who argued that natural philosophers had "fallen away from God" because they occupied themselves with what things are made of and how they work, and did not recognize God as their Creator, so for him natural philosophy had become just another source of Christian heresy.[1709] Hippolytus repeats the commonplace objection that natural philosophy could not be true because there was no agreement among natural philosophers, and in the end he exhorts his fellow Christians, "Do not pay attention to the sophisms of their crafty arguments," but embrace instead the "sacred simplicity of the humble truth" of the Gospel, for that alone will save you from Hell.[1710] Though not as hostile to natural philosophers and their activity as some, Hippolytus clearly held little respect for either. In a similar fashion, according to Eusebius (who appears to endorse the argument), Paul of Samosata attacked a sect of heretical Christians (in result of which no other trace of them now survives) because they greatly admired and devoted themselves to the study of the scientists Euclid, Aristotle, Theophrastus, and Galen—he even blamed their

1708. For this Galenic ideal see chapters 2.7 and 3.9.II. On the various Christian responses to scientific medicine, in contrast to the reception of faith healing (or the mere embrace of suffering as inevitable or morally superior to finding a cure) see: Nutton 1984a, 1985: 45–53, and 2013: 293–98 (for pagan and Jewish context: 280–93); Ferngren & Amundsen 1996; Avalos 1999 (esp.: 57–58); and Amundsen 1982, 1995, and 1996 (though Amundsen fails to distinguish an acceptance of 'traditional' or 'textbook' medical treatment from the actual scientific values required to understand, discover, and improve scientific medical care, of which there is no evidence in the early Christian period). On similarly mixed Jewish attitudes as precursors to later Christian views see Amundsen 1982: 342–43 and Newmyer 1996. But the Jewish adage that "the best of physicians are destined for hell" (discussed in Rosner 1994) was not directed at medical science *as such* but the scruples of doctors and healers (of whatever training).

1709. Hippolytus, *Refutation of All Heresies* 1 (cf. 1.26.3). For background see *OCD* 687–88 (s.v. "Hippolytus (2)") and *ODCC* 773–74 (s.v. "Hippolytus, St.").

1710. Hippolytus, *Refutation of All Heresies* 10.8 and 10.34.2.

heresy on this very devotion.[1711] Thus, again, any evident respect for natural philosophy was derided as a characteristic of *heretics*, not true believers, and consequently purged.

5.8 Assessment of Christian Hostility

Everything we have surveyed so far corroborates the same overall picture, which we already saw evidence of in earlier chapters. When Celsus, Christianity's earliest known critic, attempted to investigate the claims and doctrines of Christians, he kept running into a wall: Christians would simply exclaim "do not question, just believe!" Apparently, many expected converts to simply trust in Jesus, without evidence or demonstration—and in his rebuttal to Celsus, the Christian scholar Origen does not deny it. To the contrary, Origen defends what Celsus observed, confessing that "we admit that we teach those men to believe without reasons."

Origen does say Christians believe in inquiry into the meaning of their prophetical writings, the parables of the Gospels, and "other things narrated or enacted with a symbolic meaning," but that's it. Indeed, not only is this "study of scripture" the only inquiry ordinary Christians should engage in, Origen says most people work such long hours just to get by that they do not even have the time for *that*, and therefore the Christian exhortation to "simply believe" becomes good policy. "Is it not better for them," Origen insists, "to believe without a reason, and then become reformed and improved" rather than "not to have allowed themselves to be converted on the strength of mere faith, but to have waited until they could devote themselves to a thorough examination of the reasons?" Origen says it is indeed better to 'just believe', because most people could never complete such an examination in their short lifetime, and so would remain wicked and die unsaved. Therefore, it is better that they simply have faith, and not waste time checking the facts.[1712] Similarly, in responding to the

1711. Eusebius, *History of the Church* 5.28 (Carrier 2016: 150). See discussion in Walzer 1949: 75–86 and for background see *ODCC* 1242 (s.v. "Paul of Samosata").

1712. All of this from: Origen, *Against Celsus* 1.9–10. Galen made the same observation as Celsus regarding Christian education and epistemology in *On the Different Pulses* 2.4 and 3.3 (= Kühn 8.579 and 8.657) and in several lost works (see

Celsean charge that Christians all say "the wisdom in the world is evil but foolishness is good," Origen also does not deny that either. He only argues that "wisdom in the world" means only philosophy, while true wisdom is the revealed knowledge of God, which philosophers only *think* is foolish.[1713]

This is coming from the only Christian leader we have found in the first three centuries who definitely included any natural philosophy in the education of Christian students.[1714] So if even Origen approved of this upside-down epistemology, where faith and revelation must precede, even substitute for, empirical knowledge, how much more must other Christian intellectuals have agreed, who did not have as much patience or respect for the study of natural philosophy? Hence the most hostile Lactantius makes the very same argument, explaining that to study philosophy requires an extensive education in reading, writing, mathematics, musical theory, and astronomy, which most people (especially women, slaves, and the poor) have no time for.[1715] But wisdom should belong to everyone, educated and ignorant alike, therefore "divine instruction" can be the only real wisdom, since it is simple and easily taught and requires no literacy or scientific knowledge, and it alone produces genuine moral reform.[1716] Philosophers, meanwhile, by studying nature and attributing everything to nature, stumble into ignorance and immorality and are therefore damned.[1717]

In just the same way, Clement of Alexandria produced a lengthy defense of faith as an infallible criterion and a fully justified starting point requiring no demonstration, and insists that this infallible faith includes an absolute trust that the Scriptures are the inerrant voice of God, therefore nothing is true that does not agree with scripture, and everything scripture says is

---

fragments and discussion in Walzer 1949: 10–74, updated in Gero 1990). A similar observation may be reflected in a comment by Marcus Aurelius, *Meditations* 11.3 (compare Walzer 1949: 15–16, 57–74).

1713. Origen, *Against Celsus* 1.9 and 1.13.

1714. Discussed here in chapter 4.9 and in Carrier 2016: 149–55.

1715. As quoted and discussed in Carrier 2016: 160–63.

1716. Lactantius, *Divine Institutes* 3.26 (the Christian gospel is also superior to philosophy because both the certainty of faith *and* threats of heaven and hell are necessary to belief: 3.27).

1717. Lactantius, *Divine Institutes* 3.27–28.

true.[1718] And though Clement does at least believe the study of math and astronomy raises our mind toward God, even these must be subordinated to faith and scripture. When he observes that most Christians avoid studying any philosophy at all, his solution is to tell them to embrace what agrees with, and reject everything that contradicts, scripture and faith.[1719] He gives the same answer to his fellow Christians who ask, "What's the use in knowing the causes of how the sun moves, for instance, or the other celestial bodies, or in reviewing geometric theorems, or dialectics, and all the other sciences?" since they are not only useless, but born of human wisdom and thus always uncertain, while God and faith give us all the wisdom we really need.[1720] Though Clement answers by arguing for the utility of these studies in grasping and defending the faith, the fact that he even has to make this argument indicates that many Christians regarded science as vain and useless, while the fact that Clement's only defense is that some science is of use in interpreting scripture and defending orthodoxy indicates that even he placed faith firmly ahead of empirical discovery in both moral *and* epistemic value.

The Christian's epistemic values in our period of interest were therefore the reverse of those of natural philosophers and of the dominant pagan elite. It was most generally the case that the Christian did not have much interest in the study of nature, and even regarded such interest with suspicion. The Christian also did not believe in the empirical or rational methods that prevailed among natural philosophers, but instead belittled those methods as either useless or all but certain to lead to strife or folly, or even damnation. Instead, the Christian embraced an epistemology of the Holy Spirit, wherein all useful knowledge comes from the Word of God, whether through scripture or inspiration or private revelation, and is confirmed by miracles and shows of confidence and sincerity, rather than by the evidence of the natural world. And as we have seen here, and throughout sections 5.6 and 5.7 and more, the evidence overall confirms the underlying psychology proposed in section 5.5.[1721]

1718. Clement of Alexandria, *Stromata* 2.4, 5.13, 7.16 (cf. 2.2, 5.1, 6.9.78–79, 8.3; and *Prophetic Eclogues* 4.2); and *Exhortation to the Greeks* (= *Protrepticus*) 8.

1719. Clement of Alexandria, *Stromata* 6.10 (cf. 1.9). See Carrier 2016: 156–57.

1720. Clement of Alexandria, *Stromata* 6.11.92–95.

1721. Supporting evidence can be found in sections 5.1 through 5.4 and chapter

5.9 Exceptions That Prove the Rule

As we observed before (in chapter 3.8.IV), Dionysius the Great composed a treatise *On Nature* to refute atomist philosophy, but the one large section of this that survives shows him arguing only from scripture and commonplace observations of 'intelligent design' in nature—rather than conducting any careful investigations of his own (as Galen did to prove the same point) or even employing a survey of supporting work in natural philosophy. In such a way, Dionysius presages a later trend to treat natural philosophy as a logical exercise rather than an empirical one, and as a means to refute heresies and illuminate scriptural wisdom, rather than to understand and predict the natural world.[1722] Likewise, the Christian architect Julius Africanus exhibited a definite interest in mathematics, history, chronology, astronomy, and several other subjects in natural philosophy, as seen in various fragments of his lost works on the history of the world (the *Chronographies*) and his encyclopedia (the *Kestoi*), but from extant fragments it is clear these same works mingled such subjects with a passion for apocalypticism and supernaturalism. What learning Africanus had or sought apparently consisted of knowledge others had acquired, which he deployed in the defense of scripture and the Christian gospel, rather than pursuing astronomy or natural philosophy on his own or as ends in themselves. Though Africanus says he built the new library in the Pantheon for emperor Severus, which would make him the only (known) Christian engineer in the first three centuries, he still preferred the fabulous and the supernatural over scientific research, having written nothing on the latter.[1723]

It is notable that both Dionysius and Africanus were connected with Origen as students (Dionysius directly, Africanus through another of Origen's students), just like Gregory Thaumaturgus, another student of

---

4.9 here and chapters nine and ten of Carrier 2016.

1722. This is more or less confirmed in P. Harrison 1998 for the entire middle ages.

1723. As architect of the Pantheon library: Julius Africanus, *Kestoi* frg. 5.1. His preference for the fabulous is shown in R.M. Grant 1952: 109–12. For background see *OCD* 755; *ODCC* 915; *EANS* 450; Meissner 1999: 258–61. His only discernible works in science are brief and unoriginal (e.g. a summary weights and measures: *EANS* 39).

Origen, and the only Christian author we can find who offered direct praise of natural philosophy (which I've examined in chapter nine of Carrier 2016 and chapter 4.8 here).[1724] In other words, almost the entirety of any evidence of Christian respect for natural philosophy in the early Roman Empire appears to be linked to one eventually-condemned heretical teacher and his pupils. These men would seem to represent a disadvantaged minority view within the early centuries of the Christian intellectual movement. Even though Origen's students became prominent and respected leaders in the Church, there is no evidence this owed anything to their education in philosophy, rather than their mastery of scriptural exegesis, which was always the central objective of Origen's school. Hence Origen's knowledge of scripture continued to win wide praise, while at the same time Origen had to defend himself to his fellow Christians for his study of pagan philosophy.[1725] And yet even Origen did not explicitly defend the methods and activity of scientists. And even he was eventually declared a heretic and his special school did not survive another century beyond him.

Though Origen appears to be the only Christian author in the first three centuries to actually defend and praise human curiosity about the nature and causes of things—in fact, arguing that this questioning thirst for knowledge must have been implanted by God and is therefore approved by God—he does not argue this to defend the pursuit of science or natural philosophy, but rather to explain to "simple" Christians that scriptural references to "eating in heaven" did not mean actual food, but a diet of "wisdom," since we will have our curiosity about all scientific facts satisfied *in the afterlife*—not by any efforts in the present—because only after death will we be able to perceive things clearly enough to understand them, and only then will we be taught by God and his angels what the nature and purpose of everything really had been.[1726] In fact, contrary to advocating a scientific method,

1724. Carrier 2016: 150–54. See: Jerome, *Lives of Illustrious Men* 63, 65, 69; and Eusebius, *History of the Church* 6.3.1–2, 6.15, 6.19.13, 6.29–31. Besides exchanging scholarly letters with Origen, Africanus was a student of Heraclas, who was in turn a student of Origen, who converted Heraclas from his previous devotion to Greek philosophy (in fact, Eusebius claims Africanus went to Alexandria specifically to study under the 'renowned' Heraclas).

1725. Eusebius, *History of the Church* 6.19.12–14.

1726. Origen, *On the First Principles* 2.11.4–7. This material should perhaps be received with caution. We only have this work in a Latin translation which "revised"

Origen had already argued shortly before this that in the present life we must trust the Holy Spirit to teach us "how or why things happen."[1727] Thus, even when he seems close to embracing scientific values, Origen still failed to defend them, and instead advocated epistemic values quite the contrary.

Nevertheless, it cannot be said that leading Christians avoided the study of natural philosophy. It is evident, for example, that even its deriders had some training in the subject, and used their studies to defend Christian doctrine against attacks and heresies, and even to argue *against* the value of studying pagan philosophy. When Ellspermann claims Tertullian "was awake to the results arrived at from natural sciences" and "aware of the advantages that a thinker could derive from them," the evidence he offers confirms only that Tertullian bothered to read the works of his enemies in order to use their own teachings against them.[1728] Tertullian exhibits no other use for natural philosophy. Instead, as we saw in section 5.2, whenever Tertullian declares his own views, natural philosophy, and the epistemic values required to successfully pursue it, are consistently condemned.

The practical utility of science was also recognized by some Christians, though in the first three centuries this recognition was not very developed or pronounced, and was never transferred into any evident respect for the theory and research required to improve or understand it. In fact, it seems Christians failed to recognize any connection between the practical success of applied science and the epistemic values necessary to develop and perfect

---

the original "in directions his fourth-century editor Rufinus considered to be more orthodox" (Bynum 1995: 63–64), enough to launch a bitter enmity between Rufinus and Jerome, e.g. in Jerome's *Apology for Himself against the Books of Rufinus* he says Origen's teachings "had been changed by the translator so as to give them a more orthodox meaning" (1.6) and that Rufinus had "taken away words that existed" and "put in those that did not exist" (1.7; cf. also 2.11). Jerome's outrage at this moved him to produce his own faithful translation into Latin of Origen's *On the First Principles*, but this (being heretical) was not preserved. Though in other cases we can sometimes reconstruct or suspect what Rufinus changed, there is no way of knowing if or in what way Rufinus "revised" anything in the section of our concern, but since it is consonant with Origen's thinking elsewhere, it has probably not been much distorted.

1727. Origen, *On the First Principles* 2.7.4 (though see previous note). On how this inverts the Stoic version of the same expectation in Cicero, see chapter 4.2.

1728. Ellspermann 1949: 38–39.

it. To the contrary, as we have seen, even when some begrudging respect was shown for the practice of medicine or astronomy or engineering, the methods and values actually required for developing and perfecting those applications were vehemently denounced.

For example, while Arnobius, a Christian author of the late 3rd century, said "life" has been "built up and refined for the common good" by "medicine, philosophy, music, and all the other arts" and "it may be laudable to know what medicine or art is practical for curing the ill," he immediately goes on to argue that faith healing is far superior to medicine, and that a doctor can only sometimes cure people with drugs, diet, or exercise, because he is "an animal born from the dirt, not relying on a true science, but on a system founded on a suspect art, teetering on calculations of pure conjecture." In fact, "men endowed with such great genius as orators, teachers of grammar or rhetoric, lawyers and doctors, even those who pry into the obscure teachings of philosophy," have pursued Christian doctrines and "come to despise what they trusted only a little earlier," which implies, and praises the fact, that scholars who converted were giving up the values and principles underlying their former pursuits.[1729]

Likewise, astronomy commanded a wider respect among Christian intellectuals for its use in regulating the Christian calendar, but again without any understanding or praise of the epistemic values the development of astronomy required. This is shown in the case of Anatolius of Alexandria (Bishop of Syrian Laodicea in the late 3rd century), whom Jerome praised as "a man of amazing learning in arithmetic, geometry, astronomy, grammar, rhetoric, and dialectic" yet "the magnitude of his genius" was proven not by new discoveries or theories in astronomy, but merely because he applied mathematics and existing astronomical facts and theories to solving the problems of the Christian calendar.[1730] Of the same man, Eusebius says:

1729. Arnobius, *Against the Nations* 2.69.1 ("life is built up and refined" by the arts, though when he gives examples all he offers is divination and astrology); 1.48.6 ("it *may* be laudable to know," *sit laudabile scire*); 1.49 (faith healing superior to medicine), argued further in 1.64; 1.48.5 (medicine is a suspect art); 3.23.5 (doctors often fail); 2.5.4 (converts abandon former beliefs). Notably, Arnobius was an accomplished pagan educator converted very late in life, who allegedly wrote this treatise to prove to his bishop that he was a sincere convert (cf. Ellspermann 1949: 54–55). See *OCD* 168 (s.v. "Arnobius") and *ODCC* 109 (s.v. id.).

1730. Jerome, *Lives of Illustrious Men* 73, who says the "greatness of his genius"

> Anatolius, because of his skill and education in the philosophy of the Greeks, obtained for himself the top accomplishments of the very best men of our time, such as arithmetic, geometry, astronomy, and the rest, whether dialectics or physics, and the theory of rhetoric, once again riding to the top in learning. Also, because of this, the story has it that he was deemed worthy to be appointed to the school of Aristotle's successors in Alexandria by its citizens.[1731]

Eusebius does not say whether this appointment to the Aristotelian school preceded or followed his conversion to Christianity, or whether he resigned after converting. Nor do we have any idea what opinion Anatolius held regarding the value of his studies, but he had clearly mastered natural philosophy, as had other Christian intellectuals before and after him, and Eusebius does mention this fact with pride. Yet we know from his more direct assessments of the subject that Eusebius did not think highly of natural philosophy, or natural philosophers, but made a point of the fact that his rejection of it was an informed decision, of one who had bothered to study it before concluding it was rubbish.[1732] His praise of Anatolius probably reflects the same opinion: he had beaten the pagans at their own game, and yet still chose a life of God instead.

---

is proven "by the volume which he wrote *On the Passover* and his ten books *On the Principles of Arithmetic*." For problems crediting accomplishments to Anatolius see R.M. Grant 1952: 116. For more on Anatolius generally: *DSB* 1.148–49, *ODCC* 58–59, *EANS* 73.

1731. Eusebius, *History of the Church* 7.32.6–12. See discussions in chapter 4.7 here and in Carrier 2016: 128–30. Eusebius does not say (as some translations have it) that Anatolius was asked to *found* this school. Since Eusebius says "the school of Aristotle's successors in Alexandria" he probably means a school that was already there (further evidence of which is included in the relevant notes in Carrier 2016: 128–31). The word *sustêsasthai* can mean "to be appointed to" or "to be recommended to" or "to be joined to" or "to establish." Eusebius could have chosen a wording that more clearly indicates founding a school, but instead chose a word that most commonly means "joining membership with." The word for "school" (*diatribê*) also suggests an occupation more than an institution. So Eusebius probably means Anatolius was appointed a member or even the municipal chair in Aristotelian philosophy at Alexandria (so Marrou 1964: 286, 573 n. 13 = Marrou 1956: 190, 410).

1732. Eusebius, *Preparation for the Gospel* 14.2.7, 14.3.6, 14.13.9 (and see section 5.4).

In a similar fashion, after surveying some of the achievements of astronomers up to and including Ptolemy (fifty to a hundred years earlier), Hippolytus concludes that all these accomplishments are nothing more than the all-but-useless products of "a soul puffed up and laboring in vain" such that "it is unbelievable to think that, of all people, anyone regards Ptolemy as wise among those who work in the same field."[1733] Though Hippolytus himself exhibits considerable knowledge of the field and its major contributors, he clearly describes all this as an outsider—a disinterested reader of astrological, magical, and heretical works.[1734] Moreover, he conflates astronomy with astrology and condemns both together, and then implies his only reason for surveying the field at all is that it formed the basis of some of the heresies he plans to refute.[1735] In other words, just as his wide and acute reading of magical treatises certainly did not indicate any value or appreciation for magic, so, too, his wide and acute reading of astronomical treatises. Since Hippolytus argues that anyone is a heretic who bases their doctrines on the teachings of philosophers (including natural philosophers) rather than scripture, his only apparent motive for even having studied astronomy in the first place (just as for astrology and magic) was his desire to refute heretics who employ it.[1736]

Such was probably the most widespread view among Christian intellectuals: apart from trivial applications like fixing the calendar or selecting the right herbs or procedures to 'cure' some real or imagined ailment, the study of natural philosophy was only worthwhile as a tool for refuting natural philosophers—and those who would rely on them to support heresy or unbelief. For such Christians, natural philosophy was not the pursuit of natural causes or knowledge of the natural world, but the use of the enemy's own "scriptures" against him. Meanwhile, the epistemic values required for a successful pursuit of science and natural philosophy—curiosity, empiricism, and a belief in scientific progress—came increasingly

1733. Hippolytus, *Refutation of All Heresies* 4.12 (Hippolytus even says the 'only' use for Ptolemy's calculations of celestial distances would have been in explaining to the builders of the Tower of Babel that heaven was too far to reach). For more on Hippolytus and his attitude to science see near the end of section 5.7.

1734. Hippolytus, *Refutation of All Heresies* 4.8–11.

1735. Hippolytus, *Refutation of All Heresies* 4.1–27 and 5.1.

1736. Hippolytus, *Refutation of All Heresies* 1.pr.8–11 (and cf. 1.26.3–4 and 10.2–4).

to be condemned and rejected by Christians, even more so in late antiquity and the middle ages than in the first three centuries. The Christian intellectuals who later ushered in the Scientific Revolution had to work hard to dismantle this hostility among their peers.[1737]

5.10 Medieval Christianity

As noted in the first chapter, some contemporary scholars attempt to deny for the middle ages the conclusion I just implied. This is not directly relevant, since our present concern is only with the early Roman empire, and as we have surveyed all the relevant evidence and its results are conclusive, no assertions to the contrary can hold any merit for that period. But something should be said about the subsequent centuries of Christian domination of the entire intellectual sphere of Western civilization, since our findings for the first three centuries are so stark that it would seem odd to imply that there immediately followed an astounding revival of interest in the scientific enterprise in the early middle ages. In actual fact, there was nothing of the kind.[1738]

David Lindberg is a representative example of the apologetic trend, which often involves making hyperbolic assertions in defense of medieval Christianity without a proper basis in evidence, or often in contradiction to it. As a paradigmatic example, Lindberg asserts:

> How did the dominance of Christianity affect knowledge of, and attitudes toward, nature? The standard answer, developed in the eighteenth and nineteenth centuries and widely propagated in the twentieth, maintains that Christianity presented serious obstacles to the advancement of science and, indeed, sent the scientific enterprise into a tailspin from which it did not recover for more than a thousand years. The truth, as we shall see, is far different and much more complicated.[1739]

1737. This growth of Christian hostility and its eventual defeat (by tracking attitudes toward 'curiosity') is surveyed in P. Harrison 2001 (supported by Neil 2004: 99–138).

1738. See relevant discussion in chapter 3.8.IV.

1739. Lindberg 1992: 149. Lindberg 2007 updated 1992, but made no significant changes to the text pertaining to this (2007: 148) or the discussion to follow.

Lindberg produces no relevant facts in support of this assertion.[1740] For he offers no evidence of any Christian enthusiasm or support for the 'scientific enterprise', which is a curiosity-driven empirical quest for progress in the knowledge of nature.[1741] He can only find what we have already seen: a selective and unscientific plundering of pagan philosophy in the service of defending and interpreting Christianity, which is not at all scientific. And again (just as in section 5.9) the exceptions prove the rule. For instance, Lindberg's own evidence shows that though the medieval Augustine was sometimes uncommonly positive toward the utility of natural philosophy, he was still no advocate of "the scientific enterprise" and was in essential ways hostile to its interests.[1742] Likewise, Roger Bacon was clearly, even from Lindberg's own quotations, battling against a hostile zeitgeist—and this almost a thousand years after Constantine—while his namesake, Francis Bacon, was *still* battling against this Christian hostility and opposition four hundred years later (as discussed in chapter 1.1).

More importantly, Lindberg offers no evidence that "the scientific enterprise" was not essentially put on hold for a thousand years (and we know of no scholar to date who has proven otherwise, as also noted in chapter 1.1). Except for some small achievements within a brief period of Islamic (not Christian) history, and some continuation of education in the West (which was never in doubt), he fails to locate a single significant scientific achievement before the 14th century. Although he correctly documents how the recovery of *pagan* science and philosophy from the 12th century onward eventually contributed to a revival of "the scientific enterprise" by the 14th century, since Christianity had ruled the West since the 4th century, this still marks a thousand years of scientific stagnation *even by his own account*.[1743] Apparently having nothing to say against that,

---

1740. In fact he cites only himself: Lindberg 1986 (repeating material from Lindberg 1983) and Lindberg 1987.

1741. Which is the result in practice of the "scientific spirit" as discussed in chapter 3.8.IV.

1742. See the analysis in the early chapters of P. Harrison 1998 (especially regarding Augustine, *City of God* 22.24), and in Beck 2006: 170–75.

1743. Lindberg 1992: 151–90 and 317–25 (2007: 150–203, 321–28) presents his survey of medieval science before the 12th century; and of Islamic developments: 175–80 (2007: 163–92; on the tenuous and ultimately doomed status of science

Lindberg cites only two facts in defense of his claim that the truth is "far different" from the view that Christianity presented "serious obstacles to the advancement of science." But neither of his two facts argue against that view *at all*, much less demonstrate the truth was "far different" from it. We should look at them, because these arguments are typical of like-minded scholars.

First, Lindberg challenges the accusation that the medieval church was "broadly anti-intellectual" and that church leaders "preferred faith to reason and ignorance to education." Though we have already seen plenty of evidence of this for the first three centuries, it is true that in late antiquity the Church gradually warmed up to the very erudition that Eusebius and Lactantius insisted it should abandon, and came to prefer instead Clement's more moderate plan of subordinating knowledge and reason to faith and dogma. The anti-intellectualism that launched and sustained much of the early Christian movement thus came to be tempered and replaced with what it had caused: a more dogmatic hostility or indifference to scientific values. Yet even at their most anti-intellectual, many elite Christians were certainly, even in the early Roman empire, advocates of logic and some measure of usable erudition (at least among themselves). That still tells us nothing about Christian attitudes towards "the advancement of science" or "the scientific enterprise" or the epistemic values necessary to both.

One could say many Christian intellectuals were "enthusiastic" about natural philosophy in the sense that they valued adopting or formulating theories of nature in agreement with scripture and dogma, but that is exactly the opposite of science—it was a return to what intellectuals had been doing *before* Aristotle, which is a backwards step *away* from 'science' in any meaningful sense. Under Christian tenure, science and natural philosophy became a mere 'tradition' passed on from antiquity, often erroneously or imperfectly, and usually modified as suggested or required by scripture, dogma, or fancy, rather than in accordance with new observations, or improvements in testable theories and methods.[1744] This is not something to

---

under Islam, cf. Cohen 1994: 384–417); and of the early Renaissance (12th to 14th century): 190–244, 325–54, 355–68 (2007: 193–203, 329–56, 357–68). This agrees with the findings of other scholars, e.g. Crombie 1959 and 1994 and Clagett 1955: 135–36, 146–82. As to what should count as a "significant" scientific achievement, see chapter 1.1.

1744. Hence Lindberg 2002a (also Lindberg 1986 and 1983) fails to produce a single counter-example to the argument of this chapter from the first three

praise. But it is exactly what we should have expected given the prevailing attitudes toward natural philosophy among the earlier Christian intellectuals surveyed in this chapter.

It is therefore disingenuous of Lindberg to argue that Christianity did not produce "serious obstacles to the advancement of science," by citing Christianity's embrace of an erudite theology rooted in the borrowed logic and ideas of pagan philosophy (which in fact is all Lindberg has to offer).[1745] Theology and scholasticism are not science—they are the antithesis of it. Since there *was* no "advancement of science" under Christianity for a thousand years, that entails there *must* have been serious, in fact decisive obstacles to such advancement. And since whether science is valued and pursued, or stagnates as a merely transmitted (or even dogmatically adjustable) ideology, is a question of disposition that will always be determined by the intellectual zeitgeist of an era—and since Christianity *was* the dominant intellectual zeitgeist of the era—Christianity necessarily must have presented "serious obstacles to the advancement of science." Nothing Lindberg says or presents offers any challenge to that conclusion.

Lindberg's only other "evidence" in his defense is the equally disingenuous claim that Christianity did not produce "serious obstacles to the advancement of science" because it at least preserved a lot of what the ancients produced. The *non sequitur* here between "advancement" and "preservation" apparently escapes Lindberg's notice. Hence he argues that "the contribution of the religious culture of the early Middle Ages to the scientific movement was thus one of preservation and transmission," as medieval "monasteries served as the transmitters of literacy and a thin version of the classical tradition (including science and natural philosophy) through a period when literacy and scholarship were severely threatened."[1746] How does this argue against

---

centuries, and only proposes two from the 4th century, Basil and Augustine, yet neither of whom present a particularly good counter-example (consider Augustine, *Handbook on Faith, Hope, and Charity* 9 and *Confessions* 10.35; Basil, *Hexameron* 1.8–11; etc.; and see Lindberg 1983: 521–30 and R.M. Grant 1952: 115, 118–19), nor was either entirely representative of their peers. In Lindberg 2007: 363 he claims John Philopon conducted experiments, but there is actually no evidence that's true (see note in Chapter 1.1). He was also condemned as a heretic; and was wholly unrepresentative of his era.

1745. Lindberg 1992: 149–51 (2007: 148–50).

1746. Lindberg 1992: 157; cf. 151–59 (2007: 156–57; cf. 150–62).

there being obstacles to scientific *advancement*? And how does it imply there was any support for the scientific *enterprise*? Indeed, the very fact that "literacy and scholarship were severely threatened" under Christianity argues directly against Lindberg's thesis. But even if that threat is credited to other factors, the mere preservation and transmission of a tradition is not science—it is, again, the antithesis of science. Science only exists in a meaningful sense when its 'tradition' is constantly tested, corrected, and improved upon through rigorous empirical investigation. There was a lot of that going on in antiquity—but none in the early middle ages.

Lindberg has not come up with any new arguments. In later work he offers only the same two irrelevant observations we just examined, though he does introduce a bizarre analogy between the medieval use of old science and the modern use of new science to develop the atomic bomb, or by modern corporations to make money.[1747] That analogy is invalid, for in those cases *actual scientific research* is involved, unlike the early middle ages when "science" almost entirely consisted of simply repeating, often incorrectly, what someone else wrote centuries before.[1748] As well as having forgotten almost all of it. And then just making the rest up.

This kind of desperate rationalization, deploying false analogies and ignoring evident facts, is not uncommon among Medievalists who just can't accept the plain reality of what happened to science under Christian tenure in the Middle Ages. I'm reminded of a peer reviewer who claimed of this very book you are now reading that it overlooked positive Christian support for science in the first three centuries—then named no instance of there being one. In fact I have dealt in this chapter with literally every relevant reference

1747. Lindberg 2002b: 62–63. Not even Lindberg 2007 (the new edition of Lindberg 1992) added any new evidence or arguments on this point. Nor did Lindberg 2009, which rests only on two false assertions bordering on flat out lies: that "Alexandrian science and mathematics prospered for decades" after the murder of Hypatia in the early 5th century (p. 9) and "some of the most celebrated achievements of the Western scientific tradition were made by religious scholars" (p. 17)—when in fact not a single significant advance would be made in mathematics or science for almost a thousand years after her death (much less any of "the most celebrated"). It's as if Lindberg wants to pretend those thousand years didn't even exist and aren't evidence of anything. Compare Efron 2009 in the very same volume; and Carrier 2010 and 2014b.

1748. See examples in Clagett 1955: 136–82.

to the sciences in Christian authors of the period; including the heretics condemned by Paul of Samosata. It is apparently necessary to believe in non-existent testimonies. Likewise to believe that Clement and Tertullian were unrepresentative (also asserted by that reviewer, again without evidence), when they were the only authors on this subject so revered as to be preserved at all, and there is no evidence of any contrary view being representative of the Christianity that came to dominate the West.

But that's not even the worst of it. This same peer reviewer then insisted all the sciences were taught in Christian-run state universities at Alexandria in the fifth century, and with lab work no less—and the only evidence they presented for such an astonishing claim is that some bathtubs were excavated there (at Kom el-Dikka). I'm not joking. They actually claimed that tubs found in Alexandria prove science classes, a university, and state support. Sorry, but no. All that the excavators found were ordinary baths, fountains, and latrines.[1749] I wonder if the peer reviewer was Lindberg himself; though the delusionality here seems to exceed even his. In any event, no evidence for these claims exists. As I've already noted, science education did continue in the Christian East in much better detail than it did in the West, but even in the East it stagnated, seeing no progress and a steady decline in interest, and increasingly treating ancient work as an immutable tradition. It cannibalized ancient science. It did not advance it.

Lindberg's sly insertion of the word "thin" in his remarks conceals another truth about this fact that is also devastating to his argument: that, in actual fact, the vast and overwhelming majority of the ancient scientific tradition was *not* transmitted, and very little was transmitted accurately or in sufficient detail in the West—and even in the East, where the preservation of scientific knowledge fared much better, even there what was preserved remained extraordinarily thin, and often survived only in a fragmentary and disconnected way. It was in the East, after all, that Christians scraped the ink off the last known manuscript of the scientific treatises of Archimedes and wrote hymns to God over them.[1750] As Marshall Clagett put it, "in comparison with the scientist of Islam the early medieval natural philosopher in the Latin West had only meager scraps of the Hellenistic

1749. No scientific function for these baths, fountains, or latrines was claimed by their excavators, nor plausible to imagine: Majcherek 1999, 2010, 2013.

1750. Netz & Noel 2007.

corpus of scientific writing," and yet even the Muslims had only a fraction of what the ancients had written.[1751] Consider a typical example in the West. As one team of scholars explains, "through most of the Middle Ages, Ptolemy's *Geography* was a rare and little-read text, a situation paralleled in the history of other ancient scientific and technical works." Then, "the fortunes of the *Geography* changed abruptly around the year 1300," when copies began to proliferate.[1752] This story is nearly the same for the whole of extant ancient science and philosophy, and their associated ideals; and what's extant, is still the tiniest fraction of what was produced.[1753] And even when things did start to change, it was not Christianity itself, but the recovery and renewed enthusiasm for *pagan* discoveries, values, and ideals, which sparked a revival of science that eventually led to the Scientific Revolution.[1754]

In other words, the achievements of ancient scientists only *barely* survived the middle ages, and not very well, and for a whole millennium were not substantially corrected or improved upon. Even Christian doctors, astronomers, and engineers who continued using what little was preserved of pagan science did so with limited comprehension and no thought to improving it. That is not a record to be proud of. But it is a record we could have predicted, given our observations of Christian views during the first three centuries A.D.

1751. Clagett 1955: 156.

1752. Berggren & Jones 2000: 43.

1753. See, for example, chapter 3.8.IV. Consulting the manuscript history for nearly any ancient science book usually reveals the same result.

1754. Argued in Lloyd 1981, Russo 2003: 332–36 (with support in 336–97), and (for medicine) Nutton 2013: 323–24.

6. Conclusion

This study will now conclude with a summary of our results, a look at the big picture, and a closing suggestion for explaining the slow pace of scientific progress in antiquity.

6.1 Results

In *Science Education in the Early Roman Empire* (Carrier 2016), I established that science was not a major educational value but did have a respected place in the Roman educational system, at least among the most advanced students. Here, chapter two established that during the early Roman empire the natural philosopher was the nearest analogue to the scientist; chapter three established that many among the Roman elite were aware of the reality and value of scientific progress; chapter four established their respect and appreciation for science and scientists; and chapter five found exactly the contrary among the Christians of the same period.

All this has demonstrated that the natural philosopher and his activity was held in high esteem by many among the educated elite and that they were not marginalized in high society, except by Christians, who had no use or regard for natural philosophers and sometimes expressed outright contempt for them. In fact, among the pagan elite we have found significant support for scientific values in the early Roman empire. Many Romans were passionate about studying nature, valued curiosity and empiricism, believed learning the secrets of nature was a moral good, thought scientific progress was possible and valuable, and had some good ideas about how to make such progress. Ancient scientists also enjoyed sufficient liberty and socio-economic support to carry on their research without impediment. There was no significant interference from the government, the public, or any church, and they could secure for themselves lucrative careers and pensions, public or private. Though they typically came from what we might call the upper middle class, they were generally respected by those of higher social station, many of whom also showed considerable interest in their work.

In contrast, the Christians of the same period abandoned empiricism for intuitive or mystical ways of knowing, dependent on what they believed to be revelations from God and inspiration through the Holy Spirit, combined with a supreme confidence in a select body of Sacred Scriptures. Apart from rarefied heretics all of whose literature was destroyed, we found no evidence that any Christians valued observing, investigating, and explaining the natural world, or embraced curiosity about natural causes as a moral or valuable concern, or believed in the value (or in some cases even the possibility) of what we (or even they) would call scientific progress. Instead, we found a generally pervasive condemnation or subversion of the values necessary to advance the sciences. And so although there were many pagan proponents of science and scientific values, there were essentially none among the Christians, who often promoted exactly contrary values or concerns.

As the Christian educator Lactantius argued, the only reasons to pursue science are "subsistence, glory, or pleasure," things he said even base animals enjoy, so the scientist was essentially a beast who abdicates his status as a human being. For him, only those who do not aspire to know anything but God are wise. So in one fell swoop he was rejecting the entire pagan value system that supported scientific writing, study, and research. Cicero praised the immense pleasure of discovery that science brings. Apuleius believed the glory of scientific achievements granted scientists an immortality worthy of eternal praise. Galen passionately defended the practical applications of scientific theories—not only in medicine, but in astronomy and engineering as well. And we have seen many of their peers agreed with these motives, as lofty and appropriate reasons to pursue scientific inquiry. But Lactantius rejects them all as beastly aims, unworthy of men. Though he was the most extreme of the antagonists, other Christians shared the same general outlook, even if not to the same degree. None believed glory or pleasure were morally appropriate motives—many even thought them suspect or wicked. None recognized or acknowledged the practical benefits of either scientific research or of the epistemic values it required; and often enough, they rejected both. As they saw it, science had no evident use for the only two pursuits they regarded as of any real value: the moral perfection of man, and the requisite recognition and understanding of God's will.

Christians probably did not invent this attitude. It is likely they fed on and reflected a parallel thread of hostility or indifference to science and natural philosophy within the pagan population and from like-minded sectarian Jews. But this attitude did not have wide support or influence among the Roman elite, for we have seen ample evidence among them of quite the contrary set of values. So the indifferent masses, the Cynics, the more superstitious or curmudgeonly aristocrats, were only an impotent voice in the intellectual arena, lacking the power to advance or oppose scientific research no matter what their opinions may have been. It was only with the collapse of the early Roman empire, and the consequent rise of a Christian and Neoplatonic society in political and economic decline, that the dominant values among the elite finally shifted in a way that actually put an end to scientific progress for a thousand years—until the values and ideas of the old pagan elite were rediscovered and rekindled in the early Renaissance. By then Christianity had changed, and from then on would change even more, evolving on into the modern world.

6.2 Applications

As I explained in chapter one, the primary purpose of this study has been to assist historians in evaluating the causes of the Scientific Revolution. Though key developments preceded and followed, it was more or less between 1450 and 1660 A.D. that the methods and social role of the scientist underwent a revolutionary transformation. The period thus bracketed began with the launching of the first printing press in Europe, and ended with the founding of the Royal Society of London, which essentially marked the triumph of the Scientific Revolution by realizing the ideals promoted by Francis Bacon earlier in that same century. By that point the subsequent work of Newton was merely an inevitable consequence, no less than of Lavoisier, Maxwell, Darwin, or Einstein. Though the printing press did not begin this revolution, it had a definite effect, coinciding with and assisting the conceptual revolution that had been started, perhaps not even consciously, by a wide array of intellectuals in the 15th century, who in turn had been inspired by their Renaissance predecessors. Such is the general picture. Historians are still debating and working out the details.

The fundamental differences between ancient and modern science are exactly those changes that would be brought about by the Scientific Revolution. In fact, I would argue, that revolution is by definition this very shift. So any causal explanation of the Scientific Revolution cannot appeal to *these* differences between the ancient and modern world, since those are the changes that were caused, not the causes of those changes. Historians want to know why those changes took place, when and where they did, and at no other time or place. Accordingly, any hypothesis as to what caused them must explain not only the Scientific Revolution, but its previous absence. For if the alleged causes were all in place at another point in history, then they cannot have constituted a sufficient cause of the Scientific Revolution, or else they would have caused it then. Consequently it is essential to correctly identify the relevant differences that *actually* obtained between the ancient and early modern periods. And that, in turn, requires a correct understanding of the relevant conditions in antiquity. When this is combined with a correct understanding of the relevant conditions in later periods, we can then ask whether any contrasts between them could have contributed to the Scientific Revolution.

One such alleged contrast is the social status of the scientist and the value a society places on their work. We thus surveyed attitudes toward the 'scientist' in the early Roman empire, after which scientific progress ceased and Christianity prevailed. Within that period we studied the dominant attitude of the influential elite, and compared it with what would replace it: the dominant attitude of the Christian elite. From our results it appears very unlikely that a positive shift in social attitudes can be a sufficient cause of the Scientific Revolution, and it can only be a necessary or contributing cause when considered in contrast to the Christian attitudes of the middle ages. In other words, if there was such a shift preceding the Scientific Revolution, it constituted the *recovery* of an ancient social ideal, not a new development in human history, and certainly not a predictable outcome of Christianity in the absence of a pagan tradition to recover. Most of the changes credited with causing the Scientific Revolution are similar to this: they were only changes with respect to what society had become *after* the decline of the Roman empire, and *after* a thousand-year dominance of Christian ideology. In other words, they were only changes with respect to the medieval mindset, changes which were often a return, more or less, to some element of the *ancient* mindset. At least in the one respect we have examined, the social and intellectual atmosphere that Roman science inhabited was much closer to what preceded the Scientific Revolution, than to what prevailed over the centuries in between.

This analysis can be applied to a wide array of hypotheses. For example, though Joseph Ben-David correctly argues that a scientific revolution required destroying "the rule of traditional authority in intellectual life," which indeed did not happen under Christendom until certain events of the Renaissance (like the Reformation), Ben-David seems unaware of the fact that this only marked the beginning of a return to the way things had already been in antiquity, when no "traditional authority" ruled *anyone's* intellectual life.[1755] All ancient scientists were legally, politically, and ideologically free from institutional and dogmatic pressures on their work, so they could pick, choose, tinker, debate, and innovate as they pleased, and that's exactly what they did.[1756] Likewise, although Ludwig Edelstein said "the failure of the empirical trend to establish itself securely" was "due to the lack of a social integration of science," in the early Roman Empire the empirical trend *was* being established and science *had* achieved a respectable degree of social integration.[1757] Other trends certainly ran parallel to these, but the presence of alternative epistemologies and antiscientific interests is a social reality that has not changed even to this day.

On the other hand, if Edelstein means by "established" and "integrated" something more than had occurred in antiquity, his argument becomes circular, for a *decisive* establishment of empirical methods through a *greater* social integration of science *is* the Scientific Revolution. What we want to know is why this had not happened before. Pointing to the mere fact that it had not happened does not constitute an explanation. Likewise, as A.C. Crombie says, "in Greek science no set of generally accepted aims,

1755. Ben-David 1991: 306.

1756. Political freedom was a different matter. See sources on intellectual freedom in chapter 1.1.

1757. Edelstein 1952: 600.

methods, and criteria of cogency in scientific argument had yet become established for the whole scientific community," but again this merely repeats the obvious without explaining it.[1758] For the "establishment" of both a "general agreement on basic principles" and "a scientific community with conditions of communication and education within which agreement could be reached" is exactly what constituted the Scientific Revolution. I think a good case could be made that Hero, Galen, Ptolemy, and several others, were arguing for exactly this kind of general agreement within the sciences, and were on their way to achieving it. The course of events they might have set in motion was simply interrupted in the third century by a destructive fifty-year-long civil war, a series of plagues, and an economic collapse that scientists like them had nothing to do with.[1759]

There may still be differences between antiquity and the Renaissance relating to social status, though if there are, they are not obvious. As I have noted several times, during the Roman empire the largely unscientific focus of ancient education, the consequently small number of scientists in every generation, and the lack of direct institutional support for scientific research, are together an indication of science's lower social prestige relative to fields that were typically more attractive to brilliant minds of the time, and thus more prone to absorb them. These more "attractive" fields of intellectual achievement included, primarily, rhetoric and philosophy (in the broader analytical sense)—but this was no different from anytime before or during the Renaissance, when intellects were predominately being pipelined into other studies as well: the only difference being scripture and theology replaced rhetoric and philosophy. And educational content was even lighter on the sciences then than it had been in antiquity.[1760] Likewise, in antiquity perhaps the second most frequent draw was to the pursuit of *practical* careers in the "sciences" of medicine or engineering without any great interest in original research. But that, too, has always been the case.

1758. Crombie 1963: 7.

1759. As I argue in chapter 3.1 and also discuss in chapter 1.1. Commonly called "The Crisis of the Third Century," this conjoined a fifty-years-long civil war (from 235 to 284 A.D.) with the Plague of Cyprian (likely smallpox, wiping out a quarter to a half of the population between 250 and 270 A.D.), concluding in a Great Depression (with the collapse of the fiduciary economy by 270 A.D.).

1760. As demonstrated in Carrier 2016.

Differences thus elude us.

Indeed, all of this is true even today, when the vast bulk of those trained in medicine and engineering spend no significant time advancing their fields with their own research, and careers in entertainment, business, and law draw more income and prestige, and hence far more interest, than even a *practical* career in the sciences, much less in scientific research. Nevertheless, no one would claim that modern doctors, engineers, or even research scientists hold anything like a low social status. To the contrary, such careers are still quite prestigious and desirable. Thus, as I have also argued, the fact that theoretical science in antiquity was overshadowed by more prestigious outlets for creative genius does not indicate that scientists held a low social status, then any more than now.[1761] The question remains a more difficult one of degree.

Likewise, though it is true that "the economic, social, and ideological framework of ancient science differs profoundly from the modern situation," it is not true that "there was, indeed, no place in ancient society for science or the scientist as such."[1762] It is true that despite there having been some collaborative research centers, "most scientists worked in isolation and without support from either individual patrons or institutions," but as even Lloyd admits, this was no less true of "the Middle Ages and early Renaissance." Nor does this fact entail the conclusion that science and scientists had no place in ancient society. Roman doctors, engineers, and astronomers certainly had their patrons and dinner clubs, as well as the functional equivalent of universities, and routinely spoke to and interacted with each other, and were often widely known and recognized.[1763] Ancient society not only had a name for them (the *physicus*, the "natural philosopher," as well as even more specialized terms), it also gave their activity a recognized place in the social order, one of rather higher status than most occupations enjoyed, and one well paid.[1764] So at least in these respects, the extent or significance

1761. As argued in chapter 1.1 and the introduction to Carrier 2016.

1762. Lloyd 1981: 261–62.

1763. Ancient scientific societies: Carrier 2016: 124–30 (cf. 109 n. 286) Ancient universities: Carrier 2016: 130–33.

1764. Specialized labels included: *astronomos* or *astrologus* ("astronomer"), *architectus* or *mechanicus* ("engineer"), *iatros* or *medicus* ("doctor"), *mathematicus* or *geometres* ("mathematician"), etc., some with even more specializations

of any differences between antiquity and the Renaissance remain to be demonstrated.

6.3 Speculations

Nevertheless, Lloyd's remarks remain true in two respects. First, Roman patrons of science do not appear to have funded research, but only applications. As a result, scientists generally had to find their own time, and often their own resources, to conduct scientific research. Hence the *advancement* of science was the pastime of capable and passionate men and women of means, and very often an occupation of their retirement. Science education, or indeed education in general, also received less attention than it has in the past two centuries, so capable scientists were definitely much scarcer then than now. These two factors would explain the slow pace of progress in ancient science and technology, relative to the pace of more recent times. But progress nevertheless continued, and contributing scientists continued to enjoy a well-respected place in the social system, replenishing their ranks with every generation, until everything went to hell over the course of the 3rd century A.D. After that, or perhaps because of that, ancient society took a sharp turn away from scientific values and into the arms of diverse superstitions and mystical armchair philosophies. There it remained for over a thousand years, enjoying (at most) the occasional fruits of a patchwork of simplified, regurgitated science haphazardly preserved from old.

This theory of small numbers has been proposed before, and I have discussed it before.[1765] That the pace of scientific progress in any society will be in some way proportional to the number of research scientists is a reasonable hypothesis.[1766] Such a difference in numbers could operate through two effects: a greater number of scientific investigators can accelerate progress through increased division and collaboration of labor, *and* the subsequent

---

distinguished within these.

1765. In chapter 1.1 here and in Carrier 2016: 90–91 (for estimates of the number of scientists: 29–31).

1766. This is partly the thesis of Sawyer 2007 (though his emphasis is on the role of collaboration) and Collins 1998 (who emphasizes interaction networks).

pace of results can impact social consciousness by eventually impressing upon a single generation the magnitude of possibilities that such numbers can produce.[1767] But a third consequence could be even more important: the greater the number of investigators debating and attempting to persuade each other, the faster the progress toward the discovery and refinement of superior methods. So perhaps certain changes in educational institutions during the early Renaissance eventually flooded society with a critical mass of trained and curious scientists. But without any objective assessment of relative numbers, the two eras cannot reasonably be compared on this score, not even with regard to educational access to science, since the formation of universities was largely the result of certain peculiarities in medieval economics and politics, so their existence did not entail more students were attending lectures in science and natural philosophy than had attended them in antiquity, when whole cities could effectively serve as *ad hoc* universities. Even actual universities existed in the Roman world, in such diverse cities as Rome, Athens, and Alexandria, with dedicated buildings and state-funded professorships in multiple subjects.[1768] So there does not seem to be any identifiable difference here, either.

Obviously a major factor in slowing progress was also the lack of a consensus on method, the very consensus that would come to define the Scientific Revolution. Every natural philosopher had an 'epistemological toolbox', filled with a variety of methods to use when answering questions about nature, and the proper contents of that toolbox remained an object of continued debate. The most scientific among them had the right tools in their box, but these were thrown into a jumble with others as well, and pride of place was not given to the right tools, which were instead used haphazardly alongside the rest. The 'right tools' were the ones some scholars have claimed the ancients didn't have or use, though in fact they did, just not exclusively: the innovative use of instruments, controlled and repeatable experiments, mathematical descriptions, and logically cautious arguments constructed from well-confirmed empirical premises, all directed toward seeking a consensus within a community of acknowledged experts. The

1767. I proposed an example of this 'impact' thesis in chapters 3.10 and 4.7, though as shown there, a scientific revolution can also be the result of accident rather than a consequence of greater numbers.

1768. Carrier 2016: 130–33.

Scientific Revolution did not consist in discovering these methods. Ancient scientists were already using them. What was revolutionary was throwing out all the other tools (or handing them on to philosophy as a separate enterprise) and treating these five as the only tools worthy to employ when answering questions about nature. Once that happened, correct answers started pouring out like gangbusters everywhere a scientist could look, answers that could be defended with repeatable and thus effectively irrefutable evidence, proving the correctness and utility of the new, leaner toolbox. Rapid progress was then inevitable. The question that remains is why this trimming of the toolbox took place only when it did, and I suspect the answer has something to do with relative numbers (as suggested above), or recent earth-shattering discoveries (the 'impact' thesis explored earlier), or both.

Whatever the cause of the slow pace of ancient science, could such a pace ever have arrived at the same destination, eventually producing its own Gilbert, Galileo, Harvey, Boyle, Newton, or Lavoisier? Would there ever have been anything like an Imperial Society of Rome for Improving Natural Knowledge? We may never know. But it's not unreasonable. By the 2nd century A.D. everything seemed to be heading in the required direction. Numerous lay and scientific authors were calling for something like that very transformation. Apart from the historically trivial, or the entirely accidental (like the preceding invention of the compass or cannon or telescope), there is nothing in the writings of Gilbert or Harvey or Galileo that would appear at all out of character had it been written in the early Roman empire. If many generations had continued to test and challenge Galen, Ptolemy, and Hero using the very methods they promoted, progress would seem inevitable, and there is no obvious barrier to how far that progress could go—unless rapidity of development is essential to the effect. For it may have taken longer, their scientific revolution spanning perhaps four centuries rather than two. But apart from time or disaster, what else could have stopped it?

APPENDIX A
ON ANCIENT EXPLORATION

I have not discussed explorers as scientists in this book. But those interested in that subject may benefit from this reference summary.

Eratosthenes is one of the first scientists to accept the reports of the explorer Pytheas of Massalia, who was possibly the first Greek to engage in large scale exploration for the sake of mere knowledge, going as far as Britain and beyond in the late 4th century B.C. (*DSB* 11.225–26; *EANS* 711–12; *OCD* 1247). On Pytheas and other ancient explorers see Cary & Warmington 1963, Hawkes 1977, Henze 1998, Russo 2003: 112–14, Roller 2006, and Kowalski 2012. A brief survey of Roman-era exploration is included in Berggren & Jones 2000: 23–30, 145–62, with discussion of the values that motivated it in Beagon 1992: 180–91. See also *OCD* 32–33, 611–12, 752, 1108–09, 1353 (s.v. "Africa (Libya), exploration," "geography," "itineraries," "*periploi*," "Seres") and *EANS* 447 (s.v. "Itineraries") and 999–1002 (s.v. "geography"), with Dilke 1985: 130–44 and the brief survey in Strabo, *Geography* 1.2.1. On geographical writers in general see Dilke 1985: 55–71.

In addition, for uncertain examples, see *OCD* 950 (Metrodorus of Skepsis; cf. *EANS* 555), 1092 (Patrocles of Macedon; cf. *EANS* 628), and 1335 (Scylax of Carvanda; cf. *EANS* 745–46, both actual and pseudo); and the 4th century traveler Anaxicrates (*EANS* 74). Likewise see *OCD* 229 (s.v. "bematists") for Eratosthenes' use of the official surveyors of Alexander the Great. See also the personal explorations of Alexander's compatriots (Androsthenes: *OCD* 86, *EANS* 82; Nearchus: *OCD* 1004, *EANS* 568–69;

Onesicritus: *OCD* 1039, *EANS* 591–92) and those of subsequent kings (cf. Geminus, *Introduction to Astronomy* 16.24 and Strabo, *Geography* 17.1.5). Later explorers who contributed significantly to geography include Eratosthenes' student Mnaseas of Patara (*OCD* 965, *EANS* 559) and the historian Polybius (*OCD* 1174–75, *EANS* 680–81; though his geographical works are now lost, cf. Geminus, *Introduction to Astronomy* 16.32–33), then at the end of the 2nd century B.C., Posidonius (on whom see chapter 3.3 here), Eudoxus of Cyzicus (*OCD* 546, third entry; *EANS* 314), Artemidorus of Ephesus (*OCD* 175–76, second entry; *EANS* 165), and others (see Polybius, *Histories* 3.58–59), as well as expeditions financed by King Juba a century later (*OCD* 777, s.v. "Juba (2) II"; cf. *EANS* 441–42), and countless reports published by Roman magistrates and commanders (e.g. *OCD* 755, s.v. "Iulius Agricola, Gnaeus").

Detailed travel accounts were also written, e.g., in the 2nd century B.C. (Agatharchides of Cnidus: *OCD* 35, *EANS* 40–41; Scymnus: *OCD* 1335, *EANS* 746), in the 1st century B.C. (*OCD* 933, s.v. "Menippus (2)"; *EANS* 548–49), and in the 2nd century A.D. (*OCD* 169, s.v. "Arrian (Lucius Flavius Arrianus)"; cf. *EANS* 330). Many more explorations and accounts existed, most serving military or mercantile purposes, though some for tourists (e.g. *OCD* 1097, s.v. "Pausanias (3)"; cf. *EANS* 630–31). But there were also amateur armchair geographers who have no claim to being scientific, e.g. Pomponius Mela in the 1st century A.D. (*DSB* 11.74–76, *EANS* 685–86, *OCD* 1182), probably comparable to the lost *Geography* by Cornelius Nepos a century earlier (*OCD* 380, *EANS* 219–20, *NDSB* 2.81–84). There were other nonscientific 'descriptive geographies' from the 3rd century B.C. on (e.g. Polemon of Ilium: *OCD* 1169, third entry; *EANS* 678) including hack literary efforts (e.g. Gaius Iulius Solinus: *OCD* 764, *EANS* 455–56; Iulius Titianus: *OCD* 764, *EANS* 456).

Appendix B
On Science before Aristotle

For good general summaries of Greek science and natural philosophy up to Aristotle see Lloyd 1970, J. Barnes 1982, Kirk et al. 1983, Allen 1991, Warren 2007, Vamvacas 2009, and Graham 2006 and 2013; also the valuable comments in Russo 2003: 22–24, 33–38, 48–49 and background in *OCD* 1207–08 (s.v. "Presocratic philosophy"). For ancient perspective see the relevant sections of Diogenes Laertius, *Lives and Opinions of Eminent Philosophers*. After Aristotle, natural philosophy tracks the major philosophical sects (directly or eclectically).

Although all of the following listed philosophers wrote books, some of them a great many books, almost all their writings are lost. We have to reconstruct our knowledge of them and their writings and accomplishments through other authors and evidence. The following names are chosen here for their importance to the history of science and the availability of their discussion in standard references (many more could be named; a more complete list is available in *EANS*). All dates are B.C.

Pre-Aristotelian natural philosophers include the **Ionians** Thales (7th to 6th century: *DSB* 13.295–98, 4.463; *EANS* 779; *OCD* 1448), Anaximander (early 6th century: *DSB* 1.150–51; *EANS* 75–76; *OCD* 83), Anaximenes (of Miletus, early 6th century: *DSB* 1.151–52; *EANS* 76; *OCD* 83), Heraclitus of Ephesus (late 6th century: *DSB* 6.289–91; *EANS* 372–73; *OCD* 665), Xenophanes (late 6th and early 5th century: *DSB* 14.536–37; *EANS* 839; *OCD* 1580), Anaxagoras (early 5th century: *DSB* 1.149–50; *EANS* 73–74; *OCD* 82–83), Hecataeus (of Miletus, early 5th century: *DSB* 6.212–13;

EANS 361; *OCD* 649), and Scythinus (late 5th and early 4th century: *OCD* 1336; *EANS* 746); the **Sicilians** Empedocles (early 5th century: *DSB* 4.367–69; *NDSB* 2.395–98; *EANS* 283–84; *OCD* 503–04), and Ecphantus (4th century: *EANS* 280–82) and Hicetas (5th century: *DSB* 6.381–82; *EANS* 397; *OCD* 682) of Syracuse; the **Italians** Pythagoras (late 6th century: *DSB* 11.219–25; *EANS* 714–15; *OCD* 1245–46), Parmenides of Elea (early 5th century: *DSB* 10.324–25; *EANS* 626; *OCD* 1082), Zeno of Elea (5th century: *DSB* 14.607–12; *EANS* 844–45; *OCD* 1587), Menestor (5th century: *OCD* 932; *EANS* 547–48), Philolaus of Crotona (late 5th century: *DSB* 10.589–91; *EANS* 651–52; *OCD* 1133), and Philistion of Locri (late 5th and early 4th century: *OCD* 1130; *EANS* 649–50); the **Athenians** Archelaus (early 5th century: *OCD* 138–39; *EANS* 158), Hippon (5th century: *OCD* 689; *EANS* 421), Antiphon (late 5th century: *DSB* 1.170–72; *EANS* 99; *OCD* 108, first or second entry), Meton (late 5th century: *DSB* 9.337–40; *EANS* 551–52; *OCD* 942–43), Antisthenes (late 5th and early 4th century: *OCD* 109; *EANS* 99–100), and of course Aristotle's own teacher, Plato (late 5th and early 4th century: *DSB* 11.22–31; *EANS* 667–70; *OCD* 1155–58) and Plato's other pupils, e.g. Xenocrates of Chalcedon (*DSB* 14.534–36; *EANS* 838; *OCD* 1580). On the pupils also of Socrates (*OCD* 1378–79) see Carrier 2014a, p. 290 n. 19.

There were also the early **atomists** Democritus (who wrote on many subjects, from mathematics to biology: *DSB* 4.30–35; *EANS* 235–36; *OCD* 437–38) and his teacher Leucippus (*DSB* 8.269; *EANS* 506; *OCD* 824, third entry), together spanning the 5th century (*OCD* 200–01, s.v. "atomism"); their contemporary and more eclectic Diogenes of Apollonia (*OCD* 456; *EANS* 252); and Democritus' pupils Metrodorus of Chios (*OCD* 950; *EANS* 554) and Nausiphanes of Teos (*OCD* 1002; *EANS* 568), both early 4th century. Of special importance to the pre-Aristotelian development of scientific medicine is Hippocrates of Cos (late 5th century to early 4th century: *DSB* 6.418–31; *EANS* 404–05, with 406–20; *OCD* 687–88), more of whose works have been preserved than for anyone else listed here, although many are believed written or edited by his pupils and successors (for an extensive discussion of Hippocrates and early Hippocratism see Nutton 2013: 37–103, with Lloyd 1978). Lesser known today (since nothing he wrote survives), but in antiquity a famous contributor to medical science, was Democedes of Croton in the 6th century (*OCD* 434; *EANS* 234).

Pre-Aristotelian **mathematicians** (some of whom studied harmonics, others astronomy) include Pythagoras (above) and Hippasus (6th century: *OCD* 686; *EANS* 399–400), Democritus (above), Bryson of Heraclea (5th century: *DSB* 2.549–50; *EANS* 199–200; *OCD* 254), Euctemon of Athens (5th century: *DSB* 4.459–60; *EANS* 317; *OCD* 545), and Oenopides of Chios (5th century: *DSB* 10.179–82; *EANS* 587; *OCD* 1034); from the late 5th century: Hippias of Elis (*DSB* 6.405–10; *EANS* 400; *OCD* 687, first or third entry), Hippocrates of Chios (*DSB* 6.410–18; *EANS* 401–03; *OCD* 689, third entry), Theodorus of Cyrene (*DSB* 13.314–19, 15.503; *EANS* 785–86; *OCD* 1458, second entry), Eratocles (*OCD* 533; *EANS* 297); and (contemporary with Aristotle): Theatetus of Athens (late 5th to early 4th century: *DSB* 13.301–07; *EANS* 780–81; *OCD* 1449), Theudius of Magnesia (early 4th century: *DSB* 13.334; *EANS* 805), Archytas of Tarentum (early 4th century: *DSB* 1.231–34; *EANS* 161–62; *OCD* 145), Aristoxenus of Tarentum (early 4th century: *DSB* 1.281–83; *EANS* 153–55; *OCD* 163–64), Thymaridas of Paros (early 4th century: *DSB* 13.399–400; *EANS* 808–09), Eudoxus of Cnidus (early 4th century: *DSB* 4.465–67; *EANS* 310–13; *OCD* 545–46, (1)), Leo of Athens (early 4th century: *DSB* 8.189–90; *EANS* 502), Leodamas of Thasos (early 4th century: *DSB* 8.192; *EANS* 502), Menaechmus (4th century: *DSB* 9.268–77; *EANS* 542–43; *OCD* 929, second entry), Heraclides Ponticus (4th century: *DSB* 15.202–05; *EANS* 368–69; *OCD* 664), Aristaeus (4th century: *DSB* 1.245–46; *EANS* 130–31), Speusippus (4th century: *DSB* 12.575–76; *EANS* 756–57; *OCD* 1393), Dinostratus (4th century: *DSB* 4.103–05; *EANS* 229–30), Philippus of Opus (late 4th century: *OCD* 1130; *EANS* 647), and Callippus (late 4th century: *DSB* 3.21–22; *EANS* 464–65; *OCD* 267–68).

Scientific dissection may have begun as early as the late 6th century under the Sicilian natural philosopher Alcmaeon of Croton, but this is disputed (*DSB* 1.103–04; *EANS* 61; *OCD* 54). We only hear of it a thousand years later, and Aristotle does not appear to have known of it, even though he extensively studied the work of his predecessors, yet Alcmaeon's alleged achievements would have greatly altered Aristotle's conclusions on fundamental elements of physiology. There were also scientists writing treatises on mechanics and engineering during or before Aristotle's time, e.g. *Diades* (4th century; EANS 243–44).

Appendix C on the Books of Sextus Empiricus

Two books survive from Sextus Empiricus: *Outlines of Pyrrhonism* and *Against the Professors*. However, it is now believed the treatise entitled *Pros Mathêmatikous* ("Against the Professors") is actually a modern merging of two separate works or more (e.g. J. Barnes 1988: 53). Of this the first six books comprise the original contents of *Against the Professors*, but the remaining books belong to something else, some of which may be missing. I use the traditional numeration for the latter books because it is simpler and less confusing. But to convert them use the following table:

Against the Professors	= *Against the Dogmatists*	= *Against the ...*
7	1	... Logicians 1
8	2	... Logicians 2
9	3	... Natural Philosophers 1
10	4	... Natural Philosophers 2
11	5	... Moral Philosophers

Bibliography

Abbreviations commonly used in notes:

DSB = Gillispie 1980
EANS = Keyser & Irby-Massie 2012
Kühn = Kühn 1821-1833[1769]
LSG = Liddell & Scott 1996
LSL = Lewis & Short 1879
NDSB = Koertge 2008
OCD = Hornblower & Spawforth 2012
ODCC = Cross & Livingstone 1997
OLD = Glare 1996

Absmeier, Robert. 2015. *Der Holzbau in der Antike: Überlegungen zum vormittelalterlichen Holzhausbau.* Computus.

Acerbi, Fabio. 2011. "The Geometry of Burning Mirrors in Greek Antiquity: Analysis, Heuristic, Projections, Lemmatic Fragmentation." *Archive for History of Exact Sciences* 65.5: 471–97.

Aczel, Amir D. 2001. *The Riddle of the Compass: The Invention that Changed the World.* Harcourt.

1769. There is no consistent numbering system for passages in Galen other than (in most cases) Kühn 1821-1833, which I give whenever possible. If I provide any other numeration it will follow the scheme used in the most recent English translation of the given work (in most cases that means before 2008; e.g. Mark Grant 2000, P.N. Singer 1997, Iskandar 1988, Walzer & Frede 1985, M.T. May 1968, Walzer 1944, etc.).

Adams, Colin. 2007. *Land Transport in Roman Egypt: A Study of Economics and Administration in a Roman Province*. Oxford University Press.

———. 2012. "Transport." *The Cambridge Companion to the Roman Economy*. Walter Scheidel, ed. Cambridge University Press: 218–40

Adamson, James. 1976. *The Epistle of James*. W.B. Eerdmans.

———. 1989. *James: The Man and His Message*. W.B. Eerdmans.

Africa, Thomas W. 1967. *Science and the State in Greece and Rome*. Wiley.

Alexander, Loveday. 1990. "The Living Voice: Scepticism towards the Written Word in Early Christian and in Graeco-Roman Texts." *The Bible in Three Dimensions: Essays in Celebration of Forty Years of Biblical Studies in the University of Sheffield*. David Clines, Stephen Fowl and Stanley Porter, eds. Sheffield, England: Journal for the Study of the Old Testament: 221–47.

———. 2002. "'Foolishness to the Greeks': Jews and Christians in the Public Life of the Empire." *Philosophy and Power in the Graeco-Roman World: Essays in Honour of Miriam Griffin*. Gillian Clark & Tessa Rajak, eds. Oxford University Press: 229–49.

Allen, Reginald, ed. 1991. *Greek Philosophy: Thales to Aristotle*, 3rd ed. Maxwell Macmillan International.

Amundsen, Darrel. 1974. "Romanticizing the Ancient Medical Profession: The Characterization of the Physician in the Graeco-Roman Novel." *Bulletin of the History of Medicine* 48.3 (Fall): 320–37.

———. 1977. "Images of Physicians in Classical Times." *Journal of Popular Culture* 11.3 (Winter): 642–55.

———. 1982. "Medicine and Faith in Early Christianity." *Bulletin of the History of Medicine* 56.3 (Fall): 326–50.

———. 1995. "Tatian's 'Rejection' of Medicine in the Second Century." [in van der Eijk, Horstmanshoff and Schrijvers 1995: 2.377–92]

———. 1996. *Medicine, Society and Faith in the Ancient and Medieval Worlds*. Johns Hopkins University Press.

Anderson, Graham. 1993. *The Second Sophistic: A Cultural Phenomenon in the Roman Empire*. Routledge.

André, Jacques. 1987. *Etre Médecin à Rome*. Les Belles Lettres.

André, Jean-Marie, ed. 1987. "La rhétorique dans les préfaces de Vitruve: Le statut culturel de la science." *Filologia e Forme Letterarie: Studi Offerti a Francesco della Corte*, vol. 3. Università degli Studi di Urbino: 265–89.

———. 2003. "La réflexion sur la technique à l'époque néronienne." [in Lévy et al. 2003: 143–56]

Andreau, Jean. 1999. *Banking and Business in the Roman World*. Cambridge University Press.

———. 2015. *The Economy of the Roman World*. Michigan Classical Press.

Ankarloo, Bengt, and Stuart Clark. 1999. *Witchcraft and Magic in Europe: Ancient Greece and Rome*. University of Pennsylvania Press.

Apel, Willi. 1948. "Early History of the Organ." *Speculum* 23.2 (April): 191–216.

Argoud, Gilbert, and Jean-Yves Guillaumin, eds. 1998. *Sciences exactes et sciences appliquées à Alexandrie*. Publications de l'Université de Saint-Étienne.

Argyrakis, Vaios. 2011. "The Clepsydra Experiment: Clepsydra's Functioning and the Related Devices in Heron's *Pneumatics*." *Almagest: International Journal for the History of Scientific Ideas* 2.2: 16–27.

Arnaldi, Mario, and Karlheinz Schaldach. 1997. "A Roman Cylinder Dial: Witness to a Forgotten Tradition." *Journal for the History of Astronomy* 28: 107–17.

Arnander, Christopher. 2007. "A Plum Eater among the Gophers: Teaching Classics Fifty Years Ago in Minnesota." *Amphora* 6.1 (Spring): 10, 19.

Arnaud, Pascal. 1983. "L'Affaire Mettius Pompusianus ou le crime de cartographie." *Mélanges de l'École Française de Rome: Antiquité* 95.2: 677–99.

———. 1984. "L'Image du globe dans le monde romain." *Mélanges de l'École Française de Rome: Antiquité* 96.1: 53–116.

Asper, Markus. 2007. *Griechische Wissenschaftstexte: Formen, Funktionen, Differenzierungsgeschichten*. Franz Steiner.

Atkins, Margaret, and Robin Osborne, eds. 2006. *Poverty in the Roman World*. Cambridge University Press.

Aubrion, Etienne. 1996. "*Humanitas* et *Superstitio* dans la littérature latine du début de l'époque antonine." *Culture antique et fanatisme*. Jeanne Dion, ed. Éditions de Boccard: 76–94.

Aujac, Germaine. 1966. *Strabon et la science de son temps*. Les Belles Lettres.

———. 1993. *La sphère: Instrument au service de la découverte du monde: d'Autolycos de Pitanè à Jean de Sacrobosco*. Paradigme.

Austen, Jane. 1996 [orig. 1811]. *Sense and Sensibility*. Courage Books.

Authier, Michel. 1995. "Archimedes: A Scientist's Canon." *A History of Scientific Thought: Elements of a History of Science*. Michel Serres, ed. Blackwell: 124–59, 726. [tr. of *Éléments d'Histoire des Science*. Bordas. 1989]

Avalos, Hector. 1999. *Health Care and the Rise of Christianity*. Hendrickson Publishers.

Bacon, Francis. 2001 [orig. 1605]. *The Advancement of Learning*. Stephen Jay Gould, ed. Modern Library.

Bagnall, Roger, ed. 2009. *Oxford Handbook of Papyrology*. Oxford University Press.

Balabanes, Panos. 1999. *Hysplex: The Starting Mechanism in Ancient Stadia: A Contribution to Ancient Greek Technology*. University of California Press.

Ballér, Piroska. 1992. "Medical Thinking of the Educated Class in the Roman Empire: Letters and Writings of Plutarch, Fronto and Aelius Aristides." *From Epidaurus to Salerno: Symposium Held at the European University Centre for Cultural Heritage, Ravello, April, 1990*. Antje Krug, ed. PACT Belgium: 19–24.

Barker, Andrew. 1989. *Greek Musical Writings II: Harmonic and Acoustic Theory*. Cambridge University Press.

———. 1994. "Greek Musicologists in the Roman Empire." [in T. Barnes 1994: 53–74]

———. 2000. *Scientific Method in Ptolemy's Harmonics*. Cambridge University Press.

Barnes, Jonathan. 1982. *The Presocratic Philosophers*, rev. ed. Routledge.

———. 1988. "Scepticism and the Arts." *Apeiron: A Journal for Ancient Philosophy and Science* 21.2 (Summer): 53–77.

———. 1993. "Galen and the Utility of Logic." [in Kollesch & Nickel 1993: 33–52]

———, ed. 1995. *The Cambridge Companion to Aristotle*. Cambridge University Press.

———. 1997. *Logic and the Imperial Stoa*. Brill.

———. "Ancient Philosophers." [in Clark & Rajak 2002: 293–306]

Barnes, Jonathan, Jacques Brunschwig, Myles Burnyeat, and Malcolm Schofield, eds. 1982. *Science and Speculation: Studies in Hellenistic Theory and Practice*. Cambridge University Press.

Barnes, Timothy, ed. 1994. *The Sciences in Greco-Roman Society*. Edmonton, Alberta: Academic. [= *Apeiron: A Journal for Ancient Philosophy and Science* 27.4 (December)]

Barnett, Paul. 1997. *The Second Epistle to the Corinthians*. W.B. Eerdmans.

Barow, Horst. 2013. *Roads and Bridges of the Roman Empire*. Ed. Menges.

Barrera-Osorio, Antonio. 2006. *Experiencing Nature: The Spanish American Empire and the Early Scientific Revolution*. University of Texas Press.

Barton, Tamsyn. 1994a. *Power and Knowledge: Astrology, Physiognomics, and Medicine under the Roman Empire*. University of Michigan Press.

———. 1994b. *Ancient Astrology*. Routledge.

Basch, Lucien. 1987. *Le musée imaginaire de la marine antique*. Athens: Institut Hellenique pour la Preservation de la Tradition Nautique.

Bastomsky, S.J. 1972. "The Emperor Nero: A Forerunner of Salvino degli Armato?" *Apeiron: A Journal for Ancient Philosophy and Science* 6.2 (Summer): 19–23.

Batty, Roger. 2002. "A Tale of Two Tyrians." *Classics Ireland* 9: 1–18.

Beagon, Mary. 1992. *Roman Nature: The Thought of Pliny the Elder*. Clarendon Press.

Beck, Roger. 1994. "Cosmic Models: Some Uses of Hellenistic Science in Roman Religion." [in T. Barnes 1994: 99–117]

———. 2006. *The Religion of the Mithras Cult in the Roman Empire: Mysteries of the Unconquered Sun*. Oxford University Press.

Beckmann, Johann, William Johnston, William Francis, and J.W. Griffith. 1846. *A History of Inventions, Discoveries, and Origins*, 4th ed. in 2 vols. H.G. Bohn.

Bekker-Nielsen, Tønnes. 2002. "Nets, Boats and Fishing in the Roman World." *Classica et Mediaevalia* 53: 215–234.

———. 2004. "The Technology and Productivity of Ancient Sea Fishing." *Ancient Fishing and Fish Processing in the Black Sea Region*. Tønnes Bekker-Nielsen, ed. Aarhus University Press, 2004. 83–95. [http://www.pontos.dk/publications/books/black-sea-studies-2]

Bellamy, Peter, and R.B. Hitchner. 1996. "The Villas of the Vallée des Baux and the Barbegal Mill: Excavations at la Mérindole Villa and Cemetery." *Journal of Roman Archaeology* 9: 154–76.

Ben-David, Joseph. 1984. *The Scientist's Role in Society: A Comparative Study* [with a new introduction]. University of Chicago Press.

———. 1991. *Scientific Growth: Essays on the Social Organization and Ethos of Science*, ed. by Gad Freudenthal. University of California Press.

Ben-Dov, Jonathan, and Seth Sanders, eds. 2014. *Ancient Jewish Sciences and the History of Knowledge in Second Temple Literature*. New York University Press.

Benedum, Jost. 1974. "Zeuxis Philalethes und die Schule der Herophileer in Menos Kome." *Gesnerus* 31.3/4: 221–36.

Berggren, J.L., and Alexander Jones. 2000. *Ptolemy's Geography: An Annotated Translation of the Theoretical Chapters*. Princeton University Press.

Berryman, Sylvia. 1996. *Rethinking Aristotelian Teleology: The Natural Philosophy of Strato of Lampsacus*. Dissertation (Ph.D.), University of Texas (Austin).

———. 2002. "Galen and the Mechanical Philosophy." *Apeiron: A Journal for Ancient Philosophy and Science* 35.3 (September): 235–53.

———. 2003. "Ancient Automata and Mechanical Explanation." *Phronesis* 48.4 (November): 344–69.

———. 2009. *The Mechanical Hypothesis in Ancient Greek Natural Philosophy*. Cambridge University Press.

Bignami, F., and E. Salusti. 1990. "Tidal Currents and Transient Phenomena in the Strait of Messina: A Review." *The Physical Oceanography of Sea Straits*. L.J. Pratt, ed. Kluwer Academic: pp. 95–124.

Binfield Kevin, ed. 2004. *Writings of the Luddites*. Johns Hopkins University Press.

Blagg, T.F.C. 1987. "Society and the Artist." [in Wacher 1987: 2.717–42]

Blair, Carl. 1999. "The Iron Men of Rome." *Life of the Average Roman: A Symposium*. Mary Demaine and Rabun Taylor, eds. PZA Publishing: 51–65.

Bliquez, Lawrence. 2010. "Gynecological Surgery from the Hippocratics to the Fall of the Roman Empire." *Medicina nei secoli: arte e scienza* 22.1–3: 25–64.

———. 2015. *The Tools of Asclepius: Surgical Instruments in Greek and Roman Times*. Brill.

Boeselager, Dela von. 1983. "Solunt." *Antike Mosaiken in Sizilien: Hellenismus und Romische Kaiserzeit, 3. Jahrhundert v. Chr.-3. Jahrhundert n. Chr.* Rome: G. Bretschneider: 55–60, 208, and Taf. XV.

Bol, P.S. 1983. "Mosaik mit Tod des Archimedes." *Bildwerke aus Stein und aus Stuck von archaischer Zeit bis zur Spätantike*. Verlag Gutenberg: 342–43.

Bolton, Robert. 1991. "Aristotle's Method in Natural Science: *Physics* I." *Aristotle's Physics: A Collection of Essays*. Lindsay Judson, ed. Clarendon Press: 1–29.

Boutot, Alain. 2012. "Modernité de la *Catoptrique* de Héron d'Alexandrie." *Philosophie antique: problèmes, renaissances, usages* 12: 157–96.

Bowen, Alan. 2002. "The Art of the Commander and the Emergence of Predictive Astronomy." [in Tuplin & Rihll 2002: 76–111]

Bowen, Alan, and Robert Todd. 2004. *Cleomedes' Lectures on Astronomy: A Translation of* The Heavens *with an Introduction and Commentary*. University of California Press.

Bowersock, G.W. 1969. *Greek Sophists in the Roman Empire*. Clarendon Press.

———. 1974. *Approaches to the Second Sophistic: Papers Presented at the 105th Annual Meeting of the American Philological Association*. American Philological Association.

Bowman, Alan, and Andrew Wilson, eds. 2013. *The Roman Agricultural Economy: Organization, Investment and Production*. Oxford University Press.

Boylan, Michael. 1983. *Method and Practice in Aristotle's Biology*. University Press of America.

Brandon, C.J., et al. 2014. *Building for Eternity: The History and Technology of Roman Concrete Engineering in the Sea*. Oxbow Books.

Breebaart, A.B. 1976. "The Freedom of the Intellectual in the Roman World." *Talanta: Proceedings of the Dutch Archaeological and Historical Society* 7: 55–75.

Brende, Eric. 2004. *Better Off: Flipping the Switch on Technology*. HarperCollins.

Breidbach, Olaf. 2015. *Geschichte der Naturwissenschaften I: Die Antike*. Springer Spektrum.

British Museum. 1908. *A Guide to the Exhibition Illustrating Greek and Roman Life*, 3rd ed. Trustees of the British Museum (Department of Greek and Roman Antiquities).

Brown, P.R.L. 1992. *Power and Persuasion in Late Antiquity: Towards a Christian Empire*. University of Wisconsin Press.

Bruce, F.F. 1988. *The Book of the Acts*, rev. ed. W.B. Eerdmans.

———. 1990. *The Epistle to the Hebrews*, rev. ed. W.B. Eerdmans.

Brun, Jean-Pierre. 2012. "Techniques et économies de la Méditerranée antique." *Annuaire du Collège de France* 112: 465–90.

Brunt, P.A. 1980. "Free Labour and Public Works at Rome." *Journal of Roman Studies* 70: 81–100.

———. 1987. "Labour." [in Wacher 1987: 2.701–16]

———. 1994. "The Bubble of the Second Sophistic." *Bulletin of the Institute of Classical Studies* 39: 25–52.

Brush, Stephen. 1995. "Scientists as Historians." *Osiris* 10 (2nd series): 214–31.

Bryant, Joseph M. 1996. *Moral Codes and Social Structure in Ancient Greece: A Sociology of Greek Ethics from Homer to the Epicureans and Stoics*. State University of New York Press.

Buchheim, Thomas. 2001. "The Functions of the Concept of *Physis* in Aristotle's *Metaphysics*." *Oxford Studies in Ancient Philosophy* 20 (Summer): 201–34.

Burford, Alison. 1960. "Heavy Transport in Classical Antiquity." *The Economic History Review* 13.1, n.s.: 1–18.

———. 1972. *Craftsmen in Greek and Roman Society*. Cornell University Press.

Burgen, Arnold, Peter McLaughlin, and Jürgen Mittelstrass, eds. 1997. *The Idea of Progress*. Walter de Gruyter.

Burkert, Walter. 1997. "Impact and Limits of the Idea of Progress in Antiquity." [in Burgen et al. 1997: 19–46]

Burnet, John. 1930. "Appendix on the Meaning of *Physis*." *Early Greek Philosophy*. Adam & Charles Black: 363–64 (with pp. 10–13).

Burstein, Stanley. 1984. "A New *Tabula Iliaca*: The Vasek Polak Chronicle." *The J. Paul Getty Museum Journal* 12: 153–62.

Buxton, Richard, ed. 1999. *From Myth to Reason? Studies in the Development of Greek Thought*. Oxford University Press.

Byl, Simon. 1997. "Controverses antiques autour de la dissection et de la vivisection." *Revue Belge de Philologie et d'Histoire* 75.1: 113–20.

Bynum, Caroline Walker. 1995. *The Resurrection of the Body in Western Christianity, 200–1336*. Columbia University Press.

Calame, Claude. 2010. *Prométhée généticien: profits techniques et usages de métaphores*. La Encre marine.

Caley, Earle, and John Richards. 1956. [Theophrastus] *On Stones*. Ohio State University Press.

Cameron, Averil. 1993. *The Later Roman Empire: A.D. 284–430*. Harvard University Press.

Camerota, Filippo. 2002. "Optics and the Visual Arts: The Role of Skhnograf€a." [in Renn & Castagnetti 2002: 121–42]

Campbell, Brian. 2000. *The Writings of the Roman Land Surveyors: Introduction, Text, Translation and Commentary*. Society for the Promotion of Roman Studies.

Campbell, Duncan. 2011. "Ancient Catapults: Some Hypotheses Reexamined." *Hesperia: The Journal of the American School of Classical Studies at Athens* 80.4: 677–700.

Campbell, Gordon. 2006. *Strange Creatures: Anthropology in Antiquity*. Duckworth.

Capitani, Umberto. 1991. "I Sesti e la medicina." *Les Ecoles medicales à Rome: Actes du 2ème Colloque international sur les textes médicaux latins antiques: Lausanne, septembre 1986*. Philippe Mudry and Jackie Pigeaud, eds. Librairie Droz: 95–123.

Caplan, Harry. 1944. "The Decay of Eloquence at Rome in the First Century." *Studies in Speech and Drama in Honor of Alexander M. Drummond*. Cornell University Press: 295–325.

Carrier, Richard. 2005a. *Sense and Goodness without God: A Defense of Metaphysical Naturalism*. AuthorHouse.

———. 2005b. "The Spiritual Body of Christ and the Legend of the Empty Tomb." *The Empty Tomb: Jesus Beyond the Grave*. Robert Price and Jeffery Jay Lowder, eds. Prometheus: 105–232.

———. 2007. "Progress versus Directionality." *Historically Speaking* 8.4 (March–April): 44.

———. 2009. *Not the Impossible Faith: Why Christianity Didn't Need a Miracle to Succeed*. Lulu.

———. 2010. "Christianity Was Not Responsible for Modern Science." *The Christian Delusion: Why Faith Fails*. John Loftus, ed. Prometheus: 396–419.

———. 2011. "Christianity's Success Was Not Incredible." *The End of Christianity*. John Loftus, ed. Prometheus: 53–74, 372–75.

———. 2014a. *On the Historicity of Jesus: Why We Might Have Reason for Doubt*. Sheffield-Phoenix.

———. 2014b. "The Dark Ages." *Christianity is Not Great: How Faith Fails*. John Loftus, ed. Prometheus: 209–21, 509–12.

———. 2016. *Science Education in the Early Roman Empire*. Pitchstone.

Carriker, Andrew. 2003. *The Library of Eusebius of Caesarea*. Brill.

Carroll, Diane Lee. 1985. "Dating the Foot-Powered Loom: The Coptic Evidence." *American Journal of Archaeology* 89.1 (January): 168–73.

Cary, M., and E.H. Warmington. 1963. *The Ancient Explorers*. Penguin Books.

Casson, Lionel. 1950. "The Isis and Her Voyage." *Transactions and Proceedings of the American Philological Association* 81: 43–56.

———. 1956. "The Isis and Her Voyage: A Reply." *Transactions and Proceedings of the American Philological Association* 87: 239–40.

———. 1971. *Ships and Seamanship in the Ancient World*. Princeton University Press.

———. 1978. "Unemployment, the Building Trade, and Suetonius *Vesp*. 18." *Bulletin of the American Society of Papyrologists* 15.1–2: 43–51.

———. 1980. "Rome's Trade with the East: The Sea Voyage to Africa and India." *Transactions of the American Philological Association* 110: 21–36.

———. 1991. "Ancient Naval Technology and the Route to India." *Rome and India: The Ancient Sea Trade*. Vimala Begley and Richard Daniel De Puma, eds. University of Wisconsin Press: 8–11.

———. 2001. *Libraries in the Ancient World*. Yale University Press.

Cauderlier, Patrice. 1978. "Sciences pure et sciences appliquées dans l'Égypte romaine: Essai d'inventaire Antinoïte." *Recherches sur les artes à Rome*. Belles Lettres: 47–76.

Cave, J.F. 1977. "A Note on Roman Metal Turning." *History of Technology* 2: 77–94.

Charlesworth, James H. 1978. "Rylands Syriac Ms. 44 and a New Addition to the Pseudepigrapha: The Treatise of Shem, Discussed and Translated." *Bulletin of the John Rylands University Library of Manchester* 60: 376–403.

Chaumartin, F.-R. 2003. "Les sciences de la nature dans la pensée de Sénèque et son rapport avec le stoïcisme." [in Lévy et al. 2003: 157–65]

Cherniss, Harold. 1957. "Concerning the Face which Appears in the Orb of the Moon (De Facie quae in Orbe Lunae Apparet)." *Plutarch: Moralia XII*. Harold Cherniss and William Helmbold, eds. Harvard University Press: 1–223.

Chevallier, Raymond. 1976. *Roman Roads*. University of California Press. [tr. by N.H. Field]

———. 1993. *Sciences et techniques à Rome*. Presses universitaires de France.

Christ, John Ernest. 1974. *An Analysis of Atrial and Ventricular Beats in Conditions of Pathological and Experimental Disruption of Normal Control together with an Historical Introduction and Translation from the Original Greek of the Only Extant Reference to Enumeration of the Pulse*. Dissertation (Ph.D.), Baylor College of Medicine.

Christesen, Paul. 2003. "Economic Rationalism in Fourth-Century BCE Athens." *Greece & Rome* 50.1 (April): 31–56.

Ciarallo, Annamaria, and Ernesto De Carolis, eds. 1999. *Homo Faber: Natura, scienza e tecnica nell'antica Pompei*. Electa.

Clagett, Marshall. 1955. *Greek Science in Antiquity*. Ayer.

Clark, Gillian and Tessa Rajak, eds. 2002. *Philosophy and Power in the Graeco-Roman World: Essays in Honour of Miriam Griffin*. Oxford University Press.

Clarke, John R. 2003. *Art in the Lives of Ordinary Romans: Visual Representation and Non-Elite Viewers in Italy, 100 B.C.–A.D. 315*. University of California Press.

Clarke, M.L. 1971. *Higher Education in the Ancient World*. University of New Mexico Press.

Cloudsley-Thompson, J.L. 1980. *Biological Clocks: Their Functions in Nature*. Weidenfeld and Nicolson.

Codoñer, C. 1989. "La physique de Sénèque: Ordonnance et structure des 'Naturales quaestiones.'" *Aufstieg und Niedergang der römischen Welt* 2.36.3: 1779–1822.

Cohen, H. Floris. 1994. *The Scientific Revolution: A Historiographical Inquiry*. University of Chicago Press.

Cohen, Morris, and I.E. Drabkin. 1948. *A Source Book in Greek Science*. Oxford University Press.

Cohn-Haft, L. 1956. *The Public Physicians of Ancient Greece*. Smith College Publications in History.

Cohon, Robert. 2010. "Tools of the Trade: A Rare, Ancient Roman Builder's Funerary Plaque." *Antike Kunst* 53: 94–100.

Collins, Randall. 1998. *The Sociology of Philosophies: A Global Theory of Intellectual Change*. Harvard University Press.

Connolly, Peter, and Hazel Dodge. 1998. *The Ancient City: Life in Classical Athens & Rome*. Oxford University Press.

Connolly, R. Hugh. 1929. *Didascalia Apostolorum*. Clarendon Press.

Constantelos, Demetrios. 1998. *Christian Hellenism: Essays and Studies in Continuity and Change*. Aristide D. Caratzas.

Corcoran, Thomas H. 1964. "Fish Treatises in the Early Roman Empire." *The Classical Journal* 59.6 (March): 271–274.

Coulston, Jon, and Hazel Dodge. 2000. *Ancient Rome: The Archaeology of the Eternal City*. Oxford University School of Archaeology.

Coulton, John. 2002. "The Dioptra of Hero of Alexandria." [in Tuplin & Rihll 2002: 150–64]

Courland, Robert. 2011. *Concrete Planet: The Strange and Fascinating Story of the World's Most Common Man-Made Material.* Prometheus Books.

Courtney, Edward. 2001. *A Companion to Petronius.* Oxford University Press.

Creese, David. 2010. *The Monochord in Ancient Greek Harmonic Science.* Cambridge University Press.

Crombie, A.C. 1953. *Robert Grosseteste and the Origins of Experimental Science, 1100–1700.* Clarendon Press.

———. 1959. *The History of Science from Augustine to Galileo*, 2nd. rev. ed. Dover. [reprinted in 1995 with corrections and combining both volumes into one]

———. 1963. *Scientific Change: Historical Studies in the Intellectual, Social, and Technical Conditions for Scientific Discovery and Technical Invention, from Antiquity to the Present (Symposium on the History of Science, Oxford, 9–15 July 1963).* Basic Books.

———. 1994. *Styles of Scientific Thinking in the European Tradition: The History of Argument and Explanation Especially in the Mathematical and Biomedical Sciences and Arts.* 3 vols. Duckworth.

———. 1997. "Philosophical Commitments and Scientific Progress" [in Burgen et al. 1997: 47–63]

Crönert, Wilhelm. 1900. "Der Epikureer Philonides." *Mathematische und Naturwissenschaftliche Mittheilungen aus den Sitzungsberichten der Königlich Preussichen Akademie der Wissenschaften zu Berlin* 61.2: 942–59.

Cross, F.L., and E.A. Livingstone, eds. 1997. *The Oxford Dictionary of the Christian Church*, 3rd ed. Oxford University Press.

Crouch, Laura. 1975. *The Scientist in English Literature: Domingo Gonsales (1638) to Victor Frankenstein (1817).* Dissertation (Ph.D.), University of Oklahoma (Norman).

Crouwel, Joost. 2012. *Chariots and Other Wheeled Vehicles in Italy before the Roman Empire.* Oxbow Books.

Crouzel, Henri. 1969. *Remerciement à Origène Suivi de la Lettre d'Origène à Grégoire.* Cerf.

———. 1979. "Faut-il voir trois personnages en Grégoire le Thaumaturge?" *Gregorianum* 60: 287–320.

———. 1989. *Origen*. San Francisco: Harper & Row. [Translated by A.S. Worral from Henri Crouzel, *Origène*, 1985]

Cuomo, Serafina. 2000. *Pappus of Alexandria and the Mathematics of Late Antiquity*. Cambridge University Press.

———. 2001. *Ancient Mathematics*. Routledge.

———. 2002. "The Machine and the City: Hero of Alexandria's *Belopoeica*." [in Tuplin & Rihll 2002: 165–77]

———. 2007. *Technology and Culture in Greek and Roman Antiquity*. Cambridge University Press.

———. 2011. "A Roman Engineer's Tales." *Journal of Roman Studies* 101: 143–65.

Cüppers, Heinz. 1964. "Gallo-Römische Mähmaschine auf einer Relief in Trier." *Trierer Zeitschrift für Geschichte und Kunst des Trierer Landes und seiner Nachbargebiete* 27: 151–53.

Dalley, Stephanie, and John Peter Oleson. 2003. "Sennacherib, Archimedes, and the Water Screw: The Context of Invention in the Ancient World." *Technology and Culture* 44.1 (January): 1–26.

Damerow, Peter, Jürgen Renn, Simone Rieger, and Paul Weinig. 2002. "Mechanical Knowledge and Pompeiian Balances." [in Renn & Castagnetti 2002: 93–108]

D'Arms, John. 1981. *Commerce and Social Standing in Ancient Rome*. Harvard University Press.

Davies, R. W. 1970. "The Roman Military Medical Service." *Saalburg-Jahrbuch* 27: 84–104.

Davis, Danny Lee. 2009. *Commercial Navigation in the Greek and Roman World*. Dissertation (Ph.D.), University of Texas (Austin).

Deacy, Susan. 2008. *Athena*. Routledge.

Dear, Peter. 2005. "What Is the History of Science the History Of? Early Modern Roots of the Ideology of Modern Science." *Isis* 96.3 (September): 390–406.

Debru, Armelle. 1994. "L'expérimentation chez Galien." *Aufstieg und Niedergang der römischen Welt* 2.37.2: 1718–56.

———. 1995. "Les démonstrations médicales à Rome au temps de Galien." [in van der Eijk, Horstmanshoff and Schrijvers 1995: 1.69–82]

De Falco, V. 1923. *L'epicureo Demetrio Lacone*. Naples: Achille Cimmaruta. [repr. in *Epicureanism: Two Collections of Fragments and Studies*, Garland: 1987]

DeLaine, Janet. 1997. *The Baths of Caracalla: A Study in the Design, Construction, and Economics of Large-Scale Building Projects in Imperial Rome*. Journal of Roman Archaeology (Supplementary Series no. 25).

———. 2000. "Building the Eternal City: The Construction Industry in Imperial Rome." [in Coulston & Dodge 2000: 119–41]

Delattre, Joëlle. 1998. "Théon de Smyrne: modèles mécaniques en astronomie." [in Argoud & Guillaumin 1998: 371–95]

Deppert-Lippitz, Barbara. 1995. *Die Schraube zwischen Macht und Pracht: Das Gewinde in der Antike*. Thorbecke.

Derbyshire, John. 2006. *Unknown Quantity: A Real and Imaginary History of Algebra*. Joseph Henry Press.

Derks, Hans. 2002. "'The Ancient Economy': The Problem and the Fraud." *The European Legacy* 7.5 (October): 597–620.

Desclos, Marie-Laurence, and William Fortenbaugh, eds. 2011. *Strato of Lampsacus: Text, Translation, and Discussion*. Transaction Publishers.

Desmond, William. 2008. *Cynics*. Acumen.

De Ste. Croix, G.E.M. 2006. *Christian Persecution, Martyrdom, and Orthodoxy*. Michael Whitby and Joseph Streeter, eds. Oxford University Press.

DeVoto, James. 1996. *Philon and Heron: Artillery and Siegecraft in Antiquity*. Ares.

Dickie, Matthew. 2001. *Magic and Magicians in the Greco-Roman World*. Routledge.

Dickinson, H.W. 1939. *A Short History of the Steam Engine*. Cambridge University Press.

Dicks, D.R. 1959. "Thales." *The Classical Quarterly* 9.2 (November): 294–309.

Diederich, Silke. 1999. "Zur Rezeption der Naturwissenschaften in der römischen Schule der Kaiserzeit." *Antike Naturwissenschaft und ihre Rezeption* 9: 45–68.

Diels, Hermann. 1920. *Antike Technik: Sechs Vortrage von Hermann Diels*. 2nd ed. Teubner.

Dilke, O.A.W. 1971. *The Roman Land Surveyors: An Introduction to the* Agrimensores. David & Charles.

———. 1985. *Greek and Roman Maps*. Cornell University Press.

Dillon, John. 1970. "Lenses and Ancient Engraving." *SAN* (*Journal of the Society of Ancient Numismatics*) 2.2: 24–25.

———. 1993. *Alcinous: The Handbook of Platonism*. Clarendon.

Dillon, John, and A.A. Long, eds. 1988. *The Question of 'Eclecticism': Studies in Later Greek Philosophy*. University of California Press.

Di Pasquale, Giovanni. 2002. "The Fabrication of Roman Machines." [in Renn & Castagnetti 2002: 75–82]

———. 2004. *Tecnologia e Meccanica: Transmissione dei saperi tecnici dall'età ellenistica al mondo romano*. Leo S. Olschki.

Disney, Alfred, Cyril Hill and Wilfred Watson Baker, eds. 1928. *Origin and Development of the Microscope*. Royal Microscopical Society.

Distelzweig, Peter. 2013. "The Intersection of the Mathematical and Natural Sciences: The Subordinate Sciences in Aristotle." *Apeiron* 46.2: 85–105.

Dixon, Karen, and Pat Southern. 1992. *The Roman Cavalry: From the First to the Third Century A.D.* Routledge.

Dobson, J.F. 1918. "The Posidonivs Myth." *The Classical Quarterly* 12.3/4 (July/October): 179–195.

Dodds, E.R. 1951. *The Greeks and the Irrational*. University of California Press.

———. 1973. "The Ancient Concept of Progress." *The Ancient Concept of Progress and Other Essays on Greek Literature and Belief*. Clarendon Press: 1–25.

Dodge, Hazel. 2000. "'Greater than the Pyramids': The Water Supply of Ancient Rome." [in Coulston & Dodge 2000: 166–209]

Donderer, Michael. 1996. *Die Architekten der späten römischen Republik und der Kaiserzeit*. Universitätsbund Erlangen-Nürnberg.

D'Ooge, Martin Luther, Frank Egleston Robbins and Louis Charles Karpinski. 1926. *Nicomachus of Gerasa: Introduction to Arithmetic*. Macmillan.

Dougherty, Carol. 2006. *Prometheus*. Routledge.

Dover, K.J. 1976. "The Freedom of the Intellectual in Greek Society." *Talanta: Proceedings of the Dutch Archaeological and Historical Society* 7: 25–54.

Drachmann, A.G. 1963. *The Mechanical Technology of Greek and Roman Antiquity: A Study of the Literary Sources*. University of Wisconsin Press.

———. 1973. "The Crank in Graeco-Roman Antiquity." *Changing Perspectives in the History of Science: Essays in Honour of Joseph Needham*. Mikulás Teich and Robert Young, eds. D. Reidel: 33–51.

Drake, Stillman. 1989. "Hipparchus-Geminus-Galileo." *Studies in History and Philosophy of Science* 20: 47–56.

Draycott, Jane. 2013. "Glass Lenses in Roman Egypt: Literary, Documentary and Archaeological Evidence." *Bulletin of the Scientific Instrument Society* 117: 22–24.

Drinkwater, John. 2005. "Maximinus to Diocletian and the 'Crisis.'" *The Cambridge Ancient History, Volume 12: The Crisis of Empire, AD 193–337*, 2nd ed. Alan Bowman, Averil Cameron, Peter Garnsey, eds. Cambridge University Press: 28–66.

Drury, Clare. 2001. "The Pastoral Epistles." *Oxford Bible Commentary*. John Barton and John Muddiman, eds. Oxford University Press: 1220–33.

Duckworth, W.L.H., M..C. Lyons and B. Towers. 1962. *Galen On Anatomical Procedures: The Later Books*. 1962. Cambridge University Press.

Dueck, Daniela. 2012. *Geography in Classical Antiquity*. Cambridge University Press.

Duff, J. Wight, and Arnold M. Duff. 1935. *Minor Latin Poets: Volume I*. Loeb Classical Library. Harvard University Press.

Duncan-Jones, Richard P. 1977. "Giant Cargo-Ships in Antiquity." *The Classical Quarterly* 27.2 (n.s.): 331–32.

———. 1982. *The Economy of the Roman Empire: Quantitative Studies*. 2nd ed. Cambridge University Press.

Dunn, James. 1996. *The Epistles to the Colossians and to Philemon*. W.B. Eerdmans.

Dunsch, Boris. 2012. "*Arte rates reguntur*: Nautical Handbooks in Antiquity?" *Studies in History and Philosophy of Science (Part A)* 43.2: 270–83.

Durling, Richard. 1995. "Medicine in Plutarch's Moralia." *Traditio: Studies in Ancient and Medieval History, Thought, and Religion* 50: 311–14.

Eastwood, Bruce. 1997. "Astronomy in Christian Latin Europe c. 500 – c. 1150." *Journal for the History of Astronomy* 28.3 (August): 235–58.

Economou, N.A. 2000. *Astronomical Measurement Instruments from Ancient Greek Tradition*. Thessaloniki: Technology Museum of Thessaloniki.

Edelstein, Emma, and Ludwig Edelstein. 1945. *Asclepius: Collection and Interpretation of the Testimonies*, 2 vols. The Johns Hopkins Press.

Edelstein, Ludwig. 1935. "The Development of Greek Anatomy." *Bulletin of the Institute of the History of Medicine* 3.4 (April): 235–48.

———. 1952. "Recent Trends in the Interpretation of Ancient Science." *Journal of the History of Ideas* 13.4 (October): 573–604.

———. 1963. "Motives and Incentives for Science in Antiquity." [in Crombie 1963: 15–41]

———. 1967. *The Idea of Progress in Classical Antiquity*. Johns Hopkins Press.

Edelstein, Ludwig, and I.G. Kidd. 1989. *Posidonius I: The Fragments*. 2nd ed. Cambridge University Press.

Edlin, Herbert. 1976. *The Natural History of Trees*. Weidenfeld and Nicolson.

Edmunds, Michael. 2011. "An Initial Assessment of the Accuracy of the Gear Trains in the Antikythera Mechanism." *Journal for the History of Astronomy* 42.3: 307–20.

Edwards, Don Raymond. 1984. "Ptolemy's *Peri Analemmatos*: An Annotated Transcription of Moerbeke's Latin Translation and of the Surviving Greek Fragments with an English Version and Commentary." Dissertation (Ph.D.), Brown University (Providence, Rhode Island).

Efron, Noah. 2009. "That Christianity Gave Birth to Modern Science." [in Numbers 2009: 79–89]

Eggert, Gerhard. 1995. "Ancient Aluminum? Flexible Glass? Looking for the Real Heart of a Legend." *Skeptical Inquirer* 19.3 (May–June): 37–40.

Eibner, Clemens. 2013. "Astronomisches aus Salzburg und Heron von Alexandria." *Calamus: Festschrift für Herbert Grassl zum 65. Geburtstag*. Rupert Breitwieser, Monika Frass and Georg Nightingale, eds. Harrassowitz: 177–83.

Eijk, Philip J. van der. 2005. "Between the Hippocratics and the Alexandrians: Medicine, Philosophy, and Science in the Fourth Century BCE." [in Sharples 2005: 72–109]

Eijk, Philip J. van der, H. Horstmanshof, and P. Schrijvers. 1995. *Ancient Medicine in its Socio-Cultural Context: Papers Read at the Congress Held at Leiden University, 13–15 April 1992*. Rodopi. [= *Clio Medica* 27 and 28]

Eisenstein, Elizabeth L. 1980. *The Printing Press as an Agent of Change*. Cambridge University Press.

Ellspermann, Gerard. 1949. *The Attitude of the Early Christian Latin Writers toward Pagan Literature and Learning*. Catholic University of America Press.

Elsner, Jas. 1995. *Art and the Roman Viewer: The Transformation of Art from the Pagan World to Christianity*. Cambridge University Press.

Engels, Donald. 1985. "The Length of Eratosthenes' Stade." *American Journal of Philology* 106.3 (Autumn): 298–311.

Enoch, Jay. 1998. "The Enigma of Early Lens Use." *Technology and Culture* 39.2 (April): 273–91.

Erdkamp, Paul, and Koenraad Verboven, eds. 2015. *Structure and Performance in the Roman Economy: Models, Methods and Case Studies*. Peeters.

Evans, James. 1998. *The History and Practice of Ancient Astronomy*. Oxford University Press.

———. 1999. "The Material Culture of Greek Astronomy." *Journal for the History of Astronomy* 30.3 (August): 237–307.

Evans, James, and J. L. Berggren. 2006. *Geminos's Introduction to the Phenomena: A Translation and Study of a Hellenistic Survey of Astronomy*. Princeton University Press.

Everett, Nicholas. 2012. *The Alphabet of Galen: Pharmacy from Antiquity to the Middle Ages*. University of Toronto Press.

Falcon, Andrea. 2016. *Aristotelianism in the First Century B.C.E.* Cambridge University Press.

Fales, Evan. 1996a. "Scientific Explanations of Mystical Experiences, Part I: The Case of St. Teresa." *Religious Studies* 32: 143–163.

———. 1996b. "Scientific Explanations of Mystical Experiences, Part II: The Challenge to Theism." *Religious Studies* 32: 297–313.

———. 1999. "Can Science Explain Mysticism?" *Religious Studies* 35: 213–227.

Farrington, Benjamin. 1946. *Head and Hand in Ancient Greece: Four Studies in the Social Relations of Thought*. Watts & Company.

———. 1965. *Science and Politics in the Ancient World*, 2nd ed. Unwin University Books.

Fee, Gordon. 1987. *The First Epistle to the Corinthians*. W.B. Eerdmans.

Feeney, D.C. 2007. *Caesar's Calendar: Ancient Time and the Beginnings of History*. University of California Press.

Feke, Jacqueline. 2011. "Ptolemy's Defense of Theoretical Philosophy." *Apeiron: A Journal for Ancient Philosophy and Science* 45.1: 61–90.

———. 2014. "Meta-Mathematical Rhetoric: Hero and Ptolemy against the Philosophers." *Historia Mathematica* 41.3: 261–76.

Feldhaus, Franz. 1954. *Die Maschine im Leben der Volker: Ein Uberblick von der Urzeit bis zur Renaissance*. Birkhauser.

Ferngren, Gary. 1982. "A Roman Declamation on Vivisection." *Transactions & Studies of the College of Physicians of Philadelphia* 4.4 (December): 272–90.

———. 1985. "Roman Lay Attitudes towards Medical Experimentation." *Bulletin of the History of Medicine* 59.4: 495–505.

Ferngren, Gary, and Darrel Amundsen. 1996. "Medicine and Christianity in the Roman Empire: Compatibilities and Tensions." *Aufstieg und Niedergang der römischen Welt* 2.37.3: 2957–80.

Ferrua, Antonio. 1991. *The Unknown Catacomb: A Unique Discovery of Early Christian Art*. Geddes and Grosset.

Finley, Moses. 1981. "Technical Innovation and Economic Progress in the Ancient World." *Economy and Society in Ancient Greece*. Brent Shaw and Richard Saller, ed. Chatto & Windus: 176–95, 273–75. [a partly updated reprint of Finley's original article of the same title in *The Economic History Review* 18.1 (1965): 29–45]

———. 1985. *The Ancient Economy*, 2nd ed. Hogarth Press.

Fischer, Klaus-Dietrich. 1984. "Seneca d. Ä, *Controversiae* 10,5,17: Ein übersehenes Zeugnis zur Geschichte der antiken Anatomie." *Sudhoffs Archiv für Geschichte der Medizin und der Naturwissenschaften* 68.1: 110–12.

Fleming, Stuart J. 1999. *Roman Glass: Reflections on Cultural Change*. University of Pennsylvania Museum of Archaeology and Anthropology.

Foley, Vernard, and Werner Soedel. 1981. "Ancient Oared Warships." *Scientific American* 244.4 (April): 148–63.

Forbes, R.J. 1950. *Metallurgy in Antiquity: A Notebook for Archaeologists and Technologists*. Brill.

Ford, Brian. 1985. *Single Lens: The Story of the Simple Microscope*. Harper & Row.

Fortenbaugh, William, Pamely Huby, Robert Sharples, and Dimitri Gutas. 1992–2007. *Theophrastus of Eresus: Sources for His Life, Writings, Thought, and Influence*. E.J. Brill. [multiple volumes in progress]

Fox, Robin Lane. 1987. *Pagans and Christians*. Alfred A. Knopf.

Fraser, A.D. 1934. "Recent Light on the Roman Horseshoe." *The Classical Journal* 29. 9 (June): 689–69.

Fraser, Peter. 1972. *Ptolemaic Alexandria*, 3 vols. Clarendon Press.

Frede, Michael. 1981. "On Galen's Epistemology." *Galen: Problems and Prospects*. Vivian Nutton, ed. Wellcome Institute for the History of Medicine: 65–86.

Frede, Michael, and Gisela Striker. 1996. *Rationality in Greek Thought*. Clarendon Press.

Freeth, T. 2002a. "The Antikythera Mechanism I: Challenging the Classic Research" *Mediterranean Archaeology and Archaeometry* 2.1 (June): 21–35.

———. 2002b. "The Antikythera Mechanism II: Is It Posidonius' Orrery?" *Mediterranean Archaeology and Archaeometry* 2.2 (December): 45–58.

Freeth, T., Y. Bitsakis, X. Moussas, J.H. Seiradakis, A. Tselikas, H. Mangou, M. Zafeiropoulou, R. Hadland, D. Bate, A. Ramsey, M. Allen, A. Crawley, P. Hockley, T. Malzbender, D. Gelb, W. Ambrisco, and M.G. Edmunds. 2006. "Decoding the Ancient Greek Astronomical Calculator Known as the Antikythera Mechanism." *Nature* 444 (November 30): 587–91.

French, Roger. 1994. *Ancient Natural History: Histories of Nature*. Routledge.

French, Roger, and Frank Greenaway, eds. 1986. *Science in the Early Roman Empire: Pliny the Elder, His Sources and Influence*. Croom Helm.

Frere, Sheppard. 1987. *Britannia: A History of Roman Britain*, 3rd ed. Routledge: 287–89.

Freudenthal, Hans. 1977. "What is algebra and what has been its history?" *Archive for History of Exact Sciences* 16.3 (September): 189–200.

Fried, Michael, and Sabetai Unguru. 2001. *Apollonius of Perga's* Conica: *Text, Context, Subtext*. Brill.

Frontisi-Ducroux, Françoise. 2000. *Dédale: Mythologie de l'artisan en Grèce ancienne*, new ed. La Découverte.

Fulford, Michael, David Sim, and Alistair Doig. 2004. "The production of Roman ferrous armour: a metallographic survey of material from Britain, Denmark, and Germany and its implications." *Journal of Roman Archaeology* 17a: 197–220.

Fung, Ronald. 1988. *The Epistle to the Galatians*. W.B. Eerdmans.

Furley, David, and J.S. Wilkie. 1984. *Galen on Respiration and the Arteries*. Princeton University Press.

Gain, D.B. 1976. *The Aratus Ascribed to Germanicus Caesar*. Athlone Press.

Gaitzsch, Wolfgang, and Hertmut Matthäus. 1981a. "Schreinerwerkzeuge aus dem Kastell Altstadt bei Miltenberg." *Antike Welt: Zeitschrift für Archäologie und Kulturgeschichte* 12.3: 21–30.

———. 1981b. "Runcinae—Römische Hobel." *Bonner Jahrbucher des rheinischen Landesmuseums in Bonn* 181: 205–47.

Gallazzi, Claudio, and Salvatore Settis. 2006. *Le tre vite del Papiro di Artemidoro: voci e sguardi dall'Egitto greco-romano*. Electa.

Galloway, Elijah. 1837. *History and Progress of the Steam Engine with a Practical Investigation of Its Structure and Application*. Thomas Kelly.

Garnsey, Peter. 1970. *Social Status and Legal Privilege in the Roman Empire*. Clarendon.

Garzya, Antonio. 1999. "Le Vin dans la Littérature Médicale de l'Antiquité Tardive et Byzantine." *Filologia antica e moderna* 9.17: 13–25.

Gero, Stephen. 1990. "Galen on Jews and Christians: A Reappraisal of the Arabic Evidence." *Orientalia Christiana Periodica* 56: 371–411.

Gibbs, Sharon. 1976. *Greek and Roman Sundials*. Yale University Press.

Gies, Frances, and Joseph Gies. 1994. *Cathedral, Forge, and Waterwheel: Technology and Invention in the Middle Ages*. Harper Collins.

Gillispie, Charles Coulston, ed. 1980. *Dictionary of Scientific Biography*. Vols. I–XVI. Charles Scribner's Sons.

Giovacchini, Julie. 2012. *L'Empirisme d'Épicure*. Classiques Garnier.

Glare, P.G.W. 1996. *Oxford Latin Dictionary*, corrected edition. Clarendon Press.

Goethert, Friedrich Wilhem. 1931. "Das sogenannte Archimedesmosaik." *Zur Kunst der römischen Republik*. Bark & Schröter: 56–62.

Goldstein, Bernard R. 1967. "The Arabic Version of Ptolemy's *Planetary Hypotheses*." *Transactions of the American Philosophical Society* 57.4 (new ser.): 3–55.

Goodyear, F.R.D. 1965. *Incerti auctoris Aetna*. Cambridge University Press.

———. 1984. "The 'Aetna': Thought, Antecedents, and Style." *Aufstieg und Niedergang der römischen Welt* 2.32.1: 344–63.

Gorrie, Averilda. 1970. "Some Reflections about Geography in the Hellenistic Age." *Prudentia* 2.1: 11–18.

Gottschalk, H.B. 1987. "Aristotelian Philosophy in the Roman World from the Time of Cicero to the End of the Second Century A.D." *Aufstieg und Niedergang der römischen Welt* 2.36.2: 1079–1174.

Gourevitch, Danielle. 1970. "Some Features of the Ancient Doctor's Personality as Depicted in Epitaphs." *Nordisk Medicinhistorisk Årsbok* 1970: 38–49.

Graf, Fritz. 2001. "Athena and Minerva: Two Faces of One Goddess?" *Athena in the Classical World*. Susan Deacy and Alexandra Villing, eds. Brill: 127–39.

Graham, Daniel. 2006. *Explaining the Cosmos: The Ionian Tradition of Scientific Philosophy*. Princeton University Press.

———. 2013. *Science before Socrates: Parmenides, Anaxagoras, and the New Astronomy*. Oxford University Press.

Grandjouan, Clairève. 1961. "Terracottas and Plastic Lamps of the Roman Period." *The Athenian Agora* 6: 1–106.

Grant, Mark. 2000. *Galen on Food and Diet*. Routledge.

Grant, Michael. 1999. *Collapse and Recovery of the Roman Empire*. Routledge.

Grant, Robert M. 1952. *Miracle and Natural Law in Graeco-Roman and Early Christian Thought*. North-Holland.

Graves, Robert, and Michael Grant. 1979. *Suetonius: The Twelve Caesars*, rev. ed. Penguin Books.

Green, Charles. 1966. "The Purpose of the Early Horseshoe." *Antiquity* 40.160 (December): 305–08.

Green, Peter. 1986. "Hellenistic Technology: Eye, Hand, and Animated Tool." *Southern Humanities Review* 20.2 (Spring): 101–13.

———. 1990. *Alexander to Actium: The Historical Evolution of the Hellenistic Age*. University of California Press.

Green, Robert Montraville. 1955. *Asclepiades: His Life and Writings—A Translation of Cocchi's Life of Asclepiades and Gumpert's Fragments of Asclepiades*. E. Licht.

Greene, Kevin. 1990. "Perspectives on Roman Technology." *Oxford Journal of Archaeology* 9.2 (July): 209–19.

———. 1992. "How Was Technology Transferred in the Western Provinces?" *Current Research on the Romanization of the Western Provinces*. Mark Wood and Francisco Queiroga, eds. Tempvs Reparatvm: 101–05.

———. 1994. "Technology and Innovation in Context: The Roman Background to Mediaeval and Later Developments." *Journal of Roman Archaeology* 7: 22–33.

———. 2000. "Technological Innovation and Economic Progress in the Ancient World: M. I. Finley Re-Considered." *Economic History Review* 53.1, New Series (February): 29–59.

Grewe, Klaus. 2009. "Chorobat und Groma: neue Gedanken zur Rekonstruktion und Handhabung der beiden wichtigsten Vermessungsgeräte antiker Ingenieure." *Bonner Jahrbücher des Rheinischen Landesmuseums in Bonn und des Rheinischen Amtes für Bodendenkmalpflege im Landschaftsverband Rheinland und des Vereins von Altertumsfreunden im Rheinlande* 209: 109–28.

———. 2010. "Ingenieurkunst und die Kraft des Wassers." 2010. *Antike Welt: Zeitschrift für Archäologie und Kulturgeschichte* 3: 37–42.

Griffin, Miriam. 1976. *Seneca: A Philosopher in Politics*. Clarendon Press.

———. 1994. "The Intellectual Developments of the Ciceronian Age." *The Cambridge Ancient History, Volume 9: The Last Age of the Roman Republic, 146-43 BC*, 2nd ed. J.A. Crook, Andrew Lintott and Elizabeth Rawson, eds. Cambridge University Press: 689–728.

Grmek, Mirko, and Danielle Gourevitch. 1998. *Les Maladies dans l'Art Antique*. Fayard.

Grose, D. F. 1977. "Early Blown Glass." *Journal of Glass Studies* 19: 9–29.

Gummerus, Herman Gregorius. 1932. *Der Ärztestand im römischen Reiche nach den Inschriften*. Akademische Buchhandlung.

Habermehl, Peter. 2006. *Petronius,* Satyrica *79–141: Ein philologisch-literarischer Kommentar*, vol. 1. Walter de Gruyter.

Habinek, Thomas. 1989. "Science and Tradition in *Aeneid* 6." *Harvard Studies in Classical Philology* 92: 223–55.

Hadot, Pierre. 1979. "Les divisions des parties de la philosophie dans l'Antiquité." *Museum Helveticum* 36.4: 201–23.

Hague, Douglas, and Rosemary Christie. 1975. *Lighthouses: Their Architecture, History and Archaeology*. Gomer Press.

Hahn, Johannes. 1991. "Plinius und die griechischen Ärzte in Rom: Naturkonzeption und Medizinkritik in der Naturalis Historia." *Sudhoffs Archiv für Geschichte der Medizin und der Naturwissenschaften* 75: 209–39.

Hall, Stuart George. 1993. *Gregory of Nyssa: Homilies on Ecclesiastes: An English Version with Supporting Studies: Proceedings of the Seventh International Colloquium on Gregory of Nyssa (St. Andrews, 5–10 September 1990)*. Walter de Gruyter.

Hamilton, J.S. 1986. "Scribonius Largus on the Medical Profession." *Bulletin of the History of Medicine* 60.2 (Summer): 209–16.

Hamilton, N.T., N.M. Swerdlow, and G.J. Toomer. 1987. "The Canobic Inscription: Ptolemy's Earliest Work." *From Ancient Omens to Statistical Mechanics: Essays on the Exact Sciences Presented to Asger Aaboe*. J.L. Berggren and B.R. Goldstein, eds. University Library: 55–73.

Hankinson, R.J. 1988. "Galen Explains the Elephant." *Philosophy & Biology*. Mohan Matthen and Bernard Linsky, eds. Calgary (Alberta, Canada): University of Calgary Press: 135–57.

———. 1991a. *Galen: On the Therapeutic Method*. Clarendon Press.

———. 1991b. "Galen on the Foundations of Science." *Galeno, obra, pensamiento e influencia: coloquio internacional celebrado en Madrid, 22–25 de Marzo de 1988*. J.A. López Férez, ed. Universidad Nacional de Educación a Distancia: 15–29.

———. 1992. "Galen's Philosophical Eclecticism." *Aufstieg und Niedergang der römischen Welt* 2.36.5: 3505–22.

———. 1994a. "Galen's Concept of Scientific Progress." *Aufstieg und Niedergang der römischen Welt* 2.37.2: 1776–89.

———. 1994b. "Galen's Anatomical Procedures: A Second-Century Debate in Medical Epistemology." *Aufstieg und Niedergang der römischen Welt* 2.37.2: 1834–55.

———. 1995a. "Philosophy of Science." [in J. Barnes 1995: 109–39]

———. 1995b. "Science." [in J. Barnes 1995: 140–67]

———. 1998. *Cause and Explanation in Ancient Greek Thought*. Clarendon Press.

———, ed. 2008. *The Cambridge Companion to Galen*. Cambridge University Press.

Hannah, Robert. 2009. *Time in Antiquity*. Routledge.

Hanson, A.E, and M.H. Green. 1994. "Soranus of Ephesus: Methodicorum princeps." *Aufstieg und Niedergang der römischen Welt* 2.37.2: 968–1075.

Hardy, Edmund. 1884. *Der Begriff der Physis in der griechischen Philosophie*. Weidmann.

Hardy, Gavin, and Laurence Totelin. 2016. *Ancient Botany*. Routledge.

Harig, G. 1971. "Zum Problem « Krankenhaus » in Der Antike." *Klio* 53: 179–95.

Harris, H.A. 1968. "The Starting-Gate for Chariots at Olympia." *Greece & Rome* 15.2 (October): 113–26.

Harris, William V. 1980. "Roman Terracotta Lamps: The Organization of an Industry." *Journal of Roman Studies* 70: 126–45.

Harris, William V., and K. Iara, eds. 2011. *Maritime Technology in the Ancient Economy: Ship-design and Navigation*. Journal of Roman Archaeology.

Harrison, Peter. 1998. *The Bible, Protestantism, and the Rise of Natural Science*. Cambridge University Press.

———. 2001. "Curiosity: Forbidden Knowledge, and the Reformation of Natural Philosophy in Early Modern England." *Isis* 92.2 (June): 265–90.

———. 2007. "Moral Progress and Early Modern Science." *Historically Speaking* 9.1 (September/October): 13–14.

Harrison, S.J. 2000. *Apuleius: A Latin Sophist*. Oxford University Press.

Hart, Gerald D. 2000. *Asclepius: The God of Medicine*. Royal Society of Medicine.

Hasaki, Eleni. 2012. "Workshops and Technology." *A Companion to Greek Art*. Tyler Jo Smith and Dimitris Plantzos, eds. Wiley-Blackwell: 255–72.

Hawkes, C.F.C. 1977. *Pytheas: Europe and the Greek Explorers*. Blackwell.

Healy, John. 1978. *Mining and Metallurgy in the Greek and Roman World*. Thames and Hudson.

———. 1999. *Pliny the Elder on Science and Technology*. Oxford University Press.

Heath, Thomas. 1913. *Aristarchus of Samos, the ancient Copernicus: A History of Greek Astronomy to Aristarchus together with Aristarchus' Treatise on the Size and Distances of the Sun and Moon*. Clarendon Press.

Heidarzadeh, Tofigh. 2004. *Theories of Comets to the Age of Laplace*. Dissertation (Ph.D.), University of Oklahoma (Norman).

Heinemann, Gottfried. 2000. "Physik." *Der neue Pauly Enzyklopädie der Antike* (J. B. Metzler) Bd. 9: 990–7.

———. 2001. *Studien zum griechischen Naturbegriff, Teil I: Philosophische Grundlegung: Der Naturbegriff und die 'Natur'*. W.V.T. Wissenschaftlicher Verlag Trier.

———. 2005a. "Die Entwicklung des Begriffs *physis* bis Aristoteles." [in Schürmann 2005: 16–60]

———. 2005b. "Platon, Aristoteles und die Kunsttheorie der griechischen Antike." *Klassiker der Kunstphilosophie: von Platon bis Lyotard*. Stefan Majetschak, ed. Beck: 14–36.

Heinz, Werner. 2009. "Medizin und Religion in der Spätantike." *Mystik und Natur: zur Geschichte ihres Verhältnisses vom Altertum bis zur Gegenwart*. Peter Dinzelbacher, ed. Walter de Gruyter: 7–37.

Heisel, Joachim. 1993. *Antike Bauzeichnungen*. Wissenschaftliche Buchgesellschaft.

Hellevang, Kenneth J. 1998. "Temporary Grain Storage." *Publications of the North Dakota NDSU Extension Service* AE-84 (Revised). Fargo, North Dakota: North Dakota State University (College of Agriculture, Food Systems, and Natural Resources). [www.ag.ndsu.edu/pubs/ageng/grainsto/ae84-1.htm]

Henig, Martin. 1994. *Classical Gems: Ancient and Modern Intaglios and Cameos in the Fitzwilliam Museum, Cambridge*. Cambridge University Press.

Henry, John. 2008. *The Scientific Revolution and the Origins of Modern Science*, 3rd ed. St. Martin's.

Henze, D. 1998. *Enzyklopädie der Entdecker und Erforscher der Erde*. ADEVA.

Hermanns, Marcus Heinrich. 2010. "*Abacus* und *tabulae lusoriae*: geometrische Ritzzeichnungen im Gebiet der antiken Stadt Selinunt (Sizilien)." *Archäologisches Korrespondenzblatt: Urgeschichte, Römerzeit, Frühmittelalter* 40.3: 401–09.

Hicks, Alastair. 1997. "Power and Food Security." [paper delivered by the Senior Regional Agricultural Engineering and Agro Industries Officer at the International Solar Energy Society's 1997 Solar World Congress, Taejon, Republic of Korea, 24–30 August 1997: www.fao.org/sd/EGdirect/EGan0006.htm]

Hill, Donald Routledge. 1976. *On the Construction of Waterclocks*. Turner & Devereux.

———. 1984. *A History of Engineering in Classical and Medieval Times*. Open Court.

———. 1993. *Islamic Science and Engineering*. Edinburgh University Press.

Hillert, Andreas. 1990. *Antike Ärztedarstellungen*. Peter Lang.

Hills, Richard L. 1989. *Power from Steam: A History of the Stationary Steam Engine*. Cambridge University Press.

Hodge, A. Trevor. 2000. "Reservoirs and Dams." [in Wikander 2000c: 331–42]

Hofstadter, Richard. 1962. *Anti-Intellectualism in American Life*. Vintage.

Hohlfelder, Robert L. 1997. "Building Harbours in the Early Byzantine Era: The Persistance of Roman Technology." *Byzantinische Forschungen* 24: 368–80.

Hohlfelder, Robert, John Oleson, Avner Raban, and R. Lindley Vann. 1983. "Sebastos: Herod's Harbor at Caesarea Maritima." *The Biblical Archaeologist* 46.3 (Summer): 133–43.

Hopkins, Keith. 1998. "Christian Number and Its Implications." *Journal of Early Christian Studies* 6.2 (Summer): 185–226.

Horky, Phillip Sidney. 2013. *Plato and Pythagoreanism*. Oxford University Press.

Hornblower, Simon, and Antony Spawforth, eds. 2012. *The Oxford Classical Dictionary*. 4th ed. Oxford University Press. (Assistant ed., Esther Eidinow.)

Horsfall, Nicholas. 1979. "Doctus Sermones Utriusque Linguae?" *Echos du Monde Classique* 23.3 (October): 79–95.

Horstmannshoff, Hermann Freder. 1990. "The Ancient Physician: Craftsman or Scientist?" *Journal of the History of Medicine and Allied Sciences* 45.2 (April): 176–97.

Houston, George W. 1988. "Ports in Perspective: Some Comparative Materials on Roman Merchant Ships and Ports." *American Journal of Archaeology* 92: 553–64.

———. 2014. *Inside Roman Libraries: Book Collections and Their Management in Antiquity*. The University of North Carolina Press.

Hubner, Wolfgang. 2000. *Geographie und verwandte Wissenschaften*. Franz Steiner.

Huby, Pamela. 2004. "Elementary Logic in the Ancient World." *Bulletin of the Institute of Classical Studies* 47 (2004): 119–28.

Huby, Pamela, and Gordon Neal, eds. 1989. *The Criterion of Truth: Essays Written in Honour of George Kerferd Together with a Text and Translation (with Annotations) of Ptolemy's* On the Kriterion *and* Hegemonikon. Liverpool University Press.

Humphrey, John, John Oleson, and Andrew Sherwood, eds. 1998. *Greek and Roman Technology: A Sourcebook*. Routledge.

Hunsberger, Bruce, and Bob Altemeyer. 2006. *Atheists: A Groundbreaking Study of America's Nonbelievers*. Prometheus Books.

Hussey, Edward. 2002. "Aristotle and Mathematics." [in Tuplin & Rihll 2002: 217–29]

Hutchinson, D. S. 1988. "Doctrines of the Mean and the Debate Concerning Skills in Fourth-Century Medicine, Rhetoric and Ethics." *Apeiron: A Journal for Ancient Philosophy and Science* 21.2 (Summer): 17–52.

Hyland, Ann. 1990. *Equus: The Horse in the Roman World*. Yale University Press.

Irby-Massie, Georgia, ed. 2016. *A Companion to Science, Technology, and Medicine in Ancient Greece and Rome*. Wiley Blackwell.

Irby-Massie, Georgia, and Paul Keyser, ed. 2002. *Greek Science of the Hellenistic Era: A Sourcebook*. Routledge.

Iskandar, Albert. 1988. *Galen on Examinations by Which the Best Physicians Are Recognized*. Akademie-Verlag.

Israelowich, Ido. 2015. *Patients and Healers in the High Roman Empire*. Johns Hopkins University Press.

Jackson, Ralph. 1988. *Doctors and Diseases in the Roman Empire*. University of Oklahoma Press.

———. 1993. "Roman Medicine: The Practitioners and Their Practices." *Aufstieg und Niedergang der römischen Welt* 2.37.1: 79–101.

———. 1995. "The Composition of Roman Medical *Instrumentaria* as an Indicator of Medical Practice: A Provisional Assessment." [in van der Eijk, Horstmanshoff and Schrijvers 1995: 1.189–207]

———. 2010. "Cutting for Stone: Roman Lithotomy Instruments in the Museo Nazionale Romano." *Medicina nei secoli: arte e scienza* 22.1–3: 393–418.

Jacoby, Susan. 2008. *The Age of American Unreason*. Vintage Books.

Jaeger, Mary. 2002. "Cicero and Archimedes' Tomb." *Journal of Roman Studies* 92: 49–61.

James, Peter, and Nick Thorpe. 1994. *Ancient Inventions*. Ballantine.

Jansen, Gemma, et al. 2011. *Roman Toilets: Their Archaeology and Cultural History*. Peeters.

Janssen, L.F. 1979. "'Superstitio' and the Persecution of the Christians." *Vigiliae Christianae* 33.2 (June): 131–59.

Jex-Blake, K., E. Sellers, Heinrich Urlichs, and Raymond Schoder. 1968. *The Elder Pliny's Chapters on the History of Art*. Argonaut.

Johansen, Karsten Friis. 1998. *A History of Ancient Philosophy: From the Beginnings to Augustine*. Routledge.

Jones, Alexander. 1994. "The Place of Astronomy in Roman Egypt." [in T. Barnes 1994: 25–51]

———. 1999. *Astronomical Papyri from Oxyrhynchus (P. Oxy. 4133–4300a)*. American Philosophical Society.

———. 2012. "Ptolemy's *Geography*: Mapmaking and the Scientific Enterprise." [in Talbert 2012: 109–28]

———. 2016. *Inscriptions of the Antikythera Mechanism*. The Hellenistic Society of History, Philosophy and Didactics of Science.

———. 2017. *A Portable Cosmos: Revealing the Antikythera Mechanism, Scientific Wonder of the Ancient World*. Oxford University Press.

Jones, L.J. 1979. "The Early History of Mechanical Harvesting." *History of Technology* 4: 101–48.

Jones, Steven. 2006. *Against Technology: From the Luddites to Neo-Luddism*. Routledge.

Jones, W.H.S. 1947. *The Medical Writings of Anonymus Londinensis*. Cambridge University Press.

Joost-Gaugier, Christiane. 2006. *Measuring Heaven: Pythagoras and His Influence on Thought and Art in Antiquity and the Middle Ages*. Cornell University Press.

Joshel, Sandra Rae. 1992. *Work, Identity, and Legal Status at Rome: A Study of the Occupational Inscriptions*. University of Oklahoma Press.

Jost, John, and Jack Glaser, Arie Kruglanski, and Frank Sulloway. 2003. "Political Conservatism as Motivated Social Cognition." *Psychological Bulletin* 129.3: 339 –375.

Jouanna, Jacques. 1996. "Le Vin et la Médecine dans la Grèce Ancienne." *Revue des Études Grecques* 109.2: 410–34.

Judge, E.A. 1983. "The Reaction against Classical Education in the New Testament." *Journal of Christian Education* 77.1: 7–14.

Kalligas, Paul. 2004. "Platonism in Athens during the First Two Centuries A.D.: An Overview." *Rhizai: A Journal for Ancient Philosophy and Science* 2.2: 37–56.

Kambitsis, Jean. 1972. L'Antiope *d'Euripide: Édition commentée des fragments*. E. Hourzamanis.

Kapr, Albert. 1996. *Johann Gutenberg: The Man and His Invention*. Scolar Press.

Kelhoffer, James. 2000. *Miracle and Mission: The Authentication of Missionaries and Their Message in the Longer Ending of Mark*. Mohr Siebeck.

Kellaway, Peter. 1946. "The Part Played by Electric Fish in the Early History of Bioelectricity and Electrotherapy." *Bulletin of the History of Medicine* 20: 112–37.

Kelly, Jack. 2004. *Gunpowder: Alchemy, Bombards, and Pyrotechnics: The History of the Explosive That Changed the World*. Basic Books.

Keyser, Paul. 1988. "Suetonius *Nero* 41.2 and the Date of Heron Mechanicus of Alexandria." *Classical Philology* 83.3: 218–20.

———. 1992. "A New Look at Heron's 'Steam Engine.'" *Archive for History of Exact Sciences* 44.2: 107–24.

———. 1994. "On Cometary Theory and Typology from Nechepso-Petosiris through Apuleius to Servius." *Mnemosyne* 47.5 (November): 625–51.

Keyser, Paul, and Georgia Irby-Massie, eds. 2012. *The Encyclopedia of Ancient Natural Scientists*. Routledge.

Khanikoff, N. 1860. "Analysis and Extracts of [...] [the] Book of the Balance of Wisdom, An Arabic Work on the Water-Balance, Written by 'Al-Khâzinî in the Twelfth Century." *Journal of the American Oriental Society* 6: 1–128.

Kidd, I. G. 1978. "Philosophy and Science in Posidonius." *Antike und Abendland* 24: 7–15.

———. 1988. *Posidonius II: The Commentary*. Cambridge University Press.

———. 1999. *Posidonius III: The Translation of the Fragments*. Cambridge University Press.

Kieffer, John Spangler. 1964. *Galen's Institutio Logica: English Translation, Introduction, and Commentary*. The Johns Hopkins Press.

King, Helen. 1986. "Agnodike and the Profession of Medicine." *Proceedings of the Cambridge Philological Society* 212: 53–77.

Kirk, Geoffrey Stephen, John Earle Raven, and Malcolm Schofield. 1983. *The Presocratic Philosophers: A Critical History with a Selection of Texts*, 2nd ed. Cambridge University Press.

Kisa, Anton. 1908. *Das Glas im Altertume*. Hiersemanns.

Klauck, Hans-Josef. 2003. *Apocryphal Gospels: An Introduction*. T&T Clark.

Kneissl, Peter. 1981. "Die Utriclarii: Ihr Rolle im gallo-römischen Transportwesen und Weinhandel." *Bonner Jahrbucher des rheinischen Landesmuseums in Bonn* 181: 169–203.

Knorr, Wilbur. 1989. *Textual Studies in Ancient and Medieval Geometry*. Birkhauser.

———. 1993. "*Arithmêtikê stoicheiôsis*: On Diophantus and Hero of Alexandria." *Historia Mathematica* 20.2 (May): 180–92.

Koertge, Noretta, ed. 2008. *The New Dictionary of Scientific Biography*. Charles Scribner's Sons.

Kolb, Anne. 2015. "Epigraphy as a Source on Ancient Technology." [in Erdkamp & Verboven 2015: 223–38]

Kollesch, Jutta, and Diethard Nickel, eds. 1993. *Galen und das hellenistische Erbe: Verhandlungen des IV. internationalen Galen-Symposiums veranstaltet vom Institut für Geschichte der Medizin am Bereich Medizin (Charité) der Humboldt-Universität zu Berlin, 18–20. September 1989*. Franz Steiner.

Korpela, Jukka. 1987. *Das Medizinalpersonal im antiken Rom: Eine sozialgeschichte Untersuchung*. Suomalainen Tiedeakatemia.

Kowalski, Jean-Marie. 2012. *Navigation et géographie dans l'antiquité gréco-romaine : la terre vue de la mer*. Picard.

Kreitzer, Larry. 1994. *Prometheus and Adam: Enduring Symbols of the Human Situation*. University Press of America.

Kreutz, B.M. 1973. "Mediterranean Contributions to the Medieval Mariner's Compass." *Technology & Culture* 14.3 (July): 367–83.

Krug, Antje. 1987. "Neros Augenglas: Realia zu einer Anekdote." *Archéologie et Médecine*. Association pour la promotion et la diffusion des connaissances archéologiques: 459–75.

Kudlien, Fridolf. 1968. "The Third Century A.D., a Blank Spot in the History of Medicine?" *Medicine, Science, and Culture: Historical Essays in Honor of Owsei Temkin*. Lloyd Stevenson and Robert Multhauf, eds. Johns Hopkins Press: 25–34.

———. 1970. "Medical Education in Classical Antiquity." *The History of Medical Education*, ed. by C.D. O'Malley. University of California Press: 3–37.

———. 1976. "Medicine as a 'Liberal Art' and the Question of the Physician's Income." *Journal of the History of Medicine and Allied Sciences* 31.4 (October): 448–459.

———. 1986a. "Überlegungen zu einer Sozialgeschichte des frühgriechischen Arztes und seines Berufs." *Hermes* 114: 129–146.

———. 1986b. *Die Stellung des Arztes in der römischen Gesellschaft: freigeborene Römer, Eingebürgerte, Peregrine, Sklaven, Freigelassene als Ärzte*. Steiner.

Kühn, Karl Gottlob. 1821–1833. *Claudii Galeni Opera Omnia*. [= *Medicorum Graecorum Opera Quae Exstant* 1–20]

Künzl, Ernst. 1995. "Ein archäologisches Problem: Gräber römischer Chirurginnen." [in van der Eijk, Horstmanshoff and Schrijvers 1995: 1.309–19]

———. 1996. "Forschungsbericht zu den antiken medizinischen Instrumenten." *Aufstieg und Niedergang der römischen Welt* 2.37.3: 2433–2639.

Kuriyama, Shigehisa. 1999. *The Expressiveness of the Body and the Divergence of Greek and Chinese Medicine*. Zone Books.

Landels, John. 1967. "Assisted Resonance in Ancient Theatres." *Greece & Rome* 14.1 (April): 80–94.

———. 1999. *Music in Ancient Greece and Rome*. Routledge.

———. 2000. *Engineering in the Ancient World*, rev. ed. University of California Press.

Lang, Helen. 2005. "Aristotle's Science of Nature." [in Schürmann 2005: 77–92]

Lang, Philippa, ed. 2004. *Re-Inventions: Essays on Hellenistic & Early Roman Science*. Kelowna, British Columbia: Academic Publishing. [= *Apeiron: A Journal for Ancient Philosophy and Science* 37.4 (December)]

La Regina, Adriano. 1999. *L'Arte dell'assedio di Apollodoro di Damasco*. Electa.

Larrain, Carlos. 1992. *Galens Kommentar zu Platons Timaios*. Teubner.

Lausberg, Marion. 1989. "*Senecae operum fragmenta*: Überblick und Forschungsbericht." *Aufstieg und Niedergang der römischen Welt* 2.36.3: 1879–1961.

Lawson, R.P. 1957. *Origen: The Song of Songs, Commentary and Homilies*. The Newman Press.

Lee, Benjamin Todd. 2005. *Apuleius' Florida: A Commentary*. Walter de Gruyter.

Lee, Desmond. 1973. "Science, Philosophy and Technology in the Greco-Roman World." *Greece and Rome* (2nd ser.) 20.1 (April) and 20.2 (October): 65–78 and 180–93.

Leeuwen, Joyce van. 2014. "Thinking and Learning from Diagrams in the Aristotelian Mechanics." *Nuncius* 29.1: 53–87.

Lefevre, Wolfgang. 2002. "Drawings in Ancient Treatises on Mechanics." [in Renn & Castagnetti 2002: 109–20]

Lehar, Hannes. 2012. *Römische Hypokaustheizung: Berechnungen und Überlegungen zu Aufbau, Leistung und Funktion*. Shaker.

Lehoux, Daryn. 2007a. "Observers, Objects, and the Embedded Eye; or, Seeing and Knowing in Ptolemy and Galen." *Isis* 98.3 (September): 447–67.

———. 2007b. *Astronomy, Weather, and Calendars in the Ancient World: Parapegmata and Related Texts in Classical and Near Eastern Societies.* Cambridge University Press.

———. 2012. *What Did the Romans Know? An Inquiry into Science and Worldmaking.* University of Chicago Press.

Leisegang, H. 1941. "Physik." *Paulys Real-Encyclopaedie der classischen Altertumwissenschaft: Neue Bearbeitung*, ed. by Georg Wissowa, Wilhelm Kroll, Karl Mittelhaus (J. B. Metzler), Bd. 20.1: 1034–63.

Lennox, James. 1994. "The Disappearance of Aristotle's Biology: A Hellenistic Mystery." [in T. Barnes 1994: 7–24]

———. 2005. "The Place of Zoology in Aristotle's Natural Philosophy." [in Sharples 2005: 55–71]

Leveau, Philippe. 1996. "The Barbegal Water Mill in Its Environment: Archaeology and the Economic and Social History of Antiquity." *Journal of Roman Archaeology* 9: 137–53.

Levin, Flora. 2009. *Greek Reflections on the Nature of Music.* Cambridge University Press.

Lévy, Carlos, ed. 1996. *Le Concept de nature à Rome—la physique: Actes du seminaire de philosophie romaine de l'Université de Paris XII–Val de Marne (1992–1993).* Presses de l'Ecole normale superieure.

———. 2003. "Cicero and the *Timaeus*." [in Reydams-Schils 2003: 95–110]

Lévy, Carlos, B. Besnier, and A. Gigandet, eds. 2003. *Ars et Ratio: Sciences, art et métiers dans la philosophie hellénistique et romaine: Actes du Colloque international organisé à Créteil, Fontenay et Paris du 16 au 18 Octobre 1997.* Éditions Latomus.

Lewis, Charlton, and Charles Short. 1879. *A Latin Dictionary.* Clarendon Press.

Lewis, I.M. 2003. *Ecstatic Religion: A Study of Shamanism and Spirit Possession.* 3rd ed. Routledge.

Lewis, M.J.T. 1992. "The South-Pointing Chariot in Rome: Gearing in China and the West." *History of Technology* 14: 77–99.

———. 1993. "The Greeks and the Early Windmill." *History of Technology* 15: 141–89.

———. 1994. "The Origins of the Wheelbarrow." *Technology and Culture* 35.3 (July): 453–75.

———. 1997. *Millstone and Hammer: The Origins of Water Power*. University of Hull.

———. 2000. "Theoretical Hydraulics, Automata, and Water Clocks." [in Wikander 2000c: 343–70]

———. 2001a. "Railways in the Greek and Roman world." *Early Railways: A Selection of Papers from the First International Early Railways Conference*. A. Guy and J. Rees, eds. Newcomen Society: 8–19.

———. 2001b. *Surveying Instruments of Greece and Rome*. Cambridge University Press.

———. 2012. "Greek and Roman Surveying and Surveying Instruments." [in Talbert 2012: 129–62]

Lewis, Naphtali, and Meyer Reinhold, eds. 1990. *Roman Civilization: Selected Readings*, 3rd ed. 2 vols. Columbia University Press.

Lewit, Tamara. 2012. "Oil and Wine Press Technology in Its Economic Context: Screw Presses, the Rural Economy and Trade in Late Antiquity." *Antiquité tardive: revue internationale d'histoire et d'archéologie (IVe–VIIIe s.)* 20: 137–49.

Liddell, Henry, and Robert Scott. 1996. *A Greek-English Lexicon*. 9th ed., with a revised supplement. Clarendon Press.

Lierke, Rosemarie. 1999. *Antike Glastöpferei: Ein vergessenes Kapitel der Glasgeschichte*. Philipp von Zabern.

Lindberg, David. 1983. "Science and the Early Christian Church." *Isis* 74.4 (December): 509–30.

———. 1986. "Science and the Early Church." *God and Nature: Historical Essays on the Encounter between Christianity and Science*. David Lindberg and Ronald Numbers, eds. University of California Press: 19–48.

———. 1987. "Science as Handmaiden: Roger Bacon and the Patristic Tradition." *Isis* 78.4 (December): 518–36.

———. 1992. *The Beginnings of Western Science: The European Scientific Tradition in Philosophical, Religious, and Institutional Context, 600 B.C. to A.D. 1450*. University of Chicago Press.

———. 2002a. "Early Christian Attitudes toward Nature." *Science and Religion: A Historical Introduction*. Gary Ferngren, ed. Johns Hopkins University Press: 47–56.

———. 2002b. "Medieval Science and Religion." *Science and Religion: A Historical Introduction*. Gary Ferngren, ed. Johns Hopkins University Press: 57–72.

———. 2007. *The Beginnings of Western Science: The European Scientific Tradition in Philosophical, Religious, and Institutional Context, 600 B.C. to A.D. 1450*. 2nd ed. University of Chicago Press.

———. 2009. "That the Rise of Christianity Was Responsible for the Demise of Ancient Science." [in Numbers 2009: 8–18]

Lindsay, Jack. 1974. *Blast-Power and Ballistics: Concepts of Force and Energy in the Ancient World*. Barnes & Noble.

Lirb, Huib J. 1993. "Partners in Agriculture: The Pooling of Resources in Rural *Societates* in Roman Italy." *De Agricultura: In Memoriam Pieter Willem de Neeve (1945–1990)*. F.J.A.M. Meijer and H.W. Pleket, eds. J.C. Glieben: 263–95.

Littman, Robert. 1996. "Medicine in Alexandria." *Aufstieg und Niedergang der römischen Welt* 2.37.3: 2678–708.

Lloyd, G.E.R. 1970. *Early Greek Science: Thales to Aristotle*. W.W. Norton.

———. 1973. *Greek Science After Aristotle*. W.W. Norton.

———. 1978. *Hippocratic Writings*. Penguin.

———. 1979. *Magic, Reason, and Experience*. Cambridge University Press.

———. 1981. "Science and Mathematics." *The Legacy of Greece: A New Appraisal*. Moses Finley, ed. Clarendon Press: 256–300.

———. 1982. "Observational Error in Later Greek Science." [in J. Barnes et al. 1982: 128–64]

———. 1983. *Science, Folklore, and Ideology: Studies in the Life Sciences in Ancient Greece*. Cambridge University Press.

———. 1987. *The Revolutions of Wisdom: Studies in the Claims and Practice of Ancient Greek Science*. University of California Press.

———. 1990. *Demystifying Mentalities*. Cambridge University Press.

———. 1991. *Methods and Problems in Greek Science*. Cambridge University Press.

———. 1992a. "The Theories and Practices of Demonstration in Aristotle." *Proceedings of the Boston Area Colloquium in Ancient Philosophy* 6: 371–412. [latter pages include commentary by William Wians]

———. 1992b. "Methods and Problems in the History of Ancient Science: The Greek Case." *Isis* 83.4 (December): 564–77.

———. 1996a. *Adversaries and Authorities: Investigations into Ancient Greek and Chinese Science*. Cambridge University Press.

———. 1996b. *Aristotelian Explorations*. Cambridge University Press.

———. 2001. "Is There a Future for Ancient Science?" *Proceedings of the Cambridge Philological Society* 47: 196–210.

———. 2002. *The Ambitions of Curiosity: Understanding the World in Ancient Greece and China*. Cambridge University Press.

———. 2004. "New Issues in the History of Ancient Science." [in P. Lang 2004: 9–28].

———. 2005. "Mathematics as a Model of Method in Galen." [in Sharples 2005: 110–30]

Lloyd, G.E.R., and Nathan Sivin. 2002. *The Way and the Word: Science and Medicine in Early China and Greece*. Yale University Press.

Loiseau, Christophe. 2012. "Les métaux dans les constructions publiques romaines: applications architecturales et structures de production (Ier–IIIe siécle ap. J.-C.)." *Arqueología de la construcción. 3, Los procesos constructivos en el mundo romano: la economía de las obras: (École Normale Supérieure, París, 10–11 de diciembre de 2009)*. Stefano Camporeale, Hélène Dessales, and Antonio Pizzo, eds.. Mérida: Instituto de Arqueología de Mérida: 117–29.

Long, A.A. 1988. "Ptolemy *On the Criterion*: An Epistemology for the Practicing Scientist." [in Dillon & Long 1988: 176–207]

Longrigg, James. 1981. "Superlative Achievement and Comparative Neglect: Alexandrian Medical Science and Modern Historical Research." *History of Science* 19.3 (September): 155–200.

Loseby, Simon. 2012. "Post-Roman Economies." *The Cambridge Companion to the Roman Economy*. Walter Scheidel, ed. Cambridge University Press: 334–60.

Lougee, David Gilman. 1972. *The Concept of the Natural Scientist in Seventeenth-Century Defenses of Science in England*. Dissertation (Ph.D.), University of Michigan (Ann Arbor).

Lovejoy, Arthur, and George Boas. 1935. *Primitivism and Related Ideas in Antiquity*. The Johns Hopkins Press.

Lyons, Malcolm. 1963. *Galeni in Hippocratis de Officina Medici Commentariorum Versionem Arabicam*. Academiae Scientiarum. [= *Corpus Medicorum Graecorum: Supplementum Orientale* 1]

Maass, Ernest. 1958. *Commentariorum in Aratum Reliquiae*. Weidmanns.

MacDonald, B. R. 1986. "The Diolkos." *Journal of Hellenic Studies* 106: 191–95.

MacMullen, Ramsay. 1974. *Roman Social Relations: 50 B.C. to A.D. 284*. Yale University Press.

———. 1984. *Christianizing the Roman Empire (A.D. 100–400)*. Yale University Press.

———. 1988. *Corruption and the Decline of Rome*. Yale University Press.

———. 1997. *Christianity and Paganism in the Fourth to Eighth Centuries*. Yale University Press.

Macve, Richard. 1985. "Some Glosses on 'Greek and Roman Accounting.'" *History of Political Thought* 6.1/2: 233–64.

Maher, David, and John Makowski. 2001. "Literary Evidence for Roman Arithmetic with Fractions." *Classical Philology* 96.4: 376–99.

Majcherek, Grzegorz. 1999. "Kom El-Dikka Excavations, 1998/99." *Polish Archaeology in the Mediterranean*. Warsaw University Press: 27–38.

———. 2010. "The Auditoria on Kom el-Dikka: A Glimpse of Late Antique Education in Alexandria." *Proceedings of the Twenty-Fifth International Congress of Papyrology, Ann Arbor 2007*. American Studies in Papyrology: 471–84.

———. 2013. "Alexandria: Excavations and Preservation Work on Kom El-Dikka, Preliminary Report 2009/2010." *Polish Archaeology in the Mediterranean* 22: 33–53.

Malina, Bruce. 2001. *The Social Gospel of Jesus: The Kingdom of God in Mediterranean Perspective*. Fortress Press.

Malina, Bruce, and Jerome Neyrey. 1996. *Portraits of Paul: An Archaeology of Ancient Personality*. Louisville, Ky.: Westminster John Knox Press.

Malina, Bruce, and John Pilch. 2000. *Social Science Commentary on the Book of Revelation*. Fortress Press.

Malina, Bruce, and Richard Rohrbaugh. 1998. *Social-Science Commentary on the Gospel of John*. Fortress Press.

———. 2003. *Social-Science Commentary on the Synoptic Gospels*, 2nd ed. Minneapolis: Fortress Press.

Malinowski, Roman. 1982. "Ancient Mortars and Concretes: Aspects of their Durability." *History of Technology* 7: 89–100.

Manetti, Daniela, ed. 2011. *De Medicina / Anonymus Londinensis*. Walter de Gruyter.

Manning, W.H. 1987. "Industrial Growth." [in Wacher 1987: 2.586–610]

Manning, J.G., and Ian Morris. 2005. *The Ancient Economy: Evidence and Models*. Stanford University Press.

Mansfeld, Günter. 2013. *Der Held auf dem Wagen: archäologische Belege zur technischen Entwicklung des Wagens*. Harrassowitz.

Mansfeld, Jaap. 1998. *Prolegomena Mathematica: From Apollonius of Perga to Late Neoplatonism*. Brill.

Marasco, Gabriele. 1995. "L'Introduction de la médecine grecque à Rome: une dissension politique et idéologique." [in van der Eijk, Horstmanshoff and Schrijvers 1995: 1.35–48]

———. 1998. "Cléopâtre et les sciences de son temps." [in Argoud & Guillaumin 1998: 39–53]

Marchant, Josephine. 2010. *Decoding the Heavens: A 2,000-Year-Old-Computer and the Century-Long Search to Discover Its Secrets*. Da Capo Press.

Marchis, Vittorio, and Giuse Scalva. 2002. "The Engine Lost: Hydraulic Technologies in Pompeii." [in Renn & Castagnetti 2002: 25–34]

Marganne, M.H. 1981. *Inventaire analytique des Papyrus grecs de Médecine*. Droz.

———. 1998. *La Chirurgie dans l'Égypte gréco-romaine d'après les Papyrus littéraires grecs*. Brill.

———. 2001. "Hippocrate et la médecine de l'Égypte gréco-romaine." *Revue de Philosophie ancienne* 19: 39–62.

Marrou, Hénri. 1956. *A History of Education in Antiquity*, 3rd ed. Sheed and Ward. [tr. by George Lamb of *Histoire de l'éducation dans l'antiquité*, 3rd ed. Editions du Seuil: 1955]

———. 1964. *Histoire de l'éducation dans l'antiquité*, 6th ed. Éditions du Seuil.

Marsden, E.W. 1969. *Greek and Roman Artillery: Historical Development*. Clarendon.

———. 1971. *Greek and Roman Artillery: Technical Treatises*. Clarendon.

Martelli, Matteo. 2011. "Greek Alchemists at Work: 'Alchemical Laboratory' in the Greco-Roman Egypt." *Nuncius* 26: 271–311.

Marzano, Annalisa. 2013. *Harvesting the Sea: The Exploitation of Marine Resources in the Roman Mediterranean*. Oxford University Press.

Marzano, Annalisa, and Giulio Brizzi. 2009. "Costly Display or Economic Investment? A Quantitative Approach to the Study of Marine Aquaculture." *Journal of Roman Archaeology* 22.1: 215–230.

Mattern, Susan. 2008. *Galen and the Rhetoric of Healing*. Johns Hopkins University Press.

———. 2013. *The Prince of Medicine*. Oxford University Press.

Mattingly, David. 1990. "Paintings, Presses, and Perfume Production at Pompeii." *Oxford Journal of Archaeology* 9.1 (March): 71–90.

Mattingly, David, and Greg Aldrete. 2000. "The Feeding of Imperial Rome: The Mechanics of the Food Supply System." [in Coulston & Dodge 2000: 142–65]

Mau, August. 1908. *Pompeji in Leben und Kunst*. W. Engelmann.

———. 1982. *Pompeii: Its Life and Art*, new ed. (revised and corrected). Caratzes Brothers. [tr. by Francis Kelsey]

May, James. 1987. "Seneca's Neighbour, the Organ Tuner." *The Classical Quarterly* (n.s.) 37.1: 240–43.

May, Margaret Tallmadge. 1968. *Galen: On the Usefulness of the Parts of the Body*. Cornell University Press.

Mayor, Adrienne. 2011a. *The First Fossil Hunters: Dinosaurs, Mammoths, and Myth in Greek and Roman Times*. Princeton University Press. [n.b. The original 2000 edition was subtitled, "Paleontology in Greek and Roman Times."]

———. 2011b. *The Poison King: The Life and Legend of Mithradates, Rome's Deadliest Enemy*. Princeton University Press.

Mayr, Ernst. 1990. "When Is Historiography Whiggish?" *Journal of the History of Ideas* 51.2 (April–June): 301–09.

Mazzini, Innocenzo. 2012. "La medicina postgalenica." *Medicina nei secoli: arte e scienza* 24.2: 467–92.

Mazlish, Bruce, et al. 2006. "Progress in History? A Forum." *Historically Speaking* 7.5 (May–June): 18–33.

McCann, Anna Marguerite, ed. 1987. *The Roman Port and Fishery of Cosa: A Center of Ancient Trade*. Princeton University Press.

———. 2002. *The Roman Port and Fishery of Cosa: A Short Guide*. American Academy in Rome.

McWhirr, Alan. 1987. "Transport by Land and Water." [in Wacher 1987: 2.658–70]

Meiggs, Russell. 1982. *Trees and Timber in the Ancient Mediterranean World*. Clarendon Press.

Meijer, Fik. 1986. *A History of Seafaring in the Classical World*. Croom Helm.

Mercer, Henry. 1975. *Ancient Carpenter's Tools*, 5th ed. Horizon.

Meissner, Burkhard. 1999. *Die technologische Fachliteratur der Antike: Struktur, Uberlieferung und Wirkung technischen Wissens in der Antike (ca. 400 v. Chr.–ca. 500 n. Chr.)*. Akademie Verlag.

Mertens, J. 1958. "La moissonneuse de Buzenol (Eine römische Mähmaschine)." *Ur-Schweiz—Suisse Primitive* 22.4: 49–53.

Meunier, Louise. 1997. *Le Médicin Grec dans la Cité Hellénistique*. Dissertation (M.A.), Université Laval (Québec City, Canada).

Millard, Alan. 2000. *Reading and Writing in the Time of Jesus*. New York University Press.

Milne, John Stewart. 1907. *Surgical Instruments in Greek and Roman Times*. Clarendon.

Modrak, D.K.W. 1989. "Aristotle on the Difference between Mathematics and Physics and First Philosophy." *Apeiron: A Journal for Ancient Philosophy and Science* 22.4 (December): 122–39.

Mols, Stephanus. 1999. *Wooden Furniture in Herculaneum: Form, Technique and Function*. Gieben.

Moo, Douglas. 1996. *The Epistle to the Romans*. W.B. Eerdmans.

Morales, Helen. 1996. "Torturer's Apprentice: Parrhasius and the Limits of Art." *Art and Text in Roman Culture*. Jas Elsner, ed. Cambridge University Press: 182–209.

Moreno Gallo, Isaac. 2009. "La 'dioptra' de Herón de Alejandría." *Palaia Philia: studi di topografia antica in onore di Giovanni Uggeri*. Cesare Marangio and Giovanni Laudizi, eds. Congedo: 163–70.

Morley, Neville. 2007. *Trade in Classical Antiquity*. Cambridge University Press.

Moss, Candida. 2012. *Ancient Christian Martyrdom: Diverse Practices, Theologies, and Traditions*. Yale University Press.

———. 2013. *The Myth of Persecution: How Early Christians Invented a Story of Martyrdom*. Harper One.

Mratschek-Halfmann, Sigrid. 1993. *Divites et Praepotentes: Reichtum und soziale Stellung in der Literatur der Prinzipatszeit*. Franz Steiner.

Mudry, P. 1986. "Science et conscience: Réflexions sur le discours scientifique à Rome." *Études de Lettres* 1: 75–86.

Mueller, Ian. 1982. "Geometry and Scepticism." [in J. Barnes et al. 1982: 69–95]

———. 2000. "Working with Trivial Authors." *Apeiron: A Journal for Ancient Philosophy and Science* 33.3 (September): 247–58.

———. 2004. "Remarks on Physics and Mathematical Astronomy and Optics in Epicurus, Sextus Empiricus, and Some Stoics." [in P. Lang 2004: 57–87]

Munteanu, Claudiu. 2013. "Roman Military Pontoons Sustained on Inflated Animal Skins." *Archäologisches Korrespondenzblatt: Urgeschichte, Römerzeit, Frühmittelalter* 43.4: 545–52.

Murphy, Susan. 1995. "Heron of Alexandria's *On Automaton-Making.*" *History of Technology* 17: 1–44.

Murschel, A. 1995. "The Structure and Function of Ptolemy's Physical Hypotheses of Planetary Motion." *Journal for the History of Astronomy* 26: 33–61.

Naddaf, Gérard. 1992. *L'origine et l'évolution du concept grec de 'physis'*. E. Mellen Press.

———. 2005. *The Greek Concept of Nature*. State University of New York Press.

NAVEDTRA [Naval Education and Training Program Development Center]. 1994. *Basic Machines and How They Work*, rev. ed. Dover.

Naylor, Ron. 2007. "Galileo's Tidal Theory." *Isis* 98.1 (March): 1–22.

Needham, Joseph. 1971. *Science and Civilisation in China*, 4 vols. [continued to 7 vols. by 2004]. University Press.

Neil, Kenny. 2004. *The Uses of Curiosity in Early Modern France and Germany*. Oxford University Press.

Netz, Reviel. 2002. "Greek Mathematicians: A Group Picture." [in Tuplin & Rihll 2002: 196–216]

———. 2003. "The Goal of Archimedes' *Sand Reckoner.*" *Apeiron: A Journal for Ancient Philosophy and Science* 36.4 (December): 251–90.

———. 2010. "What Did Greek Mathematicians Find Beautiful?" *Classical Philology* 105.4: 426–44.

———. 2011. "The Bibliosphere of Ancient Science (Outside of Alexandria): A Preliminary Survey." *NTM: Zeitschrift für Geschichte der Wissenschaften, Technik und Medizin* 19.3: 239–269.

———. 2015. "Were There Epicurean Mathematicians?" *Oxford Studies in Ancient Philosophy* 49: 283–319.

Netz, Reviel, and William Noel. 2007. *The Archimedes Codex: How a Medieval Prayer Book is Revealing the True Genius of Antiquity's Greatest Scientist.* Da Capo Press.

Netz, Reviel, and Fabio Acerbi and Nigel Wilson. 2004. "Towards a Reconstruction of Archimedes' Stomachion," *Sciamus: Sources and Commentaries in Exact Sciences* 5, 2004: 67–99.

Neugebauer, Otto. 1975. *A History of Ancient Mathematical Astronomy.* 3 vols. Springer-Verlag.

Newmyer, Stephen Thomas. 1996. "Talmudic Medicine and Greco-Roman Science: Cross-Currents and Resistance." *Aufstieg und Niedergang der römischen Welt* 2.37.3: 2895–2911.

Neyses, Adolf. 1983. "Die Getreidemühlen beim römischen Land- und Weingut von Lösnich." *Trierer Zeitschrift für Geschichte und Kunst des Trierer Landes und seiner Nachbargebiete* 46: 208–21.

Nicholls, Matthew. 2010. "Parchment Codices in a New Text of Galen," *Greece & Rome* 57.2: 378–86.

Nickel, Diethard. 1979. "Berufsvorstellungen über weibliche Medizinalpersonen in der Antike." Klio 61.2: 515–18.

Nicolson, Frank. 1891. "Greek and Roman Barbers." *Harvard Studies in Classical Philology* 2: 41–56.

Noble, Joseph, and Derek de Solla Price. 1968. "The Water Clock in the Tower of the Winds." *American Journal of Archaeology* 72.4 (October): 345–55.

Noll, Mark. 2008. *The Scandal of the Evangelical Mind.* Eerdmans.

Notis, Mike, and Aaron Shugar. 2003. "Roman Shears: Metallography, Composition and a Historical Approach to Investigation." *International Conference: Archaeometallurgy in Europe, 24–25–26 September 2003, Milan, Italy: Proceedings*, vol. 1. Associazione italiana di metallurgia: 109–18.

Novara, Antoinette. 1982. *Les idées romaines sur le progrès d'après les écrivains de la République: essai sur le sens latin du progrès.* 2 vols. Les Belles Lettres.

———. 1996. "Cicéron et le planétaire d'Archimède." *Les Astres: Actes du Colloque international de Montpellier 23–25 Mars 1995.* Béatrice Bakhouche and Alain Maurice Moreau, eds. Université Paul Valéry: 1.227–44.

Numbers, Ronald, ed. 2009. *Galileo Goes to Jail: And Other Myths about Science and Religion.* Harvard University Press.

Nussbaum, Martha. 1994. *The Therapy of Desire: Theory and Practice in Hellenistic Ethics*. Princeton University Press.

Nutton, Vivian. 1972. "Galen and Medical Autobiography." *Proceedings of the Cambridge Philological Society* 18 (new ser.): 50–62.

———. 1977. "*Archiatri* and the Medical Profession in Antiquity." *Papers of the British School at Rome* 45: 191–226.

———. 1983. "The Seeds of Disease: An Explanation of Contagion and Infection from the Greeks to the Renaissance." *Medical History* 27: 1–34.

———. 1984a. "From Galen to Alexander: Aspects of Medicine and Medical Practice in Late Antiquity." *Dumbarton Oaks Papers* 38: 1–14.

———. 1984b. "Galen in the Eyes of His Contemporaries." *Bulletin of the History of Medicine* 58: 315–24.

———. 1985. "Murders and Miracles: Lay Attitudes towards Medicine in Classical Antiquity." *Patients and Practitioners: Lay Perceptions of Medicine in Pre-Industrial Society*. Roy Porter, ed. Cambridge University Press.

———. 1993a. "Roman Medicine: Tradition, Confrontation, Assimilation." *Aufstieg und Niedergang der römischen Welt* 2.37.1: 49–78.

———. 1993b. "Galen and Egypt." [in Kollesch & Nickel 1993: 11–31]

———. 1995. "The Medical Meeting Place." [in van der Eijk, Horstmanshoff and Schrijvers 1995: 1.3–25]

———. 1997. "Galen on Theriac: Problems of Authenticity." *Galen on Pharmacology: Philosophy, History, and Medicine: Proceedings of the Vth International Galen Colloquium, Lille, 16–18 March 1995*. Armelle Debru, ed. Brill: 133–51.

———. 1999. *Galen: On My Own Opinions*. Akademie Verlag. [= *Corpus Medicorum Graecorum* 5.3.2]

———. 2000a. "Medicine." *The Cambridge Ancient History, Volume 11: The High Empire, AD 70–192*, 2nd ed. Alan Bowman, Peter Garnsey and Dominic Rathbone, eds. Cambridge University Press: 943–65.

———. 2000b. "Did the Greeks Have a Word for It? Contagion and Contagion Theory in Classical Antiquity." *Contagion: Perspectives from Pre-modern Societies*. Lawrence Conrad and D. Wujastyk, eds. Ashgate: 137–62.

———. 2010. "Galen in Context." *Papers of the Langford Latin Seminar. 14, Health and Sickness in Ancient Rome: Greek and Roman Poetry and Historiography*. Francis Cairns and Miriam Griffin, eds. Cairns: 1–18.

———. 2013. *Ancient Medicine*. Routledge.

Oberhelman, S.M. 1994. "On the Chronology and Pneumatism of Aretaios of Cappadocia." *Aufstieg und Niedergang der römischen Welt* 2.37.2: 941–66.

Obrist, Barbara. 2004. *La cosmologie medievale: textes et images*. 2 vols. Edizioni del Galluzzo. [vol. 1 is complete, vol. 2 in progress]

O'Connor, Colin. 1993. *Roman Bridges*. Cambridge University Press.

Oleson, John Peter. 1984. *Greek and Roman Mechanical Water-Lifting Devices: The History of a Technology*. University of Toronto Press.

———. 1986. *Bronze Age, Greek and Roman Technology: A Select, Annotated Bibliography*. Garland.

———. 1988. "The Technology of Roman Harbours." *International Journal of Nautical Archaeology and Underwater Exploration* 17: 147–57.

———. 2000. "Water-Lifting" [in Wikander 2000c: 217–302]

———. 2004. "*Well-Pumps for Dummies*: Was There a Roman Tradition of Popular Sub-Literary Engineering Manuals?" *Problemi di macchinismo in ambito Romano macchine idrauliche nella letteratura tecnica, nelle fonti storiografiche e nelle evidenze archeologiche di età imperiale*. Franco Minonzio, ed. Comune di Como, Assessorato alla Cultura, Musei Civici: 65–86.

———. 2005. "Design, Materials, and the Process of Innovation for Roman Force Pumps." *Terra Marique: Studies in Art History and Marine Archaeology in Honor of Anna Marguerite McCann on the Receipt of the Gold Medal of the Archaeological Institute of America*. John Pollini, ed. Oxbow Books.

———. 2008. *Oxford Handbook of Engineering and Technology in the Classical World*. Oxford University Press.

Olson, Richard. 1984. "Aristophanes and the Antiscientific Tradition." *Transformation and Tradition in the Sciences: Essays in Honor of I. Bernard Cohen*. Everett Mendelsohn, ed. Cambridge University Press: 441–54.

Orengo, H.A., et al. 2013. "Pitch Production during the Roman Period: An Intensive Mountain Industry for a Globalised Economy?" *Antiquity* 87: 802–14.

Ormos, István. 1993. "Bemerkungen zur editorischen Bearbeitung der Galenschrift 'Über die Sektion toter Lebewesen.'" [in Kollesch & Nickel 1993: 165–72]

Osborn, Eric. 2005. *Clement of Alexandria*. Cambridge University Press.

O'Sullivan, Lara. 2008. "Athens, Intellectuals, and Demetrius of Phalerum's *Socrates*." *Transactions of the American Philological Association* 138.2 (Autumn): 393–410.

Owens, Joseph. 1991. "The Aristotelian Conception of the Pure and Applied Sciences." *Science and Philosophy in Classical Greece*. Alan Bowen and Francesca Rochberg-Halton, eds. Garland: 31–42.

Pailler, Jean-Marie, and Pascal Payen, eds. 2004. *Que reste-t-il de l'éducation classique? Relire «le Marrou» Histoire de l'éducation dans l'Antiquité*. Presses Universitaires du Mirail.

Paisley, P.B., and D.R. Oldroyd. 1979. "Science in the Silver Age: *Aetna*, a Classical Theory of Volcanic Activity." *Centaurus* 23.1: 1–20.

Panchenko, Dmitri. 2000. "On Copernicus' Success and Aristarchus' Failure." *Antike Naturwissenschaft und ihre Rezeption* 10: 97–105.

Papadopetrakis, Eftichis, and Vaios Argyrakis. 2010. "The Demonstration of the Corporeity of Air in Heron's *Pneumatics*: A Methodological Breakthrough in Experimentation." *Almagest* 1.1: 86–102.

Parker, A.J. 1987. "Trade within the Empire and Beyond the Frontiers." [in Wacher 1987: 2.635–53]

Partington, J. R. 1960. *A History of Greek Fire and Gunpowder*. W. Heffer.

Patzer, Harald. 1993. *Physis: Grundlegung zu einer Geschichte des Wortes*. Steiner.

Paul, Eberhard. 1962. "Der Tod des Archimedes und das Heuschreckenopfer." *Die Falsche Göttin: Geschichte der Antikenfälschung*. Lambert Schneider: 132–39.

Pavlovskis, Zoja. 1973. *Man in an Artificial Landscape: The Marvels of Civilization in Imperial Roman Literature*. Brill.

Peacock, David. 2012. "The Roman Red Sea Port Network." *Rome, Portus and the Mediterranean*. Simon Keay, ed. British School at Rome: 347–53.

Penn, R.G. 1994. *Medicine on Ancient Greek and Roman Coins*. Seaby.

Perry, B. E. 1936. *Studies in the Text History of the Life and Fables of Aesop*. Pennsylvania: American Philological Association.

Pettegrew, David. 2011. "The *diolkos* of Corinth." *American Journal of Archaeology* 115.4: 549–74.

Pfeiffer, R. 1968. *History of Classical Scholarship from the Beginning to the End of the Hellenistic Age*. Clarendon Press.

Phillips, William D., and Carla Rahn Phillips. 1992. *The Worlds of Christopher Columbus*. Cambridge University Press.

Pierce, Charles. 2010. *Idiot America: How Stupidity Became a Virtue in the Land of the Free*. Doubleday.

Pigliucci, Massimo. 2002. *Denying Evolution: Creationism, Scientism, and the Nature of Science.* Sinauer Associates.

Pilch, John. 2002. "Altered States of Consciousness in the Synoptics." *The Social Setting of Jesus and the Gospels*, ed. by Wolfgang Stegemann, Bruce Malina and Gerd Theissen. Fortress Press: 103–116.

Pines, S. 1961. "*Omne quod movetur necesse est ab aliquo moveri*: A Refutation of Galen by Alexander of Aphrodisias and the Theory of Motion." *Isis* 52.1 (March): 21–54.

Pitts, Lynn, and J.K. St. Joseph, eds. 1985. *Inchtuthil: The Roman Legionary Fortress Excavations, 1952–65.* Society for the Promotion of Roman Studies.

Plant, I.M. 2004. *Women Writers of Ancient Greece and Rome: An Anthology.* University of Oklahoma Press.

Plantzos, Dimitris. 1997. "Crystals and Lenses in the Graeco-Roman World." *American Journal of Archaeology* 101.3 (July): 451–64.

Pleket, H.W. 1973. "Technology in the Greco-Roman World: A General Report." *Talanta* 5: 6–47.

———. 2001. "Banking and Business in the Roman World." *The Journal of Economic History* 61.4 (December): 1105–1106.

Podlecki, A.J. 2005. *Aeschylus: Prometheus Bound.* Aris & Phillips.

Polzer, Mark. 2008. "Toggles and Sails in the Ancient World: Rigging Elements Recovered from the Tantura B Shipwreck, Israel." *International Journal of Nautical Archaeology and Underwater Exploration* 37.2: 225–52.

Portier-Young, Anathea. 2014. "Symbolic Resistance in the Book of the Watchers." *The Watchers in Jewish and Christian Traditions.* Kelley Coblentz Bautch, John Endres, and Angela Kim Harkins, eds. Catholic Biblical Association: 39–50.

Pot, Johan Hendrik Jacob van der. 1985. *Die Bewertung des technischen Fortschritts: Eine systematische Übersicht der Theorien*, 2 vols. Van Gorcum.

Pothecary, Sarah. 1995. "Strabo, Polybius, and the Stade." *Phoenix* 49.1 (Spring): 49–67.

Potter, Paul. 1993. "Apollonius and Galen on 'Joints.'" [in Kollesch & Nickel 1993: 117–23]

Press, Ludwika. 1988. "Valetudinarium at Novae and Other Roman Danubian Hospitals." *Archeologia* 39: 69–89.

"Primavera: Never Logged in the Light of the Moon." 1994. *Wood Magazine* 73 (October): 90.

Ramsay, W.M. 1934. "Studies in the Roman Province Galatia: II. Dedications at the Sanctuary of Colonia Caesarea." *Journal of Roman Studies* 8: 107–145.

Rathbone, Dominic. 1997. "Prices and Price Formation in Roman Egypt." *Economie Antique: Prix et Formation des Prix dans les Economies Antiques*. Jean Andreau, Pierre Briant and Raymond Descat, eds. Musée archéologique départemental de Saint-Bertrand-de-Comminges: 183–244.

Rausch, Manuela. 2012. *Heron von Alexandria: die Automatentheater und die Erfindung der ersten antiken Programmsteuerung*. Diplomica Verlag.

Rawson, Elizabeth. 1985. *Intellectual Life in the Late Roman Republic*. Johns Hopkins University Press.

Reece, David. 1969. "The Technological Weakness of the Ancient World." *Greece & Rome* 16.1 (April): 32–47.

Rees, Sian. 1987. "Agriculture and Horticulture." [in Wacher 1987: 2.481–507]

Rehm, Albert. 1937. "Antike «Automobile»." *Philologus* 92: 317–30.

Remes, Pauliina. 2008. *Neoplatonism*. University of California Press.

Remus, Harold. 1983. *Pagan-Christian Conflict over Miracle in the Second Century*. Philadelphia Patristic Foundation.

Rémy, Bernard. 2010. *Les médecins dans l'Occident romain (Péninsule Ibérique, Bretagne, Gaules, Germanies)*. Éditions de Boccard.

Renehan, R. 2000. "A Rare Surgical Procedure in Plutarch." *The Classical Quarterly* 50.1: 223–29.

Renn, Jürgen. 2002. "Introduction: On the Trail of Knowledge from a Sunken City." [in Renn & Castagnetti 2002: 11–24]

Renn, Jürgen, and Giuseppe Castagnetti, eds. 2002. *Homo Faber: Studies on Nature, Technology, and Science at the Time of Pompeii: Presented at a Conference at the Deutsches Museum, Munich, 21–22 March 2000*. L'Erma di Bretschneider.

Reydams-Schils, Gretchen, ed. 2003. *Plato's* Timaeus *as Cultural Icon*. University of Notre Dame Press.

Reynolds, L.D., and N.G. Wilson. 1991. *Scribes and Scholars: A Guide to the Transmission of Greek and Latin Literature*, 3rd ed. Clarendon Press.

Reynolds, Terry. 1983. *Stronger Than a Hundred Men: A History of the Vertical Water Wheel*. Johns Hopkins University Press.

Richter, Gisela. 1926. *Ancient Furniture: A History of Greek, Etruscan and Roman Furniture*. Clarendon Press.

Richter, Will. 1963. *Aetna*. Walter de Gruyter.

Rickman, G.E. 1988. "The Archaeology and History of Roman Ports." *International Journal of Nautical Archaeology and Underwater Exploration* 17: 257–67.

Riddle, John. 1986. *Dioscorides on Pharmacy and Medicine*. University of Texas Press.

———. 1993. "High Medicine and Low Medicine in the Roman Empire." *Aufstieg und Niedergang der römischen Welt* 2.37.1: 102–20.

Rihll, T.E. 1999. *Greek Science*. Oxford University Press.

———. 2002. "Introduction: Greek Science in Context." [in Tuplin & Rihll 2002: 1–21]

———. 2007. *The Catapult: A History*. Westholme.

———. 2008. "Slavery and Technology in Pre-Industrial Contexts." Slave Systems: Ancient and Modern. Enrico Dal Lago & Constantina Katsari, eds. Cambridge University Press: 127–47.

Rihll, T.E., and J.V. Tucker. 2002. "Practice Makes Perfect: Knowledge of Materials in Classical Athens." [in Tuplin & Rihll 2002: 274–305]

Riley, Mark. 1995. "Ptolemy's Use of His Predecessors' Data." *Transactions of the American Philological Association* 125: 221–50.

Ring, James W. 1996. "Windows, Baths, and Solar Energy in the Roman Empire." *American Journal of Archaeology* 100.4 (October): 717–24.

Ritti, Tullia, Klaus Grewe, and H. Paul M. Kessener. 2007. "A Relief of a Water-Powered Stone Saw Mill on a Sarcophagus at Hierapolis and Its Implications." *Journal of Roman Archaeology* 20.1: 138–63.

Rives, J.B. 1999. "The Decree of Decius and the Religion of Empire." *Journal of Roman Studies* 89: 135–54.

Roberts, Alexander, and James Donaldson, eds. 1896. "Arabic Gospel of the Infancy of the Savior." *The Ante-Nicene Fathers: Translations of the Writings of the Fathers down to A.D. 325*. Christian Literature Company: 859–81.

Roberts, C.H. 1954. "The Codex." *Proceedings of the British Academy* 40: 168–204.

Rocca, Julius. 2003. *Galen on the Brain: Anatomical Knowledge and Physiological Speculation in the Second Century A.D.* Brill.

Roccatagliata, Giuseppe. 1986. *A History of Ancient Psychiatry*. Greenwood Press.

Rodgers, R.H. 2004. *De Aquaeductu Urbis Romae*. Cambridge University Press.

Rogers, Penelope Walton. 2001. "The Re-Appearance of an Old Roman Loom in Medieval England." *The Roman Textile Industry and Its Influence: A Birthday Tribute to John Peter Wild*. Penelope Walton Rogers, Lise Bender Jørgensen, Antoinette Rast-Eicher, and John Peter Wild, eds. Oxbow: 158–71.

Roller, Duane. 2006. *Through the Pillars of Herakles: Greco-Roman Exploration of the Atlantic*. Routledge.

Röring, Christoph Wilhelm. 1983. *Untersuchungen zu römischen Reisewagen*. Numismatischer Verlag G.M. Forneck.

Rosner, Fred. 1994. "Physicians in the Talmud." *Medicine in the Bible and the Talmud: Selections from Classical Jewish Sources*. Yeshiva University Press: 211–15.

Ross, Sydney. 1964. "*Scientist*: The Story of a Word." *Annals of Science: A Quarterly Review of the History of Science and Technology since the Renaissance* 18.2: 65–85.

Rossi, Cesare, and Andrea Unich. 2013. "A Study on Possible Archimedes' Cannon." *Rivista storica dell'Antichità* 43: 55–75.

Rossi, Paolo. 2001. *The Birth of Modern Science*. Blackwell.

Rossing, Thomas, and Neville Fletcher. 1995. *Principles of Vibration and Sound*. Springer-Verlag.

Rosumek, Peter. 1982. *Technischer Fortschritt und Rationalisierung im antiken Bergbau*. Dr. Rudolf Habelt G.M.B.H.

Rouveret, Agnès. 2003. "Parrhasios ou le peintre assassin." [in Lévy et al. 2003: 184–93]

Rowland, Ingrid, and Thomas Howe. 1999. *Vitruvius: Ten Books on Architecture*. Cambridge University Press.

Runia, David T. 1986. *Philo of Alexandria and the* Timaeus *of Plato*. Brill.

Russell, D.A. 1974. "Letters to Lucilius." *Seneca*. C.D.N. Costa, ed. Routledge: 70–95.

Russell, Josiah C. 1987. "The Ecclesiastical Age: A Demographic Interpretation of the Period 200–900 A.D." *Medieval Demography*. AMS: 99–111.

Russo, Lucio. 2003. *The Forgotten Revolution: How Science Was Born in 300 B.C. and Why It Had to Be Reborn*, 2nd ed. Springer.

Sacks, Kenneth. 1990. "Culture's Progress." *Diodorus Siculus and the First Century*. Princeton University Press: 55–82.

Salzman, Michele. 1987. "'Superstitio' in the Codex Theodosianus and the Persecution of Pagans." *Vigiliae Christianae* 41.2 (June): 172–88.

Samama, Evelyne. 2003. *Les Médecins dans le Monde grec: Sources épigraphiques sur la Naissance d'un Corps médical*. Droz.

Sambursky, S. 1962. *The Physical World of the Greeks*, 2nd ed. Collier Books.

———. 1963. "Conceptual Developments and Modes of Explanation in Later Greek Scientific Thought." [in Crombie 1963: 61–78]

Sándor, Bela I. 2012. "The Genesis and Performance Characteristics of Roman Chariots." *Journal of Roman Archaeology* 25.1: 475–85.

Sarton, George. 1952. *A History of Science I: Ancient Science through the Golden Age of Greece*. Harvard University Press.

———. 1959. *A History of Science II: Hellenistic Science and Culture in the Last Three Centuries B.C.* Harvard University Press.

Sassi, Maria Michela. 2001. *The Science of Man in Ancient Greece*. University of Chicago.

Savery, Thomas. 1702. *The Miners Friend, or, An Engine to Raise Water by Fire Described and of the Manner of Fixing it in Mines, with an Account of the Several Other Uses It Is Applicable unto, and an Answer to the Objections Made against It*. S. Crouch.

Sawday, Jonathan. 1995. "Execution, Anatomy, and Infamy: Inside the Renaissance Anatomy Theatre." *The Body Emblazoned: Dissection and the Human Body in Renaissance Culture*. Routledge: 54–65.

Sawyer, Robert Keith. 2007. *Group Genius: The Creative Power of Collaboration*. Basic Books.

Scarborough, John. 1968. "Roman Medicine and the Legions: A Reconsideration." *Medical History* 12: 254–61.

———. 1969. *Roman Medicine*. Cornell University Press.

———. 1970. "Romans and Physicians." *The Classical Journal* 65.7: 296–306.

———. 1993. "Roman Medicine to Galen." *Aufstieg und Niedergang der römischen Welt* 2.37.1: 3–48.

———. 2010. *Pharmacy and Drug Lore in Antiquity: Greece, Rome, Byzantium*. Ashgate Publishing.

———. 2012. "Pharmacology and Toxicology at the Court of Cleopatra VII: Traces of Three Physicians." *Herbs and Healers from the Ancient Mediterranean through the Medieval West: Essays in Honor of John M. Riddle*. Anne Van Arsdall and Timothy Graham, eds. Ashgate: 7–18.

Schefold, Karl. 1997. *Die Bildnisse der antiken Dichter, Redner und Denker*. Schwabe.

Scheidel, Walter, and Sitta von Reden, eds. 2002. *The Ancient Economy*. Routledge.

Scheidel, Walter, et al., eds. 2013. *The Cambridge Economic History of the Greco-Roman World*. Cambridge University Press.

Schiebold, Hans. 2010. *Heizung und Wassererwärmung in römischen Thermen: historische Entwicklung—Nachfolgesysteme—neuzeitliche Betrachtungen und Untersuchungen*. Books on Demand.

Schiefsky, Mark. 2008. "Theory and Practice in Heron's Mechanics." *Mechanics and Natural Philosophy before the Scientific Revolution*. Walter Laird & Sophie Roux, eds. Springer: 15–49.

Schiller, F. 1978–1979. "Stepmother Nature: A Medico-Historical Scan." *Clio Medica* 13: 201–18.

Schiøler, Thorkild. 1980. "Bronze Roman Piston Pumps." *History of Technology* 5: 17–38.

———. 1989. "Rekonstruktion einer römischen Feuerlöschpumpe im Antiquarium Comunale." *Leichtweiss-Institut für Wasserbau, Mitteilungen* 103: 281–321.

———. 2009. "Die Kurbelwelle von Augst und die römische Steinsägemühle." *Helvetia Archaeologica* 40.159/160: 113–24.

Schipperges, Heinrich. 1970. "Zur Unterscheidung des 'Physicus' vom 'Medicus' bei Petrus Hispanus." *Asclepio: Archivo Iberoamericano de Historia de la Medicina y Antropologia Medica* 22: 321–27.

———. 1976. "Zur Bedeutung von 'physica' und zur Rolle des 'physicus' in der abendländischen Wissenschaftsgeschichte." *Sudhoffs Archiv für Geschichte der Medizin und der Naturwissenschaften* 60.4: 354–74.

Schmid, Walter. 1998. *Plato's* Charmides *and the Socratic Ideal of Rationality*. State University of New York Press.

Schmitz, Rudolf, and Franz-Josef Kuhlen. 1998. *Geschichte der Pharmazie. Vol. 1, Von den Anfangen bis zum Ausgang des Mittelalters*. Govi Press.

Schneemelcher, Wilhelm, and R. McL. Wilson, eds. 1991. *New Testament Apocrypha*, rev. ed. in 2 vols. Westminster Press.

Schneider, Helmuth. 1992. *Einführung in die antike Technikgeschichte*. Wissenschaftliche Buchgesellschaft.

Schomberg, Anette. 2008. "Ancient Water Technology: Between Hellenistic Innovation and Arabic Tradition." *Syria: Archéologie, art et histoire* 85: 119–28.

Schürmann, Astrid. 1991. *Griechische Mechanik und antike Gesellschaft: Studien zur staatlichen Förderung einer technischen Wissenschaft.* Franz Steiner.

———. 2002. "Pneumatics on Stage in Pompeii: Ancient Automatic Devices and their Social Context." [in Renn & Castagnetti 2002: 35–56]

———, ed. 2005. *Physik / Mechanik.* Franz Steiner. [*Geschichte der Mathematik und der Naturwissenschaften in der Antike* 3]

Shields, Christopher. 2007. *Aristotle.* Routledge.

Scott, James. 1999. "The Ethics of the Physics in Seneca's *Natural Questions.*" *The Classical Bulletin* 75.1: 55–68.

———. 2002. "The *Mappamundi* of Queen Kypros." *Geography in Early Judaism and Christianity: The Book of Jubilees.* Cambridge University Press: 5–22.

Seddon, Keith. 2005. *Epictetus' Handbook and the Tablet of Cebes: Guides to Stoic Living.* Routledge.

Sedley, David. 2007. *Creationism and Its Critics in Antiquity.* University of California Press.

Segal, Alan. 2004. "Religiously-Interpreted States of Consciousness: Prophecy, Self-Consciousness, and Life After Death." *Life after Death: A History of the Afterlife in Western Religion* (Doubleday): 322–50.

Seidel, Yvonne. 2010. "Leuchttürme in der Tabula Peutingeriana." *Standortbestimmung : Akten des 12. Österreichischen Archäologentages vom 28.2.–1.3.2008 in Wien.* Marion Meyer & Verena Gassner, eds. Phoibos: 321–26.

Seigne, Jacques. 2002. "A Sixth Century Water-Powered Sawmill at Jerash." *Annual of the Department of Antiquities of Jordan* 26: 205–213.

Selinger, Reinhard. 1999. "Experimente mit dem Skalpell am menschlichen Körper in der griechisch-römischen Antike." *Saeculum* 50.1: 29–47.

Sellin, Robert. 1983. "The Large Roman Water Mill at Barbegal (France)." *History of Technology* 8: 91–109.

Serban, Marko. 2009. "Trajan's Bridge over the Danube." *International Journal of Nautical Archaeology* 38.2: 331–42.

Shapin, Steven. 1996. *The Scientific Revolution.* University of Chicago Press.

Sharples, R.W. 1987. "Alexander of Aphrodisias: Scholasticism and Innovation." *Aufstieg und Niedergang der römischen Welt* 2.36.2: 1176–1243.

———, ed. 2005. *Philosophy and the Sciences in Antiquity*. Ashgate.

Shaw, Brent. 2013. *Bringing in the Sheaves: Economy and Metaphor in the Roman World*. University of Toronto Press.

Sherwin-White, A.N. 1973. *The Roman Citizenship*, 2nd ed. Clarendon Press.

Shorter, Edward. 1992. *From Paralysis to Fatigue: A History of Psychosomatic Illness in the Modern Era*. Free Press.

Sider, David and Carl Wolfram Brunschön, eds. 2007. *Theophrastus of Eresus: On Weather Signs*. Brill.

Sideras, A. 1994. "Rufus von Ephesos und sein Werk im Rahmen der antiken Medizin." *Aufstieg und Niedergang der römischen Welt* 2.37.2: 1077–1253, 2036–62.

Siegel, Rudolph. 1968. *Galen's System of Physiology and Medicine: An Analysis of His Doctrines and Observations on Bloodflow, Respiration, Humors and Internal Diseases*. Karger.

———, 1970. *Galen on Sense Perception: His Doctrines, Observations and Experiments on Vision, Hearing, Smell, Taste, Touch and Pain, and Their Historical Sources*. Karger.

———. 1973. *Galen on Psychology, Psychopathology, and Function and Diseases of the Nervous System: An Analysis of His Doctrines, Observations and Experiments*. Karger.

Simms, D.L. 1983. "Water-Driven Saws, Ausonius, and the Authenticity of the Mosella." *Technology and Culture* 24.4 (October): 635–43.

———. 1985. "Water-Driven Saws in Late Antiquity." *Technology and Culture* 26.2. (April): 275–76.

———. 1990. "The Trail for Archimedes's Tomb." *Journal of the Warburg and Courtauld Institutes* 53: 281–86.

———. 1991. "Galen on Archimedes: Burning Mirror or Burning Pitch?" *Technology and Culture* 32.1: 91–96.

———. 1995. "Archimedes the Engineer." *History of Technology* 17: 45–111.

———. 2005. "Archimedes the Mechanikos." [in Schürmann 2005: 164–83]

Sines, George, and Yannis Sakellarakis. 1987. "Lenses in Antiquity." *American Journal of Archaeology* 91.2 (April): 191–96.

Singer, Charles Joseph. 1921. "Steps Leading to the Invention of the First Optical Apparatus." *Studies in the History and Method of Science* 2. Charles Singer, ed. Clarendon: 385–413, 533–34.

———. 1952. "Galen's Elementary Course on Bones." *Proceedings of the Royal Society of Medicine* 45: 767–76.

———. 1956. *Galen On Anatomical Procedures*. Oxford University Press.

Singer, P.N. 1997. *Galen: Selected Works*. Oxford University Press.

———, ed. 2013. *Galen: Psychological Writings*. Cambridge University Press.

Slater, Niall. 1990. *Reading Petronius*. The Johns Hopkins University Press.

Sleeswyk, André Wegener. 1981. "Vitruvius' Odometer." *Scientific American* 245.4 (October): 188–200.

Sluiter, Engel. 1997. "The Telescope before Galileo." *Journal for the History of Astronomy* 28.3 (August): 223–34.

Smith, A.M. 1996. *Ptolemy's Theory of Visual Perception: An English Translation of the* Optics *with Introduction and Commentary*. American Philosophical Society.

———. 1999. *Ptolemy and the Foundations of Ancient Mathematical Optics: A Source Based Guided Study*. American Philosophical Society.

———. 2014. *From Sight to Light: The Passage from Ancient to Modern Optics*. The University of Chicago Press.

Smith, Martin Ferguson. 1996. "The Philosophical Inscription of Diogenes of Oinoanda." *Denkschriften (Österreichische Akademie der Wissenschaften: Philosophisch-Historische Klasse)* Bd. 251. Verlag der Osterreichische Akademie der Wissenschaften.

Smith, Morton. 1981. "Superstitio." *Society of Biblical Literature Seminar Papers* 20: 349–55.

Smith, Norman. 1976. "Attitudes to Roman Engineering and the Question of the Inverted Siphon." *History of Technology* 1: 45–71.

Smith, Pamela. 2004. *The Body of the Artisan: Art and Experience in the Scientific Revolution*. University of Chicago Press.

———. 2006. "Art, Science, and Visual Culture in Early Modern Europe." *Isis* 97.1 (March): 83–100.

Solomon, Jon. 2000. *Ptolemy: Harmonics: Translation and Commentary*. Brill.

Southern, Pat. 2001. *The Roman Empire: From Severus to Constantine*. Routledge.

Spain, Robert. 1985. "Romano-British Watermills." *Archaeologia Cantiana* 1984.C: 101–28.

———. 2002. "A Possible Roman Tide Mill." *eArticles of the Kent Archaeological Society* no. 005: http://www.kentarchaeology.ac/authors/005.pdf

———. 2008. *The Power and Performance of Roman Water-mills: Hydro-mechanical Analysis of Vertical-wheeled Water-mills*. Hedges.

Spruytte, J. 1983. *Early Harness Systems: Experimental Studies*. J.A. Allen.

Staden, Heinrich von. 1975. "Experiment and Experience in Hellenistic Medicine." *Bulletin of the Institute of Classical Studies of the University of London* 22: 178–99.

———. 1989. *Herophilus: The Art of Medicine in Early Alexandria*. Cambridge University Press.

———. 1992. "The Discovery of the Body: Human Dissection and its Cultural Context in Ancient Greece." *The Yale Journal of Biology and Medicine* 65.3 (May–June): 223–41.

———. 1995. "Anatomy as Rhetoric: Galen on Dissection and Persuasion." *Journal of the History of Medicine and Allied Sciences* 50.1 (January): 47–66.

———. 1996. "Body and Machine: Interactions between Medicine, Mechanics, and Philosophy in Early Alexandria." *Alexandria and Alexandrianism: Papers Delivered at a Symposium Organized by the J. Paul Getty Museum and the Getty Center for the History of Art and the Humanities and Held at the Museum, April 22–25, 1993*. J. Paul Getty Museum: 85–106.

———. 1997a. "Galen and the 'Second Sophistic.'" *Aristotle and After*. Richard Sorabji, ed. University of 33–54.

———. 1997b. "Teleology and Mechanism: Aristotelian Biology and Early Hellenistic Medicine." *Aristotelische Biologie: Intentionen, Methoden, Ergebnisse: Akten des Symposions über Aristoteles' Biologie vom 24.–28. Juli 1995 in der Werner-Reimers-Stiftung in Bad Homburg*. Wolfgang Kullmann and Sabine Föllinger, eds. F. Steiner: 183–208.

———. 1998. "Andréas de Caryste et Philon de Byzance: médicine et mécanique à Alexandrie." [in Argoud & Guillaumin 1998: 147–72]

———. 2000. "Body, Soul, and Nerves: Epicurus, Herophilus, Erasistratus, the Stoics, and Galen." *Psyche and Soma: Physicians and Metaphysicians on the Mind-Body Problem from Antiquity to Enlightenment*. John P. Wright and Paul Potter, eds. Oxford University Press: 79–116.

———. 2002. "Division, Dissection, and Specialization: Galen's *On the Parts of the Medical Technê.*" *Bulletin of the Institute of Classical Studies* 45 (S77, January): 19–45.

Stahl, William Harris. 1962. *Roman Science: Origins, Development, and Influence to the Later Middle Ages*. Greenwood.

———. 1971. *The Quadrivium of Martianus Capella: Latin Traditions in the Mathematical Sciences, 50 B.C.–A.D. 1250*. Columbia University Press.

Stark, Rodney. 2001. *One True God: Historical Consequences of Monotheism*. Princeton University Press.

———. 2003. *For the Glory of God: How Monotheism Led to Reformations, Science, Witch-Hunts, and the End of Slavery*. Princeton University Press.

———. 2005. *The Victory of Reason: How Christianity Led to Freedom, Capitalism, and Western Success*. Random House.

Stein, Richard. 2004. "Roman Wooden Force Pumps: A Case-Study in Innovation." *Journal of Roman Archaeology* 17a: 221–50.

Stern, E. Marianne. 1999. "Roman Glassblowing in a Cultural Context." *American Journal of Archaeology* 103.3 (July): 441–84.

Stern, Jacob. 2003. "Heraclitus the Paradoxographer: *Peri Apistôn, On Unbelievable Tales.*" *Transactions of the American Philological Association* 133.1 (Spring): 51–97.

Stikas, Constantin. 2014. *Antikythera Mechanism: The Book (Unwinding the History of Science and Technology)*. Stikas Constantin.

Strohmaier, Gotthard. 1993. "Hellenistische Wissenschaft im neugefundenen Galenkommentar zur hippokratischen Schrift 'Über die Umwelt'." [in Kollesch & Nickel 1993: 157–64]

Strong, Donald, and David Brown, eds. 1976. *Roman Crafts*. New York University Press.

Stückelberger, Alfred. 1965. *Senecas 88. Brief: Über Wert und Unwert der freien Künste*. Carl Winter Universitätsverlag.

Sullivan, J.P. 1968. *The Satyricon of Petronius: A Literary Study*. Indiana University Press.

Swan, Peter Michael. 2004. *The Augustan Succession: An Historical Commentary on Cassius Dio's* Roman History *Books 55–56 (9 B.C. – A.D. 14)*. Oxford University Press.

Syme, Ronald. 1969. "Pliny the Procurator." *Harvard Studies in Classical Philology* 73: 201–36. [= Ronald Syme, *Roman Papers* 2.742–73]

Taisbak, C.M. 1974. "Posidonius Vindicated at All Costs? Modern Scholarship versus the Stoic Earth Measurer." *Centaurus* 18.4: 253–69.

Talbert, Richard, ed. 2012. *Ancient Perspectives: Maps and Their Place in Mesopotamia, Egypt, Greece and Rome*. University of Chicago Press.

Talbert, Richard, and Richard Unger, eds. 2008. *Cartography in Antiquity and the Middle Ages: Fresh Perspectives, New Methods*. Brill.

Taub, Liba. 1993. *Ptolemy's Universe: The Natural Philosophical and Ethical Foundations of Ptolemy's Astronomy*. Open Court.

———. 2002. "Instruments of Alexandrian Astronomy: The Uses of the Equinoctial Rings." [in Tuplin & Rihll 2002: 133–49]

———. 2003. *Ancient Meteorology*. Routledge.

———. 2008. *Aetna and the Moon: Explaining Nature in Ancient Greece and Rome*. Oregon State University Press.

Taylor, A.E. 1917. "On the Date of the Trial of Anaxagoras." *Classical Quarterly* 11: 81–87.

Taylor, Rabun. 2003. *Roman Builders: A Study in Architectural Process*. Cambridge University Press.

Temin, Peter. 2004. "The Labor Market of the Early Roman Empire." *The Journal of Interdisciplinary History* 34.4 (Spring): 513–38.

———. 2006. "The Economy of the Early Roman Empire." *Journal of Economic Perspectives* 20.1 (Winter): 133–51.

Temkin, Owsei. 1934. "Galen's 'Advice for an Epileptic Boy.'" *Bulletin of the History of Medicine* 2: 179–89.

Thomas, Peter. 2000. *Trees: Their Natural History*. Cambridge University Press.

Thompson, E.A. 1952. *A Roman Reformer and Inventor: Being a New Text of the Treatise* De Rebus Bellicis. Clarendon Press.

Thomssen, H. 1994. "Die Medizin des Rufus von Ephesos." *Aufstieg und Niedergang der römischen Welt* 2.37.2: 1254–92.

Tieleman, Teun. 1996. *Galen and Chrysippus on the Soul: Argument and Refutation in the* De Placitis *Books II–III*. Brill.

———. 2002. "Galen on the Seat of the Intellect: Anatomical Experiment and Philosophical Tradition." [in Tuplin & Rihll 2002: 256–73]

Tohon, Anne. 2010. "An Unpublished Astronomical Papyrus Contemporary with Ptolemy." *Ptolemy in Perspective: Use and Criticism of His Work from Antiquity to the Nineteenth Century*. Alexander Jones, ed. Springer: 1–10.

Todd, Robert. 1976. *Alexander of Aphrodisias on Stoic Physics: A Study of the* De Mixtione. Brill.

———. 1984. "Philosophy and Medicine in John Philoponus' Commentary on Aristotle's *De Anima*." *Dumbarton Oaks Papers* 38 (1984): 103–20.

Toomer, G.J. 1976. *Diocles on Burning Mirrors: The Arabic Translation of the Lost Greek Original*. Springer-Verlag.

———. 1984. *Ptolemy's Almagest*. Duckworth.

———. 1985. "Galen on the Astronomers and Astrologers." *Archive for History of Exact Sciences* 32.3–4 (September): 193–206.

Touwaide, Alain. 2000. "Vin, Santé et Médecine à Travers le « Traité de Matière Médicale » de Dioscoride." *Pallas* 53: 101–11, 276.

Towner, Philip. 2006. *The Letters to Timothy and Titus*. W.B. Eerdmans.

Toynbee, Jocelyn. 1944. "Dictators and Philosophers in the First Century A.D." *Greece & Rome* 13.38/39 (June): 43–58.

Trapp, Michael. 2007. *Philosophy in the Roman Empire: Ethics, Politics and Society*. Aldershot, England: Ashgate.

Travis, John Robert. 2008. *Coal in Roman Britain*. Tempus Reparatum.

Trigg, Joseph Wilson. 1998. *Origen*. Routledge.

Trowbridge, Mary. 1930. *Philological Studies in Ancient Glass*. University of Illinois.

Tuplin, C.J., and T.E. Rihll, eds. 2002. *Science and Mathematics in Ancient Greek Culture*. Oxford University Press.

Turner, J.H. 1951. "Roman Elementary Mathematics: The Operations." *Classical Journal* 47: 63–74, 106–08.

Tybjerg, Karin. 2003. "Wonder-Making and Philosophical Wonder in Hero of Alexandria." *Studies in History and Philosophy of Science: Part A* 34.3: 443–66.

———. 2004. "Hero of Alexandria's Mechanical Geometry." [in P. Lang 2004: 29–56]

———. 2005. "Hero of Alexandria's Mechanical Treatises: Between Theory and Practice." [in Schürmann 2005: 204–26]

Ulansey, David. 1989. *The Origins of the Mithraic Mysteries: Cosmology and Salvation in the Ancient World*. Oxford University Press.

Unguru, Sabetai. 1975. "On the Need to Rewrite the History of Greek Mathematics." *Archive for History of Exact Sciences* 15.1 (March): 67–114.

———. 1979. "History of Ancient Mathematics—Some Reflections on the State of the Art." *Isis* 70.4 (December): 555–65.

Vallance, J.T. 1990. *The Lost Theory of Asclepiades of Bithynia*. Clarendon Press.

———. 1993. "The Medical System of Asclepiades of Bithynia." *Aufstieg und Niedergang der römischen Welt* 2.37.1: 693–727.

Vamvacas, Constantine. 2009. *The Founders of Western Thought: The Presocratics: A Diachronic Parallelism between Presocratic Thought and Philosophy and the Natural Sciences*. Springer.

Van Brummelen, Glen. 2009. *The Mathematics of the Heavens and the Earth: The Early History of Trigonometry*. Princeton University Press.

Van Helden, Albert. 1977. "The Invention of the Telescope." *Transactions of the American Philosophical Society* 67.4: 1–67.

Vegetti, Mario. 1995. "L'épistémologie d'Érasistrate et la technologie hellénistique." [in van der Eijk, Horstmanshoff and Schrijvers 1995: 2.461–72]

Vendries, Christophe. 2004. "La place de la musique dans l'éducation romaine selon Marrou: la vision d'un musicologue averti." [in Pailler & Payen 2004: 257–64]

Verde, Francesco. 2013. "Epicurean Attitude Toward Geometry: The Sceptical Account." *Épicurisme et Scepticisme*. Stéphane Marchand & Francesco Verde, eds. Università di Roma-La Sapienza: 131–50.

Vernant, Jean Pierre. 1983. *Myth and Thought among the Greeks*. Routledge. [translated from the 1965 original *Myth et Pensée chez les Grecs*]

Vitrac, Bernard. 2009. "Mécanique et mathématiques à Alexandrie: le cas de Héron." *Oriens-Occidens: sciences, mathématiques et philosophie de l'antiquité à l'âge classique* 7: 155–99.

Wacher, John. 1987. *The Roman World*, 2 vols. Routledge.

Wachsmann, Shelley. 2012. "Panathenaic Ships: The Iconographic Evidence." *Hesperia: The Journal of the American School of Classical Studies at Athens* 81.2: 237–66.

Waerden, B.L., van der. 1963. "Basic Ideas and Methods of Babylonian and Greek Astronomy." [in Crombie 1963: 42–60]

———. 1976. "Defence of a 'Shocking' Point of View." *Archive for History of Exact Sciences* 15.3 (September): 199–210.

Wallace-Hadrill, Andrew. 1990. "Pliny the Elder and Man's Unnatural History." *Greece & Rome* 37.1 (April): 80–96.

Walsh, P.G. 1970. *The Roman Novel: The 'Satyricon' of Petronius and the 'Metamorphoses' of Apuleius*. Cambridge University Press.

———. 1996. *The Satyricon*. Clarendon Press.

Walton, Steven, ed. 2006. *Wind And Water in the Middle Ages: Fluid Technologies from Antiquity to the Renaissance*. Arizona: ACMRS.

Walzer, Richard. 1944. *Galen on Medical Experience*. Oxford University Press.

———. 1949. *Galen on Jews and Christians*. Oxford University Press.

Walzer, Richard, and Michael Frede. 1985. *Galen: Three Treatises on the Nature of Science*. Hackett.

Warren, James. 2007. *Presocratics: Natural Philosophers before Socrates*. University of California Press.

Watts, Edward J. 2006. *City and School in Late Antique Athens and Alexandria*. University of California Press.

Webster, Colin. 2014. *Technology and/as Theory: Material Thinking in Ancient Science and Medicine*. Dissertation (Ph.D.), Columbia University.

Werner, Walter. 1997. "The Largest Ship Trackway in Ancient Times: The Diolkos of the Isthmus of Corinth, Greece, and Early Attempts to Build a Canal." *International Journal of Nautical Archaeology and Underwater Exploration* 26.2: 98–119.

White, K.D. 1966. "The Gallo-Roman Harvesting Machine." *Antiquity* 40.157 (March): 49–50.

———. 1967a. "Gallo-Roman Harvesting Machines." *Latomus Revue d'Études Latines* 26: 634–47.

———. 1967b. *Agricultural Implements of the Roman World*. Cambridge University Press.

———. 1969. "The Economics of the Gallo-Roman Harvesting Machines." *Hommages à Marcel Renard*, vol. 2. Jacqueline Bibauw, ed. Latomus: 804–09.

———. 1970. *Roman Farming*. Cornell University Press.

———. 1975a. *Farm Equipment of the Roman World*. Cambridge University Press.

———. 1975b. "Technology in North-West Europe before the Romans." *Museum Africum* 4: 43–46.

———. 1984. *Greek and Roman Technology*. Cornell University Press.

———. 1993. "'The Base Mechanic Arts'? Some Thoughts on the Contribution of Science (Pure and Applied) to the Culture of the Hellenistic Age." *Hellenistic History and Culture*. Peter Green, ed. University of California Press: 211–37.

White, Lynn, Jr. 1962. *Medieval Technology and Social Change*. Oxford University Press. [pagination from 1964 paperback edition]

———. 1963. "What Accelerated Technological Progress in the Western Middle Ages." [in Crombie 1963: 272–91]

White, Robert. 1975. *The Interpretation of Dreams:* Oneirocritica *by Artemidorus*. Noyes.

Whitehead, David, and P.H. Blyth. 2004. *Athenaeus Mechanicus:* On Machines *(Peri Mêchanêmatôn)*. Franz Steiner Verlag.

Whitewright, Julian. 2009. "The Mediterranean Lateen Sail in Late Antiquity." *The International Journal of Nautical Archaeology and Underwater Exploration* 38.1: 97–104.

Whitmarsh, Tim. 2005. *The Second Sophistic*. Oxford University Press.

Whitmarsh, Tim, John Wilkins, and Christopher Gill, eds. 2009. *Galen and the World of Knowledge*. Cambridge University Press.

Whitney, Elspeth. 1990. "Paradise Restored: The Mechanical Arts from Antiquity through the Thirteenth Century." *Transactions of the American Philosophical Society* 80.1 (New Series): 1–169.

Wikander, Örjan. 1990. "Water-Power and Technical Progress in Classical Antiquity." *Ancient Technology: A Symposium*. Museum of Technology: 68–84.

———. 2000a. "The Water-Mill." [in Wikander 2000c: 371–400]

———. 2000b. "Industrial Applications of Water-Power." [in Wikander 2000c: 401–12]

———, ed. 2000c. *Handbook of Ancient Water Technology*. Brill.

Wild, John Peter. 1987. "The Roman Horizontal Loom." *American Journal of Archaeology* 91.3 (July): 459–71.

Wilk, Stephen. 2004. "Claudius Ptolemy's Law of Refraction." *Optics and Photonics News* 15.10: 14–17.

Williams, Stephen. 1985. *Diocletian and the Roman Recovery*. B.T. Batsford.

Wilmanns, Juliane. 1995. "Der Arzt in der römischen Armee der frühen und hohen Kaiserzeit." [in van der Eijk, Horstmanshoff and Schrijvers 1995: 1.171–88]

Wilson, Andrew. 2000. "Industrial Uses of Water" [in Wikander 2000c: 127–50]

———. 2002. "Machines, Power and the Ancient Economy." *Journal of Roman Studies* 92: 1–32.

Wilson, C. Anne. 2002. "Distilling, Sublimation, and the Four Elements: The Aims and Achievements of the earliest Greek Chemists." [in Tuplin & Rihll 2002: 306–22]

Winter, Thomas. 2007. "Who Wrote the *Mechanical Problems* in the Aristotelian Corpus?" *The Mechanical Problems in the Corpus of Aristotle*. DigitalCommons@ University of Nebraska (http://digitalcommons.unl.edu/classicsfacpub/68): iii–ix.

Wolff, Michael. 1987. "Philoponus and the Rise of Preclassical Dynamics." *Philoponus and the Rejection of Aristotelian Science*. Richard Sorabji, ed. Cornell University Press: 84–120.

———. 1988. "Hipparchus and the Stoic Theory of Motion." *Matter and Metaphysics: Fourth Symposium Hellenisticum*. Jonathan Barnes and Mario Mignucci, eds. Bibliopolis: 472–545.

Woods, Ann. 1987. "Mining." [in Wacher 1987: 2.611–34]

Woodside, M. St. A. 1942. "Vespasian's Patronage of Education and the Arts." *Transactions of the American Philological Society* 73: 123–29.

Wright, Georg Henrik von. 1997. "Progress: Fact and Fiction." [in Burgen et al. 1997: 1–18]

Wright, G.R.H. 2005. *Ancient Building Technology*, in 2 vols. Brill.

Wright, M.R. 1995. *Cosmology in Antiquity*. Routledge.

Wright, M.T. 1990. "Rational and Irrational Reconstruction: The London Sundial-Calendar and the Early History of Geared Mechanisms." *History of Technology* 12: 65–102.

———. 2007. "The Antikythera Mechanism Reconsidered." *Interdisciplinary Science Reviews* 32.1 (March): 27–43.

Yerxa, Donald, ed. 2007. "Focus: Thoughts on the Scientific Revolution." *European Review* 15.4 (October): 439–512. [including contributions from Peter Harrison, William Shea, John Heilbron, H. Floris Cohen, and Theodore Rabb]

Zerzan, John. 2005. *Against Civilization: Readings and Reflections*. Feral House.

Zhmud, Leonid. 2003. "The Historiographical Project of the Lyceum: The Peripatetic History of Science, Philosophy, and Medicine." *Antike Naturwissenschaft und ihre Rezeption* 13: 109–26.

———. 2006. *The Origin of the History of Science in Classical Antiquity*. Walter de Gruyter.

Zilsel, Edgar. 1945. "The Genesis of the Concept of Scientific Progress." *Journal of the History of Ideas* 6.3: 325–49.

Zimmermann, Linda. 2011. *Bad Science: A Brief History of Bizarre Misconceptions, Totally Wrong Conclusions, and Incredibly Stupid Theories*. Eagle Pres.

Index of Ancient Inventions

abacus (decimal), 197
abortifacients & birth control, 195
acoustic resonator, 201
anemoscope, 200, 221
Archimedean screw (see screw pump)
asbestos (fireproof) cloth, 202
astrolabe (armillary / spherical), 145–47, 151, 161, 168–69, 193, 199, 221, 401, 422–23, 433
astrolabe (plane), 141–42, 193, 199, 327–28
automata (theaters & apparatuses), 136, 159, 160, 164–66, 198, 210, 226, 234, 236, 237, 320, 338
automated door, 166, 199

banks, 194, 425
Baroulkos (see geared crane)
bottle rocket, 199
butt hinge, 200

calendrics, 129, 135, 139, 142, 148, 197, 220, 235, 532, 534,
cams & camshafts, 198
camel domestication, 191
carrier-pigeon, 196
cartographic projection, 141, 150, 152, 169, 244
Celsean surveyor, 171, 199
chronology (scientific), 135
clock (mechanical: see cuckoo & waterclock)
codex (book-binding), 200
coin-operated vending machine, 166, 199, 224
compound pulley, 161, 191, 198
computers (gear-train), 142, 143, 146–47, 151, 199, 252, 399
cranks & crankshafts, 198, 225
crop rotation (double & triple), 191, 197
cuckoo clock, 159, 198
cylinder block, 198, 217
cylinder-and-plunger pump (see piston)

diopters, 141, 168, 199, 220, 293, 438
diving bell, 201
drop-boom fishing spear, 201

electroshock therapy, 195–96, 243

floats (air-inflated: for fishing & rafting), 201
folding furniture (chairs, stands, stools, etc.), 215
folding pocket knife, 201
frame-saw, iron, 200

geared crane, 199
gears & gear trains (cogged, toothed, worm), 136, 142, 147, 165, 191, 198, 199, 202, 217, 220, 222, 223, 235, 239–40
gimbals, 200
glassblowing, 191, 218–19, 224, 236–37, 290
globes (geographical & astronomical), 137, 141, 146, 152, 168, 199, 293

heated baths, 201
heavy-beam transports, 202
hollow bronze-casting, 191
hydrometer, 200

indoor plumbing (pressurized), 201
insurance (see maritime loans)
inswinger catapult (see metal-frame)

keyboard, 211

lateen sail, 191, 204
lenses (for burning & magnifying), 144, 153, 199
logics, formal (see general index)

machine-gun (automatic catapult)
magnifiers (see lenses and mirrors)
maritime loans, 194
mannequins (anatomically articulated)
mathematics, formal (see general index)
Menelausian balance, 199
mesolabe, 200, 421
metal-frame catapult, 167, 198
meteoroscope, 169
miner's lamp, 200–01
mirrors (liquid), 195
mirrors (parabolic; for burning & magnifying), 136–37, 208, 213
multihook fishing line, 201

odometer (sea & land), 161, 166, 191, 198, 217, 423

pantograph, 121, 191, 216
parabolic mirror (see mirrors)
parchment, 200, 219
perfumer's press, 200
pile driver, 201
piston, 159, 198, 232, 243
place-notation (numerical systems), 197
plowshare (iron), 191
pneumatic (air-spring) catapult, 198
pontoon bridge (inflatable), 201, 209
postal system, 197
pressurized fountain, 201

quadrant, 193, 199
quick-wheat, 197

rack-and-pinion, 198
ratchets, 198
reciprocating pump (for continuous-stream bilge-clearing & firefighting), 159, 198, 212, 234, 240
robotics (see automata)
rotary grain mill (ox & quern), 191

sciences, formal (see general index)
screw (for fastening), 214–15
screw cutter, 198
screw press (for grapes, cloth, and olives), 191–92, 201–02, 215, 230, 236
screw pump, 136, 161, 191, 198, 240, 316
shorthand (Latin & Greek), 196
shower, 201
siphon, 192, 198, 206, 235, 239, 254
slide-rule (see mesolabe)
snorkel, 201
sphere-lathe, 200
springs (metal), 159, 192, 198, 214, 239,
star charts (scientific), 141, 199
steam engine (see general index)
steam-powered rotator, 158, 165–66, 191–92, 243
steam-powered billows, 216, 243
steam-powered cannon (speculated) 213
sundials (conical; geared; portable; etc.), 129, 134, 135, 136, 137, 141–42, 162, 168, 181, 198, 217, 220–21, 235, 238–39, 293, 375, 438, 460
surveying instruments, 191, 220

tachygraphy (see shorthand),
toilet sponge, 201
telegraphy (optical), 196–97, 313
textual criticism (as a science), 196
thermometer (thermoscope), 200, 243
torsion catapults, 160, 162, 167, 191, 192, 198, 213
transmissions (gear trains: see gears)
treadwheel (for pump, tractor, and windlass), 202, 210
trip-hammer (water-powered), 226–27

universal joint, 200

valves (beveled, flap, hinged, piston/shaft, screw, spindle, etc.), 192, 198, 212, 215, 220, 239
vellum (see parchment)
volumetric table, 200

waterclocks (mechanical; parastatic), 161, 166–67, 198, 220–21, 238–39, 249, 293, 423, 438
water-level, 200, 219–20
watermill (see waterwheel)
water organ (pedaled & automatic), 159, 161, 166, 191, 198, 209, 211–12, 423
waterwheel (water-powered rotator: for grinding grain, ore, and sand; sawing stone and wood; pounding iron in metalworking, or cloth in felting & fulling), 17, 191, 192, 196, 209, 225–28, 235–37, 240, 251
water-lifting wheel
  man-powered (see treadwheel)
  ox-powered (sakia), 207, 209
  water-powered (noria: not discussed; pre-Hellenistic)
whaling harpoon, 201
wheelbarrow, 192–93
windmill (for pumping or milling), 209

General Index

Abdaraxos, 156, 354
Academics (see Skeptics)
Académie des Sciences, 453
Achilles (writer), 73, 149
acoustics (see harmonics)
Aelian, 408
Aelius Demetrius, 372
Aelius Promotus, 122
Aemilius Paulus, 399–400
Aesop, 94–95, 460
Aetna (volcano & poem), 208, 211, 212, 216, 230–31, 260, 281–82, 323, 350, 379, 386–91
Agathocles of Atrax, 400, 441
Agesistratus, 162
agricultural science, 58, 180, 182, 229–31, 253–56, 278–81, 321, 352, 375–76, 395, 432, 435, 439, 448–49
agrimensores (see surveying)
Agrippa of Bithynia, 155
Agrippa (statesman), 221
alchemy (see chemistry)
Alcinous, 53–54
Alexander of Aphrodisias, 50, 55, 62, 66, 71, 83–86, 89, 139, 154, 174, 188, 372
Alexander of Damascus, 372–73
Alexander of Laodicea, 117
Alexander of Myndus, 180, 378
Alexander Philalethes, 116
Alexander the Great, 353, 451, 553
Alexandria (library, museum, research in), 103–04, 111–14, 123, 131–32, 134–35, 145, 150–51, 156, 158, 160, 182–83, 208, 220, 230, 233, 297–98, 301, 354, 441, 447, 449, 459, 530, 532–33, 539, 540, 551
Alexias, 110–11
algebra, 132, 172, 195, 244–45
Amafinius, Gaius, 20
Anatolius of Alexandria, 172, 532–33
anatomy & physiology, 17, 22, 37, 51–52, 65, 80, 85–90, 100–01, 105–08, 108, 110–13, 116–28, 132, 136, 154, 173–75, 178, 180–82, 195, 222, 244, 247, 260, 284, 285, 294–306, 329, 331–32, 335, 338, 363–64, 367, 372, 378, 435–37, 466, 495, 499, 557
Anaxagoras, 28, 47, 65, 132, 395, 481, 555
Anaximander, 44, 396, 555

Andreas (the Herophilean), 112
Andrias, 141
Andronicus of Cyrrhus, 162, 220
Andronicus of Rhodes, 104
Antigenes, 113
Antikythera computer, 142, 147, 252, 399
Antiochis, 115
Antipater of Thessalonica, 237
Antiphon (poet), 256
Antisthenes, 66–67, 556
Antoninus Pius (emperor), 372
Antonius Castor, 178
Antonius Mus (the Herophilean), 112
Antyllus, 122
Apellis, 162
Apion of Oasis, 179
Apollinarius of Aizani, 155
Apollo (god), 362, 392
Apollodorus of Alexandria, 114, 534
Apollodorus of Athens, 135
Apollodorus of Damascus, 171, 241
Apollonides, 370
Apollonius Mus (the Herophilean), 112
Apollonius of Alexandria, 114
Apollonius of Athens, 162
Apollonius of Citium, 114, 120, 319
Apollonius of Memphis, 113
Apollonius of Myndus, 142, 378
Apollonius of Perga, 60, 135–37, 139, 151, 159, 196–97
apothecary (see pharmacology)
Apuleius, 66, 180, 295, 357–58, 374, 397–98, 454, 490, 544
Arabic (see Muslim science)
Aratus (poet), 31–32, 73, 138, 181, 368
Arcesilas, 493
Archelaus (philosopher), 47, 556
Archigenes of Apamea, 119, 222
Archimedes, 13, 31, 59, 61, 103, 136, 138, 140, 145–46, 156, 160–64, 167, 196–98, 205, 213, 214, 217, 238, 241, 259, 306, 316, 353, 356, 365, 374, 397, 401–05, 420–24, 442, 446, 452, 517, 540
architects, 10, 30, 156, 171, 200, 220, 238, 272, 291, 433, 449, 456, 529
architectural science (see also architects, engineering, and surveying), 155–56, 166, 171, 199, 200, 205, 220, 229, 235, 238–39, 255, 256, 352, 354, 375, 432
Archytas of Tarentum, 59, 158, 199, 251, 420, 421, 557
Aretaeus of Cappadocia, 118, 217
Aristarchus of Samos, 29, 132, 134, 137, 138, 186–88, 370
Aristides, 523
aristocratic attitude, 14, 251–53, 272, 321, 409–45, 468
Aristomachus of Soli, 230
Aristotle & Aristotelians, 11, 18, 36, 37, 39–44, 46–48, 51–55, 60–63, 70, 72, 83–84, 86–87, 89, 99–104, 110, 112, 116, 121, 125–33, 138, 142–44, 147, 154, 156–58, 160, 165, 170, 175–78, 180–83, 185–89, 209, 245–46, 251, 255, 258, 261–62, 263–67, 269, 296, 309, 310, 313–14, 317, 342, 344–46, 349, 351, 365, 369–70, 372–73, 410–18, 425, 437–39, 442–43, 447, 449, 451, 452, 454, 467, 482, 517, 524–25, 533, 537, 555–57
Aristoxenus, 101, 557
Aristyllus, 140
arithmetic (see also algebra & combinatorics), 42, 70, 82, 89, 101,

135, 138, 154, 172, 361, 375, 427, 479, 532–33
Arius of Tarsus, 120
armillary spheres (see also orreries), 146, 151, 161, 168–69, 193, 199, 221, 401, 422–23, 433
Arnaud, Pascal, 293–94
Arnobius, 532
Arrian, 145
art & artists, 14–15, 23, 29–31, 144, 146, 193, 201, 212, 264, 274–76, 277, 286, 290, 300, 307, 309, 319, 353, 357, 362, 364–65, 378, 393, 416, 430
Artemidorus of Daldis, 338–39
Artemidorus of Ephesus, 148, 554
Asclepiades of Bithynia, 45–46, 99, 115–16, 217, 261, 358, 397–98
Asclepiodotus, 148
Asclepius (god), 362, 392–93
Aspasia, 124
associations, medical & scientific (see also museums), 28, 95, 447–55, 540, 549, 551
astrology, 24–25, 29, 67, 72, 86, 125, 130, 139, 147, 151, 178–79, 293, 346, 349, 363, 365, 368, 392, 427, 431, 532, 534
astronomers & astronomy, 12, 17, 22–24, 28, 30–32, 41–42, 52–92, 99, 101, 103, 105, 114, 120, 124–26, 129–56, 159–61, 163, 165, 167–68, 170–75, 178, 180–81, 184–87, 191, 193, 195, 199, 220, 235, 238, 241, 243, 246, 252, 259, 273–75, 278, 292, 317, 323, 326–28, 333–34, 343–44, 346, 353, 355, 358, 360–75, 378, 380–83, 390, 395, 397, 399–401, 423, 426–27, 429, 433, 436, 438–40, 448, 450, 457–58, 460–61, 464–66, 468, 479, 481, 490, 495, 498–99, 521, 527–29, 532–34, 544, 541, 549, 557
astrophysics, 83, 92, 129, 133, 152, 156, 333, 342, 369–70, 495
Athena (goddess), 362
Athenaeus Mechanicus, 158, 162, 203, 303, 319, 320
Athenaeus of Attaleia, 117–18
Athenaeus of Naucratis, 369, 441
Athenagoras, 523
Athenodorus, 148
Athens, 28, 47, 66, 103, 104, 112, 132, 135, 143, 147–48, 162, 175, 220, 242, 299–300, 308, 353, 355, 362, 372–73, 393–95, 414–15, 452, 487, 551, 556, 557
Attalus III (king), 115
Atticus, 54
augury (see divination)
Augustine, 66, 257, 297, 366, 536, 538
Augustus (emperor), 115, 187, 212, 220, 395, 451
Aurelius Ammianus, 393–94
Aurelius, Marcus (emperor), 55, 260, 268–69, 297, 352, 356–57, 365, 371, 393–94, 419, 527
Aurelius Victor, 371
Austen, Jane, 14
Autolycus of Pitane, 130–31, 149
Automedon, 415

Bacchius (the Herophilean), 112
Bacon, Francis, 9, 11, 13–15, 93, 277, 450, 536, 545
Bacon, Roger, 536
Balbus, 171
Balme, D.M., 102

Barbegal industrial mill, 228, 250
Basil, 538
Basilides, 60
Batty, Roger, 150
Beagon, Mary, 256, 432
Beckmann, Johann, 410–11, 419
Ben-David, Joseph, 9, 12–15, 247, 249, 467, 547
Bennett, Charles, 282–83
Billarus Sphere, 146
biology (see life sciences)
Biton, 159
Boëthius, Flavius, 372–73
botany (see also agricultural science & pharmacology), 38, 52, 85–86, 91–92, 105, 108, 110, 115, 120–21, 144, 176–82, 230–31, 261, 274, 345, 355, 357, 364, 367, 372, 377–78, 395, 439–40,
Boyle, Robert, 93, 552
Brahe, Tycho, 17, 154
Brunt, P.A., 287–88
Burkert, Walter, 310–11
Bury, J.B., 311

cadavers, human (dissection of), 32, 110–12, 123, 126–27, 292, 294–302, 321–22, 373
calendrics (see also parapegmata), 129, 135, 139, 142, 148, 197, 220, 235, 460–61, 532, 534
Caligula (emperor), 221, 379
Callimachus (the Herophilean), 112
Callippus, 130, 557
Callistratus, 158
Calpurnius Piso, 182, 372
Caplan, Harry, 270
Carpus of Antioch, 162–63, 423
cartography, 101, 121, 129, 131, 135, 137, 143, 148, 150–53, 168–69, 178, 195, 199, 221–22, 235, 244, 253, 255, 292–94, 328, 360
Cassius Dio, 217, 291, 293, 400
Cassius Dionysius, 230
Casson, Lionel, 287–88
Celsus (encyclopedist), 33, 87, 109, 182, 241, 248, 280, 295–99, 321, 375–77, 396, 435
Celsus (engineer), 171
Celsus (social critic), 526–27
Charybdis, 323
chemistry (see also mineralogy & pharmacology), 108–09, 115, 156, 174–75, 219–20, 223, 234, 242, 252–53, 260, 334,
Cherniss, Harold, 370
Chersiphron, 353
China (science & technology in), 8, 304
Christianity (attitudes in), 7, 13–14, 16, 18–20, 28–29, 56, 65, 67, 98, 104, 116, 122, 124, 132, 139, 150, 154, 157, 160, 172, 257, 262, 269, 305–07, 330, 337, 341–42, 350, 355, 359–60, 364–66, 379, 381, 384, 393, 406, 409, 428, 456, 461–66, 469, 471–541, 543–47
Chrysermus (the Herophilean), 112
Chrysippus, 54, 80, 274, 383, 437
Cicero, 50–53, 57–62, 67, 81, 84–87, 90, 93–94, 142, 145–46, 148, 178–79, 196, 221, 256, 259, 285, 317–18, 344–52, 359–60, 363, 368–69, 374, 382, 384, 397, 399–403, 415–17, 424–25, 431, 484, 490, 493, 522, 531, 544
Clagett, Marshall, 471–74, 540–41
class (socio-economic; see also aristocratic attitude), 24, 128–29,

265–66, 281, 301, 410–11, 419, 428–42, 444–45, 455–56, 460, 467–68, 544
Claudius Agathinus, 119
Claudius (emperor), 119, 196, 400
Claudius Menecrates, 124
Claudius Severus, 372
Cleanthes, 186
Clearchus, 177
Clement of Alexandria, 360, 473, 475–85, 496, 498, 503, 517, 527–28, 537, 540
Cleomedes, 141, 149, 154
Cleopatra (queen), 114, 115, 117, 120, 441
climatology (see also astronomy and meteorology), 91, 118, 144, 390, 439
Cohen, H. Floris, 9, 11
Columella, 208, 215, 228, 229–30, 253, 274, 278–79, 321–22, 352, 375, 410–11, 435
combinatorics, 138, 195
comets, 66, 142, 324, 369, 376, 380, 390, 499
compass (magnetic), 8, 17, 175, 190, 278, 337, 552
Conon, 140
Constantine (the Great), 16, 19–20, 371, 475, 489, 496–97, 536
Copernicus, 11, 17, 184
Cornelius Severus, 387
Courtney, Edward, 276
Crates, 137
Crateuas, 120
crisis (of the 3rd century), 18–20, 104–05, 124, 183, 231, 302, 307, 471–72, 548, 550
Crombie, A.C., 18, 248, 302–03, 311, 547–48
Ctesibius, 158–60, 164, 190, 212, 353
Cuomo, Serafina, 79, 238–39, 320, 434
curiosity, 13–14, 305–07, 313, 350, 355–57, 377, 383–86, 391, 398, 407–08, 432, 439, 441, 462, 464, 473, 483–84, 487, 530, 534–36, 543–44
cyclical time, 263–69
Cynics (philosophical sect), 50, 55–56, 342, 355, 457, 459, 474, 488, 520, 545

Daedalus, 272, 289, 393–94
Damigeron, 177
Darwin, Charles, 94, 189, 305, 545
decline, 20–21, 105, 164, 183, 190, 232–33, 240–307, 419, 468, 471–72, 488, 540, 545, 547
Delphi, 44, 46, 211, 370
Demeter (goddess), 362
Demetrius (the Epicurean), 60
Demetrius of Apamea, 113–14
Democlitus, 220
Democritus, 64, 108, 177, 274, 395, 401, 556, 557
Demosthenes Philalethes, 117
Demostratus, 181, 407–08
Dercyllides, 73–74, 186
Descartes, 247
de Ste. Croix, G.E.M., 473
dialectic (see logic)
Dicaearchus of Messana, 101, 131
Didymus, 170
dining clubs, academic (see associations and museums)
Dinochares, 353
Dio Chrysostom, 367–68, 372
Diocles (astronomer), 136–37
Diocles of Carystus, 110, 329

Diocletian and Maximian (emperors), 365
Diodorus of Alexandria, 73, 142, 169, 193
Diodorus Siculus, 66, 143, 282, 315–19, 327, 331, 358, 402, 407, 454
Diodotus, 179
Diogenes Aristokleides, 44, 46
Diogenes Laertius, 47, 49–50, 55, 60, 68–69, 82, 359, 555
Diogenes of Oenoanda, 44
Dionysius (botanist), 120
Dionysius the Great, 305, 529
Dionysius the Tyrant, 450
Dionysodorus, 136
Dionysus (god), 362
Diopeithes (decree of), 28–29
Diophantus of Alexandria, 172, 244
Dioscorides Pedanius, 110, 114, 116, 120–21, 196, 241, 322
Dioscorides Salvius, 155
Dioscurides Phacas, 114, 120
dissection (see also anatomy, cadavers, and vivisection), 32, 88, 101–03, 110–11, 126–28, 180–82, 252, 292, 294–301, 306, 322, 331–32, 363–64, 372–74, 414–15, 436–37, 488, 557
divination (see also astrology), 29, 62, 85–86, 295, 338, 346, 349, 352, 368, 532
Dobson, J.F., 147
doctors, 10, 12, 14, 24, 29–30, 32, 38–39, 42–46, 52–53, 63, 68, 80, 83, 86–89, 105–06, 107, 109–25, 128, 167, 177, 217, 222–23, 235, 252, 280, 284–85, 295, 297–301, 322, 330–33, 342, 353, 362–63, 365, 367, 369, 371–73, 376, 392, 394, 397–98, 408, 415–16, 429, 431, 433–35, 438–41, 445–47, 456, 458–61, 468, 473–74, 521, 524–25, 532, 541, 549
Dodds, E.R., 308–09
Domitian (emperor), 167, 220, 293–94, 451, 459
Dorion, 171–72
Dositheus, 136
Drachmann, A.G., 434
dragon boats (Chinese) 304
Drury, Clare, 519

Ecclesiastes, 269
eclecticism (in science & philosophy), 99, 107, 119, 122, 126, 188, 247–48, 342–46, 418, 457, 555, 556
Edelstein, Ludwig, 10–14, 127, 246, 308–09, 311, 442, 455, 547
education (see also universities), 7, 10–11, 14–16, 23, 25, 28, 30, 38, 46, 65, 82, 97, 104–05, 125, 134–35, 158, 245, 263, 275, 281–82, 284–85, 300–01, 339, 341–43, 348, 350, 360, 367, 370, 374, 376, 379, 394, 399, 409, 414–15, 417, 426, 429–35, 438, 440, 443, 447, 454, 456–57, 459, 461, 464–65, 468, 475, 477, 495, 501, 507–08, 511, 516, 521, 526–27, 530, 532–33, 536–37, 540, 543–44, 548, 550–51
Egnatius, 467
Einstein, Albert, 545
Ellspermann, Gerard, 531
Empedocles, 107, 467, 556
empiricism, 14, 22–27, 30, 57–59, 72, 79–80, 84, 88, 93, 95, 100–02, 119, 121–23, 125–27, 130–33, 143, 145, 147, 149, 157, 161, 163, 170, 173–

76, 180, 183–84, 186, 189, 234, 238, 246–49, 251, 253, 265, 280, 285, 305–06, 313, 321, 329–34, 338–39, 342–47, 360, 364, 366, 373, 377, 384, 386, 388, 391, 398, 403, 408, 418, 420, 433, 436–37, 439, 445, 457–58, 462, 464–65, 471–72, 474, 481, 488–89, 492, 495, 504, 509–13, 518, 520, 522, 525, 527–29, 534, 536, 539, 543–44, 547, 551
Empiricists (medical sect), 88, 107–08, 116, 329, 331, 343
engineering & engineers (see also architecture, mechanics, and physics), 12, 14, 24, 29–30, 72, 80, 82–83, 105–06, 120–22, 124–26, 131, 136–37, 156–75, 178, 188, 190, 192, 200, 205, 207, 216–17, 223, 225, 230, 235, 239, 241, 252–60, 281, 286–88, 291–92, 303–04, 312, 318–20, 333, 353–54, 363, 372, 375, 393, 397, 401–02, 404–05, 415–16, 423–24, 429, 431–39, 445–49, 452, 459–61, 468, 529, 532, 541, 544, 548–49, 557
Enlightenment, 310
Epicrates, 171–72
Epicureanism (philosophical sect), 20, 36, 44, 59–62, 99, 116, 132–33, 186–87, 189, 261, 314–15, 342, 346–49, 366, 368, 374, 418
epigraphy (see inscriptions)
epistemology (see also scientific methods), 22–23, 101, 122–23, 126, 152, 189, 244, 247–48, 346, 349, 482–83, 493, 498, 503–20, 526–28, 547, 551
Episynthetics (medical sect), 119
Erasistratus & Erasistrateans, 87, 99, 111–13, 117, 123, 126, 132, 260–61, 294, 297, 304, 329, 335, 373
Eratosthenes of Cyrene, 31, 134–35, 137, 139, 140, 145, 148, 149, 152, 160, 162, 229, 421, 492, 553–54
Essenes, 463
ethics, 12–15, 31, 36–37, 39, 47, 49, 51, 76, 132, 143–44, 246–47, 253, 257–58, 271–72, 274–77, 279, 281, 285, 295–96, 309–10, 319, 321, 326, 330, 336–38, 341, 345–52, 355–56, 359–61, 376, 380–83, 386, 397, 424–27, 444–45, 454, 457, 459, 461–63, 467, 473, 486–87, 490, 499–500, 502–03, 512, 514–15, 519–28, 543–45, 559
Euclid, 131–32, 137, 147, 149, 163–64, 166–67, 525
Eudemus (the Aristotelian), 373
Eudemus (the Herophilean), 112, 113
Eudemus of Pergamum, 60
Eudemus of Rhodes, 101
Eudoxus, 31–32, 130, 274, 356, 420, 421, 557
Eumolpus, 274–76
Eusebius, 41, 48, 54–56, 64–65, 359, 361, 463–65, 475, 496–500, 509, 525, 530, 532–33, 537
Euterpe (muse of science), 362–63
experiments & experimentation (see also vivisection), 8, 11, 17–18, 22, 58, 61, 80, 93, 98, 101, 112–13, 115, 119, 127–28, 132, 144, 157–59, 163, 165, 168–70, 174, 181, 189, 196, 199, 203, 230–31, 243–54, 274, 302, 312, 318, 335, 358, 373–74, 389, 408–09, 414, 435–36, 440, 447, 449, 456, 488, 538, 551
exploration, 135, 149, 553–54

Facundus Novius, 220
fall of Rome (see crisis)
Farrington, Benjamin, 434–35
Feke, Jacqueline, 80
Ferngren, Gary, 302
Ferrars, Edward, 14
Finley, Moses, 193–95, 242, 410, 452
Fischer, Klaus-Dietrich, 300
Fox, Robin Lane, 501–02
freedom (intellectual), 28–29, 106, 547
French, Roger, 183, 439
Frontinus, Sexus Julius, 171, 237, 282–83, 327, 382, 441, 442

Galen, 11, 17, 20, 22–23, 27, 36, 39, 43, 49, 51–52, 59, 61–63, 65, 80, 81, 86–94, 98, 104, 106, 108, 109–11, 113, 116, 118–28, 134, 143, 144, 146, 147, 153–54, 158, 163, 173, 174, 181–84, 188–89, 195–96, 217, 222–23, 230, 238–41, 244–49, 260–61, 283–85, 294–95, 297–99, 301–02, 304–06, 327–35, 338, 341–43, 357, 359, 361, 363–65, 369–70, 372–73, 384, 397, 408, 417–18, 426, 431–33, 435–40, 445–46, 454–55, 460, 473, 488–90, 502, 511, 525–26, 529, 544, 548, 552.
Galileo, 11, 17, 38, 93, 138, 184, 188, 242, 333, 342, 552
Gallus, Gaius Sulpicius, 32, 369, 397, 399–401
Gargilius Martialis, 230
Gellius, Aulus, 25–26, 49, 199, 241, 459
Geminus, 70–71, 73, 76, 149, 368
geography (see also cartography & exploration), 22, 30, 78, 91, 114–15, 118, 121, 129, 131, 135–39, 144–45, 148, 150, 152, 168–69, 181, 199, 221, 241, 247, 252, 255, 292, 316, 328, 360, 378, 432–33, 490, 541, 553–54
geometry, 25, 41–42, 53, 59–60, 70, 73, 76–79, 82, 89, 101, 121, 130–32, 135–36, 141, 147, 149–50, 154, 160–61, 163–64, 166–67, 169, 171, 195, 238–39, 320, 353, 357–58, 361–62, 365, 375, 420–27, 464, 476, 479, 481, 496, 521, 528, 532–33, 549
Gilbert, William, 11, 17, 128, 175, 278, 552
Gottschalk, H.B. 157
Graves, Robert, 286–88
Green, Peter, 14, 123, 143, 174, 191–98, 202, 204, 209, 211, 212, 232–37, 242–43, 246–47, 250, 252, 254, 258–59, 410, 424
Greene, Kevin, 226, 287, 307, 468
Gregory Thaumaturgus, 463–65, 502, 529
gunpowder (see also incendiary combat), 8, 175, 184, 190, 213, 337, 553
Gwynn, Aubrey, 243–44

Habermehl, Peter, 276
Hadrian of Tyre, 372
Hagnodike, 394
Hankinson, R.J., 88, 100, 248, 305, 332
harmonics (and music), 15, 22, 25, 42, 49, 60, 68, 73, 78–79, 82, 85, 89, 95, 105, 131–32, 135, 152, 154–56, 163, 165, 168–70, 175, 178, 180–81, 195, 199, 201, 238, 247, 249, 259, 315, 346, 361–62, 367, 374–75, 414, 426–27, 448, 476, 479, 481, 521, 523, 527, 532, 556

Harrison, S.J., 374
Harvey, William, 17, 38, 126, 335, 552
Healy, John, 222, 377
Hegetor, 113
Heliodorus (novelist), 64, 68
Heliodorus (scientist), 122
Hephaestus (god), 362, 387–88
Heraclas, 530
Heraclides (the Herophilean), 112, 114
Heraclides of Heraclia, 163
Heraclides of Pontus, 70–71, 76, 557
Heraclides of Tarentum, 116–17
Heraclitus of Ephesus (philosopher), 47, 481, 555
Heraclitus of Rhodiapolis (scientist), 124
Herculaneum, 31, 60, 121, 202
Hermarchus, 60
Hermes (god), 363
Hermias, 498
Hermogenes of Alabanda, 162
Hermogenes of Smyrna, 124
Hero of Alexandria, 17–18, 20, 61, 72, 79–83, 120, 121, 124–25, 157–59, 163–67, 171, 173, 185, 188, 190, 198, 199, 202, 209–13, 215–20, 224–25, 230, 233, 238, 241, 243–44, 246–49, 291, 303–04, 312, 319–21, 327, 238, 345, 433–37, 452, 473, 548, 552
Herophileans (medical sect), 112, 113, 114, 116, 117
Herophilus, 42, 111–14, 116, 126, 167, 173, 240, 260, 294–95, 297, 394
Hesiod, 279, 361–62, 443–44
Hestiaeus of Perinthus, 158
Hicesius, 117
Hiero (king), 403, 420, 424
Hipparchus of Nicea, 18, 32, 74–76, 132, 137–42, 148–51, 173, 184–85, 189, 240, 265, 326, 343, 353, 356–57, 365, 370–71
Hippocrates, 107–08, 110, 118, 122–23, 132, 223, 328–30, 332, 371, 398, 436–37, 488, 556
Hippodamus, 449
Hippolytus, 525, 534
Homer, 426, 443–44
Hume, David, 14
humoral theory, 107–08, 112, 116, 121, 261, 289, 466
Humphrey, John, 442
hydrostatics, 155, 160–61, 167, 170–71, 195, 238, 254, 259, 446, 450, 452, 460
Hyginus (agrologist), 230
Hyginus (astronomer), 149
Hyginus (engineer), 171
Hyginus (fabulist), 394
Hypatia, 28, 33, 539
Hypsicles, 140

Icarus, 393
incendiary combat, 212–13
Industrial Revolution, 250, 310, 468
innovation, 8, 99, 112, 139–40, 150–51, 154, 190, 192, 195–96, 205, 211, 213, 215, 218, 228–30, 237, 239–40, 244, 255, 262, 270, 272, 282, 283, 286–92, 312, 315–20, 332, 394, 442, 449, 452, 547, 551
inscriptions, 29–30, 44–46, 124–25, 156, 193, 198, 225, 230, 404–06, 421, 430
instruments (medical & scientific; see also index of inventions), 32, 63, 78–79, 95, 109, 129, 140, 166, 168–71, 191, 199–200, 211–12, 214,

216, 219, 220–23, 234–35, 238–39, 243, 252, 259–60, 318, 326, 414, 417, 420–25, 433, 438–39, 523, 551
Isidore of Seville, 166–67
Islam (see Muslim science)
Isocrates, 414, 417, 425, 426, 452
isoperimetrics, 136, 195
Iulius Atticus, 230
Iunius Nipsus, 171

Jackson, Ralph, 222
Jerome, 531, 532
Jesus, 465–66, 477, 482, 487, 504, 507–08, 516, 521, 526
Josephus, 461
Juba (king), 115
Judaism, 139, 393, 461–63, 473, 494, 496–97, 517–18, 525
Julius Africanus, 529
Julius Bassus, 179
Julius Caesar, 197, 369, 451
Justin Martyr, 502, 505, 521–23
Juvenal, 460

Kepler, Johannes, 17, 151, 154, 242, 244
Keskinto Inscription, 124
Kidd, I.G., 71, 75–76, 143, 147, 272
Kleoxenus, 220
Kom el-Dikka excavation, 540
Kudlien, Fridolf, 118, 473

Lactantius, 57, 475, 489–96, 498, 502, 511, 527, 537, 544
Lamprias, 370
Lavoisier, Antoine, 38, 545, 552
laws (physical), 58, 64, 108, 135, 151, 160, 166, 169, 180, 243–44, 254, 258–63, 268, 277–78, 353, 380, 401–02, 452
laws (political), 28, 87, 292–94, 297–301, 315–16, 365, 370, 375, 394, 449
Leonidas of Byzantium, 181, 441
Leptines, 140
Leucippus, 108, 556
Lewis, M.J., 194, 209, 217
Licinius Lucullus, 257
Licinius Sura, 183, 386
life sciences (see also doctors and medicine; and individual sciences by name), 99–108, 110, 117, 121–22, 125–26, 131, 136, 153, 178, 182, 245, 252, 267, 305, 365, 367, 369, 505, 556
Lindberg, David, 535–40
Livy, 143, 293, 400
Lloyd, G.E.R., 9–14, 102, 153, 264, 336, 474, 488, 549–50
logic, 36, 40, 42, 47, 52, 58, 60–61, 84, 101, 103, 125, 127, 132, 238, 247–48, 274–75, 314, 347–49, 354, 356, 361, 375, 428, 462–63, 475–76, 479–80, 486, 497, 504, 511–12, 519–20, 528, 532–33, 537–38
logistics (see also algebra), 73
lost works (of ancient science), 22, 32, 73, 74, 81, 113, 117, 118, 119, 122, 123, 125, 132, 133, 138, 139, 142, 148, 152, 154, 157, 159, 161, 162, 163, 169, 171, 176, 177, 178, 179, 180, 181, 182, 184, 188, 189, 195, 221, 223, 229, 230, 249, 303–04, 307, 313, 331, 374, 375, 376, 377, 378, 379, 380, 402, 407, 423, 432, 554, 555
Lucian of Samosata, 54–55, 93, 95, 105, 205, 221, 408, 430–31, 442, 454–55, 460

Lucilius, 183, 323, 386, 388–89
Lucius Lucullus, 180
Lucretius, 20, 67, 109, 133, 187, 189, 261, 314–15, 366, 368, 374, 377, 467
ludditism, 258, 271–73, 442
Lyceum, 103, 132, 175
Lycus (anatomist), 123, 297

magic, 12, 29, 66, 114, 164, 179, 203, 263, 292, 295, 374, 457, 474, 504, 534
Malina, Bruce, 510, 520
Manilius, 368–69, 374
Manneius, Lucius (see Menekrates)
Mansfeld, Jaap, 80–81
Mantias (engineer), 170–71
Mantias (the Herophilean), 112
Marcellinus, 111
Marcellus (general), 353, 401–02, 420–21
Marcellus of Side (scientist), 114
Marinus of Alexandria, 123
Marinus of Tyre, 150, 152, 241, 328
Marius Victorinus Afer, 67
mathematics(see also individual fields by name), 22–23, 31, 36, 39, 41–42, 52–54, 60–61, 68–92, 95, 120, 129, 135–37, 141, 143–44, 149, 156, 161–63, 177, 180, 195, 238–39, 248–49, 306, 313, 344, 346, 360, 362, 365, 374, 427, 434, 448, 452, 464, 527, 529, 532, 539, 556
Maximus of Tyre, 202, 318, 424
Maxwell, James Clerk, 94, 545
McDiarmid, J.B., 177
mechanics (see also engineering and physics), 22–23, 42, 68, 73, 79–83, 92, 105, 120, 132, 135–36, 144, 150, 152, 155–68, 174–75, 184–85, 187, 195, 199, 202, 214, 223, 230, 233, 235, 238, 254–56, 259–60, 291, 320, 343, 353, 362, 365, 411, 420–23, 427, 434, 438, 450, 557
medicine, 17, 22, 29–30, 32, 38–46, 49, 52, 61, 68, 72, 83, 86–90, 101, 103–05, 107–29, 132, 134, 136, 144, 154, 172–82, 186, 195, 217, 219, 222–23, 229–30, 235, 238, 241–43, 246, 248, 252–59, 271, 280, 283–85, 295, 297–302, 306, 314, 319, 321, 328–33, 343, 358, 361–62, 365–67, 371–76, 378, 392–94, 397–98, 410, 416, 431, 433–36, 438, 440–41, 445, 457–60, 465–66, 474, 491, 521, 524–25, 532, 541, 544, 548–49, 556
Meges of Sidon, 117
Melissus, 41
Menecrates, Claudius, 124
Menekrates of Tralles, 45–46
Menelaus of Alexandria, 141, 149–50, 167, 199
Menippus, 50
Meno, 101
Menodotus, 122
Mercury (see Hermes)
Mesomedes, 181
meteorology, 66, 69, 71, 85, 91–92, 99–100, 129, 132, 144, 149, 176, 182, 253, 262, 277, 284, 356, 376, 383, 390, 499
Methodists (medical sect), 88, 107, 116
method, scientific (see scientific methods)
Metrodorus of Alexandria (botanist), 120, 181
Mettius Pompusianus, 293–94

Michler, Markwart, 122
microscope, 8, 243, 440
Middle Ages, 13, 16–18, 38, 98–99, 101, 112, 122–23, 125, 132, 150, 157, 160, 178, 186–88, 199, 228, 231–32, 241, 246, 248, 262, 267, 302, 305–07, 346, 368, 389, 428, 434, 488, 529, 535–41, 547, 549, 551
middle class (see class)
mineralogy, 52, 108, 110, 120, 177, 179, 181–82, 208, 378, 440
Minerva, 362
Mithradates VI (king), 115
moral philosophy (see ethics)
Mueller, Ian, 60, 71, 79
museums (as academic societies; see also Alexandria and Athenaeum), 95, 103, 131–32, 158, 447, 449, 450, 455
Muses (see also museums), 31, 352, 361–62, 392
music (see harmonics)
Muslim science, 17, 38, 98, 123, 131–32, 135–36, 141, 142, 150–53, 160–61, 165, 167, 169, 176, 178, 185, 188, 195, 199, 222–23, 306, 434, 465–66, 536–37, 540–41
Musonius Rufus, 418
mysteries, religious (science compared to), 325, 351, 360, 363–64, 384, 466

natural philosopher (see *physikos* and scientist)
natural philosophy (see nature and philosophy)
nature (concept & definition of) 35–37, 49, 90
Needham, Joseph, 304–05
Nero (emperor), 93, 118, 144, 210, 211, 212, 258, 271–72, 275, 277, 289, 290, 379–80, 439
Newton, Isaac, 38, 93, 129, 188, 190, 242, 333, 545, 552
New World (discovery of), 8, 175, 279, 337
Nicarchus, 93
Niceratus, 179
Nicolaus of Damascus, 178
Nicomedes, 137
Nile, exploration of, 91, 149, 257, 376, 490
Nonius Datus, 124–25, 404–05
Novara, Antoinette, 309
Numisianus (anatomist), 123
Nutton, Vivian, 107, 127, 182, 361, 433, 439–40

Occam's Razor, 186, 248
Odysseus, 444
Oleson, John Peter, 312–13, 316, 442–45
Olympus, 114
Oppian, 180, 368
Oppius, 177
optics, 22, 42, 68–69, 72–73, 89, 92, 105, 131–32, 136, 139, 144, 149, 152–56, 159, 161, 165, 168–70, 175, 178, 189–90, 195, 199, 259, 289, 319, 370, 427
Orphism, 363
Origen, 350, 463–65, 497, 502, 520, 526–27, 529–31
orreries (see also armillary spheres), 145–46, 161, 401
Ovid, 187, 387

Paccius Antiochus, 119
Paconius, 288

Pamphilus, 179
Panaetius, 256
Papirius Fabianus, 179, 271
Pappus, 79–83
papyri (sources in), 30, 44–45, 101, 121, 140, 149, 156, 306, 354
Parrhasius, 299–300
Parker, A.J., 194, 204
Parmenides, 41, 44, 59, 65, 263, 556
Paul the Apostle, 408, 477–78, 487, 504–20, 522
Paul of Samosata, 525–26, 540
Pax Romana, 20, 104–05, 182, 233
Pelling, C.B.R., 131
Pelops (anatomist), 123, 182, 297
Pericles, 400
Peripatetics (see Aristotelians)
Perseus (astronomer), 136
Petronius (satirist), 273–76, 283–84, 289–91
Petronius Musa, 179
Peutinger Table, 221
Phanias, 52, 177
pharmacology, 52, 105, 107–08, 110, 112–22, 178, 180, 195, 243, 247, 275, 280, 322, 326, 378, 398, 440, 524, 532
Pharnaces, 370
Pherecydes, 396–97
Philinus (the Herophilean), 112
Philiscus of Thasos, 230
Philo of Alexandria, 51–52, 64, 89, 461–64, 473, 479, 497
Philo of Byzantium, 120, 159–60, 164–66, 188, 190, 312, 316, 423, 449
Philo of Eleusis, 353
Philonides, 60
Philopon, John, 17–18, 538
philosophy (branches of; see also individual sects by name), 36, 39–40, 47, 49, 346–48, 351, 361, 386, 461–63, 467
Philostephanus of Cyrene, 157
Philotas, 114, 441
Philumenus of Alexandria, 122
Phylarchus, 157
physical sciences (see also individual sciences by name), 105, 134, 158, 163
physics (see mechanics)
physicus (see *physikos*)
physikos (meaning of), 13, 24–26, 30–31, 35–49, 51, 55, 69, 86–87
physiology (see anatomy)
physis (see nature)
plague, 19, 396, 548
Plato & Platonism, 28, 35–36, 39, 48, 53–54, 61, 64–65, 67, 73–75, 99–100, 103, 110, 132, 154, 158, 180, 245, 261, 263, 264, 309, 313, 318, 342, 345–46, 360, 365–70, 382–83, 414, 418–24, 436–37, 444, 448–50, 457, 461, 481–82, 488, 498, 522, 556
Pliny the Elder, 52, 116, 119, 121, 140, 143, 156, 179–83, 226–27, 229–30, 241, 252, 256–58, 276–85, 290–91, 295, 312, 326, 353–55, 365, 377–79, 394–97, 401, 404, 405–07, 410–11, 418–19, 432, 439, 450–51, 495
Pliny the Younger, 183, 212, 257, 386, 405–07, 432
Plutarch, 41, 51, 53, 58, 62–64, 66, 75, 81, 93, 109, 115, 134, 138, 143, 150, 155, 187, 259, 262–63, 346, 354–55, 361–63, 369–70, 374, 382–83, 400, 403, 410, 419–25, 440–42, 444, 457

pneumatics (science), 83, 155, 158, 160, 164–66, 188, 195, 198–99, 209–10, 212, 216, 219–20, 247, 249, 254, 319–20, 353–54, 436, 450
Pneumatists (medical sect), 117–18
Polyaenus, 59–60
Polybius, 316, 400, 554
Pompeius Trogus, 180
Pomponius Mela, 150, 554
Porcius Latro, 300
Posidonius, 53–54, 67, 69–77, 84, 99, 117–18, 142–49, 152–53, 162, 179–80, 187, 204, 210, 259, 272, 359, 383, 402–03, 410, 418, 425, 428, 554
Praxagoras, 110, 111
Presocratics, 41, 44, 47–49, 65, 189, 246, 394, 396, 397, 500, 555–57
printing press, 8, 292, 337, 545
progress, 7, 10, 17, 20, 61, 97–339, 341, 453, 457, 459, 468, 493, 519, 534, 543–46, 550–51
Prometheus, 273, 299–300, 362, 392–93
Protagoras, 28, 273
Protarchus, 60
psychology, 41, 99, 107, 119, 132, 144, 253, 528
Ptolemaïs of Cyrene (scientist), 78–79, 163, 170, 249
Ptolemies (kings), 15, 20, 103, 113, 126, 163–64, 177, 230, 447, 450, 461
Ptolemy, Claudius (scientist), 11, 17, 18, 20, 22–23, 27, 39, 61, 67, 72, 76, 78–81, 88, 98, 104, 120, 121, 124, 134, 136, 138, 140, 144, 145, 149–55, 167–70, 173, 184–87, 189, 193, 220, 240–41, 243–49, 259, 261, 327–28, 345, 359–60, 369, 433, 473, 502, 534, 541, 548, 552
Publius Nigidius, 178
Publius Septimius, 375
Pyrrhonism (see Skeptics)
Pythagoras & Pythagoreanism, 25, 39, 63, 65, 396, 556–57

Quintilian, 56, 136, 241, 295, 350, 511
Quintus (anatomist), 123

Raban, Avner, 207
Rabbis, 342, 482
Rathbone, Dominic, 231
Rawson, Elizabeth, 467
Renaissance, 9, 15, 126, 146, 178, 184, 192, 201, 242, 252, 267, 294, 301, 305, 307, 310, 384, 434, 537, 545–51
Reynolds, Joyce, 377
Reynolds, Terry, 227, 254
Rihll, Tracy, 266–67, 370, 423
Ross, William Davis, 182
Royal Society of London, 453, 545
Rowland and Howe, 318
Rufinus, 530–31
Rufus of Ephesus, 118, 123, 217, 295, 297
Russo, Lucio, 129, 138, 164, 184, 188, 208, 209, 230, 235, 240, 243, 252, 303, 424

Sallustius, 467
Sambursky, Samuel, 244–45, 247, 250, 335
sandboxes, 31, 195
Saserna, 230
Satyrus (anatomist), 123, 301
sciences (see life sciences, physical sciences, and mathematics)
Scientific Revolution, 7–16, 19, 22, 27,

59, 68, 94, 97, 109, 121, 125, 154, 174–75, 189, 190, 236, 242–47, 250, 452–53, 468, 535, 541, 545–48, 551–52
scientific methods (see also epistemology), 8–9, 17, 24–27, 38, 48, 56–95, 98, 100–01, 126, 134, 143, 147, 157, 159, 163, 173–74, 183, 196, 240–42, 246–48, 280, 303, 326–27, 330–33, 334–35, 342–43, 387, 391, 433, 445, 453, 458, 471, 487, 498, 508–09, 522–23, 528, 532, 537, 545–48, 551–52
scientific values (see curiosity, empiricism, and progress)
scientist (definition of), 10–11, 23–27
Scribonius Largus, 119, 196, 280
Scylla, 387
Second Sophistic, 20–21
Seleucus of Seleucia, 134, 137, 138, 145
Seleucus of Tarsus, 441
Seneca (the Elder), 299–300
Seneca (the Younger), 66, 76–77, 91, 143–44, 182–83, 187, 205, 207, 210, 212, 214–15, 230, 241, 248, 258, 262, 270–75, 284, 312, 322–27, 331, 334–35, 350, 355, 363, 376–89, 395, 410, 418, 424–28, 431–33, 439, 442, 467
Serapion of Antioch, 148
Sergius Paulus, 372–73
Sextius Niger, 179
Sextius, Quintus, 395–96, 401
Sextius Sulla, 369
Sextus Empiricus, 41, 47, 49, 54, 56–57, 63–64, 67, 122–23, 343, 458–59, 493, 559
Sherwood, Andrew, 442
Siculus Flaccus, 171
Silius Italicus, 401–02
Simms, D.L., 194, 195, 213–14, 225, 252, 423
Simonides, 448
Simplicius, 71, 133, 138–39, 168
Skeptics (philosophical sects), 24, 60, 242, 342–44, 370, 418, 457–59, 494, 559
Slater, Niall, 275–76
slaves, 94, 114, 119, 175, 227, 237, 250–53, 284, 293–95, 298, 300, 342, 382, 394, 413–16, 428–29, 439–40, 466, 473, 485, 520, 523, 527
Sleeswyk, André Wegener, 217
Smith, Norman, 206
social class (see class)
social status (of scientists; see also class), 9–15, 29–30, 409–47, 468, 546–49
Socrates, 28, 36, 47, 65, 67, 355, 411–12, 431, 448, 488, 491, 498, 556
Soranus, 32, 88–89, 116, 121–22, 124, 241
Sosigenes of Alexandria, 148
Sosigenes the Aristotelian, 131, 154
Sostratus of Alexandria, 117, 182
Stark, Rodney, 16–17, 244, 258, 261–66, 510
Statilius, 122
steam engine, 158, 165–66, 191–92, 213, 216, 232–34, 243
stirrup, 17, 217
Stoicism, 53, 54, 59, 61, 68–69, 76, 80, 85, 99, 116–19, 142–43, 145, 147, 256, 267, 269, 272, 342, 345–49, 356–57, 366, 368, 370, 376, 379–80, 383, 386, 418, 427–28, 437, 459, 517, 531
Strabo, 53, 67, 77–78, 142–43, 147–49, 187, 241, 254, 316–17

Strato of Lampsacus, 18, 36, 55, 66, 67, 99, 112, 132–34, 138–39, 156–59, 185–89, 261, 359
Strato (poet), 415
Stückelberger, Alfred, 427
Suetonius, 286–88, 293–94, 459
Sulla, 147
Sullivan, J.P., 275
Sura (see Licinius)
surgery, 88, 107, 110, 114, 116–19, 122, 182, 195, 199, 222–23, 295, 321, 373, 434–35
surveying, science of, 18, 30, 73, 131, 136, 165–66, 171, 191, 199, 209, 220, 235, 328, 375, 433, 441, 534, 553

Tacitus, 50, 212, 258, 270–71, 281–82, 350, 405, 407
Tatian, 523–25
telescope, 8, 190, 243, 292, 337, 552
Temple of Artemis, 353
Temple of Peace, 221
Tertullian, 56, 121, 222–23, 284, 312, 365, 475, 483–89, 494–96, 513, 524, 531, 540
Thales, 47–48, 56–57, 99, 357–59, 395, 400, 401, 481, 484, 486, 524, 555
Themison, 116
Theodosius of Bithynia, 141–42, 149
Theon (grammarian), 370
Theon of Smyrna, 73–76, 141, 149, 154
Theophrastus, 52, 85, 103, 132, 149, 175–81, 183, 312, 439, 454, 525–26
Thessalus (the Magician), 474
Thessalus (the Methodist), 116, 408
Tiberius (emperor), 119, 196, 221, 289–92, 442
Timocharis, 140
Tiro, 196
Titanic (ship), 393
Titus (emperor), 277, 281, 369, 451
Toomer, G.J., 151–53, 170, 438
Trajan (emperor), 122, 172, 257
Trebius Niger, 180
Tremelius Scrofa, 230
trigonometry (plane & spherical), 134, 139, 149–50, 195, 244
Trimalchio, 275, 289–91
Triton of Tanagra, the, 408–09, 456
Tybjerg, Karin, 79, 163–65, 435–36

universities, 28, 447, 452, 540, 549, 551
Urania (muse of astronomy), 31, 361–62

Valerius Maximus, 400, 402
Valgius, Gaius, 179
Vallance, J.T., 177
Varro, 32, 49, 58, 109, 182, 221, 230, 257, 280, 375–76
Vernant, Jean Pierre, 190–91, 235, 239, 336
Verus of Pompeii, 438
Vesalius, 17, 434–35
Vespasian (emperor), 277, 281, 286–88, 291–92, 442, 459
Vesuvius, 281, 389, 405–06
Vettius Valens, 58, 67
Vettulenus Barbarus, 372
Virgil, 352, 387, 388–89, 444
Vitruvius, 11, 65–66, 80, 92, 106, 120, 134, 156, 162, 165, 168, 191, 202, 206, 217, 227, 238, 246, 259, 276–77, 288, 312, 318–19, 352, 375, 403–04, 415, 423, 432, 435, 451–52, 454–55
vivisection (see also dissection), 101–

02, 107, 112, 127–28, 292, 295–96, 306, 321, 329, 373, 392
volcanology, 144, 148, 179–80, 260, 323, 377, 387–90
von Staden, Heinrich, 298
von Wright, Georg Henrik, 310, 337

Walsh, P.G., 275
Whiggism, 28
White, Lynn, 17, 207, 225, 229, 262
Whitney, Elspeth, 256

Xanthus, 94–95, 460
Xenarchus, 186, 187
Xenophanes, 44, 65, 555
Xenophon, 411–12, 417, 418, 425, 448–50, 498–99

Zeno of Citium, 143
Zeno of Elea, 59, 491, 556
Zeno of Sidon, 60
Zeno (the Herophilean), 112
Zenodorus, 136
Zeuxis Philalethes, 117, 371
Zhmud, Leonid, 313
Zilsel, Edgar, 243, 309–10, 449
zoology, 38, 85, 102, 105, 108, 122, 132, 144, 176–82, 372, 377–78, 419–20, 437, 451
Zosimus, 174

About the Author

Richard Carrier, PhD, is a philosopher and historian of antiquity, specializing in contemporary philosophy of naturalism and Greco-Roman philosophy, science, and religion, including the origins of Christianity. He is the author of numerous books, including *Sense and Goodness without God*, *Not the Impossible Faith*, *Hitler Homer Bible Christ*, *Proving History*, *On the Historicity of Jesus*, and *Science Education in the Early Roman Empire*. They are all also available as audiobooks, read by Dr. Carrier. For more about Dr. Carrier and his work see www.richardcarrier.info.